FOUNDATIONS OF QUANTUM CHROMODYNAMICS

Third Edition

An Introduction to Perturbative Methods in Gauge Theories

To my friends,

W.A.B., A.J.B. and D.W.D.

World Scientific Lecture Notes in Physics

Published

Vol. 61: Modern Differential Geometry for Physicists (2nd ed.)
C J Isham

Vol. 62: ITEP Lectures on Particle Physics and Field Theory (In 2 Volumes)
M A Shifman

Vol. 64: Fluctuations and Localization in Mesoscopic Electron Systems
M Janssen

Vol. 65: Universal Fluctuations: The Phenomenology of Hadronic Matter
R Botet and M Ploszajczak

Vol. 66: Microcanonical Thermodynamics: Phase Transitions in "Small" Systems
D H E Gross

Vol. 67: Quantum Scaling in Many-Body Systems
M A Continentino

Vol. 69: Deparametrization and Path Integral Quantization of Cosmological Models
C Simeone

Vol. 70: Noise Sustained Patterns: Fluctuations & Nonlinearities
Markus Loecher

Vol. 71: The QCD Vacuum, Hadrons and Superdense Matter (2nd ed.)
Edward V Shuryak

Vol. 72: Massive Neutrinos in Physics and Astrophysics (3rd ed.)
R Mohapatra and P B Pal

Vol. 73: The Elementary Process of Bremsstrahlung
W Nakel and E Haug

Vol. 74: Lattice Gauge Theories: An Introduction (3rd ed.)
H J Rothe

Vol. 75: Field Theory: A Path Integral Approach (2nd ed.)
A Das

Vol. 76: Effective Field Approach to Phase Transitions and Some Applications to Ferroelectrics (2nd ed.)
J A Gonzalo

Vol. 77: Principles of Phase Structures in Particle Physics
H Meyer-Ortmanns and T Reisz

Vol. 78: Foundations of Quantum Chromodynamics: An Introduction to Perturbation Methods in Gauge Theories (3rd ed.)
T Muta

Vol. 79: Geometry and Phase Transitions in Colloids and Polymers
W Kung

World Scientific Lecture Notes in Physics – Vol. 78

FOUNDATIONS OF QUANTUM CHROMODYNAMICS

Third Edition

An Introduction to Perturbative Methods in Gauge Theories

T Muta

Fukuyama University, Japan

World Scientific

NEW JERSEY • LONDON • SINGAPORE • BEIJING • SHANGHAI • HONG KONG • TAIPEI • CHENNAI

Published by

World Scientific Publishing Co. Pte. Ltd.
5 Toh Tuck Link, Singapore 596224
USA office: 27 Warren Street, Suite 401-402, Hackensack, NJ 07601
UK office: 57 Shelton Street, Covent Garden, London WC2H 9HE

British Library Cataloguing-in-Publication Data
A catalogue record for this book is available from the British Library.

Cover picture (galaxy M74) reproduced by courtesy of Higashi-Hiroshima Observatory, Hiroshima University.

World Scientific Lecture Notes in Physics — Vol. 78
FOUNDATIONS OF QUANTUM CHROMODYNAMICS (3RD EDITION)
An Introduction to Perturbative Methods in Gauge Theories

ISBN-13 978-981-279-353-9
ISBN-10 981-279-353-4
ISBN-13 978-981-279-354-6 (pbk)
ISBN-10 981-279-354-2 (pbk)

PREFACE TO THE THIRD EDITION

In ancient Chinese literature, it is said that time passes as if it were a galloping white horse crossing in front peered through a narrow gap. My white horse runs fast as well. Already more than 20 years have passed since the first edition of this book was published. Graduate students at that time who studied Quantum Chromodynamics (QCD) with this book are now in their late forties. Many of them may now be professors in particle physics who would be using this book for their students. I would be extremely happy if the book has helped all those people to get a deeper understanding of the basis of QCD.

In this third edition, I tried to keep the original form of the book as far as possible and so the changes are mostly minor corrections of misprints found after the second edition was published. I also included in this edition some memorable pictures that I took a long time ago.

I would like to thank Tomohiro Inagaki, who is an associate professor in Hiroshima University and was a graduate student when the first edition was released, for his full support in the whole process of preparing the third edition. The cover picture (galaxy M74) was selected from many pictures of celestial objects taken by Koji S. Kawabata at the Higashi-Hiroshima Observatory in Hiroshima University whom I would like to acknowledge. The Observatory was established when I was serving as President of Hiroshima University. I would like to thank Hiromi Sugihara for her continuous help in the preparation of the third edition and also with my work as President of Fukuyama University. I am particularly grateful to John Albert Kearley, Jr. who gave me useful advices on how to improve my English sentences.

T. Muta

Hiroshima, Japan
April 2009

PREFACE TO THE SECOND EDITION

It is to my surprise that the basis of quantum chromodynamics (QCD) has been made firmer and firmer in its applicability in the perturbative as well as nonperturbative regime since the first edition of this book was prepared in 1986. While much theoretical progress has been made in perturbative QCD, the basic structure of its formulation has remained unchanged and so the book is still of value for researchers who wish to learn fundamental items of perturbative QCD.

In preparing the second edition, I thoroughly inspected the book to make corrections of misprints as many as possible. Two graduate students, Kazutaka Kawakami and Norikazu Yamada, were of great help in this tiring and time-consuming work. I would like to thank them for their beautiful work. In the second edition, some new sentences are added corresponding to the new developments in theoretical and experimental studies of QCD. Advices given to the author by Takashi Ohsugi, Yoshio Sumi and Reisaburo Tanaka have been very useful in the process of renewing the book. The appendix D is added for the convenience of the readers who are not used to employing the Feynman rule in our book but accustomed to the traditional Feynman rule which appears in many books.

T. Muta

Hiroshima, Japan
March 1997

PREFACE TO THE FIRST EDITION

The strong interactions of fundamental particles have been successfully described by a non-Abelian gauge field theory called quantum chromodynamics (QCD). Because of its outstanding property of asymptotic freedom, a perturbative treatment makes sense for short-distance phenomena and predictions of the theory have been tested in a variety of elementary particle reactions. On the other hand, it is most important to control the nonperturbative regime of QCD in order to fully understand the strong interactions of hadrons. Unfortunately, however, no established method has surfaced to deal with the nonperturbative regime of QCD although some promising attempts are available.

In this book the techniques developed for the perturbative regime of QCD are explained in detail. To do this I have attempted to start at a level as low as that of a first-year graduate student's and continue from there to discuss the subject in a self-contained manner as far as possible. As a consequence the part of the book on background material for QCD became rather bulky. I think this part may also serve as a comprehensive introduction to gauge field theories. In the rest of the book the renormalization group method, the operator-product expansion and other technical developments are elucidated, and QCD predictions based on these techniques are derived for short-distance reactions. In the final chapter the infrared divergence is discussed in some detail since it is of relevance to the argument on the validity of the perturbative treatment.

The book grew out of lectures given in many places. I would like to thank those people who participated in the lectures and who made valuable criticisms.

During the course of writing the manuscript I have benefited from discussions and correspondences with many people as well as their assistance. I am happy to express my thanks to my friends, Bill Bardeen, Andrzej Buras and Dennis Duke for many discussions and their encouragement. Discussions with Kazuo Fujikawa, Yoichi Kazama, Noboru Nakanishi and Akio

Sugamoto have been very useful and I wish to express my gratitude to them. I am indebted to Tony Zee for valuable correspondences and to Hisao Nakkagawa and Akira Niégawa for critical comments. I would like to express my gratitude to Kiyoshi Kato, Tomoo Munehisa, Tatsumasa Tsurugai, and Sakue Yamada for providing me with experimental data and Monte Carlo results used in the book. The criticisms and comments by my colleagues and students have been undoubtedly beneficial. My hearty thanks are due to Jiro Kodaira, Toshinobu Maehara, Junji Okada, Juichi Saito, Ryuichi Najima, Masato Inoue, Kenji Hamada and other members of our laboratory.

The idea of turning my lecture notes into a book was first suggested to me four years ago by Prof. K.K. Phua. Since then he has waited for the completion of the book with great patience and kindness for which I would like to express my sincere gratitude. I would also like to thank Yasuo Hara for his concern about the completion of my book and his constant encouragement.

My friend and colleague, Yoshio Sumi, deserves many thanks for his kind help in drawing all the figures in the book in spite of his busy teaching and research schedule. It is my pleasure to thank my colleague, Masayoshi Kikugawa, for his help in computerizing the typing procedure, and Minoru Yonezawa and Seiichi Wakaizumi for their encouragement and support while I was writing the manuscript. A note of gratitude is due also to Hiroko Nishiura for her invaluable help in typing and correcting the whole manuscript. Without her persevering assistance the present book would not have materialized. I would also like to thank Emi Nakamoto for her cheerful assistance in typing the manuscript. Finally, I wish to thank my mother for her constant support of my research activities and my family for their patience during the years of preparing the manuscript.

Hiroshima, Japan T. Muta
December 1986

CONTENTS

J. D. Bjorken (left) and H. B. Thacker (right) in 1975 at SLAC

D. J. Gross at 1978 International Conference on High Energy Physics in Tokyo

H. D. Politzer at 1978 International Conference on High Energy Physics in Tokyo

From left to right, W. A. Bardeen, A. J. Buras, D. W. Duke and T. Muta in 2005 at Fermilab

CHAPTER

ONE

INTRODUCTION

1.1. GENERAL SURVEY

In the late fifties and in the sixties it was generally believed that the strong interactions of hadrons may not be described in any sense by the perturbative method of quantum field theory. In fact a naive application of the pertubative method to meson theory was recognized to fail although such a description of meson theory might have been prompted by the success of quantum electrodynamics. Accordingly, formulations based on the perturbative method were discarded in the theory of strong interactions and a suitable formulation of the theory independent of the perturbative approach was sought after. Extensive studies in this direction have brought about, among other things, the reduction technique for S-matrix elements, dispersion relations based on the analyticity of hadronic scattering amplitudes, Regge-pole theory and the dual resonance model. Although these developments were successful in many respects to describe hadronic reactions phenomenologically, they were not quite satisfactory since they did not stem from first principles.

The principal breakthrough which eventually led us to quantum chromodynamics (QCD) was put forward in 1973 when the property of asymptotic freedom of non-Abelian gauge field theories was discovered [tHo 72a, Gro 73, Pol 73]. The property of asymptotic freedom now allows perturbative treatment of strong interactions at short distances. We shall describe in the following the developments related to the discovery of asymptotic freedom.

In the late sixties studies on the classification of hadrons, hadron mass spectra and hadronic interactions strongly suggested that hadrons were made of quarks, the fundamental building blocks [Gel 64, Zwe 64].[1] It was then natural to look for the dynamics obeyed by quark systems which is responsible for the composition of hadrons as well as hadronic reactions.

[1] The term "quark" was adopted by Gell-Mann from the sentence, "Three quarks for Muster Mark," in James Joyce's *Finnegans Wake*.

In order to obtain experimental information on quark dynamics it seems to be the most sensible to probe the inside of hadrons (e.g., protons) by applying a beam of structureless particles (i.e., leptons). This method of studying the structure of target particles is essentially the same as the one utilized a long time ago by Geiger, Marsden and Rutherford in clarifying the structure of atoms. For the study of the hadronic structure we need much higher energies and larger momentum transfers to obtain the higher resolution. The first series of such experiments to probe the structure of the proton was initiated in the 1960's at SLAC (Stanford Linear Accelerator Center) and the process was called deep inelastic electron-proton scattering.

In 1969 Bjorken [Bj 69] reported that the scaling property of structure functions in electron-nucleon scatterings[2] was expected in the deep inelastic region where momentum transfer squared q^2 and energy transfer ν of electrons are very large with the ratio q^2/ν kept fixed. This scaling is called Bjorken scaling for which it is claimed that structure functions in the deep inelastic region depend only on the ratio q^2/ν rather than on two independent variables q^2 and ν. Immediately after Bjorken's proposal, experimental confirmation was found for it [Pan 68].

One of the easiest ways to understand Bjorken scaling was to assume that the projectile electrons scatter off almost-free pointlike constituents [Bj 69a] inside nucleons which were called partons[3] [Fey 69, 69a]. For deep inelastic electron-nucleon scatterings, the momentum transfer squared q^2 is large so that the spatial resolution for observing the target nucleon by projectile electrons is high. Thus Bjorken scaling implies that the constituents of the nucleon look almost free and point-like when observed with high spatial resolution. Hence, if one accepts the parton idea, the dynamics governing the parton system should have the property that the interaction between partons becomes weaker at shorter distances. The partons were later identified with quarks since experimentally it was suggested that their quantum numbers such as charges and spins were practically the same as those of quarks.

Searches for quark dynamics were initiated right after the foundation of the parton model. All the known quantum field theories at that time were surveyed as possible candidates for quark dynamics and were shown not to enjoy the above-mentioned property that the interaction between quarks gets weaker at short distances [Zee 73]. It was the non-Abelian gauge field theory[4] which was left untouched in this analysis.[5] This theory is a gauge theory

[2] For details, see Secs. 1.3 and 4.1.2.

[3] Coined by R.P. Feynman.

[4] See Sec. 2.1 for an introduction to non-Abelian gauge field theories.

[5] Symanzik suggested that $\lambda\phi^4$ theory with negative λ had the required property [Sym 73]. The theory, however, is unphysical since it has no lower bound for energy.

similar to quantum electrodynamics though differing from it in that the corresponding gauge symmetry is not Abelian, i.e., generators of the symmetry group are noncommutative. Such a theory was originally introduced by Yang and Mills [Yan 54]. 't Hooft [tHo 72a], Gross and Wilczek [Gro 73] and Politzer [Pol 73] examined non-Abelian gauge field theories by the use of the renormalization group method and found that they satisfied the desired property which is now called asymptotic freedom. Soon after it was shown that only the non-Abelian gauge field theory exhibited the property of asymptotic freedom among the known theories in four-dimensional space-time [Col 73].[6] It was quite fortunate that, by that time, the quantization of non-Abelian gauge field theories had been achieved [Fad 67] and their renormalizability had already been proven [tHo 71].

The dynamics governing quark systems therefore is to be found in non-Abelian gauge field theories. As mentioned above non-Abelian gauge field theories are generated by symmetries described by a noncommutative algebra. This means that quark systems are required to have an extra symmetry associated with the non-Abelian gauge field theory describing the quark dynamics. What is this extra symmetry among quarks?

In the meantime it had been frequently suggested that quarks must have a new quantum number called color and exhibit the color symmetry[7] in order to resolve several difficulties in the quark model. The difficulties may be summarized as follows[8]: (1) the problem of constructing baryon wave functions, (2) nonobservation of isolated quarks, (3) discrepancy between the prediction and experimental data on total cross sections of $e^+e^- \to$ hadrons and decay rates for $\pi^0 \to 2\gamma$.

Fritzsch and Gell-Mann [Fri 72, 73] proposed that the extra symmetry of the non-Abelian gauge field theory be identified with color symmetry. By this identification most of the difficulties in quark models by that time could be resolved in a natural way and thus the theory of quark dynamics was finally established. The theory was named quantum chromodynamics (QCD).[9] The term "chromo" refers to the "color" symmetry of quark systems.

6 For the space-time with dimension different from four, there are some other possibilities of asymptotically free theories: the ϕ^3 theory in six dimensions [Mac 74], the four-fermion theory in two dimensions [Gro 74a] and the Wentzel model in three dimensions [Mut 77].

7 The idea of the color quantum number may be traced back to the paper by Han and Nambu [Han 65]. Its present form was given by Gell-Mann [Gel 72] who coined the term "color." The scheme equivalent to color but based on para-fermi statistics was proposed by Greenberg [Gre 64] and the equivalence between the color and para-fermi schemes was explicitly shown by Ohnuki and Kamefuchi [Ohn 73].

8 For detailed discussions, see Sec. 1.2.

9 By M. Gell-Mann.

Just like the photon which is an Abelian gauge field mediating electromagnetic interactions between charged particles in quantum electrodynamics (QED), the non-Abelian gauge field in QCD mediates color interactions between quarks. This non-Abelian gauge field in QCD is called the gluon[10] as it is responsible for binding the quarks together. While photons have no electric charge (i.e., they are neutral), gluons carry color charges and hence interact with each other even in the absence of quarks. This property of gluons is an essential ingredient for having asymptotic freedom.[11]

In the quark model with color symmetries hadrons appear as colorless states while quarks carry color quantum numbers. It is assumed that only colorless states are physically realized and hence quarks cannot be observed in isolated states. There is a possibility of explaining this assumption as a dynamical effect in QCD. In fact serious infrared divergences due to the massless gluons (which are more serious than those in QED because of the self-couplings of gluons) may be responsible for confining quarks at long distances [Wei 73, Gro 73a]. Thus QCD has a desirable property that it enjoys the asymptotic freedom at short distances while it has a possibility of quark confinement at long distances.[12]

According to the property of asymptotic freedom of QCD, one may now safely use perturbation theory to discuss short-distance reactions. This approach in QCD of using perturbation theory is often referred to as *perturbative QCD*. The earliest application of QCD was made to deep inelastic electron-nucleon scattering [Gro 73a, 74, Geo 74] in order to see how Bjorken scaling could be recovered although it was eventually found to be logarithmically violated by the QCD corrections. In this application it is necessary to extract the purely short-distance part out of the deep inelastic cross section in order to guarantee the safe application of perturbative calculations. The operator product expansion (OPE) [Wil 69, Zim 71, 73] was, in this context, known to be the most powerful tool. Using OPE one can uniquely extract the short-distance part to which perturbative calculation is safely applied. Apparently one finds that the lowest-order results in the perturbative calculations reproduce Bjorken scaling and so the parton picture is recovered. It can, however, be easily recognized that large logs associated with the large mass scale q^2 (momentum-transfer squared) spoil the perturbative treatment and some improvement is necessary. The

[10] The term "gluon" was originally introduced within the framework of neutral vector theories without recourse to the color symmetry [Gel 62].

[11] For details, see Sec. 3.4.

[12] In this book we restrict our discussions to short-distance phenomena and will not consider the quark-confinement problem any further. For a review see, e.g., [Mar 78, Ban 81].

improvement may be attained by summing the large logs to all orders and this summation is most efficiently performed by the use of the renormalization group equations (RGE) [Chr 72]. After this resummation it was found already in the first application of QCD to deep inelastic scatterings that the Bjorken scaling was violated logarithmically [Gro 73a, 74, Geo 74]. Thus in QCD Bjorken scaling holds only in an approximate sense. The slight deviation from the Bjorken scaling in structure functions was later confirmed experimentally in deep inelastic muon-nucleon scatterings [Wat 75, Cha 75]. This observation, together with the later analyses of further experimental data, gave strong support to the foundation of QCD.

Another immediate application of QCD was made to electron-positron (e^+e^-) annihilation processes which in themselves are purely of short-distance nature. As this process is free from long-distance effects which may destroy the perturbative argument, one may directly use perturbation theory in this application. Perturbative calculations of the total cross section of e^+e^- annihilations were performed with the help of RGE immediately after the discovery of asymptotic freedom in QCD [App 73, Zee 73a]. Unfortunately the QCD effect is very small and has not yet been given full support by experimental data.

Some more applications of QCD were also made to decays of heavy quarkonia such as J/ψ's (charmonium) and Υ's (bottomium) and to nonleptonic decays of hadrons. The first series, J/ψ's, of heavy quarkonia was found right after the foundation of QCD and the first attempt to apply QCD was made to decays of these heavy quarkonia [App 75, 75b]. Here the decay width of heavy quarkonium may be decomposed into two parts: the square of the quarkonium wave function at the origin and the quark-antiquark annihilation cross section. The latter part is purely of the short-distance nature and is subject to perturbative treatment. The possibility of explaining the $\Delta I = 1/2$ rule in nonleptonic weak decays of hadrons was argued by applying OPE to the weak effective Hamiltonian [Gai 74, Alt 74]. This argument seems to be in the right track at least qualitatively but according to subsequent detailed studies its predictive power was found to be limited by the presence of the nonperturbative effects. The photon-photon scattering obtained as a subprocess of $e^+e^- \to e^+e^-X$ (X represents unobserved hadrons) is yet another example of an early application of pertubative QCD [Ahm 75, Wit 77]. This is one of the cleanest short-distance processes and has received much attention.

A formulation equivalent to the one based on the operator-product expansion but entirely phrased in parton language was proposed by Altarelli and Parisi [Par 76, Alt 77] and is called the Altarelli-Parisi formalism. This method is much more transparent in its physical meaning since it directly deals with the parton-distribution function [Bau 78, Kod 78] inside hadrons and the parton-fragmentation function [Owe 78, Uem 78], which are basic quantities for describing short-distance reactions.

In 1977 the second phase was opened in the development of perturbative QCD. There were three major ingredients which initiated the second phase: the completion of the calculation of the next-to-leading-order effect in deep inelastic scatterings [Flo 77, 78, Bar 78], the introduction of a new kind of purely short-distance processes, i.e., jets from quarks and gluons [Ste 77, Shi 78, Ein 78] and the generalization of OPE to accommodate a wider class of short-distance processes [Ell 78, 79, Ama 78, 78a, Lib 78, Mue 78].

The successful generalization of OPE as well as the discovery of a new type of short-distance processes, jets, eventually enlarged the region of applicability of perturbative QCD. In fact, since 1978 a variety of processes such as the Drell-Yan process NN $\to l^+l^-$X (N: nucleon, $l^\pm$: charged leptons, X: unobserved hadrons), inclusive e^+e^- annihilation $e^+e^- \to$ hadron + X, large-p_T hadron reactions and multi-jets in e^+e^- annihilations have come under our control. It has, however, been argued that in some processes such as the Drell-Yan process, care must be taken of the infrared effects arising from soft gluons.

The computations of higher-order effects in perturbative QCD have been performed in many processes such as e^+e^- annihilations [Che 79, 80, Din 79, Cel 80, 80a], quarkonium decays [Bar 79a, Mac 81], photon-photon scattering [Bar 79], Drell-Yan processes [Alt 78, Kub 79, Con 79, Har 79, Hum 79], inclusive e^+e^- annihilations [Cur 80, Fur 80, Fol 81, Oka 81] and jets in e^+e^- annihilations [Ell 80, 81, Fab 80, 82, Ver 81]. In dealing with higher-order effects one immediately finds [Bar 78] that the calculated results crucially depend on the way one renormalizes divergent integrals appearing in the calculation. This is usually called the renormalization-scheme dependence of perturbative predictions and is a rather general phenomenon in the perturbative treatment of renormalizable quantum field theories [Cel 79, 79a].

The renormalization-scheme dependence of perturbative predictions can be seen even in quantum electrodynamics although in QED its observable effect is negligible as the coupling constant, the expansion parameter in perturbation series, is very small ($\alpha = e^2/4\pi \sim 1/137$). In QCD the effective coupling constant at the energy scale available presently is not very small, α_S $(= g^2/4\pi)$ $\sim 1/10$, and hence nonnegligible effects may be brought about by a different choice of the renormalization scheme. This effect may obscure the meaningful tests of our perturbative predictions in the phenomenological application of QCD. Nevertheless, with some limitations, the successful comparison of the predictions with the data has been made in a variety of processes.

In this textbook we would like to give an elementary introduction to the fundamental formulation of perturbative QCD relevant to phenomenological applications. We first present a brief survey of the formulation of QCD as a gauge field theory and then explain the operator-product expansion and the renormalization-group technique as a basis for QCD calculation in the short-distance region. We demonstrate the asymptotic freedom of the theory and the use of these techniques in some typical examples. The generalization of OPE to enlarge the region of applicability of QCD is discussed and some typical applications are given.

1.2. QUARKS AND COLOR

Matter in the universe is composed of atoms[13] which consist of a nucleus and electrons. The nucleus is further composed of protons and neutrons, generically called nucleons. They are bound together through the interaction of the pion which was first found in cosmic ray experiments. Since particle accelerators were introduced, many new particles, stable and unstable, have been created in scattering experiments of nucleons, pions, photons and electrons. The number of kinds of these particles now adds up to more than several hundreds. They are classified in the following way:

Particles with strong interactions

Hadrons { Baryons (fermions), e.g., nucleons,
Mesons (bosons), e.g., pions.

Particles with no strong interactions
Leptons (fermions), e.g., electrons, muons, neutrinos.

Particles which mediate electromagnetic and weak interactions
Gauge bosons (bosons), e.g., photons, W-bosons, Z-bosons.

There are a number of different kinds of hadrons while the number of the kinds of leptons and gauge bosons is rather limited. It is then quite hard to think that all of these hadrons are ultimate building blocks of matter. Rather it is natural to expect that hadrons are made of a fewer number of fundamental particles, the quarks [Gel 64, Zwe 64].[14] On the other hand there are not many kinds of leptons and gauge bosons, and they may still be considered as

[13] There is a theoretical possibility of having matter directly made up of quarks without any intermediate steps like atoms and hadrons [Wit 84]. We, however, do not take into account this possibility in this book.

[14] The idea of the bootstrap mechanism [Che 61, 66] which states that hadrons are made of hadrons themselves had once been proposed.

elementary and pointlike. In fact there is so far no experimental evidence of the internal structure of leptons and guage bosons. Hadrons, on the contrary, exhibit the extended structure in electron-scattering experiments. This structure may be considered as a manifestation of the fundamental entities within hadrons.

Thus we are naturally led to an attempt of classifying all hadrons in terms of their fundamental building blocks, the quarks. According to the quark model, all the hadrons found in the particle data table [Par 96] may be classified in a consistent manner if one assumes that the baryons are made of three quarks, qqq, and the mesons are made of a quark-antiquark pair, $q\bar{q}$. This assumption is partly supported by the experimental fact that $\sigma_{\pi p}/\sigma_{pp} = 2/3$ at high energies where $\sigma_{\pi p}$ and σ_{pp} are the pion-proton and proton-proton scattering total cross sections respectively [Par 96]. In order to account for the conservation of the isotopic-spin, strangeness, charm, bottom and top quantum numbers in hadronic reactions, one assumes the existence of different kinds of quarks corresponding to these conserved quantum numbers in strong interactions. They are called the up(u), down(d), strange(s), charm(c), bottom(b) and top(t) quarks respectively and the quantum numbers are generically called flavor.[15] The top quark has recently been observed [CDF 95, D0 95, Cam 97].

The leptons are considered to be elementary just as the quarks are. According to their property in weak interactions it is commonly accepted that they form sequential doublets[16]

$$\begin{pmatrix} \nu_e \\ e^- \end{pmatrix}, \quad \begin{pmatrix} \nu_\mu \\ \mu^- \end{pmatrix}, \quad \begin{pmatrix} \nu_\tau \\ \tau^- \end{pmatrix}, \tag{1.2.1}$$

where e^-, μ^- and τ^- are the electron, muon and tau lepton respectively and ν_e, ν_μ and ν_τ are the corresponding neutrinos. By studying the weak interaction of hadrons one derives information on the property of quarks in weak processes and finds that the quarks exhibit the same regularity as the sequential property of the leptons, (1.2.1):

$$\begin{pmatrix} u \\ d \end{pmatrix}, \quad \begin{pmatrix} c \\ s \end{pmatrix}, \quad \begin{pmatrix} t \\ b \end{pmatrix}, \tag{1.2.2}$$

where t is the new quark (top quark) observed recently.[17] The common sequential structure of the leptons and quarks is often called the family

[15] Introduced by Y. Nambu.

[16] Only left-handed leptons form doublets while right-handed ones form singlets.

[17] The u-, d- and s-quarks were introduced by Gell-Mann [Gel 64] and Zweig [Zwe 64]. The c-quark was proposed by Maki [Mak 64], Hara [Har 64] and Bjorken and Glashow [Bj 64]. The term "charm" was adopted by Bjorken and Glashow. The t- and b-quarks were predicted by Kobayashi and Maskawa [Kob 73], and the terms "top" and "bottom" were first employed by Harari [Har 75, 75a].

structure. Note that the electric charge of the upper quarks in the series (1.2.2) is $(2/3)e$ with e the electric charge of the proton while the lower ones have charge $(-1/3)e$.

Throughout this book it is assumed that all these fundamental particles, i.e., quarks, leptons and guage bosons, are represented by local quantum fields. The present consensus in particle physics is that these local fields are described by the standard theory: electroweak theory and quantum chromodynamics.

In addition to the flavor quantum number, i.e., up, down, strange, charm, bottom and top, quarks have a hidden quantum number, color. We now explain several facts to require the necessity of the color quantum number.

1.2.1. Low-lying baryon states

In the naive quark model there is a difficulty in constructing the low-lying baryon states. As one of the simplest examples we consider the pion-nucleon resonance Δ^{++} which is of spin 3/2. On the basis of its charge, isospin and strangeness, we see that Δ^{++} is made of three u-quarks. If we consider the $J_3 = 3/2$ state with J_3 the third component of the total angular momentum for the Δ^{++} system, we find that all three u-quarks must have spins aligned up since all relative orbital angular momenta are required to vanish for the lowest state in three-quark systems. Thus the $J_3 = 3/2$ Δ^{++} state is given by

$$|\Delta^{++}, J_3 = \tfrac{3}{2}\rangle = |\,u\uparrow, u\uparrow, u\uparrow\rangle \quad , \tag{1.2.3}$$

where the arrow represents the spin aligned up. But this assignment is not acceptable because the quarks are assumed to be fermions[18] and hence the state has to be antisymmetric with respect to the exchange of the quarks. Moreover the quarks cannot occupy the same state according to the Pauli exclusion principle.

A possible way out of this difficulty may be to consider higher orbital angular momenta for the quarks. This, however, spoils the success in the prediction of baryon magnetic moments based on the S-wave three quarks. Hence we prefer to keep quarks in S-states. Then we are forced to assume the existence of a hidden degrees of freedom[19] for quarks, color [Gel 72], in order to distinguish three quarks which are otherwise identical. We need at least

[18] In order to get rid of this difficulty, Greenberg [Gre 64] regarded quarks as para-fermions. This idea is essentially equivalent to the introduction of the color quantum number [Ohn 73].

[19] Han and Nambu [Han 65] introduced an extra degree of freedom for quarks for the first time with the charge assignment different from that for the ordinary quarks.

three different colors to discriminate these three quarks. It is then easy to construct the totally antisymmetric state for Δ^{++} in place of Eq. (1.2.3),

$$|\Delta^{++}, J_3 = \tfrac{3}{2}\rangle = \varepsilon_{ijk} \,|\, u^i \uparrow, u^j \uparrow, u^k \uparrow \rangle \quad , \tag{1.2.4}$$

where indices i,j,k imply the quark colors (we assume exactly three colors, i.e., $i = 1,2,3$, the advantage of which will be seen in a moment) and ε_{ijk} is the totally antisymmetric tensor (the repeated indices are summed). The same argument applies to other baryon states and the difficulty is now circumvented with the introduction of the extra color degrees of freedom for quarks.

Since we do not observe the color degrees of freedom directly, we may assume that the hadronic phenomena be unaltered under the exchange of colors. We choose the corresponding symmetry group from the Lie groups and adopt SU(3). A single quark state is assigned to the fundamental triplet, **3**, of SU(3). The state (1.2.4) is then a singlet, **1**, of SU(3) because $\varepsilon_{ijk} \,|\, u^j \uparrow, u^k \uparrow \rangle$ constitutes the complex conjugate representation, $\mathbf{3}^*$, and gives rise to the singlet (1.2.4) when contracted with $|\, u^i \uparrow \rangle$ which belongs to **3**.

1.2.2. Quark confinement

The baryon states are in the singlet of the color symmetry SU(3) just as in Eq. (1.2.4). Group theoretically this means that in constructing the baryon state out of three quarks we have to pick out a singlet representation in the decomposition of the product of three triplets into irreducible representations:

$$\mathbf{3} \otimes \mathbf{3} \otimes \mathbf{3} = \mathbf{1} \oplus \mathbf{8} \oplus \mathbf{8} \oplus \mathbf{10}.$$

The antiquarks belong to the complex-conjugate representation, $\mathbf{3}^*$, since by charge conjugation, the representation **3** is converted to $\mathbf{3}^*$. Thus the mesons which are made of quark-antiquark pairs correspond to one of the irreducible representations in $\mathbf{3} \otimes \mathbf{3}^*$:

$$\mathbf{3} \otimes \mathbf{3}^* = \mathbf{1} \oplus \mathbf{8}.$$

In analogy with the baryon states we may take the singlet for the meson state, $|\mathrm{M}\rangle$, for which we have

$$|\mathrm{M}\rangle = \frac{1}{\sqrt{3}} \delta_{ij} \,|\, q^i \bar{q}^j \rangle \quad , \tag{1.2.5}$$

where q^i and $\bar{q}^j$ represent quarks and antiquarks. Since we have no experimental evidence of mesons which carry color quantum numbers in an explicit manner, the above choice (1.2.5) is strongly supported.

The fact that color is not directly observable may be stated in a different way: physical phenomena are invariant under the color transformation. Hence the color SU(3) has to be an exact symmetry. According to this principle all hadrons are required to be in the singlet of the color SU(3). Other states with explicit color degrees of freedom are color nonsinglets and should not be explicitly observed. For example, diquark and four-quark states (and their charge conjugation) should not be observed as they belong to color nonsinglets:

$$|qq\rangle: \quad \mathbf{3} \otimes \mathbf{3} = \mathbf{3}^* \oplus \mathbf{6} \ ,$$

$$|qqqq\rangle: \quad \mathbf{3} \otimes \mathbf{3} \otimes \mathbf{3} \otimes \mathbf{3} = \mathbf{3} \oplus \mathbf{3} \oplus \mathbf{3} \oplus \mathbf{6}^* \oplus \mathbf{15} \oplus \mathbf{15} \oplus \mathbf{15} \oplus \mathbf{15'} \ ,$$

$$|\bar{q}\bar{q}\rangle: \quad \mathbf{3}^* \otimes \mathbf{3}^* = \mathbf{3} \oplus \mathbf{6}^* \ ,$$

$$|\bar{q}\bar{q}\bar{q}\bar{q}\rangle: \quad \mathbf{3}^* \otimes \mathbf{3}^* \otimes \mathbf{3}^* \otimes \mathbf{3}^* = \mathbf{3}^* \oplus \mathbf{3}^* \oplus \mathbf{3}^* \oplus \mathbf{6} \oplus \mathbf{6} \oplus \mathbf{15}^* \oplus \mathbf{15}^* \oplus \mathbf{15}^* \oplus \mathbf{15}'^* \ .$$

It is worth noting that among the many low-lying configurations of quarks only $q\bar{q}$ and qqq states can belong to the color singlet. Thus the postulate of the nonobservability of colored states is essentially the same as that of the quark confinement which requires that quarks be confined to the inside of hadrons and are not observed as isolated states.

We recapitulate here the color nonobservability as the quark confinement postulate:

All hadron states and physical observables are color-singlets.

Here physical observables include currents, energies, momenta and masses. It should be noted that the above postulate is just a kinematical constraint to eliminate colored states. There is, however, a hope that the quark confinement may be the natural dynamical consequence of QCD.

1.2.3. Indirect experimental evidences

There are two well-known processes which bear indirect evidence for the color degrees of freedom: the decay $\pi^0 \to 2\gamma$ and e^+e^- annihilation. In these processes the number of colors appears as an extra factor for the reaction rates.

(1) $\pi^0 \to 2\gamma$ *decays*

The decay of the neutral pion into two photons takes place through the anomaly in the divergence of axial-vector currents [Adl 69, Bel 69]. This effect may also be derived by calculating the contribution of the lowest-order Feynman diagrams as shown in Fig. 1.2.1 [Fuk 49, Ste 49]. The decay rate obtained by this calculation reads

$$\Gamma(\pi^0 \to 2\gamma) = N_c^2 (Q_u^2 - Q_d^2)^2 \frac{\alpha^2 m_{\pi^0}^3}{64\pi^3 F_\pi^2} \tag{1.2.6}$$

where N_c is the number of color degrees of freedom, Q_u and Q_d the u- and d-quark charges in units of the proton charge e, m_{π^0} the neutral pion mass, $\alpha = e^2/4\pi$ and F_π the pion decay constant for $\pi \to \mu\nu$ decays ($F_\pi = 91$ MeV). In Eq. (1.2.6) the factor N_c^2 comes about since each color degree of freedom in the quark line of Fig. 1.2.1 contributes to the $\pi^0 \to 2\gamma$ decay amplitude.

Substituting $N_c = 3$, $Q_u = 2/3$ and $Q_d = -1/3$ we have

$$\Gamma(\pi^0 \to 2\gamma) = 7.6\ \text{eV} \quad , \tag{1.2.7}$$

which is in perfect agreement with the experimental data [Par 86],

$$\Gamma(\text{exp}) = 7.48 \pm 0.33\ \text{eV} \quad . \tag{1.2.8}$$

Note that the theoretical prediction with $N_c = 1$ (no color degree of freedom), $\Gamma(\pi^0 \to 2\gamma) = 0.84$ eV, is far from explaining the data.

(2) *e^+e^- annihilations*

We consider high-energy electron-positron annihilations into hadrons,

$$e^+ + e^- \to \text{hadrons} \quad . \tag{1.2.9}$$

In the quark model with asymptotic freedom, i.e., the parton model, it is assumed that quarks behave as almost-free pointlike particles when observed at short distances. Thus, as shown in Fig. 1.2.2, the process (1.2.9) proceeds through the quark-antiquark pair production by the virtual photon followed by the formation of hadrons from the quarks after final-state interactions. For annihilations at very high energies the quark final-state interaction may be

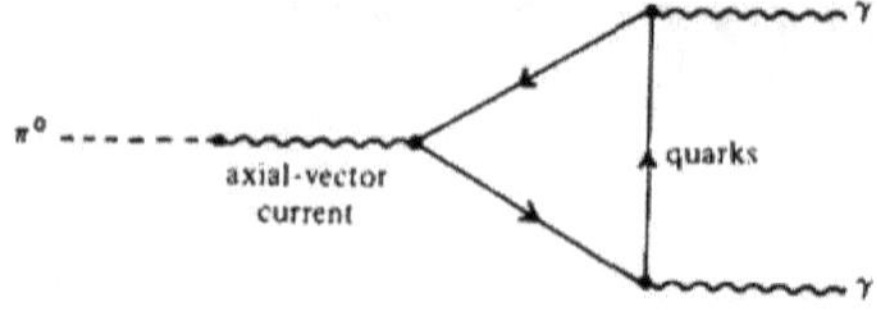

Fig. 1.2.1. The Feynman diagram relevant to the decay process $\pi^0 \to 2\gamma$.

negligible and $e^+ e^-$-annihilation total cross section $\sigma(e^+ e^- \to \text{hadrons})$ is given by[20]

$$\sigma(e^+e^- \to \text{hadrons}) = \frac{4\pi\alpha^2}{3s} N_c \sum_{i=1}^{N_f} Q_i^2 \quad , \tag{1.2.10}$$

where s is the center-of-mass total energy squared of the $e^+ e^-$ system, Q_i the charge of the i-th quark and N_f the number of flavors of quarks which may contribute to the process and is given by the condition $2m_{N_f} < \sqrt{s}$ with m_{N_f} the mass of the quark of flavor N_f. In Eq. (1.2.10) all the quark-mass effects are neglected in comparison with $\sqrt{s}$. Here again we have the explicit factor N_c which comes from the fact that each color degree of freedom in the quark line in Fig. 1.2.2 contributes to $\sigma(e^+ e^- \to \text{hadrons})$.

Instead of using Eq. (1.2.10) directly, it is more customary to define the ratio,

$$R = \sigma(e^+ e^- \to \text{hadrons})/(4\pi\alpha^2/3s) \tag{1.2.11}$$

$$= N_c \sum_{i=1}^{N_f} Q_i^2 \quad , \tag{1.2.12}$$

and compare it with experimental data. For the energy region $\sqrt{s} < 3$ GeV, only the u-, d- and s-quarks contribute since this region is below the c-quark production threshold, and hence for $N_c = 3$ we have

$$R = 3\left[\left(\frac{2}{3}\right)^2 + \left(-\frac{1}{3}\right)^2 + \left(-\frac{1}{3}\right)^2\right] = 2 \quad . \tag{1.2.13}$$

For the region $4 < \sqrt{s} < 9$ GeV there is an additional contribution from the c-quark and we obtain

$$R = 2 + 3\left(\frac{2}{3}\right)^2 = \frac{10}{3} \quad . \tag{1.2.14}$$

For $10 < \sqrt{s}$, by taking into account the b-quark contribution, we have

$$R = \frac{11}{3} \quad . \tag{1.2.15}$$

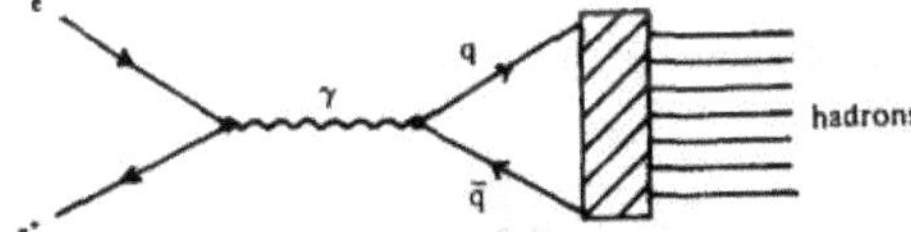

Fig. 1.2.2. The Feynman diagram contributing to the total cross section of the process $e^+e^- \to$ hadrons in the parton model.

[20] For details, see Sec. 2.3.4.

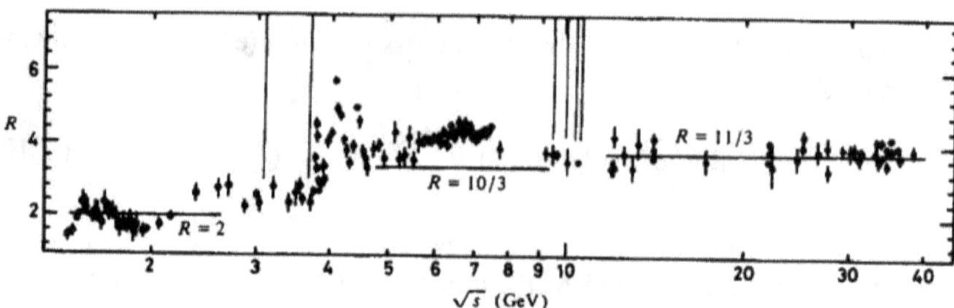

Fig. 1.2.3. Experimental data on the ratio R compared with the prediction of the parton model with $N_c = 3$. The vertical lines at $\sqrt{s} = 3.1, 3.7, 9.5, 10.0, 10.3, 10.6$ GeV represent the series of the ψ and Υ resonances.

Note that, if $N_c = 1$ (i.e., no color degree of freedom), we would have

$$R = \begin{cases} 2/3 & \text{for } \sqrt{s} < 3\,\text{GeV} \;, \\ 10/9 & \text{for } 4 < \sqrt{s} < 9 \;, \\ 11/9 & \text{for } 10 < \sqrt{s} \;. \end{cases} \tag{1.2.16}$$

The comparison of our prediction (1.2.13–15) with the existing data [Par 86] is given in Fig. 1.2.3. Obviously our prediction is consistent with the data and hence the condition $N_c = 3$ is supported here too.[21]

Digression

WHY SU(3) FOR COLOR SYMMETRY? We have not explained why we chose SU(3) as the color symmetry. In fact it could have been SO(3) or any other group. Here we explain why SU(3) is a unique candidate for color symmetry.

We restrict ourselves to compact simple Lie groups; we are only interested in simple groups as semi-simple groups are direct products of simple groups, and we require compactness so that the unitary representations are finite dimensional. The conditions that the group G for color symmetry should satisfy are summarized as follows.

(1) The number of color degrees of freedom for quarks is three, i.e., the quarks are required to belong to the triplet representation of the group G.

(2) Antiquark states are different from quark states, i.e., the triplet has to be a complex representation.

(3) Mesons and baryons are singlets of the group G.

(4) The two-quark state $|qq\rangle$ is not the singlet of the group G. The same condition holds also for other states such as $|\bar{q}\bar{q}\rangle$, $|qqqq\rangle$, ...

[21] Historically the argument with the R ratio in e^+e^- annihilations was not necessarily a strong support for $N_c = 3$ when $R = 2$ was compared with the early data. This is because errors in the data were rather large and obscured the comparison..

It is quite well-known that the compact simple Lie groups are completely classified into the following classes (see e.g., [Gil 74, Gou 67]):

$$A_N(= SU(N+1)), \quad B_N(= SO(2N+1), \quad C_N(= Sp(N)),$$

$$D_N(= SO(2N)), \quad E_6, \quad E_7, \quad E_8, \quad F_4, \quad G_2,$$

where $N = 1, 2, 3, \ldots$ (for D_N, $N > 2$) and E_6, E_7, E_8, F_4 and G_2 are exceptional groups. Among these Lie groups only SU(2), SU(3), SO(3) and Sp(1) have 3-dimensional irreducible representations. Morever three of them are isomorphic to each other:

$$SO(3) \simeq SU(2) \simeq Sp(1) \quad .$$

Hence we take SO(3) as a representative of the three. Now we are left with only two Lie groups, SU(3) and SO(3), as possible candidates for color symmetry. However, the triplet representation in SO(3) is real and hence the particle in this representation is identical to its antiparticle. On the other hand the triplet representation in SU(3) is complex. Thus only SU(3) survives under the criteria (1) and (2) among all other compact simple Lie groups. From our previous arguments it is clear that SU(3) also satisfies the criteria (3) and (4).

1.3. NEED FOR ASYMPTOTIC FREEDOM

One of the most natural ways of getting information on the structure of a particle is to use a structureless particle (like leptons) as a projectile which scatters off the particle in question. To probe the inside of the proton one naturally uses electron beams. If the proton is a point charge and the recoil effect is neglected, the differential cross section for the elastic electron-proton scattering, ep → ep, is given by the so-called Mott cross section in the laboratory frame,

$$\left(\frac{d\sigma}{d\Omega}\right)_M = \frac{\alpha^2\cos^2(\theta/2)}{4E^2\sin^4(\theta/2)} \quad , \tag{1.3.1}$$

where E and θ are the energy of the incident electron and the scattering angle in the laboratory frame respectively, and the electron mass is neglected relative to E. Equation (1.3.1) shows that there exists a considerable rate of scattering in the large angle region according to the pointlike nature of the target.

The proton is not really a point charge and the actual cross section is given instead of Eq. (1.3.1) by[22]

$$\frac{d\sigma}{d\Omega} = \left(\frac{d\sigma}{d\Omega}\right)_M \left[G_M^2(Q^2)\frac{Q^2}{2M^2}\tan^2\frac{\theta}{2} + \frac{G_E^2(Q^2) + G_M^2(Q^2)Q^2/4M^2}{1 + Q^2/4M^2}\right]\frac{E'}{E} \quad , \tag{1.3.2}$$

[22] See e.g., [Ber 68, Clo 79].

where M is the proton mass and

$$Q^2 = 4EE' \sin^2 \frac{\theta}{2} \quad , \tag{1.3.3}$$

with E' the energy of the scattered electron, and G_M and G_E are the magnetic and electric form factors respectively defined by

$$G_M(Q^2) = F_1(Q^2) + 2MF_2(Q^2)$$
$$G_E(Q^2) = F_1(Q^2) - F_2(Q^2)Q^2/2M \quad ,$$
$$\langle N|j_\mu(0)|N\rangle = \bar{u}[F_1(Q^2)\gamma_\mu + F_2(Q^2)i\sigma_{\mu\nu}q^\nu]u \quad , \tag{1.3.4}$$

with $j_\mu(x)$ and $|N\rangle$ denoting the electromagnetic current and the proton state respectively and $\sigma_{\mu\nu} = i[\gamma_\mu, \gamma_\nu]/2$. ($u$ is the Dirac spinor for the nucleon and q^ν the momentum transfered from the electron to the proton; $q^2 = -Q^2$.) The Fourier transforms of $G_M(Q^2)$ and $G_E(Q^2)$ represent the magnetic and charge spatial distributions of the proton respectively. The form factors $G_M(Q^2)$ and $G_E(Q^2)$ are decreasing functions of Q^2 if the proton is a spatially extended object. Equation (1.3.2) shows that the scattering rate at large angles is much suppressed if the proton is an extended object.

Experimentally the form factors $G_M(Q^2)$ and $G_E(Q^2)$ are known to be well approximated by the dipole form,

$$G_M(Q^2)/\mu = G_E(Q^2) = 1/(1 + Q^2/0.7)^2 \quad , \tag{1.3.5}$$

where $\mu = 2.79$ is the proton magnetic moment in units of the proton Bohr magnetons and Q^2 is measured in GeV^2. Thus the proton is really an extended object: the mean square charge radius $\langle r^2\rangle$ for the proton is given by taking the expectation value of

$$\int d^3x\, r^2 j_0(x)$$

for the proton states and is given by

$$\langle r^2 \rangle = -6G'_E(0) = 0.67 \times 10^{-26}\,\text{cm}^2 \quad , \tag{1.3.6}$$

where the prime designates the differentiation with respect to Q^2.

This extended structure of the proton may be regarded as a strong interaction effect such as that caused by pion clouds around the proton core. But for larger Q^2 it is reasonable to think that the functional form of $G_M(Q^2)$ and $G_E(Q^2)$ reflects the quark structure of the proton. In fact, according to dimensional analysis based on the quark counting (for a review, see [Siv 76]), one finds that the asymptotic behavior of the form factors is given by

$$G_M(Q^2),\quad G_E(Q^2) \sim (Q^2)^{1-n} \quad , \tag{1.3.7}$$

where n is the number of the constitutent quarks inside hadrons. In the present case of the proton, $n = 3$ and Eq. (1.3.7) is consistent with the experimental data (1.3.5) at large Q^2. Thus it is highly plausible that the extended structure of the proton observed at high Q^2 is due to the existence of quarks.

For higher incident energies of the electron the dominant process in electron-proton collisions is not the elastic but inelastic scattering: ep $\rightarrow$ e + hadrons. The process in one-photon approximation is shown in Fig. 1.3.1 and the unpolarized differential cross section for the energy and angular distribution of the scattered electrons with final hadrons unobserved is given by[23]

$$\frac{d\sigma}{d\Omega dE'} = \left(\frac{d\sigma}{d\Omega}\right)_M \left[2W_1(\nu, Q^2)\tan^2\frac{\theta}{2} + W_2(\nu, Q^2)\right]\Big/2M \quad , \qquad (1.3.8)$$

where $W_1(\nu, Q^2)$ and $W_2(\nu, Q^2)$ are relativistically invariant functions which are called structure functions (their precise definition will be given in Sec. 4.1.2), and $\nu(= E - E')$ and $-Q^2(=q^2)$ are the energy transfer (in the laboratory frame) and the momentum transfer squared of the electron respectively. The structure functions carry all the information on the proton structure relevant to the electron-proton scatterings.

The experimental data on the cross section (1.3.8) at low Q^2 reveal the resonance structure as is seen in Fig. 1.3.2. Prominent resonances seen in Fig. 1.3.2 are also observed in pion-nucleon scatterings and photon-nucleon reactions. As one observes these prominent resonance structure at low Q^2 quickly dies out as Q^2 increases, while the continuum part of the cross section persists. For example, at about $Q^2 = 2$ GeV^2 we can hardly see any of the prominent resonances and we have mainly the continuum contribution to the cross section. In fact for $Q^2 \gtrsim 2$ GeV^2 the cross section is described essentially by the continuum contribution. It is rather surprising to note that the continuum contribution is of considerable size even at high Q^2, since this fact suggests the existence of pointlike objects inside the proton.

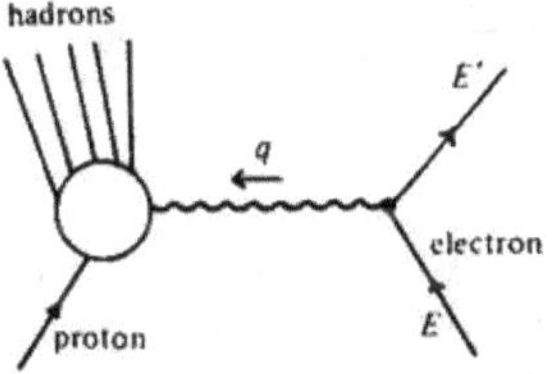

Fig. 1.3.1. The Feynman diagram relevant to the high energy electron-proton scattering in the one-photon approximation.

[23] For more details, see Sec. 4.1.2

It is this continuum region where the approximate scaling law, *Bjorken scaling*, was observed for the structure functions[24]:

$$\nu W_2(\nu, Q^2) \simeq 2MF_2(\xi) \quad , \tag{1.3.9}$$

for ν and Q^2 large with $\xi = Q^2/2M\nu$ kept fixed. A typical example of the data showing Bjorken scaling is given in Fig. 1.3.3. Bjorken scaling may be recovered in the parton model [Bj 69a]. Here we briefly recapitulate how this comes about. If the proton were pointlike, the (elastic) cross section for electron-proton scatterings would take the form

$$\frac{d\sigma}{d\Omega} = \left(\frac{d\sigma}{d\Omega}\right)_{\tilde{M}} \left(1 + \frac{Q^2}{2M^2}\tan^2\frac{\theta}{2}\right)\frac{E'}{E} \quad , \tag{1.3.10}$$

which is obtained by setting $G_M = G_E = 1$ in Eq. (1.3.2). Equation (1.3.10) can be rewritten in the following form,

$$\frac{d\sigma}{d\Omega dE} = \left(\frac{d\sigma}{d\Omega}\right)_M \left(1 + \frac{Q^2}{2M^2}\tan^2\frac{\theta}{2}\right)\delta\left(\nu - \frac{Q^2}{2M}\right) \quad . \tag{1.3.11}$$

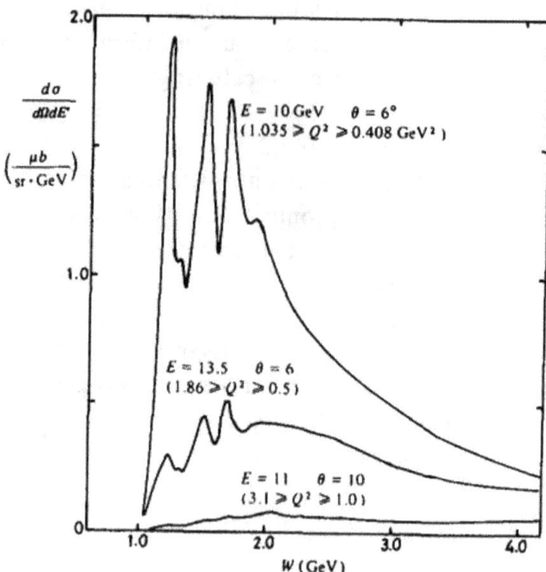

Fig. 1.3.2. A schematic view of experimental data on the cross section (1.3.8) for inelastic electron-proton scattering. Here W is an invariant mass of the final hadron system

[24] It is difficult to study the behavior of $W_1(\nu, Q^2)$ since, in Eq. (1.3.8), $W_1(\nu, Q^2)$ appears with $\tan^2(\theta/2)$ which is small in ordinary experiments.

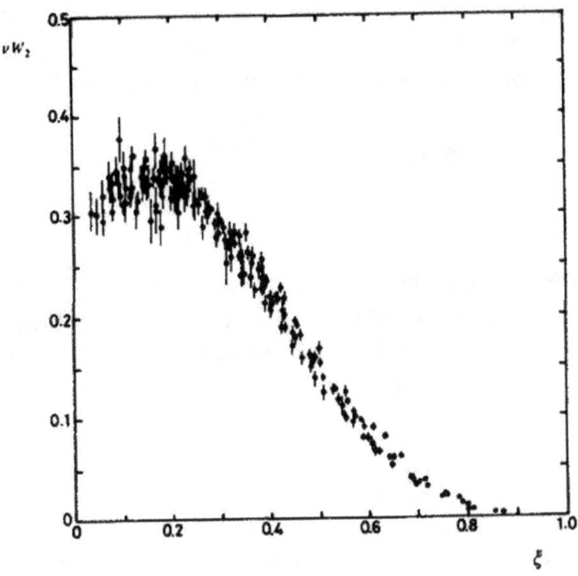

Fig. 1.3.3. Experimental data on the structure function νW_2 plotted against the Bjorken variable ξ. The concentration of data points on a curve indicates Bjorken scaling.

Thus the corresponding structure functions are

$$W_1(\nu, Q^2) = \frac{Q^2}{2M}\delta\left(\nu - \frac{Q^2}{2M}\right) \quad , \tag{1.3.12}$$

$$W_2(\nu, Q^2) = 2M\,\delta\left(\nu - \frac{Q^2}{2M}\right) \quad , \tag{1.3.13}$$

Hence we arrive at Bjorken scaling:

$$W_1(\nu, Q^2) = \xi\delta(1 - \xi) \quad , \tag{1.3.14}$$

$$\nu W_2(\nu, Q^2) = 2M\delta(1 - \xi) \quad . \tag{1.3.15}$$

Of course this is a trivial result and the actual proton is not pointlike. The above observation, however, suggests that the Bjorken scaling has something to do with the existence of the free pointlike particle. Accordingly let us assume that there are free pointlike objects (partons) inside the proton. For deep inelastic scattering where ν and Q^2 are large so that the parton mass and transverse momentum are negligible relative to ν and Q^2, the momentum of the i-th parton p_i is roughly a multiple of the proton momentum p:

$$p_i = \xi_i p \quad , \tag{1.3.16}$$

where ξ_i is the fraction of the momentum p carried by the i-th parton. Hence $\nu_i \equiv p_i{\cdot}q/m_i = \nu$ where m_i is the mass of the i-th parton and $m_i = \xi_i M$. The structure function for the i-th parton $W_1^{(i)}(\nu_i, Q^2)$ now reads

$$W_1^{(i)}(\nu_i, Q^2) = \frac{e_i^2 Q^2}{2m_i}\,\delta\left(\nu_i - \frac{Q^2}{2m_i}\right)$$
$$= e_i^2\,\delta(\xi_i - \xi) \quad , \tag{1.3.17}$$

where e_i is the charge of the i-th parton in the unit of the proton charge. The similar result may be obtained for $W_2^{(i)}(\nu_i, Q^2)$. Assuming the incoherence of electron scatterings off partons inside the proton (see Fig. 1.3.4.), we obtain for the structure function of the proton

$$W_1(\nu, Q^2) = \sum_N P(N) \sum_{i=1}^{N} \int_0^1 d\xi_i f_N(\xi_i)\, W_1^{(i)}(\nu_i, Q^2) \quad , \tag{1.3.18}$$

where $P(N)$ is the probability that the proton consists of N partons and $f_N(\xi_i)$ the probability that the i-th parton carries the momentum fraction ξ_i in the N-parton configuration. Thus we obtain Bjorken scaling:

$$W_1(\nu, Q^2) = \sum_N P(N) \sum_{i=1}^{N} e_i^2 f_N(\xi) \equiv F_1(\xi) \quad . \tag{1.3.19}$$

Similarly we have

$$\nu W_2(\nu, Q^2)/2M = \sum_N P(N) \sum_{i=1}^{N} e_i^2\, \xi f_N(\xi) \equiv F_2(\xi) \quad . \tag{1.3.20}$$

Note that in the parton model we have from Eqs. (1.3.19) and (1.3.20),

$$F_2(\xi) = \xi F_1(\xi) \quad , \tag{1.3.21}$$

which is called the Callan-Gross relation [Cal 69].

We found that Bjorken scaling is obtained by assuming the existence of free independent pointlike particles (partons) inside the proton. Conversely Bjorken scaling suggests that the quark dynamics must have the property of asymptotic freedom, i.e., the property that the quark interaction gets weaker at short distances.

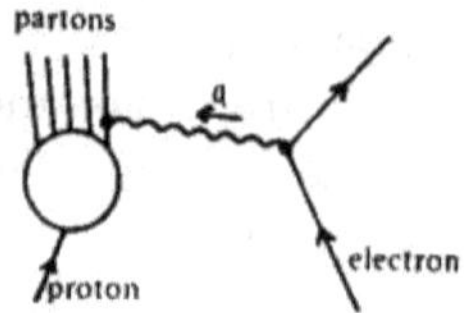

Fig. 1.3.4. The Feynman diagram exhibiting the incoherence of electron scattering off partons inside the proton.

Yet another evidence for asymptotic freedom may be found in high energy electron-positron annihilation. As was discussed in Sec. 1.2 the experimental data was well-explained by the quark-model prediction (1.2.10) in which the quark final-state interaction was neglected. This fact also suggests that the quark interaction must be weak at short distances and may be neglected in comparison with the lowest order effect.

All these evidences point to asymptotically free quark dynamics which we require to be described by quantum field theory. Such a theory is quantum chromodynamics which will be thoroughly described in the present book.

1.4. NOTATION AND CONVENTIONS

Throughout this book, unless otherwise stated, we use the natural unit $\hbar = c = 1$ where $\hbar = h/2\pi$ with h the Planck constant and c the velocity of light. Since

$$\begin{aligned} 1\ \text{cm} &= 2.84 \times 10^{37}(\hbar/c)/gr = 3.16 \times 10^{16}(\hbar c)/\text{erg} \\ &= 3.34 \times 10^{-11}\, c\ \text{sec} \quad , \end{aligned} \tag{1.4.1}$$

the units, cm, gr, erg and sec are essentially equivalent in the natural unit and we have only one independent unit. It is convenient to note that

$$\begin{aligned} 1\ \text{f} &= 10^{-13}\ \text{cm} = (\hbar c)/(0.197\ \text{GeV}) \quad , \\ 1\ \text{mb} &= (\hbar c)^2/(0.624\ \text{GeV})^2 \quad , \\ 10^{-24}\ \text{sec} &= \hbar/(0.658\,\text{GeV}) \quad , \end{aligned} \tag{1.4.2}$$

where 1 eV = 1.602×10^{-12} erg, 1 GeV = 10^3 MeV = 10^6 ke V = 10^9 eV, 1 b (barn) = 10^{-24} cm^2 and 1 mb = 10^{-3} b.

Our metric in the Minkowski space $\{x^\mu: \mu = 0, 1, 2, 3\}$ is given by $g^{\mu\nu}$ with

$$g^{00} = +1, \quad g^{11} = g^{22} = g^{33} = -1, \quad \text{otherwise} = 0 \quad . \tag{1.4.3}$$

The contravariant vectors of the space-time coordinate and energy-momentum are given by

$$x^\mu = (ct, \mathbf{r}), \quad p^\mu = (E/c, \mathbf{p}) \quad , \tag{1.4.4}$$

where t and $\mathbf{r}$ are the time and space coordinate respectively and E and $\mathbf{p}$ are the energy and momentum. The bold-faced symbols represent the three-dimensional vectors. The covariant vectors are

$$\begin{aligned} x_\mu &= g_{\mu\nu} x^\nu = (ct, -\mathbf{r}) \quad , \\ p_\mu &= g_{\mu\nu} p^\nu = (E/c, -\mathbf{p}) \quad , \end{aligned} \tag{1.4.5}$$

and hence

$$p \cdot x \equiv p^\mu x_\mu = g_{\mu\nu} p^\mu x^\nu = Et - \mathbf{p} \cdot \mathbf{r} \quad . \tag{1.4.6}$$

Here it is understood that repeated indices are summed. The contravariant vector of space and time differentiation is defined by

$$\partial^\mu \equiv \frac{\partial}{\partial x_\mu} = \left(\frac{\partial}{\partial t}, -\nabla \right) \quad , \tag{1.4.7}$$

with ∇ the gradient operation in the three-dimensional space. The d'Alembertian $\Box$ is given by

$$\Box (= \partial^2) = \partial^\mu \partial_\mu = g_{\mu\nu} \partial^\mu \partial^\nu = \frac{\partial^2}{\partial t^2} - \nabla^2 \quad . \tag{1.4.8}$$

The normalization of the one-particle state $|p, \alpha\rangle$ with momentum p and other quantum numbers α is taken to be

$$\langle p', \alpha' | p, \alpha \rangle = (2\pi)^3 2p_0 \delta^3 (p' - p) \delta_{\alpha'\alpha} \quad . \tag{1.4.9}$$

The Dirac spinors for fermion, $u_\lambda(p)$, and antifermion, $v_\lambda(p)$, with momentum p, spin component λ and mass m satisfy $(\not p - m) u_\lambda(p) = (\not p + m) v_\lambda(p) = 0$ and are normalized in such a way that

$$\begin{aligned} \bar u_\lambda(p)\, u_{\lambda'}(p) &= 2m \delta_{\lambda\lambda'} \quad , \\ \bar v_\lambda(p)\, v_{\lambda'}(p) &= -2m \delta_{\lambda\lambda'} \quad , \end{aligned} \tag{1.4.10}$$

where $\bar u_\lambda(p) = u_\lambda(p)^\dagger \gamma_0$. We use dagger† for hermitian conjugation and asterisk * for complex conjugation. Accordingly the particle and antiparticle projection operators read

$$\begin{aligned} \sum_\lambda u_\lambda(p)\, \bar u_\lambda(p) &= \not p + m \quad , \\ \sum_\lambda v_\lambda(p) \bar v_\lambda(p) &= \not p - m \quad , \end{aligned} \tag{1.4.11}$$

with $\not p = \gamma \cdot p = \gamma^\mu p_\mu$. The normalization (1.4.10) is convenient in the sense that bosons and fermions are treated in the same fashion in the Fourier expansion of the fields.

The Dirac gamma matrix $\gamma^\mu = (\gamma^0, \gamma^i)$ satisfies the anticommutation relation

$$\{\gamma^\mu, \gamma^\nu\} = 2g^{\mu\nu} \quad . \tag{1.4.12}$$

By the symbol $\{A, B\}$ we mean the anticommutator of A and B: $\{A, B\} = AB + BA$, while the commutator is given by $[A, B] = AB - BA$. The matrix γ_5 is defined by

$$\gamma_5 = i\gamma^0\gamma^1\gamma^2\gamma^3 \quad , \tag{1.4.13}$$

and anticommutes with all γ^μ:

$$\{\gamma_5, \gamma^\mu\} = 0 \quad . \tag{1.4.14}$$

The hermitian conjugate of γ^μ is taken to be

$$\gamma^{\mu\dagger} = \gamma^0\gamma^\mu\gamma^0 \quad , \tag{1.4.15}$$

so that according to the definition (1.4.13) we have

$$\gamma_5^\dagger = \gamma_5 \quad . \tag{1.4.16}$$

From the definitions (1.4.12) and (1.4.13) it follows that

$$(\gamma^0)^2 = 1, \qquad (\gamma^i)^2 = -1, \qquad \gamma_5^2 = 1 \quad , \tag{1.4.17}$$

where the Latin index i is employed to denote spatial indices 1,2,3. In the representation where γ^0 is diagonal, the explicit 4×4 form of the gamma matrices reads

$$\gamma^0 = \begin{pmatrix} 1 & 0 \\ 0 & -1 \end{pmatrix} , \quad \boldsymbol{\gamma} = \begin{pmatrix} 0 & \boldsymbol{\sigma} \\ -\boldsymbol{\sigma} & 0 \end{pmatrix} , \quad \gamma_5 = \begin{pmatrix} 0 & 1 \\ 1 & 0 \end{pmatrix} , \tag{1.4.18}$$

where by 1 and 0 we mean 2×2 matrices and σ_i is the Pauli matrix:

$$\sigma_1 = \begin{pmatrix} 0 & 1 \\ 1 & 0 \end{pmatrix} , \quad \sigma_2 = \begin{pmatrix} 0 & -i \\ i & 0 \end{pmatrix} , \quad \sigma_3 = \begin{pmatrix} 1 & 0 \\ 0 & -1 \end{pmatrix} . \tag{1.4.19}$$

CHAPTER

TWO

ELEMENTS OF QUANTUM CHROMODYNAMICS

Quantum chromodynamics is a quantized non-Abelian gauge field theory. We start with an elementary introduction to gauge field theories based on the gauge principle and then describe the method of quantizing gauge fields. We briefly review perturbation theory and present the Feynman rules for QCD. Methods of regularizing divergent integrals needed for practical higher-order calculations are explained with emphasis on dimensional regularization. Renormalization theory is discussed in some detail in consideration of its later use in the operator product expansion.

2.1. GAUGE PRINCIPLE

2.1.1. Electrodynamics

Gauge field theories are of a particular kind of field theories which are based on the gauge principle. The gauge principle is the requirement that the theory be invariant under the local gauge transformation which we shall now describe in detail.

A prototypical example of the gauge field theory is the electrodynamics of charged particles (e.g., electrons) which is an Abelian gauge field theory. In this subsection we discuss the electrodynamics of charged fermions with charge Qe and mass m (for electrons $Q = -1$) and introduce the notion of local gauge invariance.

Maxwell's equations for the electromagnetic fields $\mathbf{E}$ and $\mathbf{H}$ in the vacuum with the charge distribution $Qe\rho$ and current $Qe\mathbf{j}$ read (in the Heaviside system which is obtained by rationalizing the Gaussian CGS system by the factor $1/\sqrt{4\pi}$)

$$\nabla \times \mathbf{E} + \frac{1}{c}\frac{\partial \mathbf{H}}{\partial t} = 0 \ , \qquad \nabla\cdot\mathbf{H} = 0 \ ,$$

$$\nabla \times \mathbf{H} - \frac{1}{c}\frac{\partial \mathbf{E}}{\partial t} = \frac{1}{c}Qe\mathbf{j} \ , \quad \nabla\cdot\mathbf{E} = Qe\rho \ . \tag{2.1.1}$$

Here we shall not adopt natural units as it is more instructive at this stage to retain c and $\hbar$. Using the relativistic notation, Eq. (2.1.1) is rewritten in covariant form,

$$\partial^\mu F_{\mu\nu} = \frac{1}{c} Qej_\nu, \quad F_{\mu\nu} = \partial_\mu A_\nu - \partial_\nu A_\mu \quad , \tag{2.1.2}$$

where j^μ is the electromagnetic current four-vector,

$$j^\mu = (c\rho, \mathbf{j}) \quad , \tag{2.1.3}$$

and A^μ is the electromagnetic four-potential,

$$A^\mu = (\phi, \mathbf{A}) \quad , \tag{2.1.4}$$

related to **E** and **H** by

$$\mathbf{E} = -\nabla\phi - \frac{1}{c}\frac{\partial \mathbf{A}}{\partial t} \quad , \qquad \mathbf{H} = \nabla \times \mathbf{A} \quad . \tag{2.1.5}$$

The charged fermions may be expressed by the Dirac field ψ.[1] The relativistic Schrödinger equation for free fields, the Dirac equation, is given by

$$(i\gamma^\mu \partial_\mu - mc/\hbar)\psi = 0 \quad . \tag{2.1.6}$$

In order to obtain the Dirac equation in the presence of the electromagnetic field, let us go back to the classical electrodynamics of a charged particle of charge Qe and mass m. The classical Hamiltonian for this system has to reproduce the Lorentz force $Qe(\mathbf{E} + \mathbf{p} \times \mathbf{H}/mc)$ with $\mathbf{p}$ the momentum of the particle and is found to be [see, e.g., Pan 55]

$$H = c\sqrt{(\mathbf{p} - Qe\mathbf{A}/c)^2 + m^2c^2} + Qe\phi \quad . \tag{2.1.7}$$

It should be noted that the Hamiltonian (2.1.7) is obtained formally by replacing the momentum $\mathbf{p}$ by $\mathbf{p} - Qe\mathbf{A}/c$ and the energy (Hamiltonian) H by $H - Qe\phi$ in the free Hamiltonian $H = c\sqrt{\mathbf{p}^2 + m^2c^2}$. This means that in the quantum mechanical Hamiltonian we replace ∇ by $\nabla - iQe\mathbf{A}/\hbar c$ and ∂_0 by $\partial_0 + iQe\phi/\hbar c$ to introduce electromagnetic interactions of charged particles. Hence the prescription for obtaining the Dirac equation in the presence of the electromagnetic field is just to make the following replacement in the free Dirac equation,[2]

[1] Note that the field ψ here is the Schrödinger wave function which is yet to be quantized in the sense of the second quantization (field quantization).

[2] Note that $\nabla - iQe\mathbf{A}/\hbar c = \partial_i + iQeA_i/\hbar c$ with $i = 1, 2, 3$.

$$\partial_\mu \to \partial_\mu + iQ\frac{e}{\hbar c}A_\mu \quad . \tag{2.1.8}$$

By the above replacement we find from Eq. (2.1.6)

$$\left[i\gamma^\mu\left(\partial_\mu + iQ\frac{e}{\hbar c}A_\mu\right) - \frac{mc}{\hbar}\right]\psi = 0 \quad . \tag{2.1.9}$$

Using Eq. (2.1.9) we can easily show that $\bar{\psi}\gamma_\mu\psi(\bar{\psi} = \psi^\dagger\gamma_0)$ is a conserved current, i.e.,

$$\partial^\mu(\bar{\psi}\gamma_\mu\psi) = 0 \quad . \tag{2.1.10}$$

This conserved current is identified with the electromagnetic current j_μ in Eq. (2.1.3):

$$j_\mu = \bar{\psi}\gamma_\mu\psi \quad . \tag{2.1.11}$$

The field equations (2.1.2) and (2.1.9) with (2.1.11) describe the motion of charged particles in the electromagnetic field and are the basic equations of electrodynamics. According to the action principle these field equations have to be obtained as the Euler-Lagrange equations from a suitable Lagrangian. Such a Lagrangian (density) is given by

$$\mathcal{L} = -\frac{1}{4}F^{\mu\nu}F_{\mu\nu} + \bar{\psi}[i\gamma^\mu(\partial_\mu + iQeA_\mu) - m]\psi \quad , \tag{2.1.12}$$

where we adopted natural units so that $\hbar = c = 1$ and we understand the symmetrization $\partial_\mu \to (1/2)\overleftrightarrow{\partial}_\mu$ to guarantee the hermiticity of the Lagrangian.[3] It is an easy exercise to derive Eqs. (2.1.2) and (2.1.9) from Eq. (2.1.12) by calculating the Euler-Lagrange equations,

$$\partial_\mu\frac{\partial\mathcal{L}}{\partial(\partial_\mu A_\nu)} - \frac{\partial\mathcal{L}}{\partial A_\nu} = 0 \ , \quad \partial_\mu\frac{\partial\mathcal{L}}{\partial(\partial_\mu\bar{\psi}_\alpha)} - \frac{\partial\mathcal{L}}{\partial\bar{\psi}_\alpha} = 0 \quad , \tag{2.1.13}$$

where index α designates the four components of the Dirac field ψ.

It is immediately recognized that the Lagrangian (2.1.12) is invariant under a phase change of the field ψ:

$$\psi \to e^{-iQ\theta}\psi \quad , \tag{2.1.14}$$

with θ a real constant. The transformation (2.1.14) is called the Abelian global gauge transformation.[4] This transformation is global since the parameter θ is independent of x and it is an Abelian transformation because the charge Q is a

[3] $A\overleftrightarrow{\partial}_\mu B = A\partial_\mu B - (\partial_\mu A)B$.

[4] The term "gauge" was introduced by H. Weyl [Wey 18] in his unified theory of gravity and electromagnetism.

pure number which is commutative (Abelian). The set of transformations of the type (2.1.14) constitutes the U(1) group (the unitary group in one dimension). According to the Noether theorem[5] the invariance of the action under the transformation (2.1.14) implies the existence of the conserved current which is just equal to j^μ given by Eq. (2.1.11).

The remarkable fact is that, even when the parameter θ in the transformation (2.1.14) is made x-dependent, the Lagrangian (2.1.12) is still invariant if A^μ is transformed at the same time in an appropriate way:

$$A^\mu \to A^\mu + \frac{1}{e}\partial^\mu\theta \quad . \tag{2.1.15}$$

In fact $F^{\mu\nu} = \partial^\mu A^\nu - \partial^\nu A^\mu$ is unaltered by the transformation (2.1.15) and also the change of $\bar{\psi}\partial_\mu\psi$:

$$\bar{\psi}\partial_\mu\psi \to \bar{\psi}(\partial_\mu - iQ\partial_\mu\theta)\psi \quad ,$$

is cancelled by that of $\bar{\psi}(iQeA_\mu)\psi$:

$$\bar{\psi}(iQeA_\mu)\psi \to \bar{\psi}(iQeA_\mu + iQ\partial_\mu\theta)\psi \quad .$$

Transformations (2.1.14–15) with θ depending on x are called the *Abelian local gauge transformation*. This is local because the parameter θ depends on the local coordinate variable x. It should be noted that the transformation law for A^μ, (2.1.15), is independent of the charge Q of the fermion. As we have seen the combination $(\partial_\mu + iQeA_\mu)\psi$ behaves under the gauge transformations (2.1.14–15) as

$$(\partial_\mu + iQeA_\mu)\psi \to e^{-iQ\theta}(\partial_\mu + iQeA_\mu)\psi \quad . \tag{2.1.16}$$

The transformation property (2.1.16) is just the same as the U(1) transformation rule (2.1.14) and accordingly the operator

$$D_\mu \equiv \partial_\mu + iQeA_\mu \quad , \tag{2.1.17}$$

is called the *covariant derivative* with respect to the Abelian local gauge transformations (2.1.14–15). Note that the covariant derivative D_μ satisfies

$$[D_\mu, D_\nu] = iQeF_{\mu\nu} \quad . \tag{2.1.18}$$

We started with the Lagrangian for electrodynamics and found that it was invariant under the Abelian local gauge transformations. Let us now see what happens if we reverse the argument, i.e., we first require that the Lagrangian be invariant under the Abelian local gauge transformations and observe the

[5] We do not go into the details of this theorem here. It is explained in any texbook on field theory.

allowed form of the Lagrangian. We start with the gauge transformations (2.1.14–15) with θ depending on x. Here it is, of course, assumed that the Lagrangian is invariant under Lorentz transformation, space inversion and time reversal. The quantities obeying these properties are

$$\begin{aligned} &\partial^\mu A_\mu, \quad A^\mu A_\mu, \quad \partial^\mu A^\nu \partial_\mu A_\nu, \quad \partial^\mu A^\nu \partial_\nu A_\mu, \quad \cdots \quad , \\ &\bar{\psi}\psi, \quad \bar{\psi}\gamma^\mu \partial_\mu \psi, \quad \bar{\psi}\gamma_\mu A_\mu \psi, \quad \bar{\psi}\sigma^{\mu\nu}\partial_\mu A_\nu \psi, \quad \cdots \quad . \end{aligned} \tag{2.1.19}$$

Except for $\bar{\psi}\psi$, however, all other quantities in Eq. (2.1.19) are not gauge invariant. Recombining some of these quantities one can form gauge invariant ones, i.e.,

$$F^{\mu\nu}F_{\mu\nu}, \quad F^{\mu\nu}F_\nu^{\ \lambda}F_{\lambda\mu}, \cdots, \quad \bar{\psi}\gamma^\mu D_\mu \psi, \quad \psi\sigma^{\mu\nu}F_{\mu\nu}\psi, \quad \bar{\psi}\gamma^\mu D_\mu \psi \bar{\psi}\psi, \quad \cdots \tag{2.1.20}$$

Hence as a Lagrangian invariant under Abelian local gauge transformations, Lorentz transformations, space inversion and time reversal, we find

$$\begin{aligned} \mathscr{L} = a_1 F^{\mu\nu}F_{\mu\nu} + a_2 \bar{\psi}\gamma^\mu D_\mu \psi + a_3 \bar{\psi}\psi \\ + a_4 \bar{\psi}\sigma^{\mu\nu}F_{\mu\nu}\psi + a_5 F^{\mu\nu}F_\nu^{\ \lambda}F_{\lambda\mu} + \quad \cdots \quad , \end{aligned} \tag{2.1.21}$$

where the constants $a_i (i = 1, 2, 3, \ldots)$ are to be adjusted by rescaling the fields. The first three terms in Eq. (2.1.21) may reproduce the Lagrangian for electrodynamics (2.1.12) while the other terms are foreign. The critical difference between the first three terms and the other terms lies in their mass dimensions. To see this we need to know the mass dimensions of the fields A_μ and ψ. Since the action is dimensionless, the mass dimension of the Lagrangian in natural units is

$$\dim[\mathscr{L}] = D \quad , \tag{2.1.22}$$

where D is the space-time dimension, which is four. Hence, by observing the kinetic terms for A_μ and ψ in the Lagrangian, we find

$$\begin{aligned} \dim[A_\mu] &= \frac{D-2}{2} \quad , \\ \dim[\psi] &= \frac{D-1}{2} \quad . \end{aligned} \tag{2.1.23}$$

Thus we have

$$\begin{aligned}
&\dim[\bar{\psi}\psi] = D - 1 = 3 \quad , \\
&\dim[\bar{\psi}\gamma^{\mu} D_{\mu}\psi] = D = 4 \quad , \\
&\dim[F^{\mu\nu}F_{\mu\nu}] = D = 4 \quad , \\
&\dim[\bar{\psi}\sigma^{\mu\nu}F_{\mu\nu}\psi] = D + 1 = 5 \quad , \\
&\dim[F^{\mu\nu}F_{\nu}{}^{\lambda}F_{\lambda\mu}] = \frac{3}{2}D = 6 \quad , \ldots
\end{aligned} \tag{2.1.24}$$

According to Eqs. (2.1.22) and (2.1.24) we see that the coefficients a_1, a_2 and a_3 have positive or vanishing mass dimension while a_4, a_5, ... have negative mass dimension. As will be explained later in Sec. 2.5.2, the sign of the mass dimension of the coefficient a_i is related to the renormalizability of the theory governed by the Lagrangian (2.1.21): the term with $\dim[a_i] \geq 0$ is renormalizable while the term with $\dim[a_i] < 0$ is nonrenormalizable. Thus, if we require the renormalizability of the theory, the terms with a_4, a_5, ... in Eq. (2.1.21) should be discarded.

Summing up, we find that the Lagrangian for electrodynamics results if we impose on the theory Abelian local gauge invariance, Lorentz invariance, invariance under the space inversion and time reversal, and renormalizability.

Exercises

1. Derive the relativistic Hamiltonian for a charged particle with charge Qe in the presence of both vector and scalar potentials, A and ϕ. Confirm directly the replacement rule,

$$p^{\mu} \to p^{\mu} - QeA^{\mu}/\hbar c \quad .$$

2. Obtain the mass dimension of the electric charge e in natural units for D-dimensional space-time through the use of Lagrangian (2.1.12).

2.1.2. Yang-Mills theories

In the previous section we found that the Lagrangian for electrodynamics is obtained from the Abelian gauge principle supplemented by other obvious conditions. There, electric charges Q obey the commutative (Abelian) algebra corresponding to the U(1) group. We now try to extend the algebra to the more general one, the noncommutative (non-Abelian) algebra and see what kind of the theory results under the gauge principle. The theory obtained in this manner is the so-called *Yang-Mills theory* (or non-Abelian gauge theory).

We consider the fermion field $\psi(x)$ with mass m (the quark field, to be more specific) which belongs to the N-dimensional fundamental representation of the group G. Thus the field $\psi(x)$ has N components: $\psi_i(x)$, $i = 1, 2, \ldots, N$. The group G is not specified for the moment although we have in mind the color SU(3) group for G. For simplicity we restrict our argument to semi-simple Lie groups. The Lie algebra corresponding to the group G is generated by n generators T^a, $a = 1, 2, \ldots, n$, which are subject to the commutation relations (e.g., $T^a = \tau^a/2$, $a = 1, 2, 3$ for $G = SU(2)$ when τ^a are the Pauli matrices and $f^{abc} = \varepsilon^{abc}$ is the totally antisymmetric tensor)

$$[T^a, T^b] = if^{abc}\,T^c \quad , \tag{2.1.25}$$

where the summation on the repeated indices should be understood as usual and f^{abc} are the structure constants characterizing the algebra of the group G. The transformation property of $\psi(x)$ under the operation of the group element U of G reads

$$\psi_i' = U_{ij}\psi_j \quad , \qquad U = \exp(-iT^a\theta^a) \quad , \tag{2.1.26}$$

where θ^a are the parameters which may depend on x. The transformation (2.1.26) corresponds to the previous one in the Abelian case, (2.1.14). Note that in Eq. (2.1.26) we have T^a in place of Q in the Abelian case.

The Lagrangian for the free fermion field $\psi(x)$ is readily given by

$$\mathscr{L} = \bar{\psi}_i(i\gamma^\mu\partial_\mu - m)\psi_i \quad . \tag{2.1.27}$$

The Lagrangian (2.1.27) is clearly invariant under the transformation (2.1.26) provided the parameters θ^a are independent of x. For θ^a depending on x the Lagrangian (2.1.27) is no longer invariant under the transformation (2.1.26). According to our previous experience in Sec. 2.1.1, we suspect that the Lagrangian (2.1.27) may be made invariant under the non-Abelian local gauge transformation (2.1.26) if the derivative in Eq. (2.1.27) is replaced by the covariant derivative similar to the one in Eq. (2.1.17). Hence we try the following form for the covariant derivative,

$$D_\mu = \partial_\mu - igT^aA_\mu^a \quad . \tag{2.1.28}$$

where A_μ^a are the gauge fields and g is the constant representing the coupling strength between ψ and A_μ^a (corresponding to the unit of the charge e in the Abelian case). In component form, Eq. (2.1.28) reads

$$(D_\mu)_{ij} = \delta_{ij}\partial_\mu - i\,g\,T_{ij}^a\,A_\mu^a \quad , \tag{2.1.29}$$

where T_{ij}^a is the representation of T^a in the fundamental representation. We now replace the Lagrangian (2.1.27) by

$$\mathscr{L} = \bar{\psi}_i\left(i\gamma^\mu(D_\mu)_{ij} - m\delta_{ij}\right)\psi_j \quad ,$$
$$= \bar{\psi}(i\gamma^\mu D_\mu - m)\psi \quad , \tag{2.1.30}$$

and show that it is invariant under the non-Abelian local gauge transformation (2.1.26) provided $A^a_\mu(x)$ obey the transformation rule

$$T^a A'^a_\mu = U(T^a A^a_\mu - \frac{i}{g}U^{-1}\,\partial_\mu U)U^{-1} \quad . \tag{2.1.31}$$

The proof goes as follows: First, in order for D_μ to be the covariant derivative, we must have

$$(D_\mu\psi)' = U(D_\mu\psi) \quad . \tag{2.1.32}$$

Noting that

$$(D_\mu\psi)' = (\partial_\mu - igT^a A'^a_\mu)\psi'$$
$$= U(\partial_\mu + U^{-1}\partial_\mu U - igU^{-1}T^a U A'^a_\mu)\psi \quad , \tag{2.1.33}$$

we find that A^a_μ have to satisfy the transformation rule (2.1.31). Now, because D_μ is the covariant derivative, $\bar{\psi}D_\mu\psi$ is invariant under the non-Abelian local gauge transformation (2.1.26) and hence the Lagrangian (2.1.30) is also invariant.

The Lagrangian (2.1.30) describes the fermion fields $\psi_i(x)$ in interaction with the gauge fields $A^a_\mu(x)$. It is, however, still to be supplemented by the kinetic term consisting purely of the gauge fields $A^a_\mu(x)$. Inspired by the previous example of electrodynamics we may try the form,

$$-\frac{1}{4}(\partial_\mu A^a_\nu - \partial_\nu A^a_\mu)(\partial^\mu A^{a\nu} - \partial^\nu A^{a\mu}) \quad , \tag{2.1.34}$$

for such a term. We, however, easily see that this is not the right one, i.e., Eq. (2.1.34) is not invariant under (2.1.31). To see this it is more convenient to use an infinitesimal transformation, i.e., to make parameters θ^a infinitesimal. Keeping only terms linear in θ^a we have $U = 1 - iT^a\theta^a$, and Eq. (2.1.31) may be rewritten as

$$\delta A^a_\mu = f^{abc}\,\theta^b\,A^c_\mu - \frac{1}{g}\partial_\mu\theta^a \quad , \tag{2.1.35}$$

where $\delta A^a_\mu = A'^a_\mu - A^a_\mu$. In deriving Eq. (2.1.35) use has been made of the commutation relation (2.1.25). It should be noted here that the transformation rule for the gauge fields A^a_μ, (2.1.35), is expressed in terms only of the structure constants f^{abc} and hence we realize that, with θ^a being constant, the gauge field A^a_μ belongs to an adjoint representation A of G which is of dimension n. (Note that $(T^a)_{bc} = -if^{abc}$ in the adjoint representation.) Let us now examine how $\partial_\mu A^a_\nu - \partial_\nu A^a_\mu$ transforms under the infinitesimal transformation (2.1.35):

$$\delta(\partial_\mu A^a_\nu - \partial_\nu A^a_\mu) = \partial_\mu \delta A^a_\nu - \partial_\nu \delta A^a_\mu$$
$$= f^{abc}\,\theta^b(\partial_\mu A^c_\nu - \partial_\nu A^c_\mu) + f^{abc}\,[(\partial_\mu \theta^b)\,A^c_\nu - (\partial_\nu \theta^b)\,A^c_\mu] \quad . \tag{2.1.36}$$

Note that, if it were not for the second term at the right-hand side of Eq. (2.1.36), the form (2.1.34) would have been invariant under the infinitesimal transformation (2.1.35). Thus Eq. (2.1.34) is not invariant and we can not choose it as a part of the Lagrangian consisting of gauge fields only.

This result suggests that our choice of the field strength $\partial_\mu A^a_\nu - \partial_\nu A^a_\nu$ for the non-Abelian gauge fields is somewhat wrong and it must probably be supplemented by some additional terms so that the second term in Eq. (2.1.36) will be cancelled out. In order to get an idea on the correct expression of the field strength for the non-Abelian gauge fields, we rely on an analogy with Eq. (2.1.18) which was obtained in the Abelian case. We calculate the commutator of the covariant derivatives D_μ:

$$[D_\mu\,, D_\nu] = -igT^aF^a_{\mu\nu}\,; \quad F^a_{\mu\nu} \equiv \partial_\mu A^a_\nu - \partial_\nu A^a_\mu + gf^{abc}A^b_\mu A^c_\nu \quad . \tag{2.1.37}$$

Equation (2.1.37) is in complete analogy with the previous one (2.1.18) in the Abelian case, i.e., we have the correspondence $-g \leftrightarrow e$, $T^a \leftrightarrow Q$, $F^a_{\mu\nu} \leftrightarrow F_{\mu\nu}$. Thus $F^a_{\mu\nu}$ may be regarded as the field strength for the non-Abelian gauge fields A^a_μ. In order to obtain the transformation property of $F^a_{\mu\nu}$, we examine how the term $gf^{abc}A^b_\mu A^c_\nu$, transforms under the gauge transformation (2.1.35):

$$\delta(gf^{abc}A^b_\mu A^c_\nu) = gf^{abc}(A^b_\mu \delta A^c_\nu + A^c_\nu \delta A^b_\mu)$$
$$= f^{abc}\,[g(f^{ckl}A^b_\mu A^l_\nu + f^{bkl}A^c_\nu A^l_\mu)\,\theta^k$$
$$- A^b_\mu \partial_\nu \theta^c - A^c_\nu \partial_\mu \theta^b] \quad . \tag{2.1.38}$$

Using the Jacobi identity

$$f^{abc}f^{ckl} + f^{acl}f^{ckb} = f^{ack}f^{bcl} \quad , \tag{2.1.39}$$

which is obtained by taking the matrix element of the commutation relation

(2.1.25) in the adjoint representation $[(T^a)_{kl} = -if^{akl}]$, we have

$$\delta(gf^{abc}A^b_\mu A^c_\nu) = f^{abc}(gf^{ckl}A^k_\mu A^l_\nu\theta^b - A^b_\mu\partial_\nu\theta^c - A^c_\nu\partial_\mu\theta^b). \qquad (2.1.40)$$

According to Eqs. (2.1.36) and (2.1.40) we obtain

$$\delta F^a_{\mu\nu} = f^{abc}\theta^b F^c_{\mu\nu} \quad , \qquad (2.1.41)$$

which is precisely the transformation property in the adjoint representation. Hence $F^a_{\mu\nu}F^{a\mu\nu}$ is invariant. In fact

$$\begin{aligned}\delta(F^a_{\mu\nu}F^{a\mu\nu}) &= 2F^a_{\mu\nu}\,\delta F^{a\mu\nu} \\ &= 2f^{abc}\theta^b F^a_{\mu\nu}F^{c\mu\nu} = 0 \quad . \end{aligned} \qquad (2.1.42)$$

The last equality in Eq. (2.1.42) follows from the antisymmetry of the structure constant f^{abc}.

Finally we arrive at the general form of the Lagrangian invariant under the *non-Abelian local gauge transformations* (2.1.26) and (2.1.31),

$$\mathscr{L} = -\frac{1}{4}F^a_{\mu\nu}F^{a\mu\nu} + \bar{\psi}(i\gamma^\mu D_\mu - m)\psi \quad . \qquad (2.1.43)$$

It should be noted that in the above Lagrangian there exists only one arbitrary parameter g owing to gauge invariance. This universal constant g is called the *gauge coupling constant*. In deriving the above Lagrangian we based our argument on the gauge principle. Of course we have in mind Lorentz invariance and the invariances under the space and time reversal. Unfortunately the Lagrangian (2.1.43) is not unique in the sense that we may add to it other terms of higher powers of $F^a_{\mu\nu}$ and ψ within the requirement of the local gauge invariance, the Lorentz invariance and the invariance under the space inversion and time reversal. The additional requirement of renormalizability may eliminate all the irrelevant terms and fix the Lagrangian (2.1.43) just as in the case of electrodynamics.

A remarkable feature of the Lagrangian (2.1.43) is that it includes self-interactions among the gauge fields A^a_μ through the term $gf^{abc}A^b_\mu A^c_\nu$ in $F^a_{\mu\nu}$. This feature was not present in the case of electrodynamics and is entirely new. As will be seen later in Sec. 3.4 this self-interaction of the gauge fields (gluons in quantum chromodynamics (QCD)) is the main soure of asymptotic freedom in QCD. In QCD the gauge group G is the color SU(3), and quarks ψ and gluons A^a_μ belong to the fundamental and adjoint representations, respectively. The classical Lagrangian of QCD is given by

$$\mathscr{L} = -\frac{1}{4}F^a_{\mu\nu}F^{a\mu\nu} + \sum_{k=1}^{N_f}\bar{\psi}^k(i\gamma^\mu D_\mu - m_k)\psi^k \quad , \qquad (2.1.44)$$

where the summation on k runs over all quark flavors, and

$$F^a_{\mu\nu} = \partial_\mu A^a_\nu - \partial_\nu A^a_\mu + gf^{abc}A^b_\mu A^c_\nu \quad , \tag{2.1.45}$$

$$D_\mu = \partial_\mu - igT^a A^a_\mu \quad , \tag{2.1.46}$$

where $a = 1,2, \ldots, 8$, f^{abc} are the SU(3) structure constants and T^a the SU(3) generators.

Digression

SOME TERMINOLOGY REGARDING LIE GROUPS: The Lie group is a continuous group, i.e., a group in which each group element is labelled by continuous parameters. A group element g of the Lie group G may be written as

$$g = e^{-iT^a\theta^a} \quad , \tag{2.1.47}$$

where θ^a's are parameters necessary (and sufficient) for labelling the group elements and a runs from 1 to n. For infinitesimal θ^a one can show that the set of T^a's defined by Eq. (2.1.47) forms an algebra

$$[T^a, T^b] = if^{abc}T^c \quad , \tag{2.1.48}$$

f^{abc} being the constants peculiar to the group. The set of T^a's satisfying Eq. (2.1.48) is called the Lie algebra (or ring) of the Lie group G. Here we call

T^a	the *generators of the Lie group G*,
f^{abc}	the *structure constants* of the Lie group G,
n	the dimension of the Lie group G.

Consider a subgroup H of the group G. If, for an arbitrary group element h in H, ghg^{-1} still belongs to H for all g in G, we say that H is an invariant subgroup of the group G. In other word the invariant subgroup is a part of G which is unaltered by the transformations in G. The group which contains no (proper) invariant subgroup is called a simple group and the group which has no Abelian invariant subgroup is called a semi-simple group. Thus U(N) is neither simple nor semi-simple since U(N) may be written as SU(N) × U(1) while SU(N) is simple and SO(4) which is isomorphic to SU(2) × SU(2) is semi-simple.

A representation $D(g)$ of group element g in the group G is a matrix realization of g which obeys the same multiplication rule as that of g which defines the group G. Consider the matrices $A^a(a = 1, 2, \ldots, n)$ whose matrix elements are given by

$$(A^a)_{bc} = -if^{abc} \quad . \tag{2.1.49}$$

According to the Jacobi identity we find that A^a satisfy the same commutation relation as Eq. (2.1.48):

$$[A^a, A^b] = if^{abc}A^c \quad . \tag{2.1.50}$$

Thus $A^a = D(T^a)$, the representation of the generators T^a. The dimension of this representation is n, the dimension of the Lie group G. This representation is called the adjoint representation. In defining the Lie group G we express the generators T^a in terms of matrices. This matrix expression of T^a is by itself a representation of T^a and is called the fundamental representation. In the present book we sometimes write T^a for $D(T^a)$ if no confusion occurs.

2.2. QUANTIZATION

2.2.1. Trouble with the canonical quantization

In Sec. 2.1 we introduced non-Abelian gauge field theories in a classical form. Now we would like to quantize the gauge field theories and obtain a consistent quantum theory of gauge fields.

For simplicity we consider the case without fermions in the Lagrangian (2.1.43), i.e.,

$$\mathscr{L} = -\frac{1}{4} F^a_{\mu\nu} F^{a\mu\nu}, \tag{2.2.1}$$

with $F^a_{\mu\nu}$ given in Eq. (2.1.37). The Lagrangian (2.2.1) defines the classical theory of non-Abelian gauge fields interacting among themselves. We try to quantize the theory in the canonical formalism, i.e., we construct canonical momenta Π^a_μ conjugate to the fields A^a_μ, regard them as operators and set up canonical commutation relations between A^a_μ and Π^a_μ. The canonical momentum conjugate to A^a_μ is given by

$$\Pi^a_\mu = \frac{\partial \mathscr{L}}{\partial \dot{A}^{a\mu}} = -F^a_{0\mu} \quad , \tag{2.2.2}$$

where $\dot{A}^{a\mu} = \partial_0 A^{a\mu}$. The canonical commutation relation would be given by

$$[A^a_\mu(x)\, ,\, \Pi^b_\nu(y)]_{x_0=y_0} = i\delta_{ab} g_{\mu\nu} \delta^3(\mathbf{x} - \mathbf{y}) \quad . \tag{2.2.3}$$

Setting $\mu = \nu = 0$ and $a = b\,(= 1$, say) in Eq. (2.2.3) we obtain

$$[A^1_0(x)\, ,\, \Pi^1_0(y)]_{x_0=y_0} = i\delta^3(\mathbf{x} - \mathbf{y}) \quad . \tag{2.2.4}$$

On the other hand, from Eq. (2.2.2), we see that

$$\Pi^a_0(y) = 0 \quad . \tag{2.2.5}$$

Obviously Eq. (2.2.5) is inconsistent with the assumed commutation relation (2.2.4). Thus the simple-minded application of the canonical quantization to the gauge theory fails.

This problem already existed in the case of Abelian gauge field theory, quantum electrodynamics (QED), and has been discussed in many textbooks. The difficulty posed by Eq. (2.2.5) is inevitable as long as we rely on a gauge invariant Lagrangian where the field A^a_μ has the freedom of gauge transformations

$$A^a_\mu \to A^a_\mu + f^{abc}\theta^b A^c_\mu - \frac{1}{g}\partial_\mu \theta^a \quad , \tag{2.2.6}$$

with the infinitesimal parameters θ^a. One way of getting rid of the difficulty is to eliminate the freedom of the gauge transformation (2.2.6) by putting constraints on the field A^a_μ. For example, as a covariant constraint, we may choose the Lorentz condition

$$\partial^\mu A^a_\mu = 0 \quad . \tag{2.2.7}$$

By the constraint (2.2.7) the arbitrariness of the field A^a_μ due to the freedom of the gauge transformation (2.2.6) is eliminated. In this sense it is said that the gauge is fixed. Accordingly a constraint such as Eq. (2.2.7) is called a gauge fixing condition. By Eq. (2.2.7) the gauge is fixed to the Lorentz gauge (or covariant gauge). There are some variety of gauges other than the Lorentz gauge. Among them the following noncovariant gauges are frequently used: Coulomb (radiation) gauge $\partial_i A^a_i = 0$, axial gauge $A^a_3 = 0$ and temporal gauge $A^a_0 = 0$. Since here we are interested in quantizing the theory in a manifestly covariant manner, we adopt the Lorentz gauge.

It is well known in analytical dynamics that the Lagrange multiplier method [e.g., Gol 80] is the most useful in dealing with a constrained system. We have here the Lagrangian (2.2.1) with the constraint (2.2.7). According to the Lagrange multiplier method we add the term

$$\lambda(\partial^\mu A^a_\mu)^2 \quad , \tag{2.2.8}$$

to the Lagrangian (2.2.1) instead of directly imposing the constraint (2.2.7) on the field equations. Here the parameter λ is called the Lagrange multiplier. It is customary to write

$$\lambda = -\frac{1}{2\alpha} \quad , \tag{2.2.9}$$

and the parameter α is called the gauge parameter. The modified Lagrangian reads

$$\mathscr{L} = -\frac{1}{4} F^a_{\mu\nu} F^{a\mu\nu} - \frac{1}{2\alpha}(\partial^\mu A^a_\mu)^2 \quad . \tag{2.2.10}$$

The term added in Eq. (2.2.10) is called the gauge fixing term. Because of gauge fixing the Lagrangian is no longer gauge-invariant. However the physical predictions stemming from the Lagrangian (2.2.10) should, of course, be gauge-invariant and gauge-independent (i.e., independent of the gauge parameter α). As the value of α is irrelevant to the physical result, one often fixes the gauge parameter α. For example $\alpha = 1$ (Feynman gauge) and $\alpha \to 0$ (Landau gauge).

Fig. 2.2.1. The one-loop Feynman diagram contributing to the self-energy part $\Pi^{ab}_{\mu\nu}$ for the gauge field A^a_μ.

With the Lagrangian (2.2.10) the difficulty of having the vanishing canonical momentum (2.2.5) is circumvented since now

$$\Pi^a_\mu = -F^a_{0\mu} - \frac{1}{\alpha} g_{0\mu} (\partial^\nu A^a_\nu) \quad . \tag{2.2.11}$$

In the case of Abelian gauge field theories like quantum electrodynamics the above procedure enables us to make a consistent quantization. There is, however, a further problem in the case of non-Abelian gauge field theories. The problem is closely connected with the existence of the three-body and four-body interactions in the Lagrangian (2.2.10). Let us start with the Lagrangian (2.2.10) and perform perturbative calculations. By simply calculating the one-loop gauge-field contribution to the self-energy part $\Pi^{ab}_{\mu\nu}(q)$ for gauge fields A^a_μ (wavy line) as shown in Fig. 2.2.1 one may show that $\Pi^{ab}_{\mu\nu}(q)$ does not satisfy the requirement of the gauge invariance[6]

$$q^\mu \Pi^{ab}_{\mu\nu}(q) = 0 \quad . \tag{2.2.12}$$

Moreover, if one calculates the gauge-particle scattering cross section corresponding to the tree diagram shown in Fig. 2.2.2, and applies a simple-minded polarization sum in the final state, one fails to obtain a correct expression for the cross section. The above difficulty is related to the fact that we did not properly extract the physical polarization for the gauge field even with the Lagrangian (2.2.10).

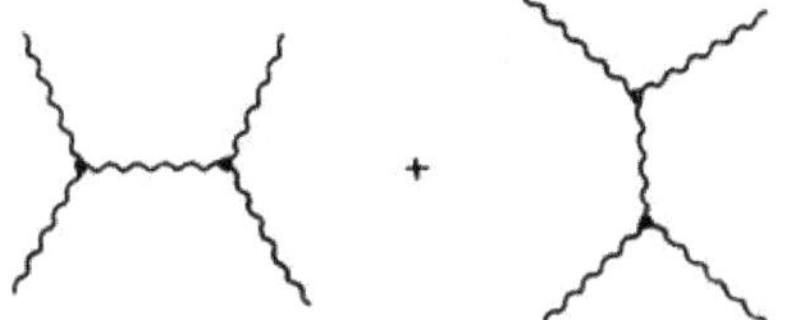

Fig. 2.2.2. Tree diagrams relevant to the elastic scattering of gauge particles.

[6] This is shown explicitly in Appendix A.

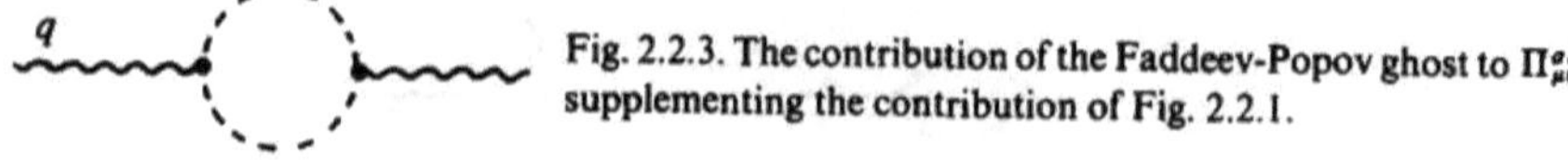

Fig. 2.2.3. The contribution of the Faddeev-Popov ghost to $\Pi^{ab}_{\mu\nu}$ supplementing the contribution of Fig. 2.2.1.

This difficulty in non-Abelian gauge field theories was first pointed out by Feynman [Fey 63] (see also [Fey 77]) in the context of the quantization of gravitational fields. The heuristic method of resolving the difficulty was suggested by Feynman and developed by DeWitt [DeW 67] in more detail (see also [Man 68]). Later the method was reformulated by Faddeev and Popov [Fad 67] in a unified manner using the Feynman path-integral (functional-integral) quantization (see also [Fra 70]). The method consists of the introduction of a new fictitious field which we now call the Faddeev-Popov ghost. This is a scalar field with a fermionic property: it is an anticommuting scalar field. The contribution of the Faddeev-Popov ghost should be added to every gauge-field loop diagram (or its discontinuity which is related to the cross sections and decay widths) in order to obtain a correct result. For example, the contributions of the Faddeev-Popov ghost (dotted line) as shown in Figs. 2.2.3 and 2.2.4 are necessary to supplement the contributions of Figs. 2.2.1 and 2.2.2, respectively.

In order to understand the origin of the Faddeev-Popov ghost it is most convenient to use the Feynman functional-integral quantization rather than canonical quantization. In the following subsections we shall discuss the quantization of non-Abelian gauge fields within the functional-integral formalism.

It should, however, be noted here that the consistent quantization of non-Abelian gauge fields based on the covariant canonical operator formalism is now available owing to the work of Kugo and Ojima [Kug 78, 79]. We shall briefly mention this formalism later in Sec. 2.3.5.

Fig. 2.2.4. The Faddeev-Popov ghost included in the final state of the gauge particle scattering diagrams to give the physical polarization sum.

2.2.2. Quantization in the Feynman functional-integral formalism

While it has become customary in most cases to use the canonical formalism in quantizing classical fields, this formalism is not the only method of field quantization. Quantization is not at all a unique procedure and a variety of quantization methods may exist which lead to the same physical prediction. There are three well-known ways of quantization which are, of course, equivalent to each other:

(1) Canonical operator formalism [Hei 29];
(2) Functional-integral formalism [Fey 48];
(3) Stochastic formalism [Par 81].

In the traditional canonical operator formalism, one regards fields as operators and sets up canonical commutation relations for them. All the Green functions which characterize the quantum theory of fields may be calculated as vacuum expectation values of the product of the field operators. In the Feynman functional-integral formalism, the fields are c-numbers and the Lagrangian is of the classical form. The Green functions are obtained by integrating the product of the fields over all of their possible functional forms with a suitable weight. As the classical form of the Lagrangian appears in the functional integral, this formalism is convenient for dealing with symmetries, such as gauge symmetries, obeyed by the classical Lagrangian. In the stochastic formalism, one notes the similarity between the functional-integral expressions of Green functions in Euclidean space and the statistical averaging, and regards the field as a stochastic variable. The Green functions are then given by the statistical average of the product of the fields in equilibrium.

We shall, in the following, describe the functional-integral formalism in more detail. For simplicity we consider a system consisting only of a neutral scalar field $\phi(x)$ with mass m. The n-point *Green function* for field $\phi(x)$ is given in the canonical operator formalism by the vacuum expectation value of the time-ordered product of n field operators[7] $\hat{\phi}(x)$: $\langle 0|\mathrm{T}[\hat{\phi}(x_1) \dots \hat{\phi}(x_n)]|0\rangle$. The same Green function may also be given by the functional integral (the proof will be given shortly):

$$\langle 0|\mathrm{T}[\hat{\phi}(x_1) \dots \hat{\phi}(x_n)]|0\rangle = \frac{\int[d\phi]\phi(x_1) \dots \phi(x_n) \exp(\mathrm{i}S)}{\int[d\phi] \exp(\mathrm{i}S)}, \tag{2.2.13}$$

where S is the classical action,

$$S = \int d^4x \mathscr{L} . \tag{2.2.14}$$

[7] We denote the field operator in the canonical operator formalism by $\hat{\phi}(x)$ and the classical field by $\phi(x)$.

Here $\mathscr{L}$ is the classical Lagrangian density:

$$\mathscr{L} = \frac{1}{2}(\partial^\mu \phi \partial_\mu \phi - m^2\phi^2) - V(\phi) \quad , \tag{2.2.15}$$

with $V(\phi)$ the part of the Lagrangian representing the self-coupling of the field ϕ. Since all the Green functions in the canonical operator formalism may be reexpressed in the form of the functional integral with the classical action, the functional-integral method gives yet another consistent description of quantum field theory and hence it provides an alternative way of quantization.

Let us first explain what is meant by the expression

$$\int [d\phi] \exp(iS) \quad , \tag{2.2.16}$$

which appeared in Eq. (2.2.13). This is one of the typical Feynman functional integrals (or path integrals) [Fey 48]. (See also [Wie 23, Dir 45, Sch 51, Mat 55].[8]) The field operator $\hat{\phi}(x)$ in the Heisenberg representation satisfies the equation of motion,

$$-i\frac{\partial\hat{\phi}(x)}{\partial t} = [\hat{H}, \hat{\phi}(x)] \quad , \tag{2.2.17}$$

and hence its time evolution is given by

$$\hat{\phi}(x) = e^{i\hat{H}t}\,\hat{\phi}(0, \mathbf{x})e^{-i\hat{H}t} \quad , \tag{2.2.18}$$

where $\hat{H}$ is the Hamiltonian of the system in the canonical operator formalism and $\hat{\phi}(0, \mathbf{x})$ is $\hat{\phi}(x)$ at $x_0 = 0$. We define $|\phi(0, \mathbf{x})\rangle$ to be an eigenstate of $\hat{\phi}(0, \mathbf{x})$ with eigenvalue $\phi(0, \mathbf{x})$:

$$\hat{\phi}(0, \mathbf{x})|\phi(0, \mathbf{x})\rangle = \phi(0, \mathbf{x})|\phi(0, \mathbf{x})\rangle \quad . \tag{2.2.19}$$

We then find according to Eqs. (2.2.18) and (2.2.19) that the state

$$|\phi(0, \mathbf{x}), t\rangle = e^{i\hat{H}t}|\phi(0, \mathbf{x})\rangle \quad , \tag{2.2.20}$$

is an eigenstate of $\hat{\phi}(x)$ with eigenvalue $\phi(0, \mathbf{x})$:

$$\hat{\phi}(x)|\phi(0, \mathbf{x}), t\rangle = \phi(0, \mathbf{x})|\phi(0, \mathbf{x}), t\rangle \quad . \tag{2.2.21}$$

We consider the transition matrix element

$$\langle \phi_f, t_f | \phi_i, t_i \rangle = \langle \phi_f | e^{-i\hat{H}(t_f - t_i)} | \phi_i \rangle \quad . \tag{2.2.22}$$

[8] For an elementary introduction to the functional integrals in field theory, see, e.g., [Abe 73, Cha 84].

where $\phi_i(\phi_f)$ and $t_i(t_f)$ are initial (final) values of $\phi(0, \mathbf{x})$ and t respectively. (The reason why we consider the quantity (2.2.22) will become clear soon.) Since the space-time coordinate x_μ is a continuous variable, the quantum field $\hat{\phi}(x)$ has infinite uncountable degrees of freedom which are inconvenient for the practical evaluation of Eq. (2.2.22). For this reason we tentatively make the volume of the space, V, finite and subdivide the space into N small cells each with volume v so that $V = Nv$. Later we let $N \to \infty$ and $v \to 0$ with V kept fixed and then $V \to \infty$ to recover the continuous Euclidean space. Now we take one representative point for each cell and represent the space by such points $\mathbf{x}_j$ $(j = 1,2, \ldots, N)$. We then have a discrete set of a finite number of fields $\phi_j \equiv \phi(0, \mathbf{x}_j)$. We also subdivide the time interval $t_f - t_i$ into $M + 1$ equal intervals of duration Δ,

$$t_0 = t_i \, , t_1 = t_i + \Delta, \ldots, t_{M+1} = t_i + (M+1)\,\Delta = t_f \quad , \tag{2.2.23}$$

and assign field ϕ^l_j to time t_l $(l = 0, 1, 2, \ldots, M+1)$ where $\phi^0_j = \phi_{ij} = \phi_i(0, \mathbf{x}_j)$ and $\phi^{M+1}_j = \phi_{fj} = \phi_f(0, \mathbf{x}_j)$. The diagrammatic expression of this subdivision is given in Fig. 2.2.5.

Using the completeness of the state $|\, \phi^l_j, t_l\rangle$,

$$\int d\phi^l_j \,|\, \phi^l_j, t_l\rangle \, \langle \phi^l_j, t_l| = 1 \quad , \tag{2.2.24}$$

we obtain

$$\begin{aligned}
\langle \phi_f, t_f|\phi_i, t_i\rangle &= \lim_{N\to\infty} \langle \phi_{f1}, \phi_{f2}, \ldots, \phi_{fN}, t_f \,|\, \phi_{i1}, \phi_{i2}, \ldots, \phi_{iN}, t_i\rangle \quad , \\
&= \lim \int \prod_{j=1}^{N} \{ d\phi^M_j \ldots d\phi^1_j \, \langle \, \phi_{fj}, t_f \,|\, \phi^M_j, t_M \,\rangle \langle \phi^M_j, t_M \,|\, \phi^{M-1}_j, t_{M-1}\rangle \\
&\quad \ldots \langle \, \phi^1_j, t_1 \,|\, \phi_{ij}, t_i\rangle \} \quad ,
\end{aligned} \tag{2.2.25}$$

where lim includes all the necessary limiting processes $N \to \infty$ keeping Nv

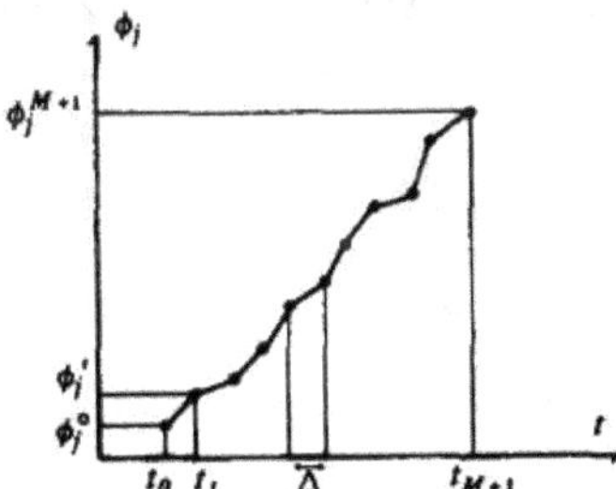

Fig. 2.2.5. The subdivision of the time interval $t_f - t_i$ into $M + 1$ sub-intervals and the field ϕ^l_j corresponding to time t_l.

finite followed by $V \to \infty$ and $M \to \infty$ with $M\Delta$ finite. The Lagrangian is discretized according to the space division,

$$\hat{L} = \int d^3x \mathscr{L} = \sum_j v \mathscr{L}_j \quad , \tag{2.2.26}$$

where $\mathscr{L}_j$ is a function only of $\dot{\hat{\phi}}_j$, $\hat{\phi}_j$ and the field at the nearest neighbor to x_j. The canonical conjugate to $\hat{\phi}_j$ is given by

$$\hat{p}_j = \frac{\partial \hat{L}}{\partial \dot{\hat{\phi}}_j} = v\dot{\hat{\phi}}_j \equiv v\hat{\pi}_j \quad . \tag{2.2.27}$$

Here each factor appearing in Eq. (2.2.25) is given, when expanded in powers of Δ, by

$$\begin{aligned} \langle \phi_j^{l+1}, t_{l+1} | \phi_j^l, t_l \rangle &= \langle \phi_j^{l+1}, t_l | e^{-i\hat{H}\Delta} | \phi_j^l, t_l \rangle \\ &= \langle \phi_j^{l+1}, t_l | \phi_j^l, t_l \rangle \\ &- i\Delta \langle \phi_j^{l+1}, t_l | \hat{H} | \phi_j^l, t_l \rangle + O(\Delta^2) \quad . \end{aligned} \tag{2.2.28}$$

The Hamiltonian may also be rewritten in the discretized form,

$$\hat{H} = \int d^3x \, \hat{\mathscr{H}} = \sum_j v \, \hat{\mathscr{H}}_j \quad , \tag{2.2.29}$$

where $\hat{\mathscr{H}}_j$ is a function only of $\hat{\pi}_j$, $\hat{\phi}_j$ and the field at the nearest neighbor to $\mathbf{x}_j$. By the use of the completeness of eigenstates $|p_j\rangle$ of $\hat{p}_j$ with eigenvalues p_j, we have for the l-th time segment,

$$\langle \phi_j^{l+1}, t_{l+1} | \phi_j^l, t_l \rangle = \int dp_j^l \langle \phi_j^{l+1}, t_l | p_j^l \rangle \langle p_j^l | \phi_j^l, t_l \rangle$$

$$- i\Delta v \int dp_j^l \langle \phi_j^{l+1}, t_l | p_j^l \rangle \langle p_j^l | \hat{\mathscr{H}}_j | p_j^l \rangle \langle p_j^l | \phi_j^l, t_l \rangle + O(\Delta^2), \tag{2.2.30}$$

$$= \int \frac{dp_j^l}{2\pi} e^{ip_j^l(\phi_j^{l+1} - \phi_j^l)} (1 - i\Delta v \mathscr{H}_j^l) + O(\Delta^2) \quad , \tag{2.2.31}$$

$$= \int \frac{dp_j^l}{2\pi} e^{ip_j^l(\phi_j^{l+1} - \phi_j^l) - \Delta v \mathscr{H}_j^l} + O(\Delta^2) \quad , \tag{2.2.32}$$

where $\mathcal{H}^l_j \equiv \langle p^l_j|\hat{\mathcal{H}}_j|p^l_j\rangle$. Here $\mathcal{H}^l_j$ is identified with the classical Hamiltonian density in the p-representation. Substituting Eq. (2.2.32) into Eq. (2.2.25) we have

$$\langle \phi_f, t_f|\phi_i, t_i\rangle$$

$$= \lim \int \prod_{j=1}^{N} \left(\prod_{l=1}^{M} d\phi_j^l \prod_{l'=0}^{M} \frac{v d\pi_j^{l'}}{2\pi} \right) \exp\left\{ i \sum_{l=0}^{M} \Delta \sum_{j=1}^{N} v \left(\pi_j^l \frac{\phi_j^{l+1} - \phi_j^l}{\Delta} - \mathcal{H}_j^l \right) \right\} . \tag{2.2.33}$$

We write the right-hand side of Eq. (2.2.33) in the following compact form

$$\langle \phi_f, t_f|\phi_i, t_i\rangle = \int [d\phi] \left[\frac{v d\pi}{2\pi} \right] \exp\left\{ i \int_{t_i}^{t_f} dt \int d^3x (\pi(x)\dot{\phi}(x) - \mathcal{H}(x)) \right\} . \tag{2.2.34}$$

Equation (2.2.34) is a typical example of a functional integral. In the case of the neutral scalar field the Hamiltonian density is given by

$$\mathcal{H} = \frac{1}{2}\pi^2 + \frac{1}{2}(\nabla\phi)^2 + \frac{1}{2}m^2\phi^2 + V(\phi) . \tag{2.2.35}$$

Replacing $\mathcal{H}$ in Eq. (2.2.34) by Eq. (2.2.35) and performing the Gaussian integrals in π in the sense of Eq. (2.2.33) we find

$$\langle \phi_f, t_f|\phi_i, t_i\rangle = C \int [d\phi] \exp\left(i \int_{t_i}^{t_f} dt \int d^3x \mathcal{L} \right) , \tag{2.2.36}$$

where $\mathcal{L}$ is the classical Lagrangian density given in Eq. (2.2.15) and C is a numerical constant. Equation (2.2.36) is almost identical to Eq. (2.2.16) but not quite. It will be shown shortly that the correspondence between them can be established by extracting the ground state contribution in Eq. (2.2.36).

We will next consider the matrix element of the field operator $\hat{\phi}(x)$, $\langle \phi_f, t_f|\hat{\phi}(x)|\phi_i, t_i\rangle$. If $t = t_l$, this matrix element may be written as

$$\begin{aligned} \langle \phi_f, t_f|\hat{\phi}(x)|\phi_i, t_i\rangle = \lim \int (\prod_{jl} d\phi_j^l) \langle \phi_{fj}, t_f|\phi_j^M, t_M \rangle \\ \ldots \langle \phi_j^{l+1}, t_{l+1}|\hat{\phi}(x)|\phi_j^l, t_l\rangle \\ \ldots \langle \phi_j^1, t_1|\phi_{ij}, t_i \rangle . \end{aligned} \tag{2.2.37}$$

According to Eq. (2.2.21) we have $\hat{\phi}(x)|\phi_j^l, t_l\rangle = \phi_j^l|\phi_j^l, t_l\rangle$ for $t = t_l$. Hence a similar procedure as that used in obtaining Eq. (2.2.36) leads to

$$\langle \phi_f, t_f | \hat{\phi}(x) | \phi_i, t_i \rangle = C \int [d\phi] \phi(x) \exp\left(i \int_{t_i}^{t_f} dt \int d^3x \mathscr{L}\right) \quad . \tag{2.2.38}$$

The matrix element of the product of two fields $\hat{\phi}(x_1)$ and $\hat{\phi}(x_2)$ may be written down in the same way. For $t_1 = t_k > t_2 = t_l$, we have

$$\begin{aligned} \langle \phi_f, t_f | \hat{\phi}(x_1) \hat{\phi}(x_2) | \phi_i, t_i \rangle = \lim \int \left(\prod_{jl} d\phi_j^l \right) \langle \phi_{fj}, t_f | \phi_j^M, t_M \rangle \\ \ldots \langle \phi_j^{k+1}, t_{k+1} | \hat{\phi}(x_1) | \phi_j^k, t_k \rangle \\ \ldots \langle \phi_j^{l+1}, t_{l+1} | \hat{\phi}(x_2) | \phi_j^l, t_l \rangle \ldots \end{aligned} \tag{2.2.39}$$

Hence we obtain

$$\begin{aligned} &\langle \phi_f, t_f | \hat{\phi}(x_1) \hat{\phi}(x_2) | \phi_i, t_i \rangle \\ &= C \int [d\phi] \phi(x_1)\, \phi(x_2) \exp\left(i \int_{t_i}^{t_f} dt \int d^3x\, \mathscr{L}\right), \quad t_f > t_1 > t_2 > t_i \end{aligned} \tag{2.2.40}$$

It is not difficult to show that for $t_2 > t_1$ the right-hand side of Eq. (2.2.40) represents the matrix element $\langle \phi_f, t_f | \hat{\phi}(x_2) \hat{\phi}(x_1) | \phi_i, t_i \rangle$. Hence for the whole region of time we obtain the formula

$$\langle \phi_f, t_f | \mathrm{T}[\hat{\phi}(x_1) \hat{\phi}(x_2)] | \phi_i, t_i \rangle = C \int [d\phi] \phi(x_1) \phi(x_2) \exp\left(i \int_{t_i}^{t_f} dt \int d^3x\, \mathscr{L}\right) \quad , \tag{2.2.41}$$

where the T-product is defined by

$$\mathrm{T}[\hat{\phi}(x_1)\hat{\phi}(x_2)] = \theta(t_1 - t_2)\hat{\phi}(x_1)\hat{\phi}(x_2) + \theta(t_2 - t_1)\hat{\phi}(x_2)\hat{\phi}(x_1), \tag{2.2.42}$$

$\theta(t)$ being the Heaviside step function. In general the following relation holds (one may prove it by induction),

$$\begin{aligned} &\langle \phi_f, t_f | \mathrm{T}\, [\hat{\phi}(x_1) \ldots \hat{\phi}(x_n)] | \phi_i, t_i \rangle \\ &\qquad = C \int [d\phi] \phi(x_1) \ldots \phi(x_n) \exp\left(i \int_{t_i}^{t_f} dt \int d^3x \mathscr{L}\right) \quad , \end{aligned} \tag{2.2.43}$$

where the T-product is the straightforward generalization of Eq. (2.2.42).

We are now in a position to extract only the ground state contribution to Eqs. (2.2.36) and (2.2.43). For this purpose we expand the eigenstate of $\hat{\phi}(0, \mathbf{x})$, $|\phi(0, \mathbf{x})\rangle$, in terms of eigenstates of the Hamiltonian (energy eigenstates) $|E_m\rangle$:

$$|\phi(0,\mathbf{x})\rangle = \sum_m |E_m\rangle\langle E_m|\phi(0,\mathbf{x})\rangle, \tag{2.2.44}$$

where $\hat{H}|E_m\rangle = E_m|E_m\rangle$ with $E_0 = 0$ and $|0\rangle$ is the vacuum state. The left-hand side of Eq. (2.2.43) is rewritten as

$$\langle\phi_f, t_f|\mathrm{T}[\hat{\phi}(x_1)\dots\hat{\phi}(x_n)]\,|\,\phi_i, t_i\rangle = \sum_{m,m'} e^{iE_m t_i - iE_{m'} t_f}\langle\phi_f|E_{m'}\rangle\langle E_m|\phi_i\rangle$$
$$\times\langle E_{m'}|\mathrm{T}[\hat{\phi}(x_1)\dots\hat{\phi}(x_n)]|E_m\rangle. \tag{2.2.45}$$

Here we set $t_i = iT$ and $t_f = -iT$, and let $T\to\infty$. We then pick up only the contribution from $E_m = E_{m'} = 0$ at the right-hand side of Eq. (2.2.45):

$$\langle\phi_f, t_f|\mathrm{T}[\hat{\phi}(x_1)\dots\phi(x_n)]|\hat{\phi}_i, t_i\rangle$$
$$\xrightarrow[T\to\infty]{}\langle\phi_f|0\rangle\langle 0|\phi_i\rangle\langle 0|\mathrm{T}[\hat{\phi}(x_1)\dots\hat{\phi}(x_n)]|0\rangle\ . \tag{2.2.46}$$

In the above manipulation t_i and t_f are imaginary. On the complex t-plane this process is shown in Fig. 2.2.6.

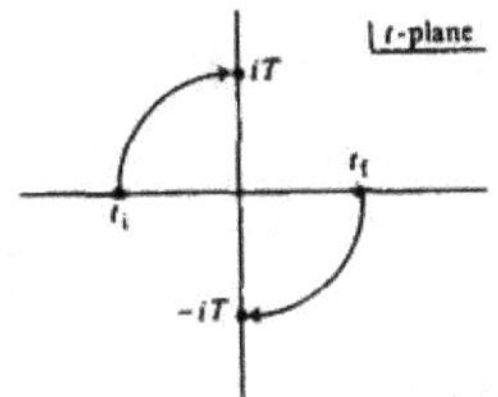

Fig. 2.2.6. The 90° rotation of the real axis in the complex t-plane needed for the analytic continuation of the matrix element (2.2.45).

The same manipulation applied to the left-hand side of Eq. (2.2. 36) leads to

$$\langle\phi_f, t_f|\phi_i, t_i\rangle\xrightarrow[T\to\infty]{}\langle\phi_f|0\rangle\langle 0|\phi_i\rangle\ . \tag{2.2.47}$$

Thus for $T\to\infty$ we find

$$\langle 0|\mathrm{T}[\hat{\phi}(x_1)\dots\phi(x_n)]|0\rangle = \frac{\int[d\phi]\phi(x_1)\dots\phi(x_n)\exp\left(i\int_{iT}^{-iT}dt\,d^3x\,\mathscr{L}\right)}{\int[d\phi]\exp\left(i\int_{iT}^{-iT}dt\,d^3x\,\mathscr{L}\right)}\ . \tag{2.2.48}$$

As there is no singularity floating around in time we can safely continue Eq. (2.2.48) analytically in time back on the real axis (by rotating the imaginary axis in Fig. 2.2.6 by +90 degrees).[8] Then we recognize that Eq. (2.2.48) is nothing but Eq. (2.2.13) which we intended to prove.

Now that the relation (2.2.13) is established, one can calculate any Green function in the functional-integral formalism without recourse to the operator language. The expression (2.2.13) may be rewritten in a more compact form by using the concept of the functional derivative with respect to an external source. This technique was first introduced by Schwinger [Sch 51a]. Let us consider an external source function $J(x)$ and introduce an artificial source term $\phi(x)J(x)$ in the functional integral,

$$Z[J] = \int [d\phi] \exp\left\{i\int d^4x(\mathscr{L} + \phi J)\right\} . \tag{2.2.49}$$

Here $Z[J]$ is a functional of $J(x)$. We define functional differentiation by

$$\frac{\delta Z[J(x)]}{\delta J(y)} = \lim_{\varepsilon\to 0}\frac{Z[J(x) + \varepsilon\delta(x-y)] - Z[J(x)]}{\varepsilon} . \tag{2.2.50}$$

According to the above definition we have

$$\frac{\delta^n Z[J]}{\delta J(x_1)\ldots\delta J(x_n)} = i^n\int [d\phi]\phi(x_1)\ldots\phi(x_n)\exp\left\{i\int d^4x(\mathscr{L}+\phi J)\right\} \tag{2.2.51}$$

Hence we obtain

$$\langle 0|\mathrm{T}[\hat{\phi}(x_1)\ldots\hat{\phi}(x_n)]|0\rangle = \frac{(-i)^n}{Z[0]}\frac{\delta^n Z[J]}{\delta J(x_1)\ldots\delta J(x_n)}\bigg|_{J=0} . \tag{2.2.52}$$

The functional integral $Z[J]$ thus generates all the Green functions. In this sense $Z[J]$ is called the *generating functional* for Green functions. The formula equivalent to Eq. (2.2.52) but in the form of the series expansion may be obtained by expanding Eq. (2.2.49) in powers of $J(x)$, i.e.,

$$\frac{Z[J]}{Z[0]} = \sum_n \frac{i^n}{n!}\int d^4x_1\ldots d^4x_n\langle 0|\mathrm{T}[\hat{\phi}(x_1)\ldots\hat{\phi}(x_n)]|0\rangle J(x_1)\ldots J(x_n) . \tag{2.2.53}$$

[8] This prescription of rotating the time axis by +90 degrees should be kept in mind when one calculates Green functions by using the functional-integral formula (2.2.13). Otherwise one would get a wrong expression for Green functions. A mnemonic aid to keeping this rule is to replace m^2 by $m^2 - i\varepsilon$ ($\varepsilon \to +0$) in the Lagrangian (2.2.15) whenever it is inserted into Eq. (2.2.13).

2.2.3. Quantization of gauge fields

We have seen in the previous subsection that in the functional-integral formalism it is enough to discuss the generating functional $Z[J]$ since it generates all the Green functions which characterize the quantum field theory under consideration. Hence we take up the generating functional for gauge fields in the present discussion on the quantization of gauge fields.

A straightforward application of Eq. (2.2.49) to the case of gauge fields suggests that the generating functional for gauge fields A_μ^a is given by

$$Z[J] = \int [dA] \exp \{ i \int d^4x (\mathscr{L} + A_\mu^a J^{a\mu}) \} \quad , \tag{2.2.54}$$

where $\mathscr{L}$ is given by Eq. (2.2.1) and $[dA]$ is understood to be shorthand notation for

$$\prod_{\mu,a} [dA_\mu^a] \quad . \tag{2.2.55}$$

In Eq. (2.2.54) $\mathscr{L}$ and $[dA]$ are invariant under the gauge transformation (2.2.6). In fact for the gauge transform $A_\mu'^a$ we find

$$\begin{aligned} [dA'] &= [dA] \det \left(\frac{\partial A_\mu'^a}{\partial A_\nu^b} \right) \\ &= [dA] \det (\delta^{ab} - f^{abc} \theta^c) \\ &= [dA] (1 + O(\theta^2)), \end{aligned} \tag{2.2.56}$$

where we used the expansion formula

$$\det (1 + L) = 1 + \operatorname{Tr} L + \ldots + \det L \quad . \tag{2.2.57}$$

Obviously, however, the source term $A^{\mu a} J_\mu^a$ is not gauge invariant. Hence $Z[J]$ as a whole is not invariant under the gauge transformation. This non-invariant nature is reflected by the gauge non-invariance of the Green functions.

If we disregard the source term, the generating functional is of course gauge invariant. We shall deal with $Z[0]$ for the moment in order to see what the main problem in connection with the choice of functional-integral measure in gauge field quantization is. The functional $Z[0]$ is given by

$$Z[0] = \int [dA] \exp (iS) \quad , \tag{2.2.58}$$

where

$$S = \int d^4x\mathscr{L} \quad , \quad \mathscr{L} = -\frac{1}{4}F^a_{\mu\nu}F^{a\mu\nu} \quad . \tag{2.2.59}$$

The action (2.2.59) is invariant under the transformation (2.1.31). In the following we set $A'^a_\mu = A_\mu^{(\theta)a}$ in order to emphasize its dependence on θ^a. Starting with a fixed A^a_μ we obtain a set of $A_\mu^{(\theta)a}$ by applying to A^a_μ all the transformations $U(\theta)$ belonging to group G. According to the above invariance the action S is constant for all $A_\mu^{(\theta)a}$ in this subset and the functional integral on this subset of $A_\mu^{(\theta)a}$ diverges as the region of the integral is infinite. Hence it is sensible to integrate only once on such $A_\mu^{(\theta)a}$ that belong to the subset and to factor out a divergent constant. For this purpose we make the following restriction on A^a_μ,

$$G^\mu A^a_\mu = B^a \quad , \tag{2.2.60}$$

which is satisfied by any $A_\mu^{(\theta)a}$ belonging to the above subset,

$$G^\mu A_\mu^{(\theta)a} = B^a \quad , \tag{2.2.61}$$

where G^μ and B^a should be chosen in an appropriate way. We require that Eq. (2.2.60) yields a unique solution θ^a for a given A^a_μ.[9] The condition (2.2.60) corresponds to the gauge fixing condition discussed previously in Sec. 2.2.1 on the canonical operator formalism. For example the Lorentz condition (2.2.7) is obtained by choosing $G^\mu = \partial^\mu$ and $B^a = 0$.

In order to take into account the constraint (2.2.61) in the functional integral (2.2.58) it is customary to use the following trick: define the functional $\Delta_G[A]$ through the equation

$$\Delta_G[A]\int[dg]\,\delta^n(G^\mu A_\mu^{(\theta)a} - B^a) = 1 \quad , \tag{2.2.62}$$

where the above functional integral is performed over the group element g in G and n is the dimension of group G so that $a = 1, 2, \ldots, n$. Here $[dg]$ is the invariant measure of the functional integral over the group space and may be given in terms of the group parameters θ^a by

$$[dg] = \prod_a [d\theta^a] \quad . \tag{2.2.63}$$

[9] This requirement is not necessarily true for Yang-Mills theories due to the phenomenon known as the Gribov ambiguity [Gri 78]. We can, however, apply the requirement to the present argument since we are only interested in the perturbative regime.

As Eq. (2.2.62) is unity, we can insert it in the right-hand side of Eq. (2.2.58) without affecting anything. We obtain

$$Z[0] = \int [dA] \prod_a [d\theta^a] \delta^n (G^\mu A^{(\theta)b} - B^b) \Delta_G[A] e^{iS} \quad . \qquad (2.2.64)$$

In Eq. (2.2.64), $[dA]$, $\prod_a [d\theta^a]$ and the action S are invariant under the gauge transformation of G. Also we can show that $\Delta_G[A]$ is invariant. In fact, according to Eq. (2.2.62) we have

$$\begin{aligned} 1/\Delta_G[A^{(\theta)}] &= \int \prod_a \{[d\theta'^a]\, \delta(G^\mu A_\mu{}^{(\theta\theta')a} - B^a)\} \\ &= \int \prod_a \{[d(\theta\theta')^a] \delta\, (G^\mu A_\mu{}^{(\theta\theta')a} - B^a)\} \\ &= 1/\Delta_G[A] \quad , \end{aligned} \qquad (2.2.65)$$

where in passing to the second line in Eq. (2.2.65) we used a property of the invariant measure (see the Digression at the end of this subsection). Thus everything except for the argument of the delta function in Eq. (2.2.64) is invariant under the gauge transformation and hence we may replace $A_\mu{}^{(\theta)a}$ by A_μ^a in the argument of the delta function:

$$Z[0] = \int [dA]\, \Delta_G[A] \int \prod_a \prod_x \{d\theta^a(x) \delta(G^\mu A_\mu^a(x) - B^a(x))\}\, e^{iS} \quad , \qquad (2.2.66)$$

where we wrote down the functional integral $[d\theta]$ in an explicit manner. Since the integrand is now independent of the group parameter $\theta^a(x)$, we can factor out the functional integral

$$\int \prod_a \prod_x d\theta^a(x) \quad , \qquad (2.2.67)$$

which is an infinite constant. We define the functional integral of $Z[0]$ for gauge fields by eliminating the factor (2.2.67), i.e.,

$$Z[0] = \int [dA] \Delta_G[A] \prod_{a,\,x} \delta(G^\mu A_\mu^a(x) - B^a(x)) e^{iS} \quad . \qquad (2.2.68)$$

We now introduce the *ansatz* that the invariant measure for functional integrals should be the one appearing in Eq. (2.2.68) for the case of gauge fields. Thus for the generating functional $Z[J]$ the proper integral measure is not that in Eq. (2.2.54) but rather

$$[dA]\Delta_G[A]\prod_{a,\,x}\delta(G^\mu A^a_\mu(x) - B^a(x)) \quad . \tag{2.2.69}$$

An explicit expression for $\Delta_G[A]$ is easily obtained by directly performing the functional integral (2.2.62):

$$\begin{aligned} 1 &= \Delta_G[A]\int\prod_a\{[d\theta^a]\delta\,(G^\mu A_\mu{}^{(\theta)a} - B^a)\} \\ &= \Delta_G[A]/\det M_G \quad , \end{aligned} \tag{2.2.70}$$

where the matrix element of M_G is given by

$$(M_G(x, y))^{ab} = \frac{\delta(G^\mu A_\mu{}^{(\theta)a}(x))}{\delta\theta^b(y)} \quad . \tag{2.2.71}$$

Using the explicit form of $\Delta_G[A]$ in the measure (2.2.69) we have an expression for $Z[J]$,

$$Z[J] = \int[dA]\,\det M_G\prod_{a,\,x}\delta(G^\mu A^a_\mu(x) - B^a(x))\exp\{i\int d^4x(\mathscr{L} + A^a_\mu J^{a\mu})\} \quad . \tag{2.2.72}$$

Since $B^a(x)$ is arbitrary, we may average $Z(J)$ over $B^a(x)$ in the sense of the functional integral, i.e., we integrate $Z[J]$ in $B^a(x)$ with a suitable weight which we choose to be

$$\exp\{-(i/2\alpha)\int d^4x(B^a(x))^2\} \quad , \tag{2.2.73}$$

where α is an arbitrary constant. Hence we obtain

$$Z[J] = \int[dA]\det M_G\exp\{i\int d^4x(\mathscr{L} - \frac{1}{2\alpha}(G^\mu A^a_\mu)^2 + A^a_\mu J^{a\mu})\} \quad . \tag{2.2.74}$$

In this way we succeeded in exponentiating the delta function which represents the constraint. The resulting exponent is the so-called gauge-fixing term with gauge parameter α.

It is $\det M_G$ which makes the quantization of gauge fields nontrivial. This factor is actually related to the origin of the difficulty in canonical quantization as discussed earlier in Sec. 2.2.1. In the following we present explicit expressions for the matrix M_G for some examples. The calculation is straightforward through the direct use of the definition (2.2.71). (Note that we neglect an irrelevant overall factor $-1/g$ in the following results.)

(1) *Coulomb gauge*: $G^\mu = (0, \nabla)$,

$$(M_G(x,y))^{ab} = (\delta^{ab}\nabla^2 - gf^{abc}\mathbf{A}^c\cdot\nabla)\delta^4(x-y) \quad . \tag{2.2.75}$$

(2) *Lorentz (covariant) gauge*: $G^\mu = \partial^\mu$,

$$(M_G(x,y))^{ab} = (\delta^{ab}\Box - gf^{abc}\partial^\mu A^c_\mu)\,\delta^4\,(x-y) \quad . \tag{2.2.76}$$

(3) *Axial gauge*: $G^\mu = n^\mu$. Here n^μ is a space-like constant 4-vector.

$$(M_G(x,y))^{ab} = (\delta^{ab} n\cdot\partial - gf^{abc} n\cdot A^c)\delta^4(x-y) \quad . \tag{2.2.77}$$

(4) *Temporal gauge*: $G^\mu = (1, 0, 0, 0)$,

$$(M_G(x,y))^{ab} = (\delta^{ab}\partial_0 - gf^{abc}A^c_0)\,\delta^4(x-y) \quad . \tag{2.2.78}$$

It should be noted here that in the cases of the axial and temporal gauges the matrix M_G is independent of the gauge field A^a_μ (conditions $n\cdot A^c = 0$ and $A^c_0 = 0$ imposed) and hence $\det M_G$ in Eq. (2.2.74) is simply a constant. Accordingly with these gauge choices we have no difficulty in the canonical quantization. Also in the case of an Abelian group G which is obtained by setting $f^{abc} = 0$, $\det M_G$ is constant. This is the reason why in quantum electrodynamics we do not have the difficulty mentioned in Sec. 2.2.1.

With the generating functional $Z[J]$ given by Eq. (2.2.74) our quantization program for gauge fields is completed. In order to develop the perturbation theory it is most convenient to use the covariant gauge. In this case, however, $\det M_G$ which is given by Eq. (2.2.76) depends on A^a_μ and hence a simple perturbative expansion of Eq. (2.2.74) is not allowed. For this purpose we need to exponentiate $\det M_G$ and regard it as a part of the effective Lagrangian. It is at this stage that we are to introduce a fictitious field called the Faddeev-Popov ghost [Fad 67]. This topic will be discussed in Sec. 2.3.3.

Digression

INVARIANT MEASURE OVER GROUP SPACE: For any element g of group G we consider an integral over G of an arbitrary integrable function $f(g)$,

$$\int f(g)\,dg \quad , \tag{2.2.79}$$

where dg is the integral measure over the group G to be defined in an appropriate way. To understand mathematical definitions it is often most instructive to start with an example. Here we take the example of the group of triangular matrices. A set of 2 × 2 triangular matrices

$$g = \begin{pmatrix} x & y \\ 0 & z \end{pmatrix} \quad , \tag{2.2.80}$$

constitutes a 3-parameter group G where x, y and z are positive real numbers. We may take as a measure over G,

$$dg = dx\, dy\, dz \quad . \tag{2.2.81}$$

Consider an element g_0 of G,

$$g_0 = \begin{pmatrix} a & b \\ 0 & c \end{pmatrix} \quad . \tag{2.2.82}$$

and apply g_0 from the left to the element g of Eq. (2.2.80) to get another element g':

$$g' = g_0 g \equiv \begin{pmatrix} x' & y' \\ 0 & z' \end{pmatrix} \quad . \tag{2.2.83}$$

For the choice of measure (2.2.81) we find

$$dg' = dx'\, dy'\, dz' = a^2 c\, dx\, dy\, dz = a^2 c\, dg \quad . \tag{2.2.84}$$

Hence the form of the integral (2.2.79) is not kept invariant under the transformation (2.2.83) even if $f(g') = f(g)$, i.e.,

$$\int f(g')\, dg' = a^2 c \int f(g_0 g)\, dg \quad . \tag{2.2.85}$$

Obviously the situation is improved if we redefine the measure dg by

$$dg = \frac{dx\, dy\, dz}{x^2 z} \quad . \tag{2.2.86}$$

In fact for the measure (2.2.86) we have $dg' = dg$ and hence

$$\int f(g')\, dg' = \int f(g_0 g)\, dg \quad . \tag{2.2.87}$$

Thus the measure (2.2.86) is much more convenient than the naive one, Eq. (2.2.81). Such a measure as (2.2.86) is called the (left) *invariant measure*. Now the precise mathematical definition: The integral measure dg is called the left invariant measure, if

$$\int f(g_0 g)\, dg = \int f(g) dg \quad . \tag{2.2.88}$$

for arbitrary elements g_0 and g of group G and arbitrary functions $f(g)$. Similarly the right invariant measure is defined through the transformation $g' = g g_0$. In the previous example, the right invariant measure is not equal to the left invariant measure. In fact the right invariant measure is given by

$$\frac{dx\, dy\, dz}{x z^2} \quad , \tag{2.2.89}$$

which disagrees with Eq. (2.2.86). Thus, in general, the left and right invariant measures are not necessarily equal. Fortunately, however, it is known that the left and right invariant measures are equal for the following kinds of groups: compact groups, simple groups, semi-simple groups, finite groups, etc... For details on the invariant measure, see, e.g., [Gil 74].

2.2.4. Fermions

The functional-integral quantization has been successfully performed for scalar and gauge fields in the last subsections 2.2.2 and 2.2.3. The case of fermion fields is dealt with in a similar manner if one properly takes into account the fact that they satisfy anticommutation relations in the canonical operator formalism.

In the functional integral formalism we have only classical c-number fields and, at first sight, there seems to be no way of considering the anticommuting nature of fermion fields. There is, however, a mathematical tool to properly describe this property known as the Grassmann algebra. This algebra is discussed in [Ber 66] with emphasis on its application to field theory (see also [Ohn 78, Fad 80]).

Ordinary numbers $\phi_j (j = 1, 2, ..., N)$, of course, commute with each other,

$$[\phi_i, \phi_j] = 0 \quad , \tag{2.2.90}$$

and constitute a commutative algebra. In contrast with the ordinary numbers, we may consider a set of anticommuting numbers $\psi_j (j = 1, 2, ..., N)$,

$$\{\psi_i, \psi_j\} = 0 \quad . \tag{2.2.91}$$

In particular $\psi_j^2 = 0$ (no summation on j). These N anticommuting numbers ψ_j generate an algebra called the Grassmann algebra. The number ψ_j is called the Grassmann number.[10] The Grassmann algebra is regarded as a linear space the basis of which is given by the set of monomials

$$1, \psi_i, \psi_i\psi_j, ..., \psi_1\psi_2 \cdots \psi_N \quad . \tag{2.2.92}$$

In the above set (2.2.92) we have only a finite number of monomials which include up to N ψ_j's because monomials with more than N ψ_j's vanish according to the property $\psi_j^2 = 0$. The total number of monomials in the set (2.2.92) is

$$1 + {}_NC_1 + {}_NC_2 + ... + {}_NC_N = 2^N \quad , \tag{2.2.93}$$

where ${}_NC_r$ is the binomial coefficient. Hence the dimension of the linear space is 2^N.

The number ϕ_j above may be identified with the discretized field ϕ_j appearing in Sec. 2.2.2 and thus ψ_j may be thought of as the corresponding anticommuting field. Then, by taking the continuum limit $N \to \infty$, we can define an anticommuting classical field $\psi(x)$ which is identified with the

[10] The Grassmann numbers commute with ordinary numbers.

fermion in the functional-integral formalism. In the following we shall introduce differentiation and integration in the finite (N)-dimensional Grassmann algebra and take the continuum limit $N \to \infty$ to define the functional integral for fermions.

As the Grassmann numbers ψ_j are anticommuting, we require some care to define the derivative. Let us consider a product $\psi_{i_1} \psi_{i_2} \cdots \psi_{i_k}$ and define the left derivative by the rule

$$\frac{\partial}{\partial \psi_j}(\psi_{i_1} \psi_{i_2} \cdots \psi_{i_k}) = \begin{cases} (-1)^{l-1} \psi_{i_1} \psi_{i_2} \cdots (\psi_{i_l}) \cdots \psi_{i_k}, & \text{if } i_l = j, \\ 0, & \text{if no } \psi_j \text{ is included,} \end{cases} \tag{2.2.94}$$

where by (ψ_{i_l}) is meant that ψ_{i_l} is missing. In other words, the left derivative is obtained by the following procedure, if $i_l = j$, first bring ψ_{i_l} to the left of ψ_{i_1} by the use of the commutation relation (2.2.91) and then drop it. The right derivative is defined in a similar manner:

$$(\psi_{i_1} \psi_{i_2} \cdots \psi_{i_k}) \frac{\overleftarrow{\partial}}{\partial \psi_j} = \begin{cases} (-1)^{k-l} \psi_{i_1} \psi_{i_2} \cdots (\psi_{i_l}) \cdots \psi_{i_k}, & \text{if } i_l = j \\ 0, & \text{if no } \psi_j \text{ is included.} \end{cases} \tag{2.2.95}$$

According to the definition (2.2.94) we obtain the following formula for the second derivative of an arbitrary element F of the Grassmann algebra,

$$\frac{\partial^2 F}{\partial \psi_i \, \partial \psi_j} = - \frac{\partial^2 F}{\partial \psi_j \, \partial \psi_i} \quad . \tag{2.2.96}$$

The formula (2.2.96) may be easily proven by observing the relations,

$$\frac{\partial}{\partial \psi_i} \frac{\partial}{\partial \psi_j} (\psi_{i_1} \cdots \psi_{i_k}) = (-1)^{m-1} (-1)^{l-1} \psi_{i_1} \cdots (\psi_{i_l}) \cdots (\psi_{i_m}) \cdots \psi_{i_k} \tag{2.2.97}$$

$$\frac{\partial}{\partial \psi_j} \frac{\partial}{\partial \psi_i} (\psi_{i_1} \cdots \psi_{i_k}) = (-1)^{l-1} (-1)^{m-2} \psi_{i_1} \cdots (\psi_{i_l}) \cdots (\psi_{i_m}) \cdots \psi_{i_k} \quad , \tag{2.2.98}$$

where $i_l = i$ and $i_m = j$. In particular, as a special case of Eq. (2.2.96), we obtain

$$\frac{\partial^2 F}{\partial \psi_j^2} = 0, \quad (\text{no summation on } j) \quad . \tag{2.2.99}$$

The definition of the integral in Grassmann algebra is more sophisticated. While the integral over ordinary numbers is defined as an inverse of differentiation, the integral in Grassmann algebra in that sense is not well-defined because of the property (2.2.99). Hence we shall rely on an indirect definition of the integral with the following operational rules

$$\int d\psi_i = 0 \quad , \tag{2.2.100}$$

$$\int d\psi_i \psi_j = \delta_{ij} \quad , \tag{2.2.101}$$

where

$$\{d\psi_i, \psi_j\} = 0, \qquad \{d\psi_i, d\psi_j\} = 0 \quad . \tag{2.2.102}$$

As a typical example of an integral over a Grassmann algebra, let us consider the following,

$$I = \int d\psi_1 d\psi_2 \ldots d\psi_N d\bar{\psi}_1 d\bar{\psi}_2 \ldots d\bar{\psi}_N \exp(\bar{\psi}_i A_{ij} \psi_j) \quad , \tag{2.2.103}$$

where summation on i and j should be understood and A_{ij} are complex numbers. In Eq. (2.2.103) we took the $2N$-dimensional Grassmann algebra and divided it into two disjoint sets $\{\psi_j\}$ and $\{\bar{\psi}_j\}$ with $j = 1, 2, \ldots, N$. Here $\bar{\psi}_j$ could be a complex conjugate of ψ_j or it may have nothing to do with ψ_j. Because of the property $\psi_i^2 = \bar{\psi}_i^2 = 0$, the integrand in Eq. (2.2.103) is expanded in a power series which terminates at order N,

$$\exp(\bar{\psi}_i A_{ij} \psi_j) = 1 + \bar{\psi}_i A_{ij} \psi_j + \ldots + \frac{1}{N!}(\bar{\psi}_i A_{ij} \psi_j)^N \quad . \tag{2.2.104}$$

We then use the rules (2.2.100) and (2.2.101) to find that only the term of order N in Eq. (2.2.104) contributes to the integral (2.2.103), i.e.,

$$I = \int \prod d\psi \prod d\bar{\psi} \frac{1}{N!} (\bar{\psi}_i A_{ij} \psi_j)^N \quad , \tag{2.2.105}$$

where

$$\prod d\psi = \prod_{i=1}^{N} d\psi_i \, , \qquad \prod d\bar{\psi} = \prod_{j=1}^{N} d\bar{\psi}_j \quad . \tag{2.2.106}$$

Applying the commutation rules for ψ_j and $\bar{\psi}_j$, we have

$$\begin{aligned}
&(\bar{\psi}_i A_{ij} \psi_j)^N \\
&= (-1)^{N(N-1)/2} \sum_{i_1 \ldots i_N j_1 \ldots j_N} A_{i_1 j_1} \cdots A_{i_N j_N} \bar{\psi}_{i_1} \cdots \bar{\psi}_{i_N} \psi_{j_1} \cdots \psi_{j_N} \quad , \\
&= (-1)^{N(N-1)/2} \sum_{i_1 \ldots i_N j_1 \ldots j_N} \varepsilon_{i_1 \ldots i_N} \varepsilon_{j_1 \ldots j_N} A_{i_1 j_1} \cdots A_{i_N j_N} \bar{\psi}_1 \ldots \bar{\psi}_N \psi_1 \ldots \psi_N
\end{aligned}$$

$$= (-1)^{N(N-1)/2}\, N! \sum_{j_1 \ldots j_N} \varepsilon_{j_1 \ldots j_N} A_{1j_1} \cdots A_{Nj_N} \bar{\psi}_1 \ldots \bar{\psi}_N \psi_1 \ldots \psi_N \quad , \tag{2.2.107}$$

where $\varepsilon_{i_1 \ldots i_N}$ is a totally antisymmetric tensor of rank N:

$$\varepsilon_{i_1 \ldots i_N} = \begin{cases} +1, & \text{if } (i_1 \ldots i_N) \text{ is an even permutation of } (1,2,\ldots,N), \\ -1, & \text{if } (i_1 \ldots i_N) \text{ is an odd permutation of } (1,2,\ldots,N), \\ 0, & \text{otherwise.} \end{cases} \tag{2.2.108}$$

Since the determinant of the matrix $A \equiv (A_{ij})$ is given by

$$\det A = \sum_{j_1 \ldots j_N} \varepsilon_{j_1 \ldots j_N} A_{1j_1} \cdots A_{Nj_N} \quad , \tag{2.2.109}$$

we finally obtain from Eqs. (2.2.105) and (2.2.107) with the integration rules (2.2.100) and (2.2.101),

$$I = (-1)^{N(N-1)/2} \det A \quad . \tag{2.2.110}$$

This result should be contrasted with that for ordinary commuting numbers. Consider an integral over complex numbers

$$J = \int d\phi_1 \ldots d\phi_N d\bar{\phi}_1 \ldots d\bar{\phi}_N \exp(-\bar{\phi}_i A_{ij} \phi_j) \quad , \tag{2.2.111}$$

where $\bar{\phi}_i$ is a complex conjugate of ϕ_i. If the matrix $A = (A_{ij})$ is diagonalizable, we can easily show that Eq. (2.2.111) reduces to the multiple Gaussian integral and find

$$J = \pi^N / \det A \quad . \tag{2.2.112}$$

Thus the integral over Grassmann algebra I gives an inverse of the integral over ordinary numbers J. Taking the continuum limit of Eq. (2.2.110) we find

$$\int [d\psi]\, [d\bar{\psi}] \exp\{ \int d^4x d^4y \bar{\psi}(x) A(x,y) \psi(y) \} = \det A, \tag{2.2.113}$$

up to a sign. The above relation (2.2.113) is very useful both in evaluating functional integrals for fermions and in exponentiating a determinant. In fact Eq. (2.2.113) will be employed in Sec. 2.3.3 to exponentiate $\det M_G$ of Eq. (2.2.74).

To close this subsection we present the generating functional including fermion fields for the sake of completeness,

$$Z[J,\eta,\bar{\eta}] = \int [dA]\, [d\psi]\, [d\bar{\psi}] \det M_G$$

$$\times \exp\{ i \int d^4x (\mathscr{L} - (1/2\alpha)(\partial^\mu A^a_\mu)^2 + A^a_\mu J^{a\mu} + \bar{\psi}\eta + \bar{\eta}\psi) \} \quad , \tag{2.2.114}$$

where η and $\bar{\eta}$ are anticommuting source functions for fermion fields $\bar{\psi}$ and ψ with $\bar{\psi}=\psi^{\dagger}\gamma_0$, and $\mathscr{L}$ is the same as Eq. (2.1.43),

$$\mathscr{L} = -\frac{1}{4} F^a_{\mu\nu} F^{a\mu\nu} + \bar{\psi}\,(i\gamma^\mu D_\mu - m)\,\psi \quad , \tag{2.2.115}$$

where m is the fermion mass. The fermion Green functions are obtained in a similar manner as in Eq. (2.2.52) by making functional derivatives with respect to η and $\bar{\eta}$. It should be stressed here that $\psi(x)$, $\bar{\psi}(x)$, $\eta(x)$ and $\bar{\eta}(x)$ are all anticommuting but classical (c-number) so that

$$\{\psi(x)\ ,\bar{\psi}(y)\} = 0\ ,\quad \{\psi(x)\ ,\ \ \eta(y)\} = 0\quad ,\text{etc.} \tag{2.2.116}$$

Hence proper account has to be taken of the sign coming from the above anticommutation relations when we define fermion Green functions. For example the fermion two-point Green function (propagator) should be defined in the following way,

$$\langle 0|\mathrm{T}[\psi_\alpha(x)\bar{\psi}_\beta(y)]|0\rangle = \frac{(-i)^2}{Z[0,0,0]}\frac{\delta^2 Z[J,\eta,\bar{\eta}]}{\delta\bar{\eta}_\alpha(x)\,\delta(-\eta_\beta(y))}\bigg|_{J=\bar{\eta}=\eta=0} . \tag{2.2.117}$$

This rule of doing the functional differentiation with $-\eta(x)$ should always be kept in mind in defining fermion Green functions.

2.3. FEYNMAN RULES

2.3.1. *S* matrix and Green functions

In elementary particle physics we are interested in physical quantities such as scattering cross sections and decay widths. As a typical example let us consider the 2-to-n-particle scattering process as depicted in Fig. 2.3.1,

$$(p_1 J_1) + (p_2 J_2) \rightarrow (k_1 j_1) + \ldots + (k_n j_n) \quad , \tag{2.3.1}$$

where p_1 and p_2 are momenta of the initial two particles with spins J_1 and J_2, respectively, while k_1, k_2, ... and k_n are momenta of the final n particles with spins j_1, j_2 ... and j_n, respectively. Here, for simplicity, we do not account for

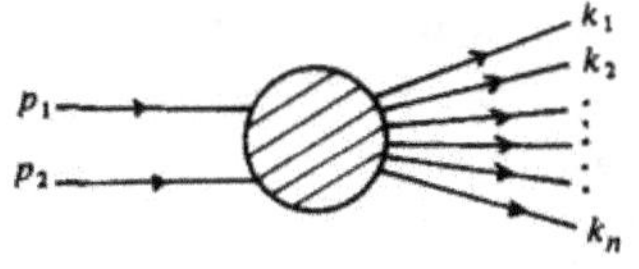

Fig. 2.3.1. Scattering process with two initial particles of momenta p_1 and p_2 producing n particles of momenta k_1, k_2, . . . , k_n.

internal degrees of freedom (such as isospin) other than spin. Inclusion of the other internal degrees of freedom, however, should be straightforward.

The unpolarized total *cross section* σ for the process (2.3.1) is given by

$$\sigma = \sum \frac{W}{FD} \quad , \tag{2.3.2}$$

where $\sum$ represents an averaging over initial particle polarizations, W is the transition probability per unit time and unit volume, F the incident particle flux and D the target-particle density. Here we work in the laboratory frame in which particle 1 is incident on target particle 2. The incident particle flux F and the target particle density D are given, with our state normalization (1.4.9), by

$$F = 2p_{10}|\mathbf{v}_1 - \mathbf{v}_2| = K(s)/m_2 \quad , \tag{2.3.3}$$

$$D = 2p_{20} = 2m_2 \quad , \tag{2.3.4}$$

where $\mathbf{v}_i = \mathbf{p}_i/p_{i0}$ $(i = 1,2)$ is the velocity of particle i $(\mathbf{v}_2 = 0)$ and

$$K(s) = \sqrt{(s-(m_1+m_2)^2)\,(s-(m_1-m_2)^2)} \quad , \tag{2.3.5}$$

with m_i $(i = 1, 2)$ the mass of the incident particle i and

$$s = (p_1 + p_2)^2 \quad . \tag{2.3.6}$$

The transition probability W per unit time and unit volume for the process (2.3.1) is calculated through the S-matrix element,

$$W = \frac{1}{VT}\int \frac{d^3k_1}{(2\pi)^3 2k_{10}} \cdots \frac{d^3k_n}{(2\pi)^3 2k_{n0}} \times \sum_{\mu_1 \cdots \mu_n} \left| \langle k_1\mu_1, \ldots, k_n\mu_n|(S-1)|p_1\lambda_1, p_2\lambda_2\rangle \right|^2 \quad , \tag{2.3.7}$$

where VT is the space-time volume in which the scattering takes place, i.e., in a symbolic notation,

$$VT = \int d^4x = \int d^4x\, e^{ip\cdot x}\bigg|_{p=0} = (2\pi)^4\delta^4(0) \quad , \tag{2.3.8}$$

S represents the S *matrix* and $\lambda_1, \lambda_2, \mu_1 \ldots, \mu_n$ denote helicities of the particles with momenta $p_1, p_2, k_1, \ldots, k_n$, respectively. Here the appearance of the factors $(2\pi)^3 2k_{10}, \ldots$ etc. in Eq. (2.3.7) is due to our choice of the state normalization (1.4.9). The S-matrix element is related to the *transition matrix element* $\langle k_1\mu_1, \ldots, k_n\mu_n|T|p_1\lambda_1, p_2\lambda_2\rangle$ by

$$\langle k_1\mu_1, \ldots, k_n\mu_n|\,(S-1)\,|p_1\lambda_1, p_2\lambda_2\rangle = (2\pi)^4\, i\delta^4\left(\sum_{j=1}^{n} k_j - p_1 - p_2\right)$$
$$\times\ \langle k_1\mu_1, \ldots, k_n\mu_n|T|p_1\lambda_1, p_2\lambda_2\rangle \quad , \tag{2.3.9}$$

where the delta function on the right-hand side represents the energy-momentum conservation. Substituting Eqs. (2.3.3), (2.3.4), (2.3.7), (2.3.8) and (2.3.9) into Eq. (2.3.2) we obtain

$$\sigma = \frac{1}{2K(s)}\frac{1}{(2J_1+1)(2J_2+1)}$$
$$\times \sum_{\lambda_1\lambda_2\mu_1\ldots\mu_n} \int \prod_{j=1}^{n} \frac{d^3k_j}{(2\pi)^3 2k_{j0}} (2\pi)^4\delta^4\left(\sum_{j=1}^{n} k_j - p_1 - p_2\right)$$
$$\times |\langle k_1\mu_1, \ldots, k_n\mu_n|T|p_1\lambda_1, p_2\lambda_2\rangle|^2 \quad . \tag{2.3.10}$$

We have seen that scattering cross sections are calculated through Eq. (2.3.10) once the transition matrix element defined in Eq. (2.3.9) is known. In the same way we can show that decay rates are calculated by the use of the formula similar to Eq. (2.3.10).

We shall now show that the transition matrix elements are related to Green functions for the relevant particles and hence all the physical quantities may be calculated by way of Green functions. The basic tool to derive this relationship is the *reduction formula* developed by Lehmann, Symanzik and Zimmermann [Leh 55]. In order to explain the reduction formula we, for simplicity, consider the case of the elastic scattering of neutral scalar particles with physical mass m as shown in Fig. 2.3.2.

The S-matrix element for this process reads

$$\langle k_1k_2|S|p_1p_2\rangle = \langle k_1k_2 \text{ out}|p_1p_2 \text{ in}\rangle \quad , \tag{2.3.11}$$

where the words "in" and "out" indicate that the states are incoming and outgoing states respectively. The incoming (outgoing) states are eigenstates of

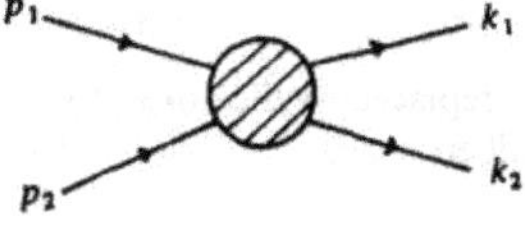

Fig. 2.3.2. Elastic scattering of two neutral spinless particles with momenta p_1 and p_2. Momenta in the final state are k_1 and k_2, respectively and the mass of the particles is m.

the total Hamiltonian with the boundary condition at the remote past (future). They are constructed by applying creation operators $\hat{a}_{\rm in}(p)^\dagger$ ($\hat{a}_{\rm out}(p)^\dagger$) to the vacuum state, e.g.,

$$|p_1p_2 \text{ in}\rangle = \hat{a}_{\rm in}(p_1)^\dagger \hat{a}_{\rm in}(p_2)^\dagger|0\rangle \quad , \tag{2.3.12}$$

where the operator $\hat{a}_{\rm in}(p)$ is a coefficient of the Fourier transform of incoming field $\hat{\phi}_{\rm in}(x)$,

$$\hat{\phi}_{\rm in}(x) = \int \frac{d^3p}{(2\pi)^3 2p_0}[\hat{a}_{\rm in}(p)e^{-ip\cdot x} + \hat{a}_{\rm in}(p)^\dagger e^{ip\cdot x}] \quad . \tag{2.3.13}$$

Note that the incoming (outgoing) field $\hat{\phi}_{\rm in}(x)$ ($\hat{\phi}_{\rm out}(x)$) is a free field, and the creation and annihilation operators satisfy the commutation relations,

$$\begin{aligned}[][\hat{a}_{\rm in}(p), \hat{a}_{\rm in}(p')^\dagger] &= (2\pi)^3 2p_0\delta^3(p'-p) \quad , \\ [\hat{a}_{\rm in}(p), \hat{a}_{\rm in}(p')] &= 0 \quad . \end{aligned} \tag{2.3.14}$$

Similar relations also hold for $\hat{a}_{\rm out}(p)$. It is easy to show that Eq. (2.3.13) is inverted to give[11]

$$\hat{a}_{\rm in}(p) = i\int d^3x\, e^{ip\cdot x}\overleftrightarrow{\partial}_0\hat{\phi}_{\rm in}(x) \quad , \tag{2.3.15}$$

where $\overleftrightarrow{\partial}_0$ is defined by

$$f\overleftrightarrow{\partial}_0 g = f\partial_0 g - (\partial_0 f)g \quad . \tag{2.3.16}$$

The incoming (outgoing) field $\hat{\phi}_{\rm in}(x)$ ($\hat{\phi}_{\rm out}(x)$) is defined by an asymptotic limit of the Heisenberg field $\hat{\phi}(x)$,

$$\hat{\phi}(x) \xrightarrow[x_0\to\pm\infty]{} \hat{\phi}_{\substack{\rm out\\ \rm in}}(x) \quad . \tag{2.3.17}$$

Here the meaning of the limiting process is that for any matrix element of $\hat{\phi}(x)$, $\langle\alpha|\hat{\phi}(x)|\beta\rangle$,

$$\lim_{x_0\to\pm\infty}\langle\alpha|\hat{\phi}(x)|\beta\rangle = \langle\alpha|\hat{\phi}_{\substack{\rm out\\ \rm in}}(x)|\beta\rangle \quad . \tag{2.3.18}$$

This defines the limit of operators in a mathematically rigorous way and is called the weak limit.

We are now ready to derive the reduction formula for scalar fields. Using Eqs. (2.3.12), (2.3.15) and (2.3.18) we have

[11] Strictly speaking the plane waves $e^{\pm ip\cdot x}$ in Eq. (2.3.13) should be replaced by the normalizable wave packets to ensure that the field operators remain in the Hilbert space.

$$\langle k_1 k_2 \text{ out}| p_1 p_2 \text{ in}\rangle = \langle k_1 | \hat{a}_{\text{out}}(k_2) | p_1 p_2 \text{ in}\rangle$$
$$= i \lim_{x_0 \to +\infty} \int d^3x\, e^{ik_2\cdot x} \overleftrightarrow{\partial}_0 \langle k_1 | \hat{\phi}(x) | p_1 p_2 \text{ in}\rangle \quad , \tag{2.3.19}$$

where we have taken into account that the single particle incoming state is identical to the single particle outgoing state: $\langle k_1 \text{ out}| = \langle k_1 \text{ in}| \equiv \langle k_1|$. Equation (2.3.19) is further rewritten as

$$\langle k_1 k_2 \text{ out}| p_1 p_2 \text{ in}\rangle = \langle k_1 | \hat{a}_{\text{in}}(k_2) | p_1 p_2 \text{ in}\rangle$$
$$+ i \int d^4x \partial_0 [e^{ik_2\cdot x} \overleftrightarrow{\partial}_0 \langle k_1 | \hat{\phi}(x) | p_1 p_2 \text{ in}\rangle] \quad . \tag{2.3.20}$$

Noting that $\partial_0 (f \overleftrightarrow{\partial}_0 g) = f \overleftrightarrow{\partial}_0^2 g$ and $\partial_0^2 e^{ik_2\cdot x} = (\nabla^2 - m^2) e^{ik_2\cdot x}$, we find[12]

$$\langle k_1 k_2 \text{ out}| p_1 p_2 \text{ in}\rangle = \langle k_1 | \hat{a}_{\text{in}}(k_2) | p_1 p_2 \text{ in}\rangle$$
$$+ i \int d^4x\, e^{ik_2\cdot x} (\Box + m^2) \langle k_1 | \hat{\phi}(x) | p_1 p_2 \text{ in}\rangle \quad . \tag{2.3.21}$$

The repeated use of the above procedure will lead us to the formula

$$\langle k_1 k_2 \text{ out}| p_1 p_2 \text{ in}\rangle$$
$$= \langle k_1 k_2 \text{ in}| p_1 p_2 \text{ in}\rangle + i^4 \int d^4x_1 d^4x_2 d^4y_1 d^4y_2 e^{i(k_1\cdot x_1 + k_2\cdot x_2 - p_1\cdot y_1 - p_2\cdot y_2)}$$
$$\times (\Box_{x_1} + m^2)(\Box_{x_2} + m^2)(\Box_{y_1} + m^2)(\Box_{y_2} + m^2)$$
$$\times \langle 0 | \mathrm{T}[\hat{\phi}(x_1)\hat{\phi}(x_2)\hat{\phi}(y_1)\hat{\phi}(y_2)] | 0\rangle \quad . \tag{2.3.22}$$

Equation (2.3.21) and Eq. (2.3.22) are the so-called *reduction formulas*. We note that $\langle k_1 k_2 \text{ out}| p_1 p_2 \text{ in}\rangle - \langle k_1 k_2 \text{ in}| p_1 p_2 \text{ in}\rangle$ is nothing but $\langle k_1 k_2 | (S-1) | p_1 p_2\rangle$ and hence we obtain

$$(2\pi)^4 i\delta(k_1 + k_2 - p_1 - p_2) \langle k_1 k_2 | T | p_1 p_2\rangle$$
$$= i^4 \int d^4x_1 d^4x_2 d^4y_1 d^4y_2 e^{i(k_1\cdot x_1 + k_2\cdot x_2 - p_1\cdot y_1 - p_2\cdot y_2)}$$
$$\times (\Box_{x_1} + m^2)(\Box_{x_2} + m^2) \times (\Box_{y_1} + m^2)(\Box_{y_2} + m^2)$$
$$\times \langle 0 | \mathrm{T}[\hat{\phi}(x_1)\hat{\phi}(x_2)\hat{\phi}(y_1)\hat{\phi}(y_2)] | 0\rangle \quad . \tag{2.3.23}$$

[12] The surface terms are discarded by replacing plane waves $e^{ik\cdot x}$ by wave packets.

Obviously Eq. (2.3.23) provides the relation between the transition matrix element and the Green function for field $\hat{\phi}(x)$. It is worth mentioning here that the delta function on the left-hand side of Eq. (2.3.23) is cancelled by the one on the right-hand side which shows up by separating the center-of-mass coordinate in the integrand using the translation invariance of the Green function.

If particles corresponding to the incoming and outgoing fields are observable as in the case of electrons in quantum electrodynamics, we can simply apply the above formalism to calculate the scattering cross sections for these particles. In quantum chromodynamics, however, quarks and gluons are most likely to be confined and may not be observed as an isolated state. Only hadrons which are composed of quarks can be observed as incoming and outgoing fields. As we have no incoming and outgoing fields for quarks and gluons, we are not allowed to apply the above formalism to the quark-gluon system. We have to exploit the method to obtain the scattering cross section for hadrons starting with the quark Green functions. It is, however, not easy to formulate the method since it requires a complete knowledge of the nonperturbative regime of quantum chromodynamics. As we are only interested in the perturbative treatment of short distance processes, we leave this problem for future developments and discuss only the Green functions for quarks and gluons.

We continue to work with the neutral scalar field $\hat{\phi}(x)$ and consider n-point Green functions for field $\hat{\phi}(x)$,

$$G_n(x_1, \ldots, x_n) = \langle 0|\mathrm{T}[\hat{\phi}(x_1) \ldots \hat{\phi}(x_n)]|0\rangle \quad . \tag{2.3.24}$$

In particular the 2-point Green function $G_2(x_1, x_2)$ is called the propagator since it represents a propagation of the particle from x_1 to x_2 (or vice versa). The Green functions defined by Eq. (2.3.24) are, in general, not the connected Green functions, since they include completely disjoint pieces. The *connected Green function* $G_n^c(x_1, \ldots, x_n)$ is the one which has no disjoint piece in it, i.e., any part of it is joined to the remaining part by at least one propagator in the sense of the Feynman diagram (see the next subsection). Diagrammatically the relation between G_n and G_n^c may be represented as in Fig. 2.3.3.

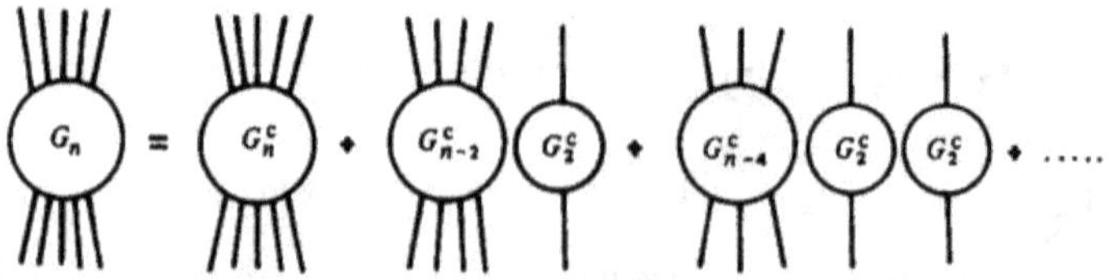

Fig. 2.3.3. The Green function G_n of Eq. (2.3.24) expanded in terms of the connected Green functions G_n^c.

It is easily observed that the disjoint pieces in the Green function G_n do not contribute to the S matrix and hence it is more convenient to work with the connected Green function G_n^c.

There is an elegant way of defining the connected Green function G_n^c in terms of the functional-integral formalism. Let us define $W[J]$ to be a generating functional which generates only the connected Green functions, i.e.,

$$G_n^c(x_1, \ldots, x_n) = (-i)^{n-1} \left. \frac{\delta^n W[J]}{\delta J(x_1) \ldots \delta J(x_n)} \right|_{J=0}, \qquad (2.3.25)$$

where the factor $(-i)^{n-1}$ has been introduced for the later convenience. We shall now show that the functional $W[J]$ is related to the generating functional $Z[J]$ for the Green functions G_n through the relation,

$$Z[J] = e^{iW[J]} \quad . \qquad (2.3.26)$$

By taking the functional differentiation of Eq. (2.3.26) and setting $J(x) = 0$, we find according to Eq. (2.2.52)

$$\left. \frac{\delta W}{\delta J(x)} \right|_{J=0} = - \left. \frac{i}{Z[0]} \frac{\delta Z}{\delta J(x)} \right|_{J=0} = \langle 0|\hat{\phi}(x)|0\rangle \quad . \qquad (2.3.27)$$

The vacuum expectation value of the field operator $\hat{\phi}(x)$ vanishes as far as the vacuum $|0\rangle$ is stable. In other words, unless spontaneous symmetry breaking takes place in the theory under consideration, $\langle 0|\hat{\phi}(x)|0\rangle$ has to be zero. (For details on this topics, see, e.g., [Tay 76]). In our perturbative approach to quantum chromodynamics we are not concerned with spontaneous symmetry breaking and hence we shall always deal with the case for which[13]

$$\langle 0|\hat{\phi}(x)|0\rangle = 0 \quad . \qquad (2.3.28)$$

With this in mind we have

$$\left. \frac{\delta^2 W}{\delta J_1\, \delta J_2} \right|_{J=0} = - \left. \frac{i}{Z[0]} \frac{\delta^2 Z}{\delta J_1\, \delta J_2} \right|_{J=0} = i\langle 0|\mathrm{T}[\hat{\phi}(x_1)\, \hat{\phi}(x_2)]|0\rangle \quad , \qquad (2.3.29)$$

$$\left. \frac{\delta^3 W}{\delta J_1\, \delta J_2\, \delta J_3} \right|_{J=0} = - \left. \frac{i}{Z[0]} \frac{\delta^3 Z}{\delta J_1\, \delta J_2\, \delta J_3} \right|_{J=0}$$

$$= -\langle 0|\mathrm{T}[\hat{\phi}(x_1)\, \hat{\phi}(x_2)\, \hat{\phi}(x_3)]|0\rangle \quad , \qquad (2.3.30)$$

[13] Or we simply apply the renormalization condition $\langle 0|\hat{\phi}(x)|0\rangle = 0$ in the case without spontaneous symmetry breaking.

where $J_i = J(x_i)$ with $i = 1, 2, 3, \ldots$ The 2- and 3-point Green functions appearing on the right-hand side of Eq. (2.3.30) cannot be split into disjoint pieces as may be recognized by the diagrammatical argument in Fig. 2.3.3. Hence the left-hand sides of Eqs. (2.3.29) and (2.3.30) represent connected Green functions. These are rather trivial examples. If we go over to the fourth functional derivative of $W[J]$, a first nontrivial example shows up,

$$\left.\frac{\delta^4 W}{\delta J_1\,\delta J_2\,\delta J_3\,\delta J_4}\right|_{J=0} =$$

$$\frac{i}{Z[0]^2}\left.\left(\frac{\delta^2 Z}{\delta J_1\,\delta J_2}\frac{\delta^2 Z}{\delta J_3\,\delta J_4} + \frac{\delta^2 Z}{\delta J_1\,\delta J_3}\frac{\delta^2 Z}{\delta J_2\,\delta J_4} + \frac{\delta^2 Z}{\delta J_1\,\delta J_4}\frac{\delta^2 Z}{\delta J_2\,\delta J_3}\right)\right|_{J=0}$$

$$- \frac{i}{Z[0]}\left.\frac{\delta^4 Z}{\delta J_1\,\delta J_2\,\delta J_3\,\delta J_4}\right|_{J=0} , \qquad (2.3.31)$$

$$= iG_2(x_1,x_2)G_2(x_3,x_4) + iG_2(x_1,x_3)\,G_2(x_2,x_4) + iG_2(x_1,x_4)\,G_2(x_2,x_3)$$
$$-iG_4(x_1, x_2, x_3, x_4) \quad . \qquad (2.3.32)$$

As can be seen from the diagrams in Fig. 2.3.3 the connected 4-point Green function G_4^c is related to the Green function G_4 by

$$G_4^c(x_1,x_2,x_3,x_4) = G_4(x_1,x_2,x_3,x_4) - G_2(x_1,x_2)\,G_2(x_3,x_4)$$
$$- G_2(x_1,x_3)\,G_2(x_2,x_4) - G_2(x_1,x_4)\,G_2(x_2,x_3). \qquad (2.3.33)$$

Inserting Eq. (2.3.33) into Eq. (2.3.32) we have the $n = 4$ case of Eq. (2.3.25). The general proof of Eq. (2.3.25) may be given by induction.

Next we introduce the notion of truncated (amputated) Green functions $G_n^t\,(x_1, \ldots, x_n)$. The *truncated Green function* is defined by eliminating propagators from all the external lines. Thus the truncated Green function is related, in momentum space, to the nontruncated Green function Eq. (2.3.24) in the following way,

$$\tilde{G}_n(p_1, \ldots, p_{n-1}) = \tilde{G}_2(p_1) \ldots \tilde{G}_2(p_n)\,\tilde{G}_n^t(p_1, \ldots, p_{n-1}) \quad , \qquad (2.3.34)$$

where

$$\tilde{G}_n(p_1, \ldots p_{n-1})\,(2\pi)^4\delta^4(p_1 + \ldots + p_n)$$
$$= \int d^4x_1 \ldots d^4x_n e^{-i(p_1\cdot x_1 + \ldots + p_n\cdot x_n)}\,G_n(x_1, \ldots, x_n) \quad , \qquad (2.3.35)$$

$$\tilde{G}_n^t(p_1, \ldots, p_{n-1})(2\pi)^4\delta^4(p_1 + \ldots + p_n)$$
$$= \int d^4x_1 \ldots d^4x_n e^{-i(p_1 \cdot x_1 + \ldots + p_n \cdot x_n)} G^t(x_1, \ldots, x_n). \quad (2.3.36)$$

In particular the truncated 2-point Green function $\tilde{G}_2^t(p)$ is an inverse propagator owing to Eq. (2.3.34),

$$\tilde{G}_2^t(p) = \frac{1}{\tilde{G}_2(p)} . \quad (2.3.37)$$

The reason why we introduced the truncated Green function is that it is this function which participates in the relation between the transition matrix element and the Green function, e.g., Eq. (2.3.23). Let us explain the situation in more detail by taking Eq. (2.3.23) as an example. Performing integrations by parts in Eq. (2.3.23), we obtain[14]

$$i\langle k_1 k_2 | T | p_1 p_2 \rangle$$
$$= i^4 \lim_{k_1^2, k_2^2, p_1^2, p_2^2 \to m^2} (m^2 - k_1^2)(m^2 - k_2^2)(m^2 - p_1^2)(m^2 - p_2^2)$$
$$\times \tilde{G}_4(-k_1, p_1, p_2). \quad (2.3.38)$$

On the other hand, near at the mass shell $p^2 = m^2$, the propagator $\tilde{G}_2$(p) behaves as

$$\tilde{G}_2(p) \sim \frac{iZ_3}{p^2 - m^2} , \quad (2.3.39)$$

where Z_3 is the field renormalization constant which will be discussed in detail in Sec. 2.5.3. Taking into account the relation (2.3.34) and using Eqs. (2.3.38) and (2.3.39), we find

$$\langle k_1 k_2 | T | p_1 p_2 \rangle = -iZ_3^{\frac{4}{3}} \tilde{G}_4^t(-k_1, p_1, p_2)\Big|_{k_1^2 = k_2^2 = p_1^2 = p_2^2 = m^2} . \quad (2.3.40)$$

The above result clearly proves the usefulness of the truncated Green functions. It should be noted here that $Z_3^{\frac{2}{3}} \tilde{G}_4^t$ is a renormalized truncated Green function which will be described in Sec. 2.5.5 and Sec. 3.1.2.

[14] It can be easily shown that disconnected pieces in $\tilde{G}_4$ do not contribute to the transition matrix element (2.3.38) and hence $\tilde{G}_4$ may be replaced by $\tilde{G}_4^c$ in Eq. (2.3.38).

Finally we introduce yet another Green function called a proper (one-particle-irreducible) Green function $G_n^{\mathrm{P}}(x_1,\ldots,x_n)$. For the definition of this Green function we have to use the language of Feynman diagrams though they will be introduced later in Sec. 2.3.2. The *proper Green function* $G_n^{\mathrm{P}}(x_1,\ldots,x_n)$ is the one whose expression in each order of the perturbative expansion is obtained by a proper Feynman diagram. The *proper diagram* is a truncated connected diagram which does not split into disjoint pieces when an arbitrary internal line is cut. When we work in perturbation theory the proper diagrams are very useful because we may construct any Feynman diagrams as an assembly of proper diagrams. In particular the notion of proper diagrams is indispensable for the discussion of renormalizability.

There is a simple expression for the generating functional of the proper Green functions $G_n^{\mathrm{P}}(x_1, \ldots, x_n)$. This generating functional is called the *effective action* and is defined by

$$\Gamma[v] = W[J] - \int d^4x\, J(x)v(x) \quad , \tag{2.3.41}$$

where the new variable $v(x)$ is given by

$$v(x) = \frac{\delta W[J]}{\delta J(x)} \quad , \tag{2.3.42}$$

which corresponds to the vacuum expectation value of $\hat{\phi}(x)$ in the presence of the source J. Equation (2.3.41) may be regarded as a change of variables from (J, W) to (v, Γ) where v is considered to be a function of J through Eq. (2.3.42) and $v = 0$ for $J = 0$ owing to Eq. (2.3.28). This type of the change of variables is often employed in solving differential equations and is called the Legendre transformation. The Legendre transformation is also used in thermodynamics in order to interchange the thermodynamical functions.

We shall now show that the effective action $\Gamma[v]$ is in fact a generating functional of the proper Green functions $G_n^{\mathrm{P}}(x_1,\ldots,x_n)$:

$$G_n^{\mathrm{P}}(x_1, \ldots, x_n) = i\frac{\delta^n \Gamma[v]}{\delta v(x_1)\ldots\delta v(x_n)}\bigg|_{v=0} \quad , \tag{2.3.43}$$

or in the form of the power expansion in $v(x)$,

$$\Gamma[v] = \sum_{n=0}^{\infty} \frac{-i}{n!} \int d^4x_1\ldots d^4x_n\, G_n^{\mathrm{P}}(x_1,\ldots,x_n)v(x_1)\ldots v(x_n) \quad . \tag{2.3.44}$$

Here again we shall be content with showing some simpler cases. The general proof may be given by induction. It follows from Eq. (2.3.42) by functional differentiation with respect to $J(x)$ that

$$\frac{\delta^2 W}{\delta J_1 \delta J_2} = \frac{\delta v_1}{\delta J_2} , \tag{2.3.45}$$

where $J_i = J(x_i)$ and $v_i = v(x_i)$ with $i = 1, 2\ 3...$ On the other hand from Eq. (2.3.41) we have

$$\frac{\delta \Gamma[v]}{\delta v(x)} = -J(x) . \tag{2.3.46}$$

Differentiating Eq. (2.3.46) with respect to $v(x)$ we obtain

$$\frac{\delta^2 \Gamma}{\delta v_1 \delta v_2} = -\frac{\delta J_1}{\delta v_2} . \tag{2.3.47}$$

Using Eqs. (2.3.45) and (2.3.47) we find that

$$\int d^4x_3 \frac{\delta^2 W}{\delta J_1 \delta J_3} \frac{\delta^2 \Gamma}{\delta v_3 \delta v_2} = -\frac{\delta v(x_1)}{\delta v(x_2)} = -\delta^4(x_1 - x_2) . \tag{2.3.48}$$

Note here that the right-hand side of Eq. (2.3.48) is obtained in conformity with the rule of functional differentiation (2.2.50). The function $-i\delta^2 W/\delta J_1 \delta J_3$ for $J = 0$ is the propagator (2-point Green function) and hence we observe through Eq. (2.3.48) that $-\ i\delta^2\Gamma/\delta v_3 \delta v_2$ for $v = 0$ is the inverse propagator. While the propagator is improper (one-particle reducible) as it includes the proper self-energy part an infinite number of times, the inverse propagator contains it only once. Thus the inverse propagator is proper although the propagator is not.[15] Thus, if we define G_2^p to be (-1) times inverse propagators, we establish

$$i\left.\frac{\delta^2 \Gamma[v]}{\delta v(x_1)\, \delta v(x_2)}\right|_{v=0} = G_2^p(x_1, x_2) , \tag{2.3.49}$$

which is the special case of Eq. (2.3.43) for $n = 2$. According to Eq. (2.3.48) we have

$$\int d^4x\, G_2(x_1, x)\, G_2^p(x, x_2) = -\delta^4(x_1 - x_2) . \tag{2.3.50}$$

[15] See Sec. 2.5.3 for the definition of the self-energy part.

We then differentiate Eq. (2.3.48) with respect to $J(x_4)$ to obtain

$$\int d^4x_3 \frac{\delta^3 W}{\delta J_1 \delta J_3 \delta J_4} \frac{\delta^2 \Gamma}{\delta v_3 \delta v_2} + \int d^4x_3\, d^4x_5 \frac{\delta^2 W}{\delta J_1 \delta J_3} \frac{\delta^3 \Gamma}{\delta v_3 \delta v_2 \delta v_5} \frac{\delta v_5}{\delta J_4} = 0. \quad (2.3.51)$$

Using Eqs. (2.3.45) and (2.3.47) and applying the formula,

$$\int d^4x \frac{\delta v(x)}{\delta J(x_1)} \frac{\delta J(x_2)}{\delta v(x)} = \delta^4(x_1 - x_2) \;, \quad (2.3.52)$$

we obtain from Eq. (2.3.51)

$$\frac{\delta^3 \Gamma}{\delta v_1 \delta v_2 \delta v_3} = -\int d^4x_4 d^4x_5 d^4x_6 \frac{\delta^2 \Gamma}{\delta v_1 \delta v_4} \frac{\delta^2 \Gamma}{\delta v_2 \delta v_5} \frac{\delta^2 \Gamma}{\delta v_3 \delta v_6} \frac{\delta^3 W}{\delta J_4 \delta J_5 \delta J_6} \;. \quad (2.3.53)$$

Here $-i\delta^2\Gamma/\delta v_1 \delta v_2$ for $v=0$ is the inverse propagator; $(-i)^2\delta^3 W/\delta J_4 \delta J_5 \delta J_6$ for $J = 0$ is the connected 3-point Green function. Obviously the right-hand side of Eq. (2.3.53) represents the proper 3-point Green function and hence

$$i \left. \frac{\delta^3 \Gamma[v]}{\delta v(x_1) \delta v(x_2) \delta v(x_3)} \right|_{v=0} = G_3^{\rm p}(x_1, x_2, x_3) \;. \quad (2.3.54)$$

We now present in Fig. 2.3.4 some examples of Feynman diagrams in ϕ^3

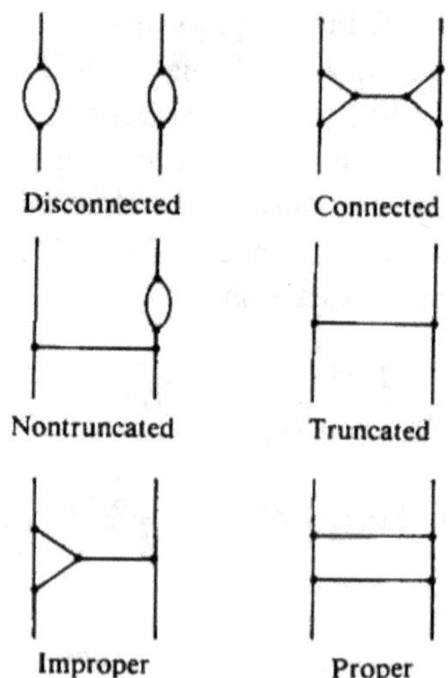

Fig. 2.3.4. Examples of disconnected and connected, nontruncated and truncated, and improper and proper diagrams. The line with dots on both ends represents the propagator.

theory representing the connected, disconnected, truncated (amputated), nontruncated (nonamputated), proper (one-particle irreducible) and improper (one-particle reducible) Green functions.

In this subsection, for simplicity, we have restricted ourselves to the case of neutral scalar fields. The argument given here, however, is quite general and may be easily generalized to incorporate particles with spin. Here, for later convenience, we write down reduction formulae for spinor and vector fields respectively. According to the reduction formula the S-matrix element for the two-body scattering of spinor fields ψ with mass m has the following form,

$$
\begin{aligned}
&\langle k_1\sigma_1, k_2\sigma_2 \text{ out} | p_1\lambda_1, p_2\lambda_2 \text{ in}\rangle = \langle k_1\sigma_1, k_2\sigma_2 \text{ in} | p_1\lambda_1, p_2\lambda_2 \text{ in}\rangle \\
&+ i^4 \int d^4x_1 d^4x_2 d^4y_1 d^4y_2 \, e^{i(k_1\cdot x_1 + k_2\cdot x_2 - p_1\cdot y_1 - p_2\cdot y_2)} \, \bar{u}^{\alpha_1}_{\sigma_1}(k_1)\, \bar{u}^{\alpha_2}_{\sigma_2}(k_2) \\
&\times (-i\partial\!\!\!/_{x_1} + m)_{\alpha_1\alpha'_1} (-i\partial\!\!\!/_{x_2} + m)_{\alpha_2\alpha'_2} \langle 0 | \mathrm{T}[\psi^{\alpha'_1}(x_1)\psi^{\alpha'_2}(x_2)\bar{\psi}^{\beta'_1}(y_1)\,\bar{\psi}^{\beta'_2}(y_2)] | 0\rangle \\
&\times (i\overleftarrow{\partial\!\!\!/}_{y_1} + m)_{\beta'_1\beta_1} (i\overleftarrow{\partial\!\!\!/}_{y_2} + m)_{\beta'_2\beta_2} \, u^{\beta_1}_{\lambda_1}(p_1)\, u^{\beta_2}_{\lambda_2}(p_2) \quad ,
\end{aligned} \tag{2.3.22a}
$$

where $\sigma_1, \sigma_2, \lambda_1$ and λ_2 represent polarizations (or helicities) of particles with momenta k_1, k_2, p_1 and p_2, respectively and $u^{\beta}_{\lambda}(p)$ is the βth component of the Dirac spinor for polarization (or helicity) λ. The definition and normalization of the Dirac spinor are found in Sec. 1.4. The transition matrix element corresponding to the above S-matrix element is given by

$$
\begin{aligned}
&i\langle k_1\sigma_1, k_2\sigma_2 | T | p_1\lambda_1, p_2\lambda_2\rangle = \\
&i^4 \lim_{k_1^2, k_2^2, p_1^2, p_2^2 \to m^2} \bar{u}^{\alpha_1}_{\sigma_1}(k_1)\, \bar{u}^{\alpha_2}_{\sigma_2}(k_2)\, (m - k\!\!\!/_1)_{\alpha_1\alpha'_1} (m - k\!\!\!/_2)_{\alpha_2\alpha'_2} \\
&\times \tilde{G}_4^{\alpha'_1\alpha'_2\beta'_1\beta'_2}(-k_1, p_1, p_2)\, (m - p\!\!\!/_1)_{\beta'_1\beta_1} (m - p\!\!\!/_2)_{\beta'_2\beta_2}\, u^{\beta_1}_{\lambda_1}(p_1)\, u^{\beta_2}_{\lambda_2}(p_2) \quad ,
\end{aligned} \tag{2.3.38a}
$$

where the Green function $\tilde{G}_4$ is defined by

$$
\begin{aligned}
&\tilde{G}_4^{\alpha_1\alpha_2\beta_1\beta_2}(-k_1, p_1, p_2)\, (2\pi)^4\, \delta^4(k_1 + k_2 - p_1 - p_2) \\
&= \int d^4x_1 d^4x_2 d^4y_1 d^4y_2 \, e^{i(k_1\cdot x_1 + k_2\cdot x_2 - p_1\cdot y_1 - p_2\cdot y_2)} \\
&\quad \times \langle 0 | \mathrm{T}[\psi^{\alpha_1}(x_1)\psi^{\alpha_2}(x_2)\bar{\psi}^{\beta_1}(y_1)\bar{\psi}^{\beta_2}(y_2)] | 0\rangle \quad .
\end{aligned} \tag{2.3.35a}
$$

As can be easily drawn from Eq. (2.3.38a), here again we have a simple relation between the transition matrix element and the truncated connected Green function $\tilde{G}_4^{\mathrm{tc}}$:

$$\langle k_1\sigma_1,k_2\sigma_2|T|p_1\lambda_1,p_2\lambda_2\rangle$$

$$= -iZ_2^4\bar{u}^{\alpha_1}_{\sigma_1}(k_1)\,\bar{u}^{\alpha_2}_{\sigma_2}(k_2)\,\tilde{G}^{tc}_{4\alpha_1\alpha_2\beta_1\beta_2}(-k_1,p_1,p_2)\,u^{\beta_1}_{\lambda_1}(p_1)\,u^{\beta_2}_{\lambda_2}(p_2) \quad , \qquad (2.3.40a)$$

which holds for $k_1^2 = k_2^2 = p_1^2 = p_2^2 = m^2$. In Eq. (2.3.40a), Z_2 is the field renormalization constant of the spinor field $\psi(x)$.

A similar reduction formula for neutral vector field $A_\mu(x)$ with mass m is obtained in an analogous way. It reads

$$\langle k_1\sigma_1,k_2\sigma_2 \text{ out}|p_1\lambda_1,p_2\lambda_2 \text{ in}\rangle = \langle k_1\sigma_1,k_2\sigma_2 \text{ in}|p_1\lambda_1,p_2\lambda_2 \text{ in}\rangle$$

$$+\, i^4\int d^4x_1d^4x_2d^4y_1d^4y_2\; e^{i(k_1\cdot x_1+k_2\cdot x_2-p_1\cdot y_1-p_2\cdot y_2)}$$

$$\times\, (\Box_{x_1}+m^2)(\Box_{x_2}+m^2)(\Box_{y_1}+m^2)(\Box_{y_2}+m^2)\,\varepsilon^{\mu_1}_{\sigma_1}(k_1)$$

$$\times\, \varepsilon^{\mu_2}_{\sigma_2}(k_2)\langle 0|\mathrm{T}[A_{\mu_1}(x_1)A_{\mu_2}(x_2)A_{\nu_1}(y_1)A_{\nu_2}(y_2)]|0\rangle\varepsilon^{\nu_1}_{\lambda_1}(p_1)\varepsilon^{\nu_2}_{\lambda_2}(p_2) \quad , \qquad (2.3.22b)$$

where $\sigma_1,\sigma_2,\lambda_1$ and λ_2 are polarizations of particles with momenta k_1, k_2, p_1 and p_2, respectively and $\varepsilon^{\mu}_{\lambda}(p)$ is the polarization vector of the vector field $A_\mu(x)$ for polarization λ and momentum p. The relation between the transition matrix element and the truncated connected Green function $\tilde{G}^{tc}_4$ is given by

$$\langle k_1\sigma_1,k_2\sigma_2|T|p_1\lambda_1,p_2\lambda_2\rangle =$$

$$-iZ_3^4\varepsilon^{\mu_1}_{\sigma_1}(k_1)\varepsilon^{\mu_2}_{\sigma_2}(k_2)\tilde{G}^{tc}_{4\mu_1\mu_2\nu_1\nu_2}(-k_1,p_1,p_2)\varepsilon^{\nu_1}_{\lambda_1}(p_1)\varepsilon^{\nu_2}_{\lambda_2}(p_2), \qquad (2.3.40b)$$

where $k_1^2 = k_2^2 = p_1^2 = p_2^2 = m^2$, Z_3 is the field renormalization constant for the vector field $A_\mu(x)$ and the truncated connected Green function $\tilde{G}^{tc}_{4\mu_1\mu_2\nu_1\nu_2}$ is defined in the same way as in Eq. (2.3.34).

Summary

We here tabulate the various Green functions and the corresponding generating functionals discussed in this subsection.

Green function		Generating functional	
Original G_n	Eq. (2.3.24)	$Z[J]$	Eq. (2.2.49)
Connected G_n^c	Eq. (2.3.25)	$W[J]$	Eq. (2.3.26)
Truncated G_n^t	Eq. (2.3.34)		
Truncated connected G_n^{tc}			
Proper G_n^p	Eq. (2.3.43)	$\Gamma[v]$	Eq. (2.3.41)

Here the terms "truncated" and "proper" may be replaced by "amputated" and "one-particle irreducible," respectively.

Exercise

1. Derive the relation between $\delta^n W/\delta J_1 \ldots \delta J_n|_{J=0}$ and $\langle 0|\mathrm{T}[\hat{\phi}(x_1)\ldots\hat{\phi}(x_n)]|0\rangle$ up to $n = 4$ when $\langle 0|\hat{\phi}(x)|0\rangle \neq 0$.

2.3.2. ϕ^3 theory

Before discussing the Feynman rules in gauge theories (i.e., quantum chromodynamics) we consider the simple case of the neutral scalar theory with ϕ^3 self-coupling and illustrate a derivation of the Feynman rules. This theory is so simple that it is free from the complications existing in gauge theories and thus is suited for explaining the derivation of the rules without any additional complexities. The theory, however, has an obvious disadvantage in that it has no stable vacuum. This corresponds to the fact that its classical energy is not bounded below owing to the ϕ^3 term. Thus any state in this theory decays without limit. In the present arguments, however, we are only interested in perturbation theory in which the problem of the stability of the vacuum does not show up. Hence we shall, from time to time, employ ϕ^3 theory as a prototype whenever the simplicity is required.

In ϕ^3 theory the Lagrangian is given by Eq. (2.2.15) with

$$V(\phi) = -\frac{g}{3!}\phi^3 \quad , \tag{2.3.55}$$

where g is a coupling constant and the factor $1/3!$ is introduced for later convenience. The generating functional $Z[J]$ is given by Eq. (2.2.49),

$$Z[J] = \int [d\phi] \exp\left\{i\int d^4x(\mathscr{L}_0 - V(\phi) + \phi J)\right\} \quad , \tag{2.3.56}$$

where $\mathscr{L}_0$ is the free Lagrangian,

$$\mathscr{L}_0 = \frac{1}{2}(\partial^\mu\phi\partial_\mu\phi - m^2\phi^2) \ . \tag{2.3.57}$$

We consider the generating functional for the free field, i.e., for the interaction being switched off ($g = 0$),

$$Z_0[J] = \int [d\phi] \exp\left\{i\int d^4x(\mathscr{L}_0 + \phi J)\right\} \ . \tag{2.3.58}$$

By applying the functional derivative with respect to $J(x)$ three times to $Z_0[J]$ we obtain

$$\left(\frac{1}{i}\frac{\delta}{\delta J(x)}\right)^3 Z_0[J] = \int [d\phi]\,(\phi(x))^3 \exp\left\{i\int d^4x'(\mathcal{L}_0 + \phi J)\right\} . \tag{2.3.59}$$

Repeating this process and forming an infinite series we find

$$\sum_{n=0}^{\infty}\frac{1}{n!}\left\{\frac{ig}{3!}\int d^4x\left(\frac{\delta}{i\delta J(x)}\right)^3\right\}^n Z_0[J]$$

$$= \int [d\phi]\sum_{n=0}^{\infty}\frac{1}{n!}\left\{\frac{ig}{3!}\int d^4x(\phi(x))^3\right\}^n \exp\left\{i\int d^4x'(\mathcal{L}_0 + \phi J)\right\}, \tag{2.3.60}$$

which is identical to

$$\exp\left\{-i\int d^4x V\left(\frac{\delta}{i\delta J(x)}\right)\right\}Z_0\,[J]$$

$$= \int [d\phi]\exp\left\{-i\int d^4x V(\phi)\right\}\exp\left\{i\int d^4x'(\mathcal{L}_0 + J)\right\} . \tag{2.3.61}$$

Equation (2.3.61) gives a convenient representation for $Z[J]$ of Eq. (2.3.56)

$$Z[J] = \exp\left\{-i\int d^4x V\left(\frac{\delta}{i\delta J(x)}\right)\right\} Z_0[J] . \tag{2.3.62}$$

Note that Eq. (2.3.62) holds for any arbitrary function $V(\phi)$.

We are interested in deriving a perturbative expansion of Eq. (2.3.62) in powers of g. This expansion has already been given in Eq. (2.3.60) and we have

$$Z[J] = \left\{1 - \frac{g}{3!}\int d^4x\left(\frac{\delta}{\delta J(x)}\right)^3 + \frac{1}{2}\left(\frac{g}{3!}\right)^2\left(\int d^4x\left(\frac{\delta}{\delta J(x)}\right)^3\right)^2 + \ldots\right\}Z_0[J] . \tag{2.3.63}$$

In order to find an explicit expression for the above formal series (2.3.63) we need to perform the functional integration for $Z_0[J]$ with respect to $\phi(x)$ and to have an explicit form of $Z_0[J]$ as a functional of $J(x)$. We first rewrite $Z_0[J]$ by applying integration by parts

$$Z_0[J] = \int [d\phi]\exp\left\{\frac{-i}{2}\int d^4x d^4y \phi(x)K(x,y)\phi(y) + i\int d^4x\phi(x)J(x)\right\} , \tag{2.3.64}$$

where

$$K(x, y) = \delta^4(x - y)\,(\Box_y + m^2), \qquad \Box_y = \frac{\partial}{\partial y^\mu}\frac{\partial}{\partial y_\mu} . \tag{2.3.65}$$

Here it should be borne in mind that m^2 in Eq. (2.3.65) is just $m^2 - i\varepsilon\,(\varepsilon > 0$ and infinitesimal) in order to ensure the convergence of the Gaussian integral in Eq. (2.3.64). To evaluate the functional integral in Eq. (2.3.64) we represent it as the continuum limit of a discrete multiple integral

$$Z_0[J] = C \lim_{N\to\infty} \int d\phi_1 \ldots d\phi_N \exp\left(\frac{-i}{2} \sum_{ij} \phi_i K_{ij} \phi_j + i \sum_i \phi_i J_i\right) \quad , \tag{2.3.66}$$

where C is an irrelevant numerical factor and K_{ij} are constants whose continuum limit is $K(x, y)$. Making the change of variables, $\phi'_i = \phi_i - \sum_j (K^{-1})_{ij} J_j$, we have

$$Z_0[J] = C \lim_{N\to\infty} \exp\left\{ i \frac{1}{2} \sum_{ij} J_i (K^{-1})_{ij} J_j \right\} I_N \quad , \tag{2.3.67}$$

where

$$I_N = \int d\phi'_1 \ldots d\phi'_n \exp\left(\frac{-i}{2} \sum_{ij} \phi'_i K_{ij} \phi'_j\right) \quad . \tag{2.3.68}$$

The J-independent factor I_N is easily calculated by diagonalizing the matrix $K = (K_{ij})$. (Since K is a symmetric matrix, it may be diagonalized by an orthogonal matrix R.) Writing $\phi' = R\phi''$ and $R^T K R = (\lambda_i \delta_{ij})$, we obtain

$$I_N = \prod_{j=1}^{N} \sqrt{\frac{2\pi}{i\lambda_j}} = \text{const.} / \sqrt{\det K}. \tag{2.3.69}$$

Equation (2.3.69) gives a numerical constant and we obtain, by neglecting an irrelevant numerical factor,

$$Z_0[J] = \exp\left\{ \frac{i}{2} \int d^4x\, d^4y\, J(x)\, \Delta(x,y)\, J(y) \right\} \quad , \tag{2.3.70}$$

where $\Delta(x, y)$ is the inverse of $K(x, y)$, i.e.,

$$\int d^4\xi\, K(x, \xi)\, \Delta(\xi, y) = \delta^4(x-y) \quad . \tag{2.3.71}$$

Replacing $K(x, \xi)$ in Eq. (2.3.71) by Eq. (2.3.65) and taking the Fourier transform of Eq. (2.3.71) we find

$$\Delta(x, y) = \int \frac{d^4k}{(2\pi)^4} \frac{e^{-ik\cdot(x-y)}}{m^2 - k^2 - i\varepsilon} \quad . \tag{2.3.72}$$

Equation (2.3.72) is nothing else but the Feynman propagator for the free field ϕ. In fact, for the free field ϕ, we have

$$G_2(x, y) = \langle 0|T[\hat{\phi}(x)\,\hat{\phi}(y)]|0\rangle$$
$$= \frac{(-i)^2}{Z_0[0]}\frac{\delta^2 Z_0[J]}{\delta J(x)\,\delta J(y)}\bigg|_{J=0} = -i\Delta(x, y) \quad . \tag{2.3.73}$$

We are now ready to evaluate the perturbative terms in Eq. (2.3.63). We insert Eq. (2.3.70) into Eq. (2.3.63) and calculate $Z[J]$ term by term for each order of g. After some calculation we obtain

$$Z[J] = Z_0[J]\,\{1 - gz_1[J] + g^2 z_2[J] + O(g^3)\} \quad , \tag{2.3.74}$$

where $z_1[J]$ and $z_2[J]$ are given by[16]

$$z_1[J] = \frac{i^2}{2!}\int dx_1 \Delta_{11}\,(\Delta_1 J_1) + \frac{i^3}{3!}\int dx_1 (\Delta_1 J_1)\,(\Delta_1 J_2)\,(\Delta_1 J_3) \quad , \tag{2.3.75}$$

$$\begin{aligned} z_2[J] = \frac{1}{2(3!)^2}\int dx_1 dx_2 \{ & 6i^3 \Delta_{12}^3 + 9i^3\,\Delta_{11}\,\Delta_{12}\Delta_{22} \\ & + 9i^4 \Delta_{11}\Delta_{22}\,(\Delta_1 J_1)\,(\Delta_2 J_2) + 18 i^4 \Delta_{11}\Delta_{12}\,(\Delta_2 J_1)\,(\Delta_2 J_2) \\ & + 18 i^4\,\Delta_{12}^2\,(\Delta_1 J_1)\,(\Delta_2 J_2) + 6i^5 \Delta_{11}\,(\Delta_1 J_1)\,(\Delta_2 J_2)\,(\Delta_2 J_3)\,(\Delta_2 J_4) \\ & + 9i^5 \Delta_{12}\,(\Delta_1 J_1)\,(\Delta_1 J_2)\,(\Delta_2 J_3)\,(\Delta_2 J_4) \\ & + i^6\,(\Delta_1 J_1)\,(\Delta_1 J_2)\,(\Delta_1 J_3)\,(\Delta_2 J_4)\,(\Delta_2 J_5)\,(\Delta_2 J_6)\} \quad , \end{aligned} \tag{2.3.76}$$

with Δ_{ij} and $(\Delta_i J_j)$ defined by

$$\Delta_{ij} = \Delta(x_i, x_j), \quad i,j = 1, 2,$$
$$(\Delta_i J_j) = \int dy_j\,\Delta(x_i, y_j)\,J(y_j), \quad i = 1, 2, \quad j = 1, 2, 3, 4. \tag{2.3.77}$$

Equations (2.3.75) and (2.3.76) correspond to what may be obtained by the use of Wick's theorem in the canonical operator formalism.

It is instructive at this point to introduce a diagrammatic interpretation for Eqs. (2.3.75) and (2.3.76). Let us make the following correspondence between a building block of Eqs. (2.3.75) and (2.3.76) and a part of the diagrams:

[16] For simplicity we write dx for d^4x.

$\int dy J(y)$	$\leftrightarrow$	—⊗ y	External source
$i\Delta(x, y)$	$\leftrightarrow$	x •—• y	Propagator
$-\int dx\, g$	$\leftrightarrow$	x (vertex)	Vertex

Then two terms of Eq. (2.3.75) may be represented by diagrams as shown in Fig. 2.3.5.

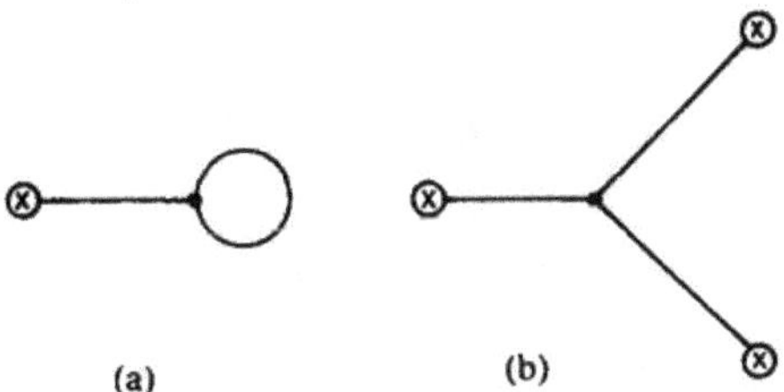

Fig. 2.3.5. Diagrammatic representation of two terms on the right-hand side of Eq. (2.3.75).

In the same way the eight terms of Eq. (2.3.76) have the graphical representation given in Fig. 2.3.6. The first two terms in Eq. (2.3.76) correspond to the so-called vacuum diagrams as seen in Fig. 2.3.6 and are eliminated when $Z[J]$ is normalized by $Z[0]$. In fact we see from Eq. (2.374) that

$$Z[0] = 1 + \frac{g^2}{2(3!)^2} \int dx_1 dx_2 \,(6i^3\Delta_{12}^3 + 9i^3\Delta_{11}\Delta_{12}\Delta_{22}) + O(g^4) \quad . \qquad (2.3.78)$$

Moreover the third, sixth and eighth terms in Eq. (2.3.76) are represented by disconnected diagrams as shown in Fig. 2.3.6 and may be rearranged into the compact expression $z_1[J]^2/2$. Hence we obtain

$$Z[J]/Z[0] = Z_0[J]\{1 - gz_1[J] + g^2\left(\frac{1}{2}z_1[J]^2 + z[J]\right) + O(g^3)\} \quad , \qquad (2.3.79)$$

where $z[J]$ consists only of the connected terms,

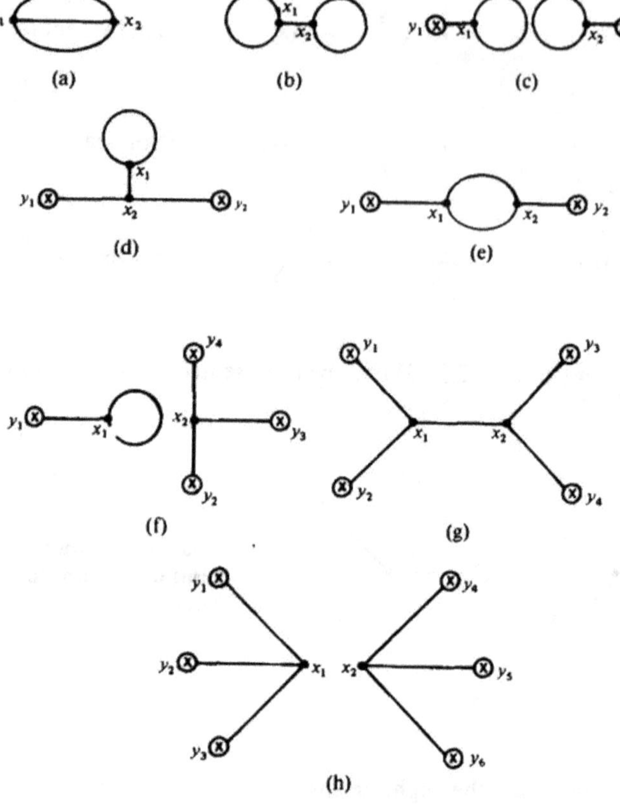

Fig. 2.3.6. Diagrammatic representation of eight terms on the right-hand side of Eq. (2.3.76).

$$z[J] = \frac{i^4}{4}\int dx_1 dx_2 dy_1 dy_2 \Delta(x_1,x_1)\,\Delta(x_1,x_2)\,\Delta(x_2,y_1)\,\Delta(x_2,y_2)\,J(y_1)\,J(y_2)$$

$$+\frac{i^4}{4}\int dx_1 dx_2 dy_1 dy_2 \Delta(x_1,x_2)^2\,\Delta(x_1,y_1)\,\Delta(x_2,y_2)\,J(y_1)\,J(y_2)$$

$$+\frac{i^5}{8}\int dx_1 dx_2 dy_1 \ldots dy_4 \Delta(x_1,x_2)\,\Delta(x_1,y_1)\,\Delta(x_1,y_2)\,\Delta(x_2,y_3)$$

$$\times\,\Delta(x_2,y_4)\,J(y_1)\ldots J(y_4) \quad . \tag{2.3.80}$$

As we have explained in Sec. 2.3.1, the generating functional $W[J]$ for connected Green functions is given by Eq. (2.3.26)

$$W[J] = -i\ln Z[J] \quad . \tag{2.3.81}$$

We substitute Eq. (2.3.79) into Eq. (2.3.81) to get

$$W[J] = -i\ln(Z_0[J]Z[0]) - i\ln\{1 - gz_1[J] + g^2\left(\frac{1}{2}z_1[J]^2 + z[J]\right) + O(g^3)\} \quad . \tag{2.3.82}$$

Re-expanding Eq. (2.3.82) in powers of g we find

$$W[J] = -i\ln Z[0] + \frac{1}{2}\int dxdyJ(x)\Delta(x,y)J(y) - i\{-gz_1[J] + g^2z[J] + O(g^3)\} \quad . \tag{2.3.83}$$

Note that the disconnected contribution $z_1[J]^2/2$ present in Eq. (2.3.82) has cancelled out, illustrating the fact that $W[J]$ is the generating functional for connected Green functions. Now we are in a position to calculate the connected Green functions G_n^c to order g^2 by the use of Eq. (2.3.25)

$$G_1^c(x) = -i\frac{g}{2}\int dy\Delta(x,y)\Delta(y,y) + O(g^3) \quad , \tag{2.3.84}$$

$$G_2^c(x,y) = -i\Delta(x,y) - \frac{g^2}{2}\int dx_1dx_2\Delta(x_1,x_1)\,\Delta(x_1,x_2)\Delta(x_2,x)\Delta(x_2,y) - \frac{g^2}{2}\int dx_1dx_2\Delta(x_1,x_2)^2\,\Delta(x_1,x)\,\Delta(x_2,y) + O(g^4) \quad , \tag{2.3.85}$$

$$G_3^c(x,y,z) = -g\int dx_1\Delta(x_1,x)\Delta(x_1,y)\,\Delta(x_1,z) + O(g^3) \quad , \tag{2.3.86}$$

$$G_4^c(x,y,z,w) = ig^2\int dx_1dx_2\Delta(x_1,x_2)\,\{\Delta(x_1,x)\Delta(x_1,y)\Delta(x_2,z)\Delta(x_2,w) + \Delta(x_1,x)\Delta(x_1,z)\Delta(x_2,y)\Delta(x_2,w) + \Delta(x_1,x)\Delta(x_1,w)\Delta(x_2,z)\Delta(x_2,y)\} + O(g^4) \quad . \tag{2.3.87}$$

Here $G_1^c(x)$ is a connected part of $\langle 0|\hat{\phi}(x)|0\rangle$ which we assumed to vanish in Eq. (2.3.28). We, however, include this contribution here for completeness. We can now read off from Eqs. (2.3.84–87) the diagrammatic rules for

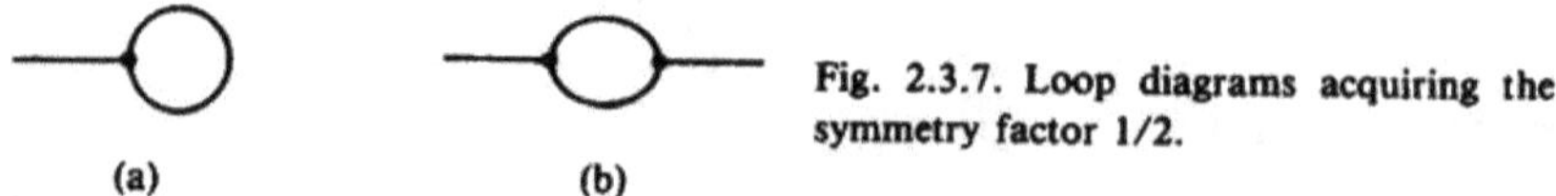

Fig. 2.3.7. Loop diagrams acquiring the symmetry factor 1/2.

calculating the connected Green function $G_n^c(x_1, \ldots, x_n)$.

Feynman rules in configuration space

(1) Make the following identification

Propagator	$-i\Delta(x,y)$	$\leftrightarrow$	x ——— y
Vertex	$i\int d^4x\, g$	$\leftrightarrow$	x (three-line vertex)

(2) To calculate an order-g^N contribution to $G_n^c(x_1, \ldots, x_n)$, draw all possible topologically-independent connected diagrams with n external lines and N vertices and apply the above correspondence.

(3) Multiply each loop contribution by the symmetry factor 1/2 for diagrams such as Fig. 2.3.7.

By using these Feynman rules we can easily and quickly reproduce the previous results (2.3.84–87). A remark is in order about the symmetry factor 1/2 in the above rule. This factor emerges from the fact that we overcount the number of loop diagrams if we make a simple-minded application of the rules (1) and (2) to the loop diagrams. Consider a single vertex which is connected to external sources. There are 3! possibilities of connecting the vertex to 3 sources and all make the same contribution. The factor 3! thus obtained is compensated by 1/3! coming from the interaction Lagrangian $-V(p) = (g/3!)\,\phi^3$. If two of the external lines are joined together to make a loop as shown in Fig. 2.3.7(a), we have only one external line left for attaching a source and there are only 3 possibilities of putting the source. Thus we have the factor $3/3! = 1/2$ which adjusts the overcounting of the diagrams. The same argument also applies to the loop diagram of order g^2 as shown in Fig. 2.3.7(b). If there were six sources to be attached to six external lines originating from two vertices, we would have $2!(3!)^2$ ways of getting the same diagram (Fig. 2.3.6(h)). Since we have a loop in Fig. 2.3.7(b), there are only 3 possibilities of attaching a source at each vertex, 2 possibilities of exchanging two vertices and 2 possibilities of joining lines to make a loop. Thus we again have the correction factor $3^2 \times 2 \times 2/2!(3!)^2 = 1/2$. A more general argument will lead to the symmetry factor $1/S$ where

$$S = p \prod_{n=2,3} (n!)^{d_n} \quad , \tag{2.3.88}$$

d_n being the number of vertex pairs connected by n lines and p the number of vertex permutations which leave the diagram unchanged with external lines held fixed.[17]

In most cases it is much more convenient to work in momentum space rather than in configuration space. Hence we rewrite Eqs. (2.3.84–87) in momentum space by taking into account the representation (2.3.72). We find[18]

$$iG_1^c(x) = \frac{g}{2m^2}\int\frac{d^4k}{(2\pi)^4}\frac{1}{m^2-k^2} + O(g^3) \quad , \tag{2.3.89}$$

$$\begin{aligned} iG_2^c(x,y) &= \int\frac{d^4k}{(2\pi)^4}e^{-ik\cdot(x-y)} \\ &\quad\times\left\{\frac{1}{m^2-k^2}+\frac{1}{(m^2-k^2)^2}\frac{g^2}{2}\left(\frac{1}{m^2}\int\frac{d^4p}{(2\pi)^4 i}\frac{1}{m^2-p^2}\right.\right. \\ &\quad\left.\left.+\int\frac{d^4p}{(2\pi)^4 i}\frac{1}{m^2-p^2}\frac{1}{m^2-(k-p)^2}\right)+O(g^4)\right\} \quad , \end{aligned} \tag{2.3.90}$$

$$\begin{aligned} iG_3^c(x,y,z) &= -ig\int\frac{d^4k}{(2\pi)^4}\frac{d^4p}{(2\pi)^4}e^{-ik\cdot(x-z)-ip\cdot(y-z)} \\ &\quad\times\frac{1}{m^2-k^2}\frac{1}{m^2-p^2}\frac{1}{m^2-(k+p)^2}+O(g^3) \quad , \end{aligned} \tag{2.3.91}$$

$$\begin{aligned} iG_4^c(x,y,z,w) &= -g^2\int\frac{d^4k}{(2\pi)^4}\frac{d^4p}{(2\pi)^4}\frac{d^4q}{(2\pi)^4}e^{-ik\cdot(x-w)-ip\cdot(y-w)-iq\cdot(z-w)} \\ &\quad\times\frac{1}{m^2-k^2}\frac{1}{m^2-p^2}\frac{1}{m^2-q^2}\frac{1}{m^2-(k+p+q)^2} \\ &\quad\times\left\{\frac{1}{m^2-(k+p)^2}+\frac{1}{m^2-(k+q)^2}+\frac{1}{m^2-(p+q)^2}\right\}+O(g^4) \quad . \end{aligned} \tag{2.3.92}$$

Judging from the expression of Eqs. (2.3.89–92) we note that it is much more convenient to employ truncated connected Green functions rather than nontruncated ones as there are many redundant propagators associated with external lines in Eqs. (2.3.89–92). Hence we eliminate the redundant

[17] Take $d_n = 1$ for the diagram of Fig. 2.3.7(a).

[18] The loop integrals in Eqs. (2.3.89) and (2.3.90) are obviously divergent. These divergences are taken care of by the renormalization procedure as will be described in Sec. 2.5.

propagators on all the external lines in Eqs. (2.3.89–92) following the definition of the truncated Green functions (2.3.34) and (2.3.37) where $\tilde{G}_2(k) = i/(k^2-m^2) + O(g^2)$. The resulting truncated Green functions will be denoted by $\tilde{G}_n^{tc}(p_1, ..., p_{n-1})$ since it is connected as well as truncated. We obtain

$$-i\tilde{G}_1^{tc} = \frac{g}{2}\int \frac{d^4k}{(2\pi)^4 i}\frac{1}{m^2-k^2} + O(g^3) \quad , \tag{2.3.89a}$$

$$-i\tilde{G}_2^{tc}(p) = m^2 - p^2 - \frac{g^2}{2}\left\{\frac{1}{m^2}\int \frac{d^4k}{(2\pi)^4 i}\frac{1}{m^2-k^2}\right.$$
$$\left. + \int \frac{d^4k}{(2\pi)^4 i}\frac{1}{m^2-k^2}\frac{1}{m^2-(k-p)^2}\right\} + O(g^4) \quad , \tag{2.3.90a}$$

$$-i\tilde{G}_3^{tc}(p,q) = g + O(g^3) \quad , \tag{2.3.91a}$$

$$-i\tilde{G}_4^{tc}(p,q,k) = g^2\left\{\frac{1}{m^2-(p+q)^2} + \frac{1}{m^2-(q+k)^2}\right.$$
$$\left. + \frac{1}{m^2-(k+p)^2}\right\} + O(g^4) \quad . \tag{2.3.92a}$$

It should be noted here that, according to the discussion leading to Eq. (2.3.40), $(-i)\tilde{G}_n^{tc}$ is exactly equal to the transition matrix element $\langle ...|T|...\rangle$ except for the renormalization constant. On the basis of the observation of Eqs. (2.3.89a–92a) we write out the Feynman rules in momentum space.

Feynman rules in momentum space

(1) To calculate an order g^N contribution to the truncated connected n-point Green function times $(-i)$, draw all possible topologically-independent connected diagrams with n external points and N vertices.
(2) Make the following identification:

Propagator $\quad \overset{k}{\bullet\!\!-\!\!-\!\!-\!\!\bullet} \quad \leftrightarrow \quad \dfrac{1}{m^2-k^2}$

Vertex $\quad -\!\!\bullet\!\!< \quad \leftrightarrow \quad g$

(3) Take into account the energy-momentum conservation at each vertex.
(4) Perform the integration for each loop with the measure

$$\int \frac{d^4k}{(2\pi)^4 i} \quad . \tag{2.3.93}$$

(5) Multiply each loop contribution by the symmetry factor $1/S$ given by Eq. (2.3.88).

It should be remarked here that our Feynman rules presented here are somewhat different from the commonly used Feynman rules. In the Feynman rules most frequently used, the propagator is $i/(k^2 - m^2)$ instead of $1/(m^2 - k^2)$, the vertex factor is ig instead of g, and the measure of the loop integrals is $d^4k/(2\pi)^4$ instead of $d^4k/(2\pi)^4 i$. Moreover the usual rules give Green functions while ours give $(-i)$ times the Green functions. Of course the two systems are essentially equivalent and so there is no reason for discriminating one from the other. It is, however, worth noting that (1) our Feynman rules produce the minimum number of i(imaginary unit) in the course of the calculation (this is useful in the practical sense because the appearance of many i's often causes a mistake in sign), (2) the propagator $1/(m^2 - k^2)$ does not change sign when we make a Wick rotation in the k_0 plane, and (3) our Feynman rules give $-i\tilde{G}_n^{tc}$ which is just equal to the transition matrix element.

Exercise

1. Calculate the $O(g^3)$ term in Eq. (2.3.91).

2.3.3. Quantum chromodynamics

In the last subsection we explained in an elementary way how to derive the Feynman rules in ϕ^3 theory on the basis of the functional-integral formalism. This formalism is actually the most useful in dealing with the complications arising in gauge theories.

In gauge theories the generating functional $Z[J]$ is much more involved than that of ϕ^3 theory. In particular the presence of $\det M_G$ in Eq. (2.2.74) or (2.2.114) prevents a direct application of the method developed in Sec. 2.3.2 to the derivation of Feynman rules in gauge theories. In the axial and temporal gauge, $\det M_G$ turns out to be constant as was seen in Eqs. (2.2.77) and (2.2.78), and hence we may make a straightforward application of this method in order to obtain the Feynman rules. In these gauges, however, the theory is not manifestly covariant and the rules become rather intricate. Hence we would like to develop Feynman rules in a manifestly covariant manner. For this reason we shall work in the covariant gauge in which $G^\mu = \partial^\mu$. (See Eq. (2.2.114).) In the covariant gauge, however, $\det M_G$ depends on the guage field $A_\mu^a(x)$ and also on the gauge coupling constant g as is seen in Eq. (2.2.76). Therefore in making a perturbative expansion of Eq. (2.2.114) we have to take into account the effect of this factor $\det M_G$. The most elegant way of dealing with $\det M_G$ in perturbation theory may be to use Eq. (2.2.113) to exponentiate $\det M_G$ eventually modifying the Lagrangian. Up to an irrelevant factor, $\det M_G$ is represented by, according to Eq. (2.2.113),

$$\det M_G = \int [d\chi]\,[d\chi^*] \exp\{-i\int d^4x d^4y \chi^a(x)^*(M_G(x,y))^{ab}\chi^b(y)\} \quad , \qquad (2.3.94)$$

where M_G is given by Eq. (2.2.76) and $\chi^a(x)$ is a complex fictitious field[19] obeying the Grassmann algebra and belonging to the adjoint representation of the gauge group $G = \mathrm{SU}(3)$. The field $\chi^a(x)$ is called the *Faddeev-Popov ghost* [Fad 67], as it has the strange property that it is fermionic (Grassmann number) as well as bosonic (its propagator is boson-like as will be seen shortly). The exponent of the integrand in Eq. (2.3.94) may be rewritten by doing an integration by parts such that

$$\int d^4x d^4y \chi^a(x)^* \,(M_G(x,y))^{ab}\chi^b(y) = -\int d^4x (\partial^\mu \chi^a(x))^* D^{ab}_\mu \chi^b(x) \quad , \qquad (2.3.95)$$

where D^{ab}_μ is the covariant derivative (Eq. (2.1.28) or (2.1.46)) in the adjoint representation (note that $(T^a)_{bc} = -if^{abc}$)

$$D^{ab}_\mu = \delta^{ab}\partial_\mu - gf^{abc}A^c_\mu \quad . \qquad (2.3.96)$$

We insert Eq. (2.3.94) together with Eq. (2.3.95) into Eq. (2.2.114) to obtain the generating functional

$$Z[J,\xi,\xi^*,\eta,\bar\eta] = \int [dA]\,[d\chi]\,[d\chi^*]\,[d\psi]\,[d\bar\psi]$$

$$\times \exp\{i\int d^4x(\mathscr{L} + AJ + \chi^*\xi + \xi^*\chi + \bar\psi\eta + \bar\eta\psi)\} \quad , \qquad (2.3.97)$$

where ξ^a and ξ^{a*} are source functions (Grassmann numbers) for the ghosts and AJ, $\chi^*\xi$ and $\xi^*\chi$ are shorthand notations so that

$$AJ = A^a_\mu J^{a\mu} \quad , \quad \chi^*\xi = \chi^{a*}\xi^a \quad , \quad \xi^*\chi = \xi^{a*}\chi^a \quad . \qquad (2.3.98)$$

[19] Here we chose χ^a and χ^{a*} as two independent fields as we prefered to have Eq. (2.3.102) later. This choice, however, is not necessary. We could have chosen, say, two independent real fields χ^a_1 and χ^a_2 instead of χ^a and $-i\chi^{a*}$. See Sec. 2.3.5 for more details.

Here $\mathcal{L}$ is an effective quantum Lagrangian which includes the effect of $\det M_G$ through Eq. (2.3.94),

$$\mathcal{L} = \mathcal{L}_{\mathrm{G}} + \mathcal{L}_{\mathrm{GF}} + \mathcal{L}_{\mathrm{FP}} + \mathcal{L}_{\mathrm{F}} \quad , \tag{2.3.99}$$

$$\mathcal{L}_{\mathrm{G}} = -\frac{1}{4} F^a_{\mu\nu} F^{a\mu\nu} , \quad F^a_{\mu\nu} = \partial_\mu A^a_\nu - \partial_\nu A^a_\mu + g f^{abc} A^b_\mu A^c_\nu \quad , \tag{2.3.100}$$

$$\mathcal{L}_{\mathrm{GF}} = -\frac{1}{2\alpha} (\partial^\mu A^a_\mu)^2 \quad , \tag{2.3.101}$$

$$\mathcal{L}_{\mathrm{FP}} = (\partial^\mu \chi^{a*}) D^{ab}_\mu \chi^b \quad , \tag{2.3.102}$$

$$\mathcal{L}_{\mathrm{F}} = \bar{\psi}^i (i\gamma^\mu D^{ij}_\mu - m\delta^{ij}) \psi^j \quad , \tag{2.3.103}$$

where we considered, for simplicity, only one species of quarks ψ belonging to the fundamental representation of $G = \mathrm{SU}(3)$ with $\bar{\psi} = \psi^\dagger \gamma_0$ and D^{ij}_μ is the covariant derivative in the fundamental representation of $G = \mathrm{SU}(3)$. The suffixes of the Lagrangians (2.3.100 – 103) stand for "gauge," "gauge fixing," "Faddeev-Popov" and "fermion" terms respectively. The above Lagrangian (2.3.99) forms the basis of quantum chromodynamics (for one species of quarks).

Since the generating functional (2.3.97) is now of the same form as the one in Eq. (2.3.56), we can straightforwardly apply the method developed in Sec. 2.3.2 to the present case. For this purpose we split up the Lagrangian (2.3.99) into the free part $\mathcal{L}_0$ and the interaction part $\mathcal{L}_1$,

$$\mathcal{L} = \mathcal{L}_0 + \mathcal{L}_1 \quad . \tag{2.3.104}$$

Here the free Lagrangian $\mathcal{L}_0$ is made of three parts each of which corresponds to the participating particles, i.e., the gluons, Faddeev-Popov ghosts and quarks,

$$\mathcal{L}_0 = \mathcal{L}^{\mathrm{G}}_0 + \mathcal{L}^{\mathrm{FP}}_0 + \mathcal{L}^{\mathrm{F}}_0 \quad , \tag{2.3.105}$$

$$\mathcal{L}^{\mathrm{G}}_0 = -\frac{1}{4}(\partial_\mu A^a_\nu - \partial_\nu A^a_\mu)(\partial^\mu A^{a\nu} - \partial^\nu A^{a\mu}) - \frac{1}{2\alpha}(\partial^\mu A^a_\mu)^2 \quad , \tag{2.3.106}$$

$$\mathcal{L}^{\mathrm{FP}}_0 = (\partial^\mu \chi^{a*})(\partial_\mu \chi^a) \quad , \tag{2.3.107}$$

$$\mathcal{L}^{\mathrm{F}}_0 = \bar{\psi}(i\gamma^\mu \partial_\mu - m)\psi \quad . \tag{2.3.108}$$

As Eq. (2.3.107) is of the form of the Lagrangian for massless charged scalar fields, we recognize that the Faddeev-Popov ghost is bosonic although it is fermionic owing to its nature as a Grassmann number. It should also be noted that in Eq. (2.3.106) the gauge fixing term $\mathscr{L}_{GF}$ is included. The remaining part of the Lagrangian $\mathscr{L}$ after subtracting $\mathscr{L}_0$ is the interaction Lagrangian $\mathscr{L}_1$ which amounts to

$$\begin{aligned}\mathscr{L}_1 &= \mathscr{L}_1(A^a, \chi^a, \chi^{a*}, \psi, \bar{\psi}) \\ &= -\frac{g}{2} f^{abc}(\partial_\mu A^a_\nu - \partial_\nu A^a_\mu) A^{b\mu} A^{c\nu} - \frac{g^2}{4} f^{abe} f^{cde} A^a_\mu A^b_\nu A^{c\mu} A^{d\nu} \\ &\quad - g f^{abc}(\partial^\mu \chi^{a*}) \chi^b A^c_\mu + g\bar{\psi} T^a \gamma^\mu \psi A^a_\mu \quad . \end{aligned} \tag{2.3.109}$$

Just as in the previous subsection we can rewrite Eq. (2.3.97) in the form of Eq. (2.3.62), i.e.,

$$\begin{aligned} & Z[J,\xi,\xi^*,\eta,\bar{\eta}] \\ & = \exp\left\{i\int d^4x\, \mathscr{L}_1\left(\frac{\delta}{i\delta J^{a\mu}}, \frac{\delta}{i\delta \xi^{a*}}, \frac{\delta}{i\delta(-\xi^a)}, \frac{\delta}{i\delta\bar{\eta}}, \frac{\delta}{i\delta(-\eta)}\right)\right\} Z_0[J,\xi,\xi^*,\eta,\bar{\eta}] \quad , \end{aligned} \tag{2.3.110}$$

where Z_0 is a generating functional for free fields,

$$Z_0[J, \ldots] = Z_0^G[J] Z_0^{FP}[\xi,\xi^*] Z_0^F[\eta,\bar{\eta}] \quad , \tag{2.3.111}$$

$$Z_0^G[J] = \int [dA] \exp\{i\int d^4x(\mathscr{L}_0^G + AJ)\} \quad , \tag{2.3.112}$$

$$Z_0^{FP}[\xi,\xi^*] = \int [d\chi][d\chi^*] \exp\{i\int d^4x(\mathscr{L}_0^{FP} + \chi^*\xi + \xi^*\chi)\} \quad , \tag{2.3.113}$$

$$Z_0^F[\eta,\bar{\eta}] = \int [d\psi][d\bar{\psi}] \exp\{i\int d^4x(\mathscr{L}_0^F + \bar{\psi}\eta + \bar{\eta}\psi)\} \quad . \tag{2.3.114}$$

In order to obtain $Z[J,\ldots]$ perturbatively we first calculate $Z_0[J,\ldots]$ for the gluon, Faddeev-Popov ghost and quark respectively. To do this we re-express the free Lagrangian by doing an integration by parts,

$$\mathscr{L}_0^G = -\frac{1}{2} A^a_\mu K^{ab\mu\nu} A^b_\nu, \quad K^{ab}_{\mu\nu} = \delta^{ab}\left(-g_{\mu\nu}\Box + \left(1 - \frac{1}{\alpha}\right)\partial_\mu\partial_\nu\right), \tag{2.3.115}$$

$$\mathscr{L}_0^{FP} = -\chi^{a*} K^{ab} \chi^b, \quad K^{ab} = \delta^{ab}\Box, \tag{2.3.116}$$

$$\mathscr{L}_0^F = -\bar{\psi}\Lambda\psi, \quad \Lambda = -i\gamma^\mu\partial_\mu + m. \tag{2.3.117}$$

If we denote the inverse of $K^{ab}_{\mu\nu}$, K^{ab} and Λ by $D^{ab}_{\mu\nu}$, D^{ab} and S, respectively, we have

$$\int d^4z\, K^{ac}_{\mu\lambda}(x-z) g^{\lambda\rho} D^{cb}_{\rho\nu}(z-y) = \delta^{ab} g_{\mu\nu} \delta^4(x-y) \quad , \tag{2.3.118}$$

$$\int d^4z\, K^{ac}(x-z) D^{cb}(z-y) = \delta^{ab}\delta^4(x-y) \quad , \tag{2.3.119}$$

$$\int d^4z\, \Lambda(x-z)\, S(z-y) = \delta^4(x-y) \quad , \tag{2.3.120}$$

where it should be understood that K's and Λ include delta functions of argument $x - z$. These functions $D^{ab}_{\mu\nu}$, D^{ab} and S are propagators of the gluon, Faddeev-Popov ghost and quark respectively. Solving the conditions (2.3.118 – 120) for Fourier coefficients, we find

$$D^{ab}_{\mu\nu}(x) = \delta^{ab} \int \frac{d^4k}{(2\pi)^4} \frac{e^{-ik\cdot x}}{k^2 + i\varepsilon} \left(g_{\mu\nu} - (1-\alpha) \frac{k_\mu k_\nu}{k^2} \right) \quad , \tag{2.3.121}$$

$$D^{ab}(x) = \delta^{ab} \int \frac{d^4k}{(2\pi)^4} \frac{-1}{k^2 + i\varepsilon} e^{-ik\cdot x} \quad , \tag{2.3.122}$$

$$S(x) = \int \frac{d^4p}{(2\pi)^4} \frac{1}{m - \not{p}} e^{-ip\cdot x} \quad . \tag{2.3.123}$$

We can now perform the functional integrations of A^a_μ, χ, χ^*, ψ and $\bar{\psi}$ in Eqs. (2.3.112–114), respectively. We obtain, apart from irrelevant constant multiples,

$$Z^G_0[J] = \exp\{\frac{i}{2} \int d^4x d^4y J^{a\mu}(x) D^{ab}_{\mu\nu}(x-y) J^{b\nu}(y)\} \quad , \tag{2.3.124}$$

$$Z^{FP}_0[\xi,\xi^*] = \exp\{i \int d^4x d^4y \xi^a(x)^* D^{ab}(x-y) \xi^b(y)\} \quad , \tag{2.3.125}$$

$$Z^F_0[\eta,\bar{\eta}] = \exp\{i \int d^4x d^4y \bar{\eta}(x) S(x-y) \eta(y)\} \quad . \tag{2.3.126}$$

We insert Eqs. (2.3.124–126) into Eq. (2.3.111) and use Eq. (2.3.110) to generate the perturbation series,

$$Z[J,...] = \left\{ 1 + i \int d^4x \mathscr{L}_1 \left(\frac{\delta}{i\delta J^{a\mu}(x)}, ... \right) + ... \right\} Z_0[J,...] \quad . \tag{2.3.127}$$

As our first example we calculate the contribution of the first term (three-gluon coupling) in Eq. (2.3.109) to the first-order term in the above expansion (2.3.127). Except for the irrelevant factor $Z_0^{FP}[\xi,\xi^*]Z_0^F[\eta,\bar{\eta}]$ we have for that contribution

$$i\int d^4x\mathscr{L}_1^{3G}\left(\frac{\delta}{i\delta J^{a\mu}}\right)Z_0^G[J] \equiv$$

$$i\int d^4x\frac{-g}{2}f^{abc}\left(\partial_\mu\frac{\delta}{i\delta J^{a\nu}}-\partial_\nu\frac{\delta}{i\delta J^{a\mu}}\right)\frac{\delta}{i\delta J_\mu^b}\frac{\delta}{i\delta J_\nu^c}Z_0^G[J] \quad . \tag{2.3.128}$$

By straightforward calculation we find that Eq. (2.3.128) reduces to (keeping only the term with three J's)

$$\begin{aligned}&-\mathrm{i}\frac{g}{2}f^{abc}\int dxdy_1dy_2dy_3\{\partial_\mu D_{\nu\lambda_1}^{aa_1}(x-y_1)\\&\qquad-\partial_\nu D_{\mu\lambda_1}^{aa_1}(x-y_1)\}D_{\lambda_2}^{ba_2\mu}(x-y_2)D_{\lambda_3}^{ca_3\nu}(x-y_3)\\&\qquad\times J^{a_1\lambda_1}(y_1)J^{a_2\lambda_2}(y_2)J^{a_3\lambda_3}(y_3)Z_0[J]\quad ,\end{aligned} \tag{2.3.129}$$

where ∂_μ refers to x. This term will make a contribution to the three-point Green function $G_3(x_1, x_2, x_3)$ for gluons to order g in such a way that

$$G_{3\mu_1\mu_2\mu_3}^{a_1a_2a_3}(x_1,x_2,x_3) = (-i)^2\frac{\delta^3}{\delta J_1\delta J_2\delta J_3}\int d^4x\mathscr{L}_1^{3G}\left(\frac{\delta}{i\delta J^{a\mu}}\right)Z_0^G[J]\bigg|_{J=0} \quad , \tag{2.3.130}$$

where $J_i = J^{a_i\mu_i}(x_i)$ with $i = 1, 2, 3$. After some calculation we get

$$\begin{aligned}G_{3\mu_1\mu_2\mu_3}^{a_1a_2a_3}(x_1,x_2,x_3) &= gf^{abc}\int d^4x\{\partial_\mu D_{\nu\mu_1}^{aa_1}(x-x_1)\\&\qquad-\partial_\nu D_{\mu\mu_1}^{aa_1}(x-x_1)\}D_{\mu_2}^{ba_2\mu}(x-x_2)\,D_{\mu_3}^{ca_3\nu}(x-x_3)\\&\qquad+(231)+(312)\quad ,\end{aligned} \tag{2.3.131}$$

where by (231) we mean the term similar to the first term with indices 1, 2 and 3 replaced by 2, 3 and 1, respectively. Substituting Eq. (2.3.121) for the gluon propagator in Eq. (2.3.131) we obtain the representation of $G_3(x_1, x_2, x_3)$ in momentum space,

$$
\begin{aligned}
-iG^{a_1a_2a_3}_{3\mu_1\mu_2\mu_3}(x_1,x_2,x_3) = & \int \frac{d^4k_1}{(2\pi)^4}\frac{d^4k_2}{(2\pi)^4}\, e^{i(k_1\cdot x_1+k_2\cdot x_2+k_3\cdot x_3)} \\
& \times \frac{d_{\mu_1\lambda_1}(k_1)d_{\mu_2\lambda_2}(k_2)d_{\mu_3\lambda_3}(k_3)}{k_1^2k_2^2k_3^2} \\
& \times g f^{a_1a_2a_3}\{(k_1-k_2)^{\lambda_3}g^{\lambda_1\lambda_2}+(k_2-k_3)^{\lambda_1}g^{\lambda_2\lambda_3} \\
& +(k_3-k_1)^{\lambda_2}g^{\lambda_3\lambda_1}\} \quad , \qquad (2.3.132)
\end{aligned}
$$

where $k_3 = -k_1 - k_2$, and

$$
d_{\mu\nu}(k) = g_{\mu\nu} - (1-\alpha)k_\mu k_\nu/k^2 \quad . \qquad (2.3.133)
$$

Equation (2.3.132) provides us with the Feynman rule for the three-gluon vertex in momentum space. Note that the above Green function (2.3.132) will be truncated by eliminating the factor $-id_{\mu\nu}(k)/k^2$ at each leg.

As our second example we consider the contribution of the second term (four-gluon coupling) in Eq. (2.3.109) to the first-order term in Eq. (2.3.127). We calculate

$$
i\int d^4x\mathscr{L}_1^{4G}\left(\frac{\delta}{i\delta J^{a\mu}}\right)Z_0^G[J] \equiv -i\int d^4x\frac{g^2}{4}f^{abe}f^{cde}\frac{\delta^4}{\delta J^{a\mu}\delta J^{b\nu}\delta J^c_\mu \delta J^d_\nu}Z_0^G[J] \quad . \qquad (2.3.134)
$$

Equation (2.3.134) is easily evaluated and results in (keeping only the terms with four J's)

$$
\begin{aligned}
& -i\frac{g^2}{4}f^{abe}f^{cde}\int dxdy_1 \ldots dy_4 \\
& \times D^{aa_1}_{\mu\mu_1}(x-y_1)\, D^{ba_2}_{\nu\mu_2}(x-y_2)\, D^{ca_3\mu}_{\mu_3}(x-y_3)\, D^{da_4\nu}_{\mu_4}(x-y_4) \\
& \times J^{a_1\mu_1}(y_1) \ldots J^{a_4\mu_4}(y_4) \quad . \qquad (2.3.135)
\end{aligned}
$$

Now we may easily deduce the contribution of Eq. (2.3.135) to the first-order term in the four-point Green function $G_4(x_1,\ldots,x_4)$ for gluons

$$
\begin{aligned}
-iG^{a_1\ldots a_4}_{4\mu_1\ldots\mu_4}(x_1,\ldots x_4) = & -g^2\Big\{(f^{b_1b_3b}f^{b_2b_4b}-f^{b_1b_4b}f^{b_3b_2b})g^{\lambda_1\lambda_2}g^{\lambda_3\lambda_4} \\
& +(f^{b_1b_2b}f^{b_3b_4b}-f^{b_1b_4b}f^{b_2b_3b})g^{\lambda_1\lambda_3}g^{\lambda_2\lambda_4} \\
& +(f^{b_1b_3b}f^{b_4b_2b}-f^{b_1b_2b}f^{b_3b_4b})g^{\lambda_1\lambda_4}g^{\lambda_3\lambda_2}\Big\} \\
& \times \int d^4x D^{b_1a_1}_{\lambda_1\mu_1}(x-x_1)\, D^{b_2a_2}_{\lambda_2\mu_2}(x-x_2)\, D^{b_3a_3}_{\lambda_3\mu_3}(x-x_3)\, D^{b_4a_4}_{\lambda_4\mu_4}(x-x_4). \qquad (2.3.136)
\end{aligned}
$$

From Eq. (2.3.136) we can read off the Feynman rule for the four-gluon vertex.

Similar arguments apply to the gluon-ghost-ghost three-point Green function $G_{3FP}(x_1, x_2, x_3)$. Here we should keep in mind the rule explained at the end of Sec. 2.2.4 for the definition of fermionic Green functions (the ghost field χ^a and its source function ξ^a are Grassmann numbers), i.e., in defining $G_{3FP}(x_1, x_2, x_3)$ through the generating functional we have to make functional differentiation with respect to $-\xi^a(x)$ instead of $\xi^a(x)$ as it stands on the right in the source term $\chi^*\xi$. Hence we have

$$G^{a_1a_2a_3}_{3FP,\mu}(x_1,x_2,x_3) = (-i)^3 \left.\frac{\delta^3 Z}{\delta\xi^{a_1}(x_1)^*\delta(-\xi^{a_2}(x_2))\delta J^{a_3\mu}(x_3)}\right|_{\xi^*=\xi=J=0} ,$$

where it is understood that $\eta(x)$ and $\bar{\eta}(x)$ are also set equal to zero. To the first order in g the relevant contribution to the Green function is given by

$$i\int d^4x\left[-gf^{abc}\left(\partial^\mu\frac{\delta}{i\delta(-\xi^a)}\right)\frac{\delta}{i\delta\xi^{b*}}\frac{\delta}{i\delta J^{c\mu}}\right]Z_0^G[J]Z_0^{FP}[\xi,\xi^*]$$

$$= -igf^{abc}\int dxdy_1dy_2dy_3\partial^\mu D^{aa_1}(y_1-x)\,D^{ba_2}(x-y_2)D^{ca_3}_{\mu\nu}(x-y_3)$$

$$\times\,\xi^{a_1}(y_1)^*\xi^{a_2}(y_2)J^{a_3\nu}(y_3)Z_0^G[J]Z_0^{FP}[\xi,\xi^*] ,$$

where we kept only the term with $\xi^*\xi J$. Inserting this result into the above definition of G_{3FP} we find in the momentum representation,

$$-iG^{a_1a_2a_3}_{3FP,\mu}(x_1,x_2,x_3)$$

$$= gf^{a_1a_2a_3}\int\frac{d^4k_1}{(2\pi)^4}\frac{d^4k_2}{(2\pi)^4}e^{i(-k_1\cdot x_1+k_2\cdot x_2+k_3\cdot x_3)}\frac{d_{\mu\nu}(k_3)}{k_1^2k_2^2k_3^2}k_1^\nu ,$$

(2.3.137)

where $k_3 = k_1 - k_2$, $d_{\mu\nu}(k)$ is given by Eq. (2.3.133) and momentum k_1 refers to the χ^{a*} field.

In the same way we can derive the expression for the gluon-quark-quark three-point Green function G_{3F} to order g,

$$-iG^a_{3F,\mu}(x_1,x_2,x_3) = igT^a \int \frac{d^4p_1}{(2\pi)^4}\frac{d^4p_2}{(2\pi)^4} e^{i(-p_1\cdot x_1+p_2\cdot x_2+k\cdot x_3)}$$

$$\times \frac{1}{m-\not{p}_1}\gamma^\nu \frac{1}{m-\not{p}_2}\frac{d_{\mu\nu}(k)}{k^2}, \tag{2.3.138}$$

where $k = p_1 - p_2$. Note here again that the above Green functions (2.3.137) and (2.3.138) can be truncated by eliminating the two-point Green functions $-id_{\mu\nu}(k)/k^2$, $-i/k^2$ and $-i/(m-\not{p})$ for the gluon, ghost and quark legs, respectively.

Accordingly we have the following Feynman rules for the Lagrangian of quantum chromodynamics:

Gluon propagator	$a\mu$ —k— $b\nu$	$\delta_{ab}\frac{d_{\mu\nu}(k)}{k^2}$
Ghost propagator	a —k— b	$\delta_{ab}\frac{-1}{k^2}$
Quark propagator	i —p— j	$\delta_{ij}\frac{1}{m-\not{p}}$
3-gluon vertex	$a_1\mu_1$, k_1; $a_2\mu_2$, k_2; $a_3\mu_3$, k_3	$-igf^{a_1a_2a_3}V_{\mu_1\mu_2\mu_3}(k_1,k_2,k_3)$
4-gluon vertex	$a_1\mu_1$; $a_2\mu_2$; $a_3\mu_3$; $a_4\mu_4$	$-g^2W^{a_1\ldots a_4}_{\mu_1\ldots\mu_4}$
Gluon-ghost vertex	$a\mu$; k; b; c	$-igf^{abc}k_\mu$
Gluon-quark vertex	$a\mu$; i; j	$g\gamma_\mu T^a_{ij}$

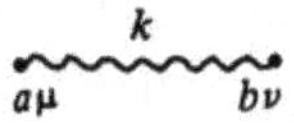

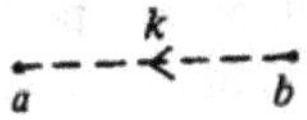

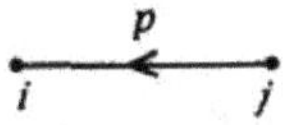

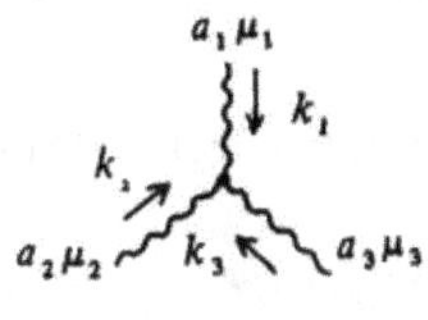

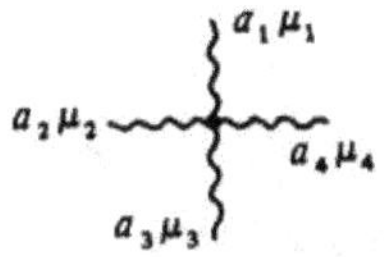

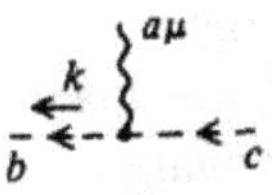

Here the wavy, dotted and solid lines represent the gluon, ghost and quark, respectively; the arrows on the ghost and quark line show a flow of the ghost and fermion number, respectively. The functions V and W are given by

$$V_{\mu_1\mu_2\mu_3}(k_1,k_2,k_3) = (k_1-k_2)_{\mu_3}g_{\mu_1\mu_2} + (k_2-k_3)_{\mu_1}g_{\mu_2\mu_3} + (k_3-k_1)_{\mu_2}g_{\mu_3\mu_1} \quad , \tag{2.3.139}$$

$$W^{a_1\ldots a_4}_{\mu_1\ldots\mu_4} = (f^{13,24}-f^{14,32})g_{\mu_1\mu_2}g_{\mu_3\mu_4} + (f^{12,34}-f^{14,23})g_{\mu_1\mu_3}g_{\mu_2\mu_4}$$

$$+ (f^{13,42}-f^{12,34})g_{\mu_1\mu_4}g_{\mu_3\mu_2} \quad , \tag{2.3.140}$$

where $f^{ij,kl}$ denotes the following combination

$$f^{ij,kl} = f^{a_i a_j a} f^{a_k a_l a} \quad . \tag{2.3.141}$$

The rules given above are still incomplete unless they are supplemented by additional rules on the loop-sign factor and the symmetry factor.

Let us first discuss the loop-sign factor which is an extra minus sign needed for each ghost and quark loop. To see the necessity of this extra minus sign when applying the above Feynman rules, we consider, as an example, the contribution of a fermion loop to the gluon two-point Green function at order g^2. For this purpose we calculate the second-order quark contribution to the generating functional keeping only the term with two source functions of gluons $J^{a\mu}(x)$,

$$\frac{1}{2!}\left\{i\int d^4x\,\mathscr{L}^{\mathrm{F}}_1\left(\frac{\delta}{i\delta(-\eta)},\frac{\delta}{i\delta\bar{\eta}},\frac{\delta}{i\delta J}\right)\right\}^2 Z^{\mathrm{F}}_0[\eta,\bar{\eta}]Z^{\mathrm{G}}_0[J] \,,$$

$$= (-1)\frac{g^2}{2}\int dx_1dx_2dy_1dy_2\mathrm{Tr}[T^a\gamma^\mu S(x_1-x_2)T^b\gamma^\nu S(x_2-x_1)]$$

$$\times D^{ac}_{\mu\lambda}(x_1-y_1)\,D^{bd}_{\nu\rho}(x_2-y_2)\,J^{c\lambda}(y_1)\,J^{d\rho}(y_2)\,Z^{\mathrm{F}}_0[\eta,\bar{\eta}]Z^{\mathrm{G}}_0[J] \,, \tag{2.3.142}$$

where the trace Tr is effective on both group space and spinor space and $\mathscr{L}^{\mathrm{F}}_1(\bar{\psi},\psi,A)$ is defined by

$$\mathscr{L}^{\mathrm{F}}_1(\bar{\psi},\psi,A) = g\bar{\psi}T^a\gamma^\mu\psi A^a_\mu \quad . \tag{2.3.143}$$

Needless to say, the factor -1 at the right-hand side of Eq. (2.3.142) comes from the anticommuting nature of the Grassmann numbers η and $\bar{\eta}$. From Eq. (2.3.142) we deduce the second-order quark contribution to the gluon two-point Green function $G^{ab}_{2\mu\nu}(x,y)$,

$$-iG^{ab}_{2\mu\nu}(x,y) = -(-1)g^2\int\frac{d^4k}{(2\pi)^4}e^{-ik\cdot(x-y)}\frac{d_{\mu\lambda}(k)}{k^2}\frac{d_{\nu\rho}(k)}{k^2}$$

$$\times\int\frac{d^4p}{(2\pi)^4 i}\mathrm{Tr}\left[T^a\gamma^\lambda\frac{1}{m-\not{p}}T^b\gamma^\rho\frac{1}{m+\not{p}-\not{k}}\right] \quad . \tag{2.3.144}$$

If we would have simple-mindedly followed the Feynman rules given previously to calculate $-iG^{ab}_{2\mu\nu}$, we would have obtained the above result without the factor -1. Thus, in order to obtain the correct Green functions, we have to amend the previous rules by adding an additional rule of multiplying by -1 each fermion-loop contribution. This additional rule comes from the Grassmann nature of the quarks and hence the same argument as above also applies to the ghost loop contribution to the Green functions. The factor -1 associated with each quark and ghost loop is the so-called loop-sign factor. Since the loop-sign factor is always accompanied by a loop integration, we may include it in the rule for loop integrations. Now the rules for loop integrations read as follows.

Gluon loop ($a\mu$, $b\nu$, $\uparrow k$) $\qquad \displaystyle\int \frac{d^4k}{(2\pi)^4 i}\,\delta^{ab}\,g^{\mu\nu}$

Ghost loop (a, b, $\uparrow k$) $\qquad \displaystyle -\int \frac{d^4k}{(2\pi)^4 i}\,\delta^{ab}$

Quark loop ($i\alpha$, $j\beta$, $\uparrow p$) $\qquad \displaystyle -\int \frac{d^4p}{(2\pi)^4 i}\,\delta^{ij}\delta^{\alpha\beta}$

In the above rules, δ^{ab} (δ^{ij}), $g^{\mu\nu}$ and $\delta^{\alpha\beta}$ stand for contractions of group, Lorentz and spinor indices, respectively, at the point where the loops close.

We next go over to a discussion of the symmetry factor which is responsible for the adjustment of the possible overcounting of diagrams caused by a simple-minded application of the previous Feynman rules. The problem of overcounting in fact arises in gluon loops while no such problem occurs in ghost and quark loops. Let us consider the diagram Fig. 2.3.8(a).

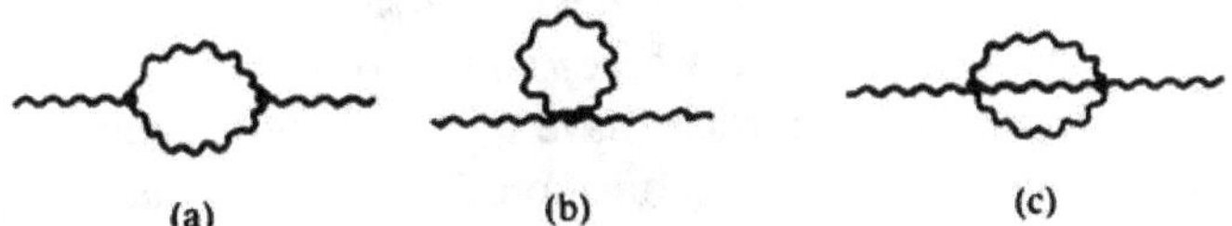

Fig. 2.3.8. Gluon one-loop diagrams which need the symmetry factor: (a) 1/2!, (b) 1/2!, (c) 1/3!.

If we apply the Feynman rules in a simple-minded way, we have 2 identical terms obtained by interchanging two vertices and 3! terms at each vertex corresponding to 3! possibilities of choosing one of three gluon lines as an external line. Thus the total number of terms divided by 2! in the second-order Green function with two three-gluon vertices may be

$$\frac{1}{2!} \times 2 \times (3!)^2 = (3!)^2 \quad .$$

In fact for the three-gluon vertex we have exactly $3! = 6$ terms as seen in Eq. (2.3.139) and so we actually have $(3!)^2$ terms for Fig. 2.3.8(a) directly following the previous Feynman rule. This, however, is obviously wrong because for the loop diagram Fig. 2.3.8(a) the number of terms divided by 2! is as discussed before in Sec. 2.3.2, i.e.,

$$\frac{1}{2!} \times 2 \times 3 \times 3 \times 2 = \frac{1}{2} \times (3!)^2 \quad .$$

In order to correct this overcounting we have to multiply each contribution of the diagram of the type Fig. 2.3.8(a) by 1/2. Although the above argument is based on a simple counting of terms, it is not difficult to show the existence of the symmetry factor 1/2 by directly calculating the second-order two-point gluon Green function. A similar argument also applies to the diagrams shown in Figs. 2.3.8(b) and (c), and we find the following symmetry factors corresponding to the diagrams (a), (b) and (c):

$$\text{(a) } 1/2! \quad , \qquad \text{(b) } 1/2! \quad , \qquad \text{(c) } 1/3! \quad . \tag{2.3.145}$$

Finally we arrive at the complete Feynman rules in quantum chromodynamics for calculating $-iG_n^{\text{tc}}$ (truncated connected Green function) as shown in the table. The truncated connected Green functions times $(-i)$ calculated through the Feynman rules will be called *Feynman amplitudes*. Note here that $d_{\mu\nu}(k)$, $V_{\mu_1\mu_2\mu_3}(k_1, k_2, k_3)$ and $W^{a_1 \ldots a_4}_{\mu_1 \ldots \mu_4}$ are given in Eqs. (2.3.133), (2.3.139) and (2.3.140), respectively. For a more general exposition of Feynman rules in the case of the theory with spontaneous symmetry breaking, see [Aok 82] where the same convention is adopted.

The Feynman rules given above provide us with truncated connected Green functions times $(-i)$. In order to obtain transition matrix elements, we add the rule that external lines should be on their mass shell. If the external particles are scalar, we simply set $p^2 = m^2$ where p and m are the momentum and mass, respectively. If the external particles have spin, we have to attach, in addition to setting $p^2 = m^2$, to the Green function the spin wave function, e.g., the Dirac spinor $u(p)$ for fermions ($v(p)$ for anti-fermions) and the polarization vector $\varepsilon^\mu(k)$ for vector particles in conformity with the reduction formulas (2.3.40a) and (2.3.40b), respectively. See the additional table.

Here the following comment is in order. In quantum chromodynamics quarks and gluons are most likely to be confined and so it seems to be meaningless to consider free on-shell states of quarks and gluons. Although the nonexistence of free isolated quark and gluon states should be respected in a rigorous argument, it is sometimes convenient as a technical means to treat quarks and gluons in approximate perturbative calculations as if they appear in free on-shell states. This technical expedient is in fact allowed owing to

QCD Feynman rules

Propagators

gluons A	 $a\mu$ — k — $b\nu$	$\delta_{ab}\dfrac{d_{\mu\nu}(k)}{k^2}$
ghosts χ	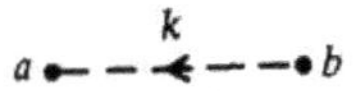 a — k — b	$\delta_{ab}\dfrac{-1}{k^2}$
quarks ψ	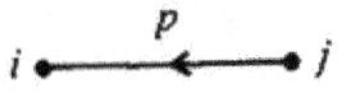 i — p — j	$\delta_{ij}\dfrac{1}{m-\not p}$

Vertices

3-gluon	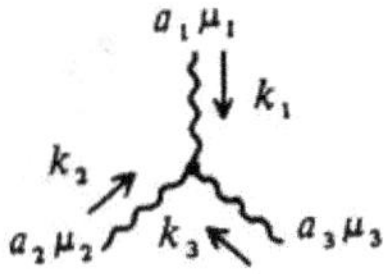 $a_1\mu_1$, k_1; $a_2\mu_2$, k_2; $a_3\mu_3$, k_3	$-igf^{a_1a_2a_3}V_{\mu_1\mu_2\mu_3}(k_1,k_2,k_3)$
4-gluon	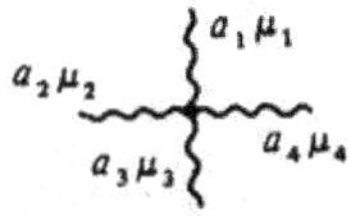 $a_1\mu_1$, $a_2\mu_2$, $a_3\mu_3$, $a_4\mu_4$	$-g^2W^{a_1\ldots a_4}_{\mu_1\ldots\mu_4}$
gluon-ghost	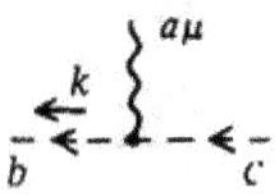 $a\mu$; k; b, c	$-igf^{abc}k_\mu$
gluon-quark	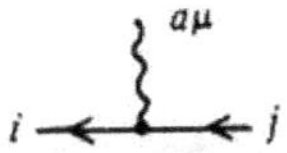 $a\mu$; i, j	$g\gamma_\mu T^a_{ij}$

Loops

gluon	 $a\mu$, $b\nu$, k	$\displaystyle\int\frac{d^4k}{(2\pi)^4 i}\,\delta^{ab}g^{\mu\nu}$
ghost	 a, b, k	$\displaystyle-\int\frac{d^4k}{(2\pi)^4 i}\,\delta^{ab}$
quark	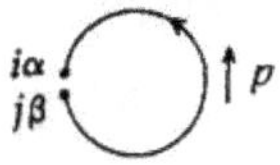 $i\alpha$, $j\beta$, p	$\displaystyle-\int\frac{d^4p}{(2\pi)^4 i}\,\delta^{ij}\delta^{\alpha\beta}$
gluon-quark	 k	$\displaystyle\int\frac{d^4k}{(2\pi)^4 i}$
gluon-ghost		

Symmetry factors

$\frac{1}{2!}$  , $\frac{1}{2!}$, $\frac{1}{3!}$

asymptotic freedom of quantum chromodynamics which guarantees that quarks and gluons are almost free at short distances. With this reasoning we shall, from time to time, calculate quark-gluon transition matrix elements following the above Feynman rules in some practical applications to short-distance QCD processes.

Additional rules for transition matrix elements

Mass shell condition for external lines		$p^2 = m^2$
Fermions (quarks)		
outgoing fermion	p, λ	$\bar{u}_\lambda(p)$
outgoing antifermion	p, λ	$v_\lambda(p)$
incoming fermion	p, λ	$u_\lambda(p)$
incoming antifermion	p, λ	$\bar{v}_\lambda(p)$
Vector fields (gluons)		
outgoing vector	k, λ	$\varepsilon^{\mu *}_\lambda(k)$
incoming vector	k, λ	$\varepsilon^\mu_\lambda(k)$

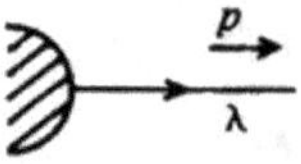

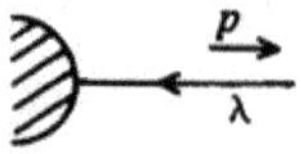

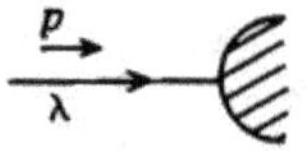

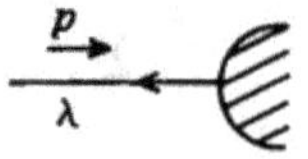

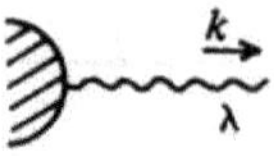

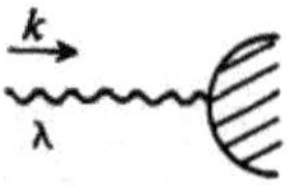

In some cases it will be necessary for us to deal with electromagnetic interactions and we need Feynman rules for quantum electrodynamics (QED). The rules are readily derived in the same way as before for the Lagrangian

$$\mathscr{L} = -\frac{1}{4}F_{\mu\nu}F^{\mu\nu} - \frac{1}{2\alpha}(\partial^\mu A_\mu)^2 + \bar{\psi}\,[i\gamma^\mu(\partial_\mu + iQeA_\mu) - m]\,\psi \quad , \qquad (2.3.146)$$

where $F^{\mu\nu}$ is given previously in Eq. (2.1.2), α the electromagnetic gauge parameter and ψ the charged fermion (lepton or quark) field with charge Qe. The resulting rules are essentially the same as those for QCD if one disregards the color index and sets $f^{abc} = 0$ and $gT^a_{ij} = -Qe$. In these rules $d_{\mu\nu}(k)$ is given by Eq. (2.3.133) with α the electromagnetic gauge parameter in Eq. (2.3.146).

QED Feynman rules

Propagators

photons A	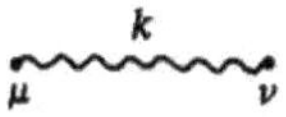	$d_{\mu\nu}(k)/k^2$
fermions ψ		$\dfrac{1}{m - \not p}$

Vertices

photon-fermion	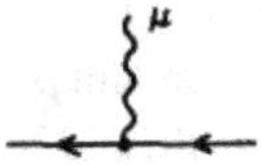	$-Qe\gamma_\mu$

Loops

fermion	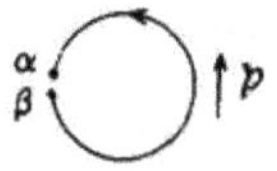	$-\displaystyle\int \frac{d^4p}{(2\pi)^4 i}\,\delta^{\alpha\beta}$
photon-fermion		$\displaystyle\int \frac{d^4k}{(2\pi)^4 i}$

2.3.4. Simple example in e^+e^- annihilation

As one of the simplest examples of the application of the Feynman rules developed in Sec. 2.3.3, we take up high-energy electron-positron annihilation processes. At high energies, electron-position pairs annihilate dominantly into hadrons as in Eq. (1.2.9),

$$e^- + e^+ \to \text{hadrons}. \tag{1.2.9}$$

In quantum chromodynamics the above process is considered to be caused through pair production of almost-free quarks,

$$e^- + e^+ \to q + \bar{q} \quad , \tag{2.3.147}$$

and subsequent hadronization processes. The possible QCD higher-order effects to the process (2.3.147) may be safely neglected as a first approximation according to the asymptotic freedom of QCD. Hence the total cross section for the process (1.2.9) is approximately equal to that for the process (2.3.147).

Let us calculate the total cross section for the process (2.3.147) where the quark is assumed to have charge Qe. The Feynman diagram for the process (2.3.147) is shown in Fig. 2.3.9. A simple application of the Feynman rules given in Sec. 2.3.3 yields the transition matrix element for the process (2.3.147),

$$\langle q\bar{q}|T|e^-e^+\rangle = \bar{u}_{\lambda_1'}(p_1')(-Qe)\gamma^\mu v_{\lambda_2'}(p_2')\frac{d_{\mu\nu}(q)}{q^2}\bar{v}^e_{\lambda_2}(p_2)e\gamma^\nu u^e_{\lambda_1}(p_1) \quad , \tag{2.3.148}$$

where $q^\mu = p_1^\mu + p_2^\mu$, $u_\lambda(p)$ $(v_\lambda(p))$ is the Dirac spinor for the quark (antiquark) with momentum p, mass m and polarization λ, and $u^e_\lambda(p)$ $(v^e_\lambda(p))$ that for the electron (positron) with momentum p, mass m_e and polarization λ. Here $d_{\mu\nu}(q)$ is defined by Eq. (2.3.133). It should be noted that the $q_\mu q_\nu$ term in $d_{\mu\nu}(q)$ does not contribute in Eq. (2.3.148) owing to the Dirac equations for $u_\lambda, v_\lambda, u^e_\lambda$ and v^e_λ and hence may be neglected. Thus we have

$$\langle q\bar{q}|T|e^-e^+\rangle = -Qe^2\bar{u}_{\lambda_1'}(p_1')\gamma_\mu v_{\lambda_2'}(p_2')\frac{1}{q^2}\bar{v}^e_{\lambda_2}(p_2)\gamma^\mu u_{\lambda_1}(p_1) \quad . \tag{2.3.149}$$

The total cross section for the process (2.3.147) is obtained by using the general formula (2.3.10). We have

$$\sigma = \frac{1}{2\sqrt{s(s-4m_e^2)}}\frac{1}{4}\sum_{\lambda_1\lambda_2\lambda_1'\lambda_2'}\int\frac{d^3p_1'}{(2\pi)^3\,2p_{10}'}\frac{d^3p_2'}{(2\pi)^3 2p_{20}'}$$
$$\times\,(2\pi)^4\,\delta^4(p_1'+p_2'-p_1-p_2)\,|\langle q\bar{q}|T|e^-e^+\rangle|^2 \quad , \tag{2.3.150}$$

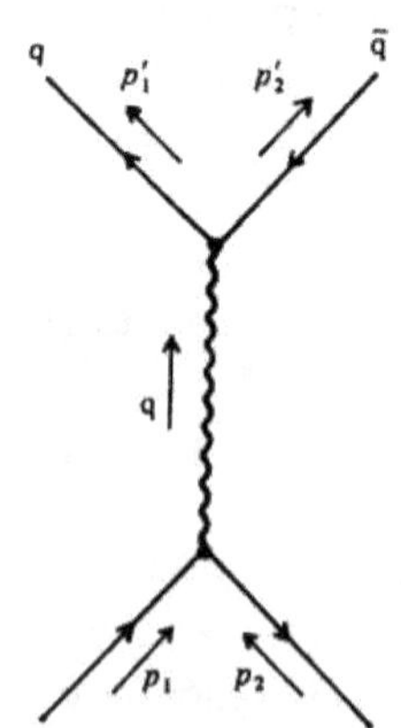

Fig. 2.3.9. The Feynman diagram for the process $e^- + e^+ \to q + \bar{q}$.

where $s = q^2$. The polarization sum in the above equation is easily performed by using Eq. (1.4.11) if one inserts Eq. (2.3.149) into Eq. (2.3.150). In fact

$$\sum_{\lambda_1,\dots\lambda_2'} |\langle q\bar{q}|T|e^-e^+\rangle|^2$$
$$= \frac{Q^2e^4}{s^2}\mathrm{Tr}[(\not p_1' + m)\gamma_\mu(\not p_2' - m)\gamma_\nu]\,\mathrm{Tr}[(\not p_2 - m_e)\gamma^\mu(\not p_1 + m_e)\gamma^\nu]$$
$$= \frac{4Q^2e^4}{s^2}(q_\mu' q_\nu' - k_\mu' k_\nu' - q'^2 g_{\mu\nu})\,(q^\mu q^\nu - k^\mu k^\nu - q^2 g^{\mu\nu}) \quad , \qquad (2.3.151)$$

where $q = p_1 + p_2, k = p_1 - p_2, q' = p_1' + p_2'$ and $k' = p_1' - p_2'$. Substituting Eq. (2.3.151) in Eq. (2.3.150) and rewriting the 3-dimensional integrals as 4-dimensional ones with the aid of

$$\int\frac{d^3p'}{2p_0'} = \int d^4p'\theta(p_0')\,\delta(p'^2 - m^2) \quad , \qquad (2.3.152)$$

we obtain

$$\sigma = \frac{2Q^2\alpha^2}{s^2\sqrt{s(s-4m_e^2)}}\int d^4p_1'd^4p_2'\theta(p_{10}')\,\delta(p_1'^2 - m^2)\,\theta(p_{20}')\,\delta(p_2'^2 - m^2)$$
$$\times\ \delta^4(q'-q)\,(q_\mu' q_\nu' - k_\mu' k_\nu' - s g_{\mu\nu})\,(q^\mu q^\nu - k^\mu k^\nu - s g^{\mu\nu}) \quad , \qquad (2.3.153)$$

where $\alpha = e^2/4\pi = 1/137.036...$ We change the integration variables from (p_1', p_2') to (q', k') and perform the q'-integration to yield

$$\sigma = \frac{Q^2\alpha^2}{2s^3\sqrt{1-4m_e^2/s}}\int d^4k'\theta(q_0 + k_0')\theta(q_0 - k_0')$$
$$\times\ \delta(k'^2 + s - 4m^2)\,\delta(k'\cdot q)\,[(k\cdot k')^2 + s(2s + k^2 + 4m^2)] \quad . \qquad (2.3.154)$$

Taking the center-of-mass system in which

$$q^\mu = (\sqrt{s}, 0, 0, 0), \quad k^\mu = (0, \mathbf{k}), \qquad (2.3.155)$$

we can easily execute the k'-integration,

$$\sigma = Q^2\frac{4\pi\alpha^2}{3s}\sqrt{\frac{s-4m^2}{s-4m_e^2}}\left(1 + \frac{2m^2}{s}\right)\left(1 + \frac{2m_e^2}{s}\right) \quad . \qquad (2.3.156)$$

If we hold the angular variables unintegrated, we obtain the following differential cross section,

$$\frac{d\sigma}{d\Omega} = Q^2 \frac{\alpha^2}{4s} \sqrt{\frac{s-4m^2}{s-4m_e^2}} \left[1 + \frac{4(m^2+m_e^2)}{s} + \left(1 - \frac{4m^2}{s}\right)\left(1 - \frac{4m_e^2}{s}\right)\cos^2\theta \right] , \tag{2.3.157}$$

where θ is the angle between the produced quark and incident electron in the e^-e^+ center-of-mass system as shown in Fig. 2.3.10.

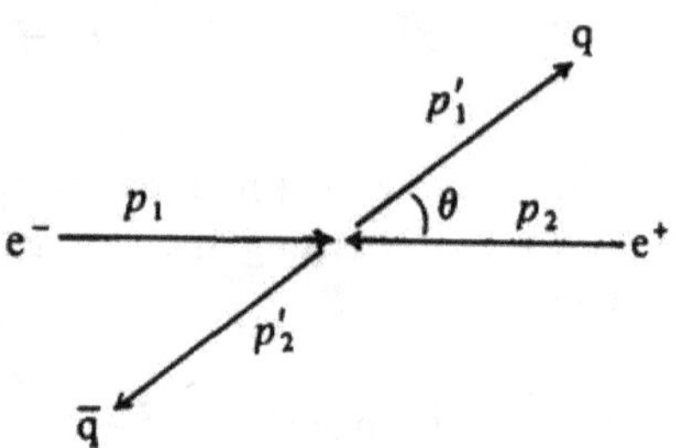

Fig. 2.3.10. Definition of the scattering angle in the center-of-mass frame for the process $e^- + e^+ \to q + \bar{q}$.

For high energies, where $s \gg m^2$, m_e^2, we can safely neglect terms of order m^2/s and m_e^2/s and hence we have

$$\sigma = Q^2 \frac{4\pi\alpha^2}{3s} , \tag{2.3.158}$$

$$\frac{d\sigma}{d\Omega} = Q^2 \frac{\alpha^2}{4s}(1 + \cos^2\theta) . \tag{2.3.159}$$

If we take into account the quark flavors and colors, we have to sum up Eq. (2.3.158) over these degrees of freedom and then we derive the previously announced result (1.2.10).

Equation (2.3.159) suggests that the quarks and antiquarks produced distribute sharply in the forward and backward region. Hence the final hadrons also tend to distribute in the same way. This property is a reflection of the spin-1/2 nature of quarks and may be tested experimentally.

2.3.5. Covariant canonical quantization of gauge fields

So far we have been working in the functional integral formalism. As was mentioned at the end of Sec. 2.2.1, the canonical operator formalism is now available for the covariant quantization of non-Abelian gauge fields. Here we shall give a brief survey on this topics.

Before going into a discussion of the canonical operator formalism, we first examine the symmetry property of the quantum Lagrangian in the functional integral formalism. The final form of the Lagrangian for quantum chromodynamics was given in Eq. (2.3.99). Let us now concentrate our attention on the ghost part (2.3.102) of the Lagrangian. There we chose a pair of the ghost fields χ^a and χ^{a*} as two independent fields. It is also possible to choose two real fields χ_1^a and χ_2^a in place of χ^a and χ^{a*}. However these two different choices of the ghost fields are essentially equivalent to each other. In fact, by setting

$$\chi^a = (\chi_1^a + i\,\chi_2^a)/\sqrt{2} \quad , \tag{2.3.160}$$

and paying attention to the Grassmann property

$$(\chi_1^a)^2 = (\chi_2^a)^2 = 0 \quad , \quad \text{(no summation on } a\text{)}, \tag{2.3.161}$$

we find that the ghost Lagrangian $\mathscr{L}_{\text{FP}}$ in Eq. (2.3.102) may be rewritten as

$$\mathscr{L}_{\text{FP}} = i(\partial^\mu \chi_1^a)\, D_\mu^{ab}\, \chi_2^b \quad , \tag{2.3.162}$$

where a total derivative is ignored. In the following discussion we shall adopt the above form for $\mathscr{L}_{\text{FP}}$ instead of Eq. (2.3.102). The reason for this will become clear soon.

As we have seen in Sec. 2.1.2, the classical part of the Lagrangian $\mathscr{L}$,

$$\mathscr{L}_{\text{CL}} = \mathscr{L}_{\text{G}} + \mathscr{L}_{\text{F}} \quad , \tag{2.3.163}$$

is invariant under the local gauge transformation

$$\delta A_\mu^a = -\frac{1}{g} D_\mu^{ab}\, \theta^b \quad , \tag{2.3.164}$$

$$\delta\psi = -\, iT^a\theta^a\psi \quad , \tag{2.3.165}$$

where D_μ^{ab} is given by Eq. (2.3.96). Since the full quantum Lagrangian $\mathscr{L}$ includes $\mathscr{L}_{\text{GF}}$ and $\mathscr{L}_{\text{FP}}$ in addition to $\mathscr{L}_{\text{CL}}$, it violates the invariance under the local gauge transformations (2.3.164) and (2.3.165). One may wonder what happens if the ghost fields χ_1^a and χ_2^a are also transformed at the same time. There is a possibility of having the invariance of the full quantum Lagrangian $\mathscr{L}$ under local gauge transformations of χ_1^a and χ_2^a as well as of A_μ^a and ψ. To examine this possibility we try the following ansatz [Bec 76]

$$\theta^a(x) = -\, g\delta\lambda\, \chi_2^a(x) \quad , \tag{2.3.166}$$

where $\delta\lambda$ is an infinitesimal Grassmann number independent of x and

$$\{\delta\lambda, \chi_2^a(x)\} = 0 \quad . \tag{2.3.167}$$

With this ansatz (2.3.166) the local gauge transformations take the following form,

$$\delta A_\mu^a = \delta\lambda D_\mu^{ab} \chi_2^b \quad , \tag{2.3.168}$$

$$\delta\psi = \delta\lambda i g T^a \chi_2^a \psi \quad . \tag{2.3.169}$$

Obviously the classical Lagrangian $\mathscr{L}_{CL}$ is still invariant under the transformations (2.3.168) and (2.3.169). We shall now see whether we can determine the transformation rule of χ_1^a and χ_2^a so that the combination $\mathscr{L}_{GF} + \mathscr{L}_{FP}$ is invariant under the local gauge transformations (2.3.168–169). Denoting the infinitesimal transformation of χ_1^a and χ_2^b by $\delta\chi_1^a$ and $\delta\chi_2^b$, we can calculate the infinitesimal change of $\mathscr{L}_{GF} + \mathscr{L}_{FP}$ under the local gauge transformation,

$$\delta(\mathscr{L}_{GF}+\mathscr{L}_{FP}) = -\frac{1}{\alpha}(\partial^\mu A_\mu^a)(\partial^\nu \delta A_\nu^a) + i(\partial^\mu \delta\chi_1^a) D_\mu^{ab}\chi_2^b + i(\partial^\mu \chi_1^a)\,\delta(D_\mu^{ab}\chi_2^b) \quad . \tag{2.3.170}$$

We note that

$$\delta(D_\mu^{ab}\chi_2^b) = \delta^{ab}\partial_\mu \delta\chi_2^b - g f^{abc}\delta\lambda(D_\mu^{cd}\chi_2^d)\chi_2^b - g f^{abc} A_\mu^c \delta\chi_2^b \quad . \tag{2.3.171}$$

Inserting Eq. (2.3.171) into Eq. (2.3.170) and rearranging terms, we obtain, up to a total derivative,

$$\begin{aligned}\delta(\mathscr{L}_{GF} + \mathscr{L}_{FP}) = &-i\left(\delta\chi_1^a - i\frac{1}{\alpha}\delta\lambda\partial^\mu A_\mu^a\right)\partial^\nu(D_\nu^{ab}\chi_2^b) \\ &+ i(\partial^\mu\chi_1^a)\,\partial_\mu(\delta\chi_2^a + \frac{1}{2}\delta\lambda g f^{abc}\chi_2^b\chi_2^c) \\ &- ig(\partial^\mu\chi_1^a) A_\mu^e (f^{abe}\,\delta\chi_2^b + \delta\lambda g f^{abc} f^{cde}\chi_2^b\chi_2^d) \quad .\end{aligned} \tag{2.3.172}$$

By direct observation of Eq. (2.3.172) we are naturally led to the following transformation rule of χ_1^a and χ_2^a which makes the right-hand side vanish:

$$\delta\chi_1^a = i\delta\lambda\frac{1}{\alpha}\partial^\mu A_\mu^a \quad , \tag{2.3.173}$$

$$\delta\chi_2^a = -\frac{1}{2}\delta\lambda g f^{abc}\chi_2^b\chi_2^c \quad . \tag{2.3.174}$$

In fact the first and second terms of the right-hand side of Eq. (2.3.172) apparently vanish under the above transformation rule. It is easy to show that the third term also vanishes according to the Jacobi identity (2.1.39)

$$f^{abe}\,\delta\chi_2^b + \delta\lambda g f^{abc} f^{cde}\,\chi_2^b\chi_2^d$$

$$= \delta\lambda\frac{g}{2}(f^{abc}f^{cde} + f^{adc}f^{ceb} + f^{aec}f^{cbd})\,\chi_2^b\chi_2^d = 0 \quad . \qquad (2.3.175)$$

Thus we find that, under the local gauge transformations (2.3.168), (2.3.169), (2.3.173) and (2.3.174), the combination $\mathscr{L}_{\rm GF} + \mathscr{L}_{\rm FP}$ remains invariant:

$$\delta(\mathscr{L}_{\rm GF} + \mathscr{L}_{\rm FP}) = 0 \quad . \qquad (2.3.176)$$

Since the classical part $\mathscr{L}_{\rm CL}$ is separately invariant under these transformations, we conclude that the full quantum Lagrangian $\mathscr{L}$ is also invariant,

$$\delta\mathscr{L} = 0 \quad . \qquad (2.3.177)$$

Summing up, we found that the full quantum Lagrangian $\mathscr{L}$ prosesses a new symmetry under the extended local gauge transformations (2.3.168), (2.3.169), (2.3.173) and (2.3.174) although it violates the classical local gauge symmetry under the transformations (2.3.164–165). The above extended local gauge transformation which includes the transformation of the ghost fields may be regarded as the quantum version of the classical local gauge transformation and is called the BRS (Becchi-Rouet-Stora) transformation [Bec 76]. The invariance of the full quantum Lagrangian under the BRS transformation is referred to as the *BRS symmetry* and is actually a basis for developing the canonical operator formalism. This transformation is also very useful for deriving the generalized Ward-Takahashi identities (Slavnov-Taylor identities). This topic will be dealt with later in Sec. 2.5.6.

It is known to be advantageous to linearize the gauge fixing term $\mathscr{L}_{\rm GF}$ of the Lagrangian $\mathscr{L}$ by introducing an auxiliary field $B^a(x)$. In the functional-integral formalism this is simply accomplished by re-expressing the generating functional $Z[J, \xi_1, \xi_2, \eta, \bar{\eta}]$ in the following way,

$$Z[J, \xi_1, \xi_2, \eta, \bar{\eta}] = \int [dA]\,[dB]\,[d\chi_1]\,[d\chi_2]\,[d\psi]\,[d\bar{\psi}]$$
$$\times \exp\left\{i\int d^4x\,(\mathscr{L}' + AJ + \chi_1\xi_1 + \chi_2\xi_2 + \bar{\psi}\eta + \bar{\eta}\psi)\right\} \quad , \qquad (2.3.178)$$

where $\mathscr{L}'$ is given by

$$\mathscr{L}' = \mathscr{L}_{\rm CL} + \mathscr{L}_{\rm FP} + B^a\partial^\mu A_\mu^a + \frac{\alpha}{2}B^aB^a \qquad (2.3.179)$$

The Lagrangian $\mathscr{L}'$ in Eq. (2.3.179) has the BRS symmetry under the following version of the BRS transformation,

$$\begin{aligned}\delta A_\mu^a &= \delta\lambda\, D_\mu^{ab}\,\chi_2^b \quad ,\\ \delta\psi &= i\delta\lambda\, gT^a\,\chi_2^a\psi \quad ,\\ \delta\chi_1^a &= -\,i\delta\lambda\, B^a \quad ,\\ \delta\chi_2^a &= -\frac{1}{2}\delta\lambda\, g f^{abc}\,\chi_2^b\,\chi_2^c \quad ,\\ \delta B^a &= 0 \quad .\end{aligned} \tag{2.3.180}$$

As can be seen above, the use of the auxiliary field B^a in the functional integral formalism is nothing but a re-expression of the same generating functional in a different form. In the canonical operator formalism, however, the introduction of the auxiliary field B^a plays an important role in imposing a subsidiary condition. This technique has provided us with a satisfactory formalism in quantum electrodynamics called the Nakanishi-Lautrup formalism [Nak 66, Lau 67].

With the Lagrangian $\mathscr{L}'$ in hand we are now ready to go over to the canonical operator formalism for the covariant quantization of non-Abelian gauge fields [Kug 78]. Our argument goes as follows: We start with the operator version of the classical Lagrangian

$$\mathscr{L}_{\mathrm{CL}} = -\frac{1}{4}\hat{F}_{\mu\nu}^a\,\hat{F}^{a\mu\nu} + \bar{\hat{\psi}}\,(i\gamma^\mu\hat{D}_\mu - m)\hat{\psi} \quad , \tag{2.3.181}$$

where $\hat{F}_{\mu\nu}^a$ consists of $\hat{A}_\mu^a$ as in Eq. (2.1.45) and

$$\hat{D}_\mu = \partial_\mu - igT^a\hat{A}_\mu^a \quad . \tag{2.3.182}$$

The Lagrangian $\mathscr{L}_{\mathrm{CL}}$ is invariant under the classical local gauge transformation. It, however, does not bring about a consistent quantization and hence we add the gauge-fixing term

$$\mathscr{L}_{\mathrm{GF}} = \hat{B}^a\partial^\mu\hat{A}_\mu^a + \frac{\alpha}{2}\,\hat{B}^a\hat{B}^a \quad . \tag{2.3.183}$$

The resulting Lagrangian

$$\mathscr{L}' = \mathscr{L}_{\mathrm{CL}} + \mathscr{L}_{\mathrm{GF}} \quad , \tag{2.3.184}$$

no longer has the classical local gauge symmetry. At this stage we require that the Lagrangian $\mathscr{L}'$ be modified by adding an extra term $\mathscr{L}_{\mathrm{X}}$ so that the full Lagrangian

$$\mathscr{L} = \mathscr{L}' + \mathscr{L}_X \quad , \tag{2.3.185}$$

is invariant under the quantized version of the local gauge transformation, i.e., the BRS transformation which is given by Eq. (2.3.180) with a hat ˆ on all fields. Here we assume that we need to introduce extra ghost fields $\hat{\chi}_1^a$ and $\hat{\chi}_2^b$ in order to obtain a quantized local gauge transformation. It is now an easy exercise to show that $\mathscr{L}_X$ has to be equal to $\mathscr{L}_{FP}$ where

$$\mathscr{L}_{FP} = i(\partial^\mu \hat{\chi}_1^a)\, \hat{D}_\mu^{ab}\, \hat{\chi}_2^b \quad . \tag{2.3.186}$$

Thus we obtain our final form of the full Lagrangian (2.3.185) in the canonical operator formalism. It should be noted here that, in contrast with the functional integral formalism, the choice of the ghost-field operators becomes an issue in the canonical operator formalism. In fact, the choice of the set $(\hat{\chi}_1^a, \hat{\chi}_2^a)$ is not equivalent to that of $(\hat{\chi}^a, \hat{\chi}^{a\dagger})$, because, in the canonical operator formalism, we have equal-time anticommutation relations for the ghost-field operators. In the canonical operator formalism, only a consistent choice of the ghost-field operators amounts to the hermitean one $(\hat{\chi}_1^a, \hat{\chi}_2^a)$ which guarantees the hermiticity of the Lagrangian and hence the unitarity of the S-matrix [Kug 78].

In order to complete the consistent quantization in the canonical operator formalism, it remains to show that a sufficient set of subsidiary conditions for physical states is imposed. It is shown in [Kug 78] that such conditions are actually given by

$$\hat{Q}_B \,|\, \mathrm{phys}\,\rangle = 0 \quad , \tag{2.3.187}$$

$$\hat{Q}_C \,|\, \mathrm{phys}\,\rangle = 0 \quad , \tag{2.3.188}$$

where $\hat{Q}_B$ and $\hat{Q}_C$ are, respectively, generators of the BRS transformation and the scale transformation for the ghost fields. According to these conditions the physical space is constrained to that with a positive semi-definite norm thus leading to the unitarity of the S-matrix.

2.4. REGULARIZATION

2.4.1. Review of regularization schemes

Now that we settled the Lagrangian for quantum chromodynamics (QCD Lagrangian (2.3.99)) and established the Feynman rules for it in Sec 2.3.3, we are free to make perturbative calculations of cross sections for arbitrary quark-gluon processes. Since quarks and gluons are most probably confined inside hadrons [see, e.g., Ban 81], the quark-gluon processes cannot be observed directly and one might feel that such calculations are useless.

Fortunately, however, we have in our hands, the operator-product expansion technique which enables us to extract information on quark-gluon processes out of hadronic reactions. This important technique will be described later in Chap. 4. Here in the present section we only deal with quark-gluon Green functions.

It will be seen later that the lowest-order calculation in quark-gluon processes generally reproduces the parton model results. Thus the parton picture corresponds to the tree approximation (i.e., no loop diagrams) in the perturbative expansion based on the QCD Lagrangian. In the tree approximation, however, the dynamical effect of QCD does not show up and the really important ingredient of QCD is hidden in the QCD radiative correction to the tree amplitudes which necessarily includes the contributions of loop diagrams. Thus it is essential to deal with QCD loop effects in quark-gluon processes. In general, however, the loop contributions to quark-gluon Green functions generate divergences as we have seen already in Eq. (2.3.90). The divergences are properly taken care of by the renormalization program which will be described in some detail in Sec. 2.5.

Although the divergences are subtracted out in the final physical answer on the basis of the renormalization program, we, in the intermediate stage, still require that divergent integrals be mathematically manageable. This procedure which makes divergent integrals tentatively finite by introducing a suitable convergence device is generically called *regularization*. While the divergences associated with loop diagrams have been studied since the onset of quantum electrodynamics, the importance of the regularization seems to have been emphasized less frequently. Regularization is a purely mathematical procedure and has no physical consequences. Accordingly it is not a unique procedure, i.e., we have a variety of the regularization schemes.

In order to explain the regularization schemes developed so far, we take a specific example of the divergent integral, the quark self-energy part $\sum_{ij}(p)$. The quark self-energy part $\sum_{ij}(p)$ is a component of the truncated connected 2-point quark Green function $\tilde{G}^{tc}_{2ij}(p)$,

$$\tilde{G}^{tc}_{2ij}(p) = i\{\delta_{ij}(m-\not{p}) - \textstyle\sum_{ij}(p)\} \quad . \tag{2.4.1}$$

In other words its relation to the quark propagator $\tilde{S}_{ij}(p)$ (the full propagator which includes all the radiative corrections) is given by

$$\tilde{S}_{ij}(p) = \delta_{ij}\Big/\Big(m - \not{p} - \textstyle\sum(p)\Big) \quad , \tag{2.4.2}$$

where $\sum_{ij}(p) = \delta_{ij}\sum(p)$ and

$$\tilde{S}_{ij}(\mathrm{p}) = i\tilde{G}^{c}_{2ij}(p) = i\int d^4x\, e^{-ip\cdot x}\ \langle 0|\mathrm{T}[\psi_i(x)\,\bar{\psi}_j(0)]\,|\,0\rangle_c \ , \tag{2.4.3}$$

with suffix c indicating the connected piece. Here the self-energy part $\sum(p)$ is proper (one-particle irreducible) while the propagator is not. This may be explicitly seen by making an expansion of Eq. (2.4.2) in powers of $\sum(p)$:

$$\begin{aligned}\tilde{S}_{ij}(p) = \delta_{ij}\{&\tilde{S}_0(p) + \tilde{S}_0(p)\textstyle\sum(p)\,\tilde{S}_0(p)\\ &+\tilde{S}_0(p)\textstyle\sum(p)\,\tilde{S}_0(p)\textstyle\sum(p)\,\tilde{S}_0(p) + \ldots\}\ ,\end{aligned} \tag{2.4.4}$$

where

$$\tilde{S}_0(p) = 1/(m-\not{p}) \ . \tag{2.4.5}$$

Obviously each term of Eq. (2.4.4) is one-particle-reducible (improper).

Now following the Feynman rules for QCD developed in Sec. 2.3.3, we can write down an expression for the quark self-energy part to order g^2 (the corresponding Feynman diagram is shown in Fig. 2.4.1),

$$\textstyle\sum_{ij}(p) = \displaystyle\int\frac{d^4k}{(2\pi)^4 i}\, g\gamma_\mu T^a_{il}\,\frac{\delta_{ln}}{m-\not{p}+\not{k}}\, g\gamma_\nu\, T^b_{nj}\,\frac{\delta_{ab}}{k^2}\, d^{\mu\nu}(k) \ , \tag{2.4.6}$$

where $d^{\mu\nu}(k)$ is given by Eq. (2.3.133). Noting that for SU(N) (we consider the case for a general N though only $N = 3$ is relevant[20])

$$T^a_{il}\,T^a_{lj} = (T^aT^a)_{ij} = \delta_{ij}C_F\,, \quad C_F = \frac{N^2-1}{2N} \ , \tag{2.4.7}$$

we obtain

$$\textstyle\sum(p) = g^2C_F\displaystyle\int\frac{d^4k}{(2\pi)^4 i}\,\frac{\gamma_\mu(m+\not{p}-\not{k})\,\gamma_\nu\, d^{\mu\nu}(k)}{k^2\,(m^2-(p-k)^2)} \ . \tag{2.4.8}$$

For simplicity we consider the Feynman gauge $\alpha = 1$ so that $d^{\mu\nu}(k) = g^{\mu\nu}$. We then have

$$\textstyle\sum(p) = g^2C_F\displaystyle\int\frac{d^4k}{(2\pi)^4 i}\,\frac{\gamma_\mu\,(m+\not{p}-\not{k})\,\gamma^\mu}{k^2\,(m^2-(p-k)^2)} \ . \tag{2.4.9}$$

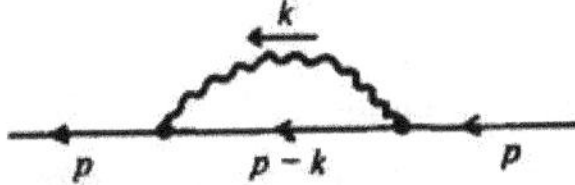

Fig. 2.4.1. One-loop diagram for the quark self-energy part in QCD.

[20] See the Digression at the end of this subsection for a derivation of Eq. (2.4.7).

The 4-dimensional integral in Eq. (2.4.9) is linearly divergent as can be easily seen by the simple power counting in k, i.e.,

$$\int d^4k \frac{\not{k}}{k^2k^2} \sim \lim_{K\to\infty} K \quad . \tag{2.4.10}$$

The divergence comes from the high-momentum region $|k| \to \infty$. To make the integral (2.4.9) mathematically meaningful, we need to write it as a suitable limit of a convergent integral. This convergent integral corresponds to the regularization of the original divergent integral (2.4.9). In the following we shall give a short review of a variety of regularization methods. It should be noted that the theory based on the regularized integral, necessarily violates some of the underlying physical requirements, i.e., the Lorentz invariance, gauge symmetry, unitarity, etc. Of course the final finite results after renormalization recover all these physical requirements. Thus the criterion to choose a good regularization is that the regularized theory should preserve as many physical laws as possible.

(1) *Cut-off Method*

One of the simplest regularizations is the cut-off method in which the high-momentum region which is the source of the divergence is cut off in the divergent integrals. The method was employed in earlier literatures in quantum electrodynamics (see, e.g., [Jau 55]). It, however, breaks translation invariance and hence the shift of momenta in the integral, in general, changes the result. In other words the so-called surface term should be carefully estimated [Jau 55]. Also gauge invariance is not maintained in this regularization procedure. A typical example is the one-loop photon self-energy part in quantum electrodynamics which violates the requirement of the gauge invariance if calculated in the cut-off method. Hence the cut-off method is not suitable for the regularization of gauge theories.

(2) *Pauli-Villars regulator method*

In the integrand of the divergent integral the propagator $1/(m^2 - k^2)$ may be replaced by

$$\frac{1}{m^2-k^2} - \frac{1}{M^2-k^2} = \frac{M^2-m^2}{(m^2-k^2)(M^2-k^2)} \quad . \tag{2.4.11}$$

The modified propagator (2.4.11) reduces to the original one as $M \to \infty$. If M is kept finite, the high-momentum behavior of the modified propagator is better ($O(1/k^4)$) than that of the original propagator ($O(1/k^2)$). Hence this modification is of use for regularization. This regularization is the Pauli-Villars regulator method [Pau 49]. The additional term in Eq. (2.4.11) is called the Pauli-Villars regulator.

This method respects translation and Lorentz invariance and hence is a good candidate for consistent regularization. In fact gauge invariance in quantum electrodynamics is preserved by this regularization method. Thus quantum electrodynamics is dealt with in a perfectly consistent way by the Pauli-Villars regulator method. The method is also applicable to the massless Yang-Mills theory such as quantum chromodynamics although the proof of the unitarity of the theory is found to be rather difficult in this regularization. In the massive Yang-Mills theory like Weinberg-Salam theory [Gla 61, Wei 67, Sal 68] the Pauli-Villars regularization does not maintain gauge invariance. In fact, for example, the self-energy part of the charged weak boson (W-boson) can not be regularized in a gauge-invariant way if the Pauli-Villars regulator is employed [tHo 71].

(3) *Analytic regularization*

Yet another method of modifying the propagator is to change the power of the propagator denominator,

$$\frac{1}{m^2-k^2} \to \frac{1}{(m^2-k^2)^\alpha} \quad . \tag{2.4.12}$$

The parameter α is complex with $\mathrm{Re}\,\alpha > 1$ ensuring the convergence of the integral. As $\alpha \to 1$, the original propagator is recovered. This regularization method is called the analytic regularization and is extensively employed by Speer [Spe 68, 69] for the proof of renormalizability. (See also [Bol 64].) This method also violates gauge invariance and we shall not use it in quantum chromodynamics.

(4) *Lattice regularization*

This regularization method is somewhat different in nature from the others discussed here in the present subsection. In lattice regularization we discretize space-time so that it is made of small cells of size a, the lattice constant. According to the lattice structure of space-time, the short-distance contribution to the space-time integration is eliminated. In momentum space this corresponds to cutting off the high momentum region thus leading to a convergent momentum integral. This regularization method obviously breaks translation and Lorentz invariance, and hence is not suitable for regularizing gauge theories. Nevertheless it is the only known regularization which allows a nonperturbative calculation, since it can be applied to the configuration-space integral that appears in the functional integral for the expression of the generating functional $Z[J]$. In this connection the lattice regularization has been frequently used in the recent works for evaluating physical quantities of hadrons (see, e.g. [Cre 83]).

(5) *Dimensional regularization*

A divergent multiple integral may be made convergent by reducing the number of multiple integrals. For example, the linearly divergent 4-dimensional integrals in Eq. (2.4.9) would be finite if the space-time were 2-dimensional. This fact is the basic idea of dimensional regularization [Bol 72, tHo 72, Ash 72, Cic 72]. In dimensional regularization we keep the space-time dimension D lower than four and replace the divergent 4-dimensional integral by a convergent D-dimensional one. By making momentum integrations explicitly we obtain an analytic expression as a function of the dimension D. We make the analytic continuation in D in this expression. Then the original divergence will show up as a pole at $D = 4$ in the above analytic expression. For a review, see [Lei 75].

Since in the dimensional regularization nothing has been violated except that space-time is not 4-dimensional, all the physical requirements are preserved. Hence this regularized theory is Lorentz invariant, gauge invariant, unitary, etc. [tHo 72, Spe 74]. In this sense dimensional regularization is the most suitable for gauge theories. We shall give a more detailed account of this topic in the next subsection.

Digression

SOME USEFUL FORMULAS FOR SU(N): In the course of QCD calculations we often encounter quantities like T^aT^a, $f^{acd}f^{bcd}$, $T^aT^bT^a$, etc, where T^a ($a = 1, 2, \ldots, N^2-1$) are SU(N) generators and f^{abc} the structure constants. We shall present here a derivation of formulas for the above quantities.

The algebra of SU(N) is defined by the commutation relations

$$[T^a, T^b] = i f^{abc} T^c \quad . \tag{2.4.13}$$

As a normalization condition we adopt

$$\mathrm{Tr}[T^a T^b] = \frac{1}{2}\delta^{ab} \quad , \tag{2.4.14}$$

where the trace refers to the $N \times N$ matrix in the fundamental representation.

We first derive the useful formula

$$T^a_{ij} T^a_{kl} = \frac{1}{2}\left(\delta_{il}\delta_{jk} - \frac{1}{N}\delta_{ij}\delta_{kl}\right) \quad . \tag{2.4.15}$$

To derive this formula we note that the identity 1 and T^a ($a = 1, 2, \ldots, N^2-1$) form the basis of a set of hermitian $N \times N$ matrices A, and thus

$$A = c^0 1 + c^a T^a \quad , \tag{2.4.16}$$

where c^0, c^1, ... are real numbers. The constants c^0 and c^a are determined by using the normalization condition (2.4.14):

$$c^0 = \mathrm{Tr}[A]/N \quad , \tag{2.4.17}$$

$$c^a = 2\mathrm{Tr}[T^a A] \quad . \tag{2.4.18}$$

Inserting Eqs. (2.4.17–18) into Eq. (2.4.16) we have

$$A_{lk}(2T^a_{ij}T^a_{kl} + \frac{1}{N}\delta_{kl}\delta_{ij} - \delta_{il}\delta_{jk}) = 0 \quad . \tag{2.4.19}$$

Since Eq. (2.4.19) holds for arbitrary hermitean matrix A, we obtain Eq. (2.4.15).

An immediate consequence of Eq. (2.4.15) is that

$$(T^a T^a)_{ij} = \frac{N^2-1}{2N}\delta_{ij} \quad , \tag{2.4.20}$$

$$(T^a T^b T^a)_{ij} = -\frac{1}{2N}T^b_{ij} \quad . \tag{2.4.21}$$

We can also calculate $f^{acd}f^{bcd}$ as follows. Using Eqs. (2.4.13) and (2.4.14) we have

$$\begin{aligned} f^{acd}f^{bcd} &= -2\mathrm{Tr}[[T^a, T^c][T^b, T^c]] \quad , \\ &= -2\mathrm{Tr}\,[2T^aT^cT^bT^c - (T^aT^b + T^bT^a)\,T^cT^c] \,. \end{aligned} \tag{2.4.22}$$

Applying Eqs. (2.4.20) and (2.4.21) to Eq. (2.4.22), we obtain

$$f^{acd}f^{bcd} = N\delta^{ab} \quad . \tag{2.4.23}$$

It should be remarked here that the T^aT^a appearing in Eq. (2.4.20) is the Casimir operator of the group SU(N) and hence commutes with all the operators in SU(N). By Schur's lemma such an operator is a multiple of the identity operator. In general the number of independent Casimir operators is equal to the rank of the group.

2.4.2. Dimensional regularization

We carry on our discussion by using our example of the one-loop quark self-energy part in the Feynman gauge (2.4.9). We now move off from 4-dimensional space-time and replace the 4-dimensional integral d^4k by the D-dimensional one d^Dk. Before performing the D-dimensional integration, we mention that some care has to be given to the algebra in D dimensions.

First of all the space-time index μ now runs from 0 to $D-1$ so that components of a momentum vector are

$$p^\mu = (p^0, p^1, \ldots, p^{D-1}) \quad , \tag{2.4.24}$$

and the contracted metric tensor is

$$g^\mu_\mu = g_{\mu\nu}g^{\mu\nu} = D \quad . \tag{2.4.25}$$

The Dirac algebra in D dimensions is unchanged except that we have to use Eq. (2.4.25) whenever we make contractions of indices. The Dirac matrix satisfies anticommutation relations as in 4 dimensions,

$$\{\gamma^\mu, \gamma^\nu\} = 2g^{\mu\nu} \quad . \tag{2.4.26}$$

When we make contractions of indices μ, we must be careful so that we use Eq. (2.4.25), i.e.,

$$\gamma_\mu \gamma^\mu = D \quad , \tag{2.4.27}$$

$$\gamma_\mu \gamma_\nu \gamma^\mu = (2 - D)\gamma_\nu \quad . \tag{2.4.28}$$

There are some ambiguities in the analytic continuation of the space-time dimension. In fact the measure of the integral, $1/(2\pi)^4$, may be replaced by $1/(2\pi)^D$ in D dimensions or it may be the same as in 4 dimensions. The only relevant requirement is that the measure in D dimensions should recover the 4-dimensional one $1/(2\pi)^4$ as $D \to 4$. The trace of the γ-matrices is also a source of the ambiguity. If one naively follows the Clifford algebra[21] in arbitrary dimensions, one would have $\mathrm{Tr}[\gamma_\mu \gamma_\nu] = 2^{D/2} g_{\mu\nu}$ for D even. This form of course, reduces to the 4-dimensional form as $D \to 4$. As we are only interested in 4-dimensional space-time, the above form for $\mathrm{Tr}[\gamma_\mu \gamma_\nu]$ need not be adopted. In particular we could alternatively use the 4-dimensional form $\mathrm{Tr}[\gamma_\mu \gamma_\nu] = 4g_{\mu\nu}$.

In order to circumvent the above ambiguity we fix our convention such that the integral measure has the form

$$\int \frac{d^D k}{(2\pi)^D} \quad , \tag{2.4.29}$$

and the trace of the γ-matrices is normalized to

$$\mathrm{Tr}[\gamma_\mu \gamma_\nu] = 4g_{\mu\nu} \quad . \tag{2.4.30}$$

It is important to note that the above convention should always be kept in mind. Otherwise an error may result in the final finite answer.

We are now in a position to evaluate Eq. (2.4.9) by using dimensional regularization. In order to make the argument simpler, we consider the case of massless quarks, $m = 0$. The regularized form of Eq. (2.4.9) reads

$$\Sigma(p) = g^2 C_F (2 - D) \int \frac{d^D k}{(2\pi)^D i} \frac{\not{k} - \not{p}}{k^2 (k - p)^2} \quad , \tag{2.4.31}$$

where we keep $D < 3$ to ensure the convergence of the above integral and we have used Eq. (2.4.28) to rewrite the numerator of the integrand of Eq. (2.4.9).

[21] This topic is discussed at the Digression.

We shall evaluate the above integral explicitly. For this purpose it seems to be the most elegant to use the Feynman parametrization[22]

$$\frac{1}{AB} = \int_0^1 \frac{dx}{\{xA + (1-x)\,B\}^2} \quad . \tag{2.4.32}$$

We apply the above formula to re-express the denominator of the integrand of Eq. (2.4.31) and obtain

$$\Sigma(p) = g^2 C_F\,(2-D) \int \frac{d^D k}{(2\pi)^D i} (\not{k} - \not{p}) \int_0^1 \frac{dx}{\{x(k-p)^2 + (1-x)k^2\}^2} \quad . \tag{2.4.33}$$

As far as $D < 3$, the k-integration is convergent and hence the k- and x-integrations are interchangeable. We have

$$\Sigma(p) = g^2 C_F\,(2-D) \int_0^1 dx \int \frac{d^D k'}{(2\pi)^D i} \frac{\not{k} - \not{p}}{\{(k-xp)^2 + x(1-x)p^2\}^2} \quad , \tag{2.4.34}$$

where we made a rearrangement in the denominator. Since dimensional regularization preserves translational invariance, we are free to make a shift of the momentum variable k,

$$k' = k - xp \quad . \tag{2.4.35}$$

We then have

$$\Sigma(p) = g^2 C_F\,(2-D) \int_0^1 dx \int \frac{d^D k'}{(2\pi)^D i} \frac{\not{k}' - (1-x)\,\not{p}}{\{k'^2 + x(1-x)\,p^2\}^2} \quad . \tag{2.4.36}$$

In dimensional regularization we break none of the symmetries of space-time and so an integral of an odd function in k vanishes,

$$\int d^D k\, k_\mu f(k^2) = 0 \quad , \tag{2.4.37}$$

where $f(k^2)$ is an integrable function of k^2. According to Eq. (2.4.37) we obtain[23]

[22] The most general form of the Feynman parametrization will be given later at the end of Sec. 2.4.3.

[23] Note that the linearly divergent piece in 4 dimensions disappeared according to Eq. (2.4.37) leaving only the logarithmically divergent integral.

$$\Sigma(p) = g^2 C_F (D-2) \not{p} \int_0^1 dx(1-x) \int \frac{d^D k'}{(2\pi)^D i} \frac{1}{\{k'^2 + x(1-x)p^2\}^2} \quad . \tag{2.4.38}$$

We wish to perform the k'-integration. However this is not so easy since the integration should be performed in the Minkowski space. To circumvent this problem, we make a rotation by $+90$ degrees in the complex k_0'-plane as shown in Fig. 2.4.2 and change the k'-integration to Euclidean integration. This rotation in the k_0'-plane is called the *Wick rotation*, which amounts to the change of the variable,

$$k_0' = i K_0 \quad , \qquad (K_0 = \text{real}) \quad . \tag{2.4.39}$$

Denoting $\mathbf{k}' = \mathbf{K}$, we have

$$d^D k' = i d^D K \; , \quad k'^2 = -K^2 \; , \quad K^2 = K_0^2 + \mathbf{K}^2 \; . \tag{2.4.40}$$

Thus after Wick rotation we are in Euclidean space and

$$\Sigma(p) = g^2 C_F (D-2) \not{p} \int_0^1 dx(1-x) \int \frac{d^D K}{(2\pi)^D} \frac{1}{(K^2+L)^2} \quad , \tag{2.4.41}$$

where

$$L = -x(1-x)\, p^2 \quad . \tag{2.4.42}$$

Note that the integrand is nonsingular for $L > 0$. According to this requirement we have to keep the momentum p in the space-like region, $p^2 < 0$. Now that the K-integral in Eq. (2.4.41) is Euclidean, it may be easily performed following the standard procedure. We work in the D-dimensional polar coordinate system,

$$\begin{aligned} K^0 &= K\cos\theta_1 \quad , \\ K^1 &= K\cos\theta_2 \sin\theta_1 \quad , \\ &\vdots \\ K^{D-1} &= K\sin\theta_{D-1} \ldots \sin\theta_1 \quad , \end{aligned} \tag{2.4.43}$$

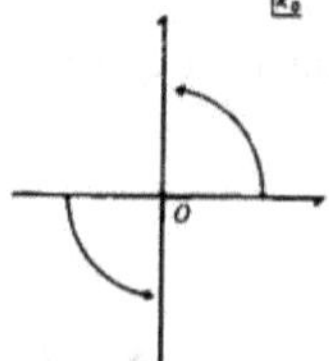

Fig. 2.4.2. Wick rotation in the complex k_0'-plane.

where $K = (K_0^2 + \mathbf{K}^2)^{1/2}$ as is defined in Eq. (2.4.40). In this system the K-integral is given by

$$d^D K = K^{D-1} dK \, d\Omega_D \quad , \tag{2.4.44}$$

where

$$d\Omega_D = \prod_{l=1}^{D-1} \sin^{D-1-l} \theta_l d\theta_l \quad . \tag{2.4.45}$$

Accordingly we find

$$\int \frac{d^D K}{(2\pi)^D} \frac{1}{(K^2+L)^2} = \frac{B(D/2, 2-D/2)}{(4\pi)^{D/2} \Gamma(D/2)} L^{D/2-2} \quad , \tag{2.4.46}$$

where we used the formulas [Erd 54, Gra 65, Mor 60]

$$\int_0^\infty \frac{t^{p-1}}{(1+t)^{p+q}} dt = B(p, q) \quad , \tag{2.4.47}$$

$$\int d\Omega_D = \int_0^\pi d\theta_1 \sin^{D-2} \theta_1 \ldots \int_0^\pi d\theta_{D-2} \sin \theta_{D-2} \int_0^{2\pi} d\theta_{D-1} = \frac{2\pi^{D/2}}{\Gamma(D/2)} \quad , \tag{2.4.48}$$

with $B(x, y)$ and $\Gamma(x)$ the Beta and Gamma functions, respectively. Taking into account the formula

$$B(p, q) = \frac{\Gamma(p)\Gamma(q)}{\Gamma(p+q)} \quad , \tag{2.4.49}$$

and inserting Eq. (2.4.46) into Eq. (2.4.41), we obtain

$$\begin{aligned} \Sigma(p) = {} & g^2 C_F (D-2) \not{p} \frac{\Gamma(2-D/2)}{(4\pi)^{D/2}} (-p^2)^{D/2-2} \\ & \times \int_0^1 dx \, x^{D/2-2} (1-x)^{D/2-1} \quad . \end{aligned} \tag{2.4.50}$$

Using another expression for the Beta function.

$$B(p, q) = \int_0^1 dx \, x^{p-1} (1-x)^{q-1} \quad , \tag{2.4.51}$$

we finally come to the following form for the quark self-energy part,

$$\Sigma(p) = \frac{2C_F g^2}{(4\pi)^{D/2}} \not{p} (-p^2)^{D/2-2} (D-1) \, B\left(\frac{D}{2}, \frac{D}{2}\right) \Gamma\left(2-\frac{D}{2}\right) \quad . \tag{2.4.52}$$

where use has been made of the relation

$$\Gamma(x+1) = x\Gamma(x) \quad . \tag{2.4.53}$$

We note that the integral representation of $\sum(p)$, Eq. (2.4.41), is valid only for $D < 3$ and $p^2 < 0$. Since in Eq. (2.4.52) $\sum(p)$ is now given as an explicit function of the space-time dimension D and the momentum p, we can analytically continue it to the region where D and p^2 are arbitrary complex numbers. According to the expression (2.4.52), we immediately recognize that there are poles at $D = 4, 6, 8, \ldots$ and a branch cut on the positive real axis in p^2-plane. The above poles in D come from the Gamma function $\Gamma(2-D/2)$, because $\Gamma(z)$ has a pole at $z = -n (n = 0, 1, 2, \ldots)$,

$$\Gamma(z) \sim \frac{(-1)^n}{n!}\frac{1}{z+n} \quad , \qquad \text{for } z \sim -n \quad . \tag{2.4.54}$$

For $D \sim 4$ we find from Eq. (2.4.52)

$$\sum(p) \sim \frac{C_F g^2}{(4\pi)^2}\frac{2}{4-D}\not{p} \quad . \tag{2.4.55}$$

Thus the logarithmic divergence in the original 4-dimensional integral shows up as a pole in D at $D = 4$.

It is worth noting here that the gauge coupling constant g is no longer dimensionless for arbitrary space-time dimensions. Recall our dimension counting in Sec. 2.1.1; we get

$$\dim[g] = 2 - D/2 \quad . \tag{2.4.56}$$

The derivation of Eq. (2.4.56) goes as follows: The action

$$S = \int d^D x \mathscr{L}, \tag{2.4.57}$$

should be dimensionless and hence $\dim[\mathscr{L}] = D$ as in Eq. (2.1.22). By observing kinetic terms in $\mathscr{L}$ we find, just as in Eq. (2.1.23),

$$\dim[A_\mu^a] = \frac{D-2}{2} \quad ,$$

$$\dim[\psi] = \frac{D-1}{2} \quad . \tag{2.4.58}$$

On the other hand an examination of the dimension of the interaction terms in Eq. (2.3.109), say the term $g\bar{\psi}T^a\gamma^\mu\psi A_\mu^a$, puts the condition

$$\dim[g] + 2\dim[\psi] + \dim[A_\mu^a] = D \quad . \tag{2.4.59}$$

Equation (2.4.59) with Eq. (2.4.58) leads us to Eq. (2.4.56).

We introduce a mass scale μ by hand and rewrite the gauge coupling constant g in the following way,

$$g = g_0\mu^{2-D/2} \quad , \tag{2.4.60}$$

where g_0 is the dimensionless gauge coupling constant. Inserting Eq. (2.4.60) into Eq. (2.4.52) and making some modification in the expression by the use of Eq. (2.4.53), we obtain

$$\Sigma(p) = \frac{g_0^2}{(4\pi)^2} C_F \not{p} \left(\frac{-p^2}{4\pi\mu^2}\right)^{-\varepsilon} (1-\varepsilon)\, B(1-\varepsilon, 1-\varepsilon)\, \Gamma(\varepsilon) \quad , \tag{2.4.61}$$

where we defined a new parameter ε by

$$\varepsilon = \frac{4-D}{2} \quad . \tag{2.4.62}$$

We make a Laurent expansion of Eq. (2.4.61) around $\varepsilon = 0$ to obtain

$$\Sigma(p) = \frac{g_0^2}{(4\pi)^2} C_F \not{p} \left(\frac{1}{\varepsilon} - \gamma + 1 - \ln\frac{-p^2}{4\pi\mu^2}\right) + O(\varepsilon) \quad , \tag{2.4.63}$$

where we have employed the following expansion formulas,

$$\Gamma(\varepsilon) = \frac{1}{\varepsilon} - \gamma + O(\varepsilon) \quad , \tag{2.4.64}$$

$$(1-\varepsilon)\, B(1-\varepsilon, 1-\varepsilon) = 1 + \varepsilon + O(\varepsilon^2) \quad , \tag{2.4.65}$$

where $\gamma = 0.57721\ldots$ is the Euler constant.

In the present example of the quark self-energy part we have not encountered the Dirac matrix γ_5. We, however, need γ_5 in some practical applications such as QCD corrections to weak processes and polarized electroproductions. In 4 dimensions γ_5 is defined by Eq. (1.4.13). It is easy to see that such a definition of γ_5 is always possible in even dimensions through

$$\gamma_5 = i\gamma^0\gamma^1\ldots\gamma^{D-1} \quad . \tag{2.4.66}$$

In odd dimensions, however, if we define γ_5 in the same way as in Eq. (2.4.66), it turns out to be a multiple of the identity as will be shown later at the end of this subsection. Hence there is no mathematically consistent generalization of γ_5 in arbitrary dimensions. We thus give up our efforts to define γ_5 in an explicit way for arbitrary dimensions.[24] Rather we require the existence of the quantity called γ_5 which is hermitian and anticommutes with all γ^μ [Cha 79],

$$\{\gamma_5, \gamma^\mu\} = 0 \quad (\mu = 0, 1, 2, \ldots, D-1) \quad . \tag{2.4.67}$$

This operational definition of γ_5 is legitimate as far as the theory is free from anomalies [Adl 69, Bel 69]. In the case where an anomaly is present dimensional regularization has to be used with care and probably the best way is to adopt the Pauli-Villars regulator method.

We now summarize our conventions for dimensional regularization in this book.

Conventions for dimensional regularization

(1) The D-dimensional space-time has the metric $g^{\mu\nu} = (+, -, \ldots, -)$.
(2) $\mathrm{Tr}[1] = 4$ in the space of the gamma matrices.
(3) The integral measure is $\int d^D k/(2\pi)^D$.
(4) γ_5 is an object which satisfies $\{\gamma_5, \gamma^\mu\} = 0$.

Our derivation of Eq. (2.4.63) was restricted to the Feynman gauge $\alpha = 1$ just for simplicity. It is, however, not difficult to perform the calculation in an arbitrary covariant gauge. Let us go back to Eq. (2.4.8) and we have for arbitrary α and for $m = 0$

$$\Sigma(p) = g^2 C_F \int \frac{d^D k}{(2\pi)^D i} \frac{1}{k^2 (k-p)^2} \left\{ \gamma_\mu (\not{k} - \not{p}) \gamma^\mu - (1-\alpha) \frac{\not{k}(\not{k}-\not{p})\not{k}}{k^2} \right\} \quad . \tag{2.4.68}$$

We need to calculate the second term in Eq. (2.4.68) which is proportional to $1-\alpha$. We write

$$\Sigma(p) = \Sigma_1(p) - (1-\alpha)\Sigma_2(p) \quad . \tag{2.4.69}$$

The first term $\Sigma_1(p)$ has already been calculated and is given in Eq. (2.4.63). We shall now calculate $\Sigma_2(p)$ which is given by

$$\Sigma_2(p) = g^2 C_F \int \frac{d^D k}{(2\pi)^D i} \frac{\not{k}(\not{k}-\not{p})\not{k}}{(k^2)^2 (k-p)^2} \quad . \tag{2.4.70}$$

Using the Feynman parametrization

$$\frac{1}{AB^2} = 2\int_0^1 dx \frac{1-x}{\{xA + (1-x)B\}^3} \quad . \tag{2.4.71}$$

[24] For other possibilities of defining γ_5 in arbitrary dimensions, see [Fuj 81] and references therein.

we rewrite Eq. (2.4.70) in such a way that

$$\Sigma_2(p) = -2g^2 C_F \int_0^1 dx(1-x) \int \frac{d^D k}{(2\pi)^D i} \frac{\not{k}(\not{k}-\not{p})\not{k}}{\{-(k-xp)^2+L\}^3} , \tag{2.4.72}$$

where L is the same as the one given by Eq. (2.4.42). We shift the momentum k as in Eq. (2.4.35) and discard terms odd in k' according to Eq. (2.4.37) so that we find

$$\Sigma_2(p) = 2g^2 C_F \int_0^1 dx(1-x) \int \frac{d^D k'}{(2\pi)^D i} \frac{(1-x)\not{k}'\not{p}\not{k}' - 2xk'^2 \not{p} - xL\not{p}}{(-k'^2+L)^3} \tag{2.4.73}$$

We next use the formula (valid for an arbitrary integrable function $f(k'^2)$)

$$\int d^D k' k'_\mu k'_\nu f(k'^2) = \frac{1}{D} g_{\mu\nu} \int d^D k' k'^2 f(k'^2) , \tag{2.4.74}$$

to obtain the following,

$$\Sigma_2(p) =$$

$$2g^2 C_F \not{p} \int_0^1 dx(1-x) \int \frac{d^D k'}{(2\pi)^D i} \frac{1}{(-k'^2+L)^3} \left\{ \left(\frac{2(1-x)}{D} - 1 - x \right) k'^2 - xL \right\}$$

$$= 2g^2 C_F \not{p} \int_0^1 dx(1-x) \int \frac{d^D K}{(2\pi)^D}$$

$$\times \left\{ \frac{1}{(K^2+L)^3} \left(\frac{2(1-x)}{D} - 1 - 2x \right) L - \frac{1}{(K^2+L)^2} \left(\frac{2(1-x)}{D} - 1 - x \right) \right\} , \tag{2.4.75}$$

where we performed a Wick rotation (2.4.39). The formula (2.4.46) may be easily generalized to give

$$\int \frac{d^D K}{(2\pi)^D} \frac{1}{(K^2+L)^a} = \frac{\Gamma(a-D/2)}{(4\pi)^{D/2}\Gamma(a)} L^{D/2-a} , \tag{2.4.76}$$

where a is an arbitrary complex parameter with Re $a > 0$. By using Eq. (2.4.76) we can perform the integrations in Eq. (2.4.75) and obtain

$$\Sigma_2(p) = \frac{2C_F g^2}{(4\pi)^{D/2}} \not{p}(-p^2)^{D/2-2} (D-1) B\left(\frac{D}{2}, \frac{D}{2} \right) \Gamma\left(2 - \frac{D}{2} \right) , \tag{2.4.77}$$

which is exactly equal to Eq. (2.4.52). Hence we have

$$\sum_2 (p) = \sum_1 (p) \quad . \tag{2.4.78}$$

Thus for the covariant gauge with α arbitrary we find

$$\begin{aligned}\sum(p) &= \alpha \frac{2C_F g^2}{(4\pi)^{D/2}} \not{p}(-p^2)^{D/2-2} (D-1)\, B\left(\frac{D}{2}, \frac{D}{2}\right) \Gamma\left(2-\frac{D}{2}\right) \\ &= \alpha \frac{g_0^2}{(4\pi)^2} C_F \not{p} \left(\frac{1}{\varepsilon} - \gamma + 1 - \ln \frac{-p^2}{4\pi\mu^2}\right) + O(\varepsilon) \quad . \end{aligned} \tag{2.4.79}$$

Digression

CLIFFORD ALGEBRA: The numbers γ_i ($i = 1, 2, ..., N$) which satisfy the relations

$$\{\gamma_i, \gamma_j\} = 2\delta_{ij} \quad , \tag{2.4.80}$$

are called the Clifford numbers [eg., Boe 70]. The Clifford numbers γ_i have some similarity with the Grassmann numbers ψ_i discussed in Sec 2.2.4. An essential difference between these two kinds of the numbers lies in the fact that $\gamma_i^2 = 1$ while $\psi_i^2 = 0$. As in the case of the Grassmann numbers, the Clifford numbers γ_i generate an algebra called the Clifford algebra, the basis of which is given by the set of monomials

$$1, \gamma_i, \gamma_i\gamma_j, ..., \gamma_1\gamma_2 ... \gamma_N \quad . \tag{2.4.81}$$

In fact each commutation relation of these monomials may be written as a linear combination of themselves, thus forming a closed algebra. Since the total number of the above monomials is 2^N, the Clifford algebra spans a linear space of dimension 2^N, i.e., an arbitrary element of the algebra is given as a linear combination of the monomials (2.4.81).

The Dirac gamma matrices γ_μ are the typical example of the Clifford numbers where $N = D$, the space-time dimension. The commutation relations

$$\{\gamma^\mu, \gamma^\nu\} = 2g^{\mu\nu}; \mu, \nu = 0, 1, 2, ..., D-1, \tag{2.4.82}$$

determine the Clifford algebra of the gamma matrices. The basis of the algebra consists of 2^D monomials

$$1, \gamma^\mu, \gamma^\mu\gamma^\nu, ..., \gamma^0\gamma^1 ... \gamma^{D-1} \quad . \tag{2.4.83}$$

Since the above basis spans a 2^D-dimensional linear space, each element of the algebra may be expressed as a matrix in this space. However special care is needed in dealing with odd D.

Let us first consider the case of even D which includes our familiar case $D = 4$. For D even all the monomials in Eq. (2.4.83) are independent as can be directly checked by observing the commutation relations. Hence we must work in the full 2^D-dimensional space spanned by the monomials (2.4.83) and so we need $2^{D/2} \times 2^{D/2}$ matrices to represent the γ^μ's. In particular we have the familiar 4×4 matrices for γ^μ's in 4 dimensions ($D = 4$). This matrix representation of γ^μ is faithful. The trace of products of γ^μ's can now be easily calculated, e.g.,

$$\mathrm{Tr}(\gamma^\mu\gamma^\nu) = 2^{D/2} g^{\mu\nu} \quad . \tag{2.4.84}$$

The case of odd D is somewhat different and should be treated with some care [Shi 85]. By examining the commutation relations for all the monomials (2.4.83) for odd D, we find that

$$[\gamma, \Gamma_j] = 0 \quad , \tag{2.4.85}$$

for any Γ_j with Γ_j representing the monomials (2.4.83) where

$$\gamma \equiv \gamma^0\gamma^1 ... \gamma^{D-1} \quad . \tag{2.4.86}$$

Since γ commutes with all the monomials in the basis, it should be a multiple of the identity according to Schur's lemma,

$$\gamma = c1 \quad , \tag{2.4.87}$$

where c is a constant to be determined. On the other hand we find by using the definition (2.4.86) and the commutation relation (2.4.82) that

$$\gamma^2 = (-1)^{D(D-1)/2} \quad . \tag{2.4.88}$$

Hence we obtain

$$c = \begin{cases} \pm 1 \text{ for } D = 4n+1, \\ \pm i \text{ for } D = 4n+3 \; (n = 0, 1, 2, ...). \end{cases} \tag{2.4.89}$$

According to Eq. (2.4.87) we realize that one of γ^μ's is expressible in terms of the other γ^μ's, e.g.,

$$\gamma^{D-1} = -c\gamma^0\gamma^1 ... \gamma^{D-2} \quad , \tag{2.4.90}$$

and hence the number of independent monomials in the basis (2.4.83) is reduced to be 2^{D-1}. Thus γ^μ is represented by $2^{(D-1)/2} \times 2^{(D-1)/2}$ matrices. Corresponding to two different values of c (± 1 for $D = 4n + 1$ or $\pm i$ for $D = 4n + 3$), we have two inequivalent representations of γ^μ. For example, in 5-dimensional space-time ($D = 5$) we have the representation of γ^μ in terms of 4×4 matrices for two different signs.

2.4.3. Preliminary remarks on renormalization schemes

Given an explicit expression for the quark self-energy part to one-loop order for massless quarks, we were offered a chance to make a preliminary argument about various renormalization schemes by using this concrete example prior to the more detailed discussions to be given in Sec. 2.5.

Here by the term "renormalization" we mean, together with the redefinition of the mass and coupling constant, the readjustment of the normalization of Green functions by suitable multiplicative factors which may eliminate possible infinities in the Green function.[25] It is important to note here that the way of eliminating divergences in perturbation theory is not unique because there exists an ambiguity in defining the divergent piece of the Green function. This ambiguity eventually leads to ambiguity in the finite piece of the Green function. In order to remove this ambiguity we have to specify how we define the divergent piece which will be subtracted out in the renormalization process. The prescription in subtracting divergences in Green functions is called the *renormalization scheme*. Two different renormalization schemes are always connected by a finite renormalization.

[25] Note that renormalization is not necessarily directed to the elimination of divergences. There may be a finite renormalization.

Let us consider the one-loop quark self-energy part $\sum(p)$ calculated in the previous subsection and explain renormalization schemes with the help of it. We insert Eq. (2.4.79) into Eq. (2.4.2) with $m = 0$ to obtain

$$\tilde{S}_{ij}(p) = -\frac{\delta_{ij}}{\not{p}}\frac{1}{1+\sigma(p^2)} \quad , \tag{2.4.91}$$

where

$$\sigma(p^2) = \alpha\frac{g_0^2}{(4\pi)^2}C_F\left(\frac{1}{\varepsilon} - \gamma + 1 - \ln\frac{-p^2}{4\pi\mu^2}\right) + O(g_0^4) \quad , \tag{2.4.92}$$

with the terms of order ε set equal to zero. Since the one-loop corrected propagator (2.4.91) has a pole at $\not{p} = 0$, we note that the massless quark remains massless after the one-loop correction. It is generally true that the massless quark does not have radiative corrections to its mass to all orders of perturbation theory.

We renormalize the quark propagator $\tilde{S}_{ij}(p)$ by a multiplicative factor Z_2 (the quark-field renormalization constant),[26]

$$\tilde{S}_{Rij}(p) = Z_2^{-1}\tilde{S}_{ij}(p) \quad , \tag{2.4.93}$$

where $\tilde{S}_{Rij}(p)$ is the renormalized (finite) propagator of quarks. We expand Z_2 in powers of g_0 and write

$$Z_2 = 1 - z_2 + O(g_0^4) \quad , \tag{2.4.94}$$

with z_2 the term of order g_0^2 which is assumed to be divergent. Substituting Eq. (2.4.94) for Z_2 in Eq. (2.4.93) and using Eq. (2.4.91) we obtain to order g_0^2,

$$\tilde{S}_{Rij}(p) = -\frac{\delta_{ij}}{\not{p}}\frac{1}{1+\sigma(p^2)-z_2} \quad , \tag{2.4.95}$$

where we keep only the terms up to order g_0^2 in the denominator (note that $Z_2\sigma(p^2) = \sigma(p^2) + O(g_0^4)$). In Eq. (2.4.95) the coupling constant g_0 should be replaced by its renormalized version. There is, however, no practical effect in Eq. (2.4.95) by this replacement to the order we are considering. Therefore we simply retain g_0 in the following arguments. Since $\tilde{S}_{Rij}(p)$ is the renormalized propagator, it should be free of divergences and hence $\sigma(p^2) - z_2$ has to be finite. Thus the divergence in $\sigma(p^2)$ should be cancelled by that of z_2. This requirement determines the constant z_2 up to a finite additive constant. In order to fix this arbitrary finite constant in z_2, we need an additional requirement which sets up a renormalization scheme (prescription). There are

[26] We follow tradition by denoting the fermion field renormalization constant by Z_2 and the boson one by Z_3 as in QED. The constant Z_1 is reserved for the vertex renormalization.

a variety of renormalization schemes depending on the choice of the above additional requirement.

(1) *On-shell subtraction*

The renormalization constant Z_2 is determined on the mass shell of quarks, i.e., by the condition

$$\tilde{S}_{\mathrm{R}ij}(p) \sim \frac{\delta_{ij}}{m-\not{p}} \quad \text{for } \not{p} \sim m \quad . \tag{2.4.96}$$

This is the traditional renormalization prescription employed in QED and is the most natural for obtaining physical quantities such as S-matrices. For the present case $m = 0$, and so $z_2 = \sigma(0)$. Unfortunately for massless quarks the mass singularity develops in $\sigma(p^2)$ and $\sigma(0)$ is not well-defined in the present example.

(2) *Off-shell subtraction*

At an unphysical (off-shell) value of p^2, say $p^2 = -\lambda^2 < 0$, we require that $\tilde{S}_{\mathrm{R}ij}(p)$ be of the form of the free (massless) propagator,

$$\tilde{S}_{\mathrm{R}ij}(p) \sim -\frac{\delta_{ij}}{\not{p}} \quad \text{for } p^2 \sim -\lambda^2 \quad . \tag{2.4.97}$$

With this renormalization condition [Wei 73a, Geo 76] we determine z_2 such that

$$z_2 = \sigma(-\lambda^2) = \alpha\frac{g_0^2}{(4\pi)^2}C_{\mathrm{F}}\left(\frac{1}{\varepsilon} - \gamma + 1 - \ln\frac{\lambda^2}{4\pi\mu^2}\right) \quad . \tag{2.4.98}$$

Hence the renormalized propagator reads

$$\tilde{S}_{\mathrm{R}ij}(p) = -\frac{\delta_{ij}}{\not{p}}\left(1 - \alpha\frac{g_0^2}{(4\pi)^2}C_{\mathrm{F}}\ln\frac{-p^2}{\lambda^2}\right)^{-1} \quad . \tag{2.4.99}$$

This renormalization scheme is often refered to as the momentum-space subtraction scheme (MOM) [Cel 79, 79a, Bar 79c].

(3) *Minimal subtraction (MS)*

This renormalization condition is due to 't Hooft [tHo 73] and is specific to dimensional regularization. In the minimal subtraction scheme we eliminate only the pole term $1/\varepsilon$ in the dimensionally regularized expression of the Green functions. As this scheme is the most economical and has some further advantages, it has been frequently used in many applications of QCD (and also other gauge-field theories). With this renormalization condition we have

$$z_2 = \alpha \frac{g_0^2}{(4\pi)^2} C_F \frac{1}{\varepsilon} \quad . \tag{2.4.100}$$

The renormalized propagator in this scheme is

$$\tilde{S}_{Rij}(p) = -\frac{\delta_{ij}}{\not p}\left\{1 - \alpha \frac{g_0^2}{(4\pi)^2} C_F\left(\gamma - 1 + \ln\frac{-p^2}{4\pi\mu^2}\right)\right\}^{-1} . \tag{2.4.101}$$

Thus in the minimal subtraction scheme the renormalization constants acquire the simplest expression while the renormalized Green functions have rather complicated forms. Since the renormalization constants in this scheme are independent of mass parameters in the theory, we shall have great facility in defining the renormalization group functions as will be seen in Sec. 3.2.1. Note here that the expression Eq. (2.4.101) may be converted to that of Eq. (2.4.99) in the MOM scheme by setting

$$\lambda^2 = 4\pi e^{1-\gamma}\mu^2 \quad .$$

(4) *Modified minimal subtraction* ($\overline{\text{MS}}$)

In the expression for $\sigma(p^2)$ in Eq. (2.4.92) the pole term is accompanied by the natural constants γ and $\ln 4\pi$ in the combination,

$$\frac{1}{\varepsilon} - \gamma + \ln 4\pi \quad . \tag{2.4.102}$$

The appearance of the above combination is a phenomenon peculiar to dimensional regularization with the convention described in Sec. 2.4.2. It can be shown that the combination (2.4.102) always appears in any calculations to one-loop order. Hence it is more convenient to eliminate, in the renormalization process, the whole of Eq. (2.4.102) instead of eliminating only the term $1/\varepsilon$ as in the MS scheme. This is the modified minimal subtraction ($\overline{\text{MS}}$) scheme [Bar 78] which is frequently used in the definition of the QCD coupling constant and also in other applications of gauge field theories. The renormalizaton constant z_2 in the $\overline{\text{MS}}$ scheme takes the following form,

$$z_2 = \alpha \frac{g_0^2}{(4\pi)^2} C_F\left(\frac{1}{\varepsilon} - \gamma + \ln 4\pi\right) \quad . \tag{2.4.103}$$

The renormalized propagator reads

$$\tilde{S}_{Rij}(p) = -\frac{\delta_{ij}}{\not p}\left\{1 - \alpha \frac{g_0^2}{(4\pi)^2} C_F\left(-1 + \ln\frac{-p^2}{\mu^2}\right)\right\}^{-1} \tag{2.4.104}$$

The $\overline{\text{MS}}$ scheme shares many advantages with the MS scheme and at the same time offers a sufficiently compact expression for the renormalized propagator.

As we have seen, the above four different renormalization schemes provide different forms for the renormalized propagator. In general the form of the Green functions varies from scheme to scheme. This dependence of the form of Green functions to the renormalization scheme is called the renormalization-scheme dependence.

Digression

THE FEYNMAN PARAMETRIZATION: It is often very useful to employ the Feynman parametrization in calculating loop integrals. Some specific examples have already appeared in Eqs. (2.4.32) and (2.4.71). Here we present the most general form of the Feynman parametrization,

$$\prod_{i=1}^{n}\frac{1}{A_i^{\alpha_i}}=\frac{\Gamma(\alpha)}{\prod_{i=1}^{n}\Gamma(\alpha_i)}\int_0^1\left(\prod_{i=1}^{n}dx_i\,x_i^{\alpha_i-1}\right)\frac{\delta(1-x)}{\left(\sum_{i=1}^{n}x_iA_i\right)^{\alpha}}\quad, \tag{2.4.105}$$

where $\alpha_i(i=1,2,\ldots,n)$ are arbitrary complex numbers and

$$\alpha=\sum_{i=1}^{n}\alpha_i\quad, \tag{2.4.106}$$

$$x=\sum_{i=1}^{n}x_i\quad. \tag{2.4.107}$$

The proof of the above formula (2.4.105) goes as follows. For $n=2$ we can easily show that the formula holds, i.e.,

$$\frac{1}{A^{\alpha}B^{\beta}}=\frac{\Gamma(\alpha+\beta)}{\Gamma(\alpha)\Gamma(\beta)}\int_0^1 dxdy\frac{x^{\alpha-1}y^{\beta-1}}{(xA+yB)^{\alpha+\beta}}\delta(1-x-y)\quad. \tag{2.4.108}$$

Now we assume that the formula (2.4.105) holds for n and prove that it is also valid for $n+1$. Let us start with the expression

$$\frac{1}{A_{n+1}^{\alpha_{n+1}}}\prod_{i=1}^{n}\frac{1}{A_i^{\alpha_i}}=\frac{\Gamma(\alpha)}{\prod_{i=1}^{n}\Gamma(\alpha_i)}\int_0^1\left(\prod_{i=1}^{n}dx_ix_i^{\alpha_i-1}\right)\frac{\delta(1-x)}{\left(\sum_{i=1}^{n}x_iA_i\right)^{\alpha}}\frac{1}{A_{n+1}^{\alpha_{n+1}}}\quad. \tag{2.4.109}$$

Applying Eq. (2.4.108) we rewrite Eq. (2.4.109) in the following way,

$$\prod_{i=1}^{n+1}\frac{1}{A_i^{\alpha_i}}=\frac{\Gamma(\alpha)}{\prod_{i=1}^{n}\Gamma(\alpha_i)}\int_0^1\left(\prod_{i=1}^{n}dx_ix_i^{\alpha_i-1}\right)\delta(1-x)\frac{\Gamma(\alpha+\alpha_{n+1})}{\Gamma(\alpha)\,\Gamma(\alpha_{n+1})}$$

$$\times \int_0^1 dx_{n+1}\, dy \frac{y^{\alpha-1} x_{n+1}^{\alpha_{n+1}-1}}{\left(y \sum_{i=1}^{n} x_i A_i + x_{n+1} A_{n+1}\right)^{\alpha+\alpha_{n+1}}} \delta(1-y-x_{n+1}) \quad . \tag{2.4.110}$$

We make the change of variables $x_i' = yx_i$ ($i = 1, 2, \ldots, n$) in Eq. (2.4.110) and then perform the y-integration. After replacing x_i' by x_i we obtain

$$\prod_{i=1}^{n+1} \frac{1}{A_i^{\alpha_i}} = \frac{\Gamma(\alpha+\alpha_{n+1})}{\prod_{i=1}^{n+1} \Gamma(\alpha_i)} \int_0^1 \left(\prod_{i=1}^{n+1} dx_i x_i^{\alpha_i - 1} \right) \delta(1-x-x_{n+1}) \frac{1}{\left(\sum_{i=1}^{n+1} x_i A_i \right)^{\alpha+\alpha_{n+1}}} \quad . \tag{2.4.111}$$

This completes the proof of Eq. (2.4.105).

2.5. RENORMALIZATION

2.5.1. Power counting

For higher orders of the perturbation series in quantum field theory, we inevitably come across divergences in the loop integrals. Those divergences usually stem from the high momentum limit of the loop integrals and hence are called the ultraviolet (UV) divergences. On the other hand, if massless particles are involved in the theory, another kind of divergence may occur which originates from the low momentum limit of the loop integrals. This is the so-called infrared (IR) divergence and will be discussed separately later in Chapter 6. In the present section we concentrate our attention on UV divergences.

As we have seen in the concrete example in the previous subsection 2.4.3, the divergences in the Green functions may be removed by multiplicative factors (such as Z_2) together with the redefinition of the parameters (such as g and m) in the original Lagrangian. At each order of perturbation theory, this process of removing the divergences turns out to be a subtraction of divergent pieces in the proper (one-particle irreducible) Green functions. The central subject of renormalization theory lies in the question of whether the above subtraction process is consistently performed to all orders with a *finite* number of multiplicative factors (renormalization constants) and parameters which will be redefined. We shall show that the renormalization procedure is successfully carried out in a restricted class of interactions. The theory for which the above renormalization procedure operates is said to be *renormalizable*. Thus throughout Sec. 2.5 we are essentially dealing with the problem of renormalizability of the theory under consideration.

Our discussion will proceed as follows. First we have to know which diagram at a given order gives a divergent Feynman amplitude and what is the condition for an amplitude at a given order to be divergence-free (Sec. 2.5.1). We then use this result to classify all interactions into renormalizable and non-renormalizable ones (Sec. 2.5.2). After the precise definition of renormalizability we present a sketch of the proof of renormalizability of the theory with renormalizable interactions. The argument is given for ϕ^3 theory in six dimensions (Sec. 2.5.3) and for QCD (Sec. 2.5.5). In a rigorous proof of renormalizability the most important point is to find a way to systematically decompose a complicated diagram into an ensemble of fundamental divergent pieces. This program is most efficiently executed by the BPHZ method which we shall describe in Sec. 2.5.4. For the proof of renormalizability of QCD the generalized version of the Ward-Takahashi identity is used in an essential manner and so we present a full account of the derivation of generalized Ward-Takahashi identities in Sec. 2.5.6.

We have applied naive power counting to Eq. (2.4.9) to see that the one-loop quark self-energy part is linearly divergent. This power-counting method is applicable to any Feynman amplitude at an arbitrary order for any interaction. For this purpose it is convenient to define a quantity which expresses an overall power of the momentum-integration variable at the high momentum region in a Feynman integral. Here by the Feynman integral we mean a loop-momentum integral appearing in the Feynman amplitudes. The quantity introduced above is called the *superficial degree of divergence d*. The reason why it is superficial will become clear soon.

Consider the two-loop gluon self-energy part $\Pi(q)$ corresponding to the diagram shown in Fig. 2.5.1 (Here for simplicity we neglect the Lorentz and group indices in $\Pi(q)$). The contribution of the large values of loop momenta k_1 and k_2 to $\Pi(q)$ is roughly given by

$$\Pi(q) \sim \int d^D k_1 d^D k_2 (k_1)^2 \left(\frac{1}{k_1^2}\right)^3 \frac{1}{K_1 + K_2} \frac{1}{K_2} , \tag{2.5.1}$$

in conformity with the Feynman rules of Sec. 2.3.3 where D is the space-time dimension. Restricting our argument to the contribution from the region $k_1 \sim k_2$ we find that $\Pi(q)$ diverges as

$$\Pi(q) \sim \lim_{K \to \infty} K^{2D+2-6-2} . \tag{2.5.2}$$

We see that the superficial degree of divergence d for the diagram in Fig. 2.5.1 is

$$d = 2D + 2 - 6 - 2 , \tag{2.5.3}$$

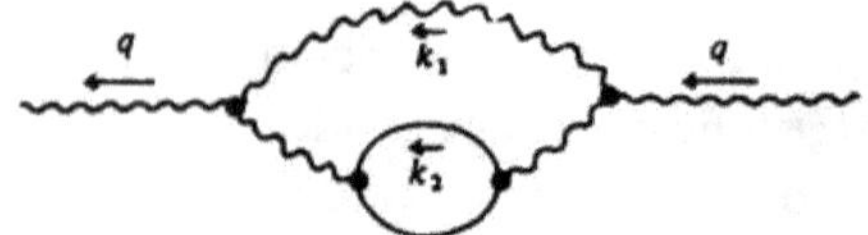

Fig. 2.5.1. A typical example of a two-loop contribution to the gluon self-energy part.

where terms $2D$, 2, -6 and -2 in the right-hand side of Eq. (2.5.3) correspond to two loop-integrals, two vertex factors for the three-gluon vertex, three gluon-propagators and two quark-propagators. For 4 dimensions, $d = 2$ and $\Pi(q)$ seems to diverge quadratically.

The generalization of Eq. (2.5.3) for an arbitrary diagram G is straightforward and the superficial degree of divergence d_G reads

$$d_G = Dl + \sum_v \delta_v - 2n_B - n_F \quad , \tag{2.5.4}$$

where l, δ_v, n_B and n_F are defined as follows,

$$\begin{aligned} l &= \text{the number of loops in } G \text{ (the number of independent integrals)}, \\ \delta_v &= \text{the number of momentum factors at vertex } v \text{ in } G, \\ n_B &= \text{the number of boson internal lines in } G,^{26} \\ n_F &= \text{the number of fermion internal lines in } G. \end{aligned} \tag{2.5.5}$$

Obviously to one-loop order the Feynman amplitude is finite for $d < 0$, logarithmically divergent for $d = 0$ and power-divergent for $d > 0$.[27] This statement, however, cannot be generalized straightforwardly to higher orders (multi-loops). For example let us first consider the two-loop quark-quark scattering diagram in QCD shown in Fig.2.5.2. According to the definition

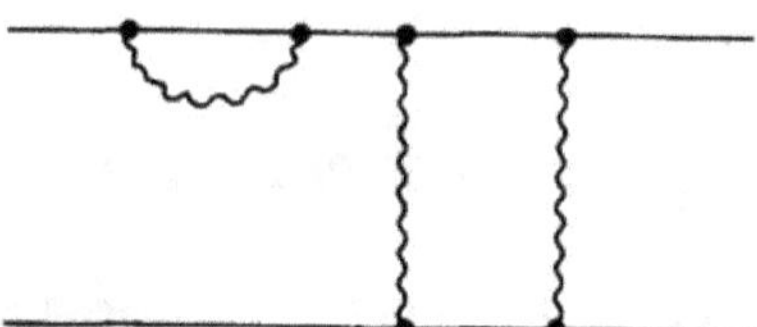

Fig. 2.5.2. An example of a quark-quark scattering diagram in the two-loop order.

[26] We do not consider the case of a massive non-gauge vector boson whose propagator behaves as $O(k_\mu k_\nu/k^2)$ instead of $O(1/k^2)$ for large momentum $k \to \infty$.

[27] In some cases the divergence may become weaker by symmetry reasons. For example the one-loop quark self-energy part $\Sigma(p)$ (2.4.9) is logarithmically divergent in 4 dimensions in spite of the fact that $d = 1$. This reduction of the divergence is due to the requirement of Lorentz invariance.

Fig. 2.5.3. An example of a quark-quark scattering diagram in the two-loop order.

(2.5.4) the superficial degree of divergence for this diagram is given by $d = 2D - 10 = -2$ for $D = 4$, but, as we obviously see, the amplitude for this diagram is divergent due to the one-loop quark self-energy part. The diagram that we considered is improper (one-particle reducible) and the divergence arises as a simple recurrence of a lower-order divergence. Hence, if we restrict an application of the notion of the superficial degree of divergence to proper (one-particle irreducible) diagrams, the above problem may be circumvented. However, we easily find that this restriction is not the solution for the whole problem. Consider as a second example the proper two-loop diagram drawn in Fig. 2.5.3. The superficial degree of divergence for this diagram reads $d = 2D - 10 = -2$ for $D = 4$. The corresponding Feynman amplitude, however, is divergent owing to the divergence of the one-loop quark self-energy part. Thus our hope to classify the degree of divergence of Feynman amplitudes in terms of the sign of d seems to be frustrated. As a reflection of this situation d is called the "superficial" degree of divergence. The Feynman integral for the diagram with $d \geq 0$ is said to be superficially divergent. Also another reason for calling d superficial is that in some Feynman integrals the degree of divergence is lower than the one expected because of symmetry. We have already encountered an example of this phenomenon in the case of the one-loop quark self-energy part $\sum(p)$ in which the divergence is logarithmic rather than linear due to the requirement of Lorentz invariance.

To go a step further, let us see why the value of d failed to give the true degree of divergence in the above example in Fig. 2.5.3. As in the discussion in Eq. (2.5.2), all the relevant loop momenta ought to be large in order that the superficial degree of divergence works as a measure of divergence. In the example of Fig. 2.5.3 the divergence does not come from the integration region where all loop momenta are large but stems from the region in which the loop momentum for the one-loop self-energy part gets large while the other loop momentum is finite. This fact suggests that the notion of superficial degree of divergence should be applied to all subdiagrams including the whole diagram where by the subdiagram we mean a part of the whole diagram which corresponds to a certain truncated connected Green function. Consider a set of subdiagrams H of the Feynman diagram G and define the superficial degree of divergence d_H for subdiagram H. We find that, if at least one of d_H is non-negative ($d_H \geq 0$), the Feynman integral I_G for the diagram G is divergent except that the special reduction of the degree of divergence occurs by symmetry reason. Stating this the other way round we have the following theorem:

Power-counting theorem (Convergence theorem): The Feynman integral I_G for diagram G is absolutely convergent if the superficial degree of divergence d_H for the subdiagram H in G is negative for all possible H (including the case H = G).

In the above theorem the term absolute convergence refers to the fact that when the absolute value of the integrand is used, the Feynman integral is bounded from above by a finite positive number. To prove the theorem, one, for example, employs the Feynman parameter representation for an arbitrary Feynman amplitude and discusses the convergence of the integral. Since the rigorous proof of the theorem is beyond the scope of the present textbook, we shall not go into the details of the proof. The theorem is, however, basically important as it plays a fundamental role in verifying renormalizability, and serious readers should look into the original papers [Wei 60, Nak 57, 63, Hah 68, Zim 68]. The power-counting theorem was first mentioned by Dyson in his second paper on quantum electrodynamics [Dys 49a] and the mathematically rigorous proof of the theorem was given in terms of Euclidean momenta by Weinberg [Wei 60] and in parametric representation by Nakanishi [Nak 57, 63]. The proof was later elaborated upon in [Hah 68, Zim 68].

2.5.2. Renormalizable interactions

We shall now see by using the notion of superficial degree of divergence what types of interactions are renormalizable. We first restrict ourselves to the case of only one interaction term $\mathscr{L}_1$ in the Lagrangian $\mathscr{L}$. (Note that we have four interaction terms in QCD.) Later a generalization to the case where more than one interaction term exist will be given.

For our purpose it is necessary to rewrite the expression for the superficial degree of divergence (2.5.4) in terms of the quantities characteristic to the interaction and those related to the external lines of the Feynman diagram under consideration. We define the following quantities relevant to the interaction Lagrangian $\mathscr{L}_1$:

$$
\begin{aligned}
b &= \text{the number of boson fields in } \mathcal{L}_1, \\
f &= \text{the number of fermion fields in } \mathcal{L}_1, \\
\delta &= \text{the number of space-time derivatives in } \mathcal{L}_1.
\end{aligned}
\tag{2.5.6}
$$

We also define the quantities specific to the Feynman diagram G as follows:

$$
\begin{aligned}
n &= \text{the number of vertices corresponding to } \mathcal{L}_1 \text{ in G}, \\
N_B &= \text{the number of boson external lines in G}, \\
N_F &= \text{the number of fermion external lines in G}.
\end{aligned}
\tag{2.5.7}
$$

These quantities obey simple relations such as

$$2n_B + N_B = nb \quad , \tag{2.5.8}$$

$$2n_F + N_F = nf \quad , \tag{2.5.9}$$

$$l = n_B + n_F - n + 1 \quad , \tag{2.5.10}$$

$$\sum_v \delta_v = n\delta \quad , \tag{2.5.11}$$

where n_B, n_F, l and δ_v have already been defined in Eq. (2.5.5). The derivation of Eqs. (2.5.8) and (2.5.9) is straightforward if one recognizes that internal boson (fermion) lines connected to a vertex contribute twice to the total number of vertices times $b(f)$ while external boson (fermion) lines contribute once. Equation (2.5.10) is obtained in the following way: originally a momentum integration is associated with each internal line and so we have $n_B + n_F$ integrations in G. Because of the energy-momentum conservation at each vertex we have n delta functions in G. One of these delta functions represents the total energy-momentum conservation in G and is irrelevant to the momentum intergration. Thus we find that the number of independent momentum integrals is $n_B + n_F - (n - 1)$.

Using Eqs. (2.5.8) and (2.5.9) we can eliminate internal quantities n_B and n_F in Eq. (2.5.10).

$$l = \left(\frac{b+f}{2} - 1\right)n - \frac{N_B + N_F}{2} + 1 \quad . \tag{2.5.12}$$

Inserting Eqs. (2.5.11) and (2.5.12) into Eq. (2.5.4) and eliminating n_B and n_F by the use of Eqs. (2.5.8) and (2.5.9), we find

$$d = rn - \frac{D-2}{2} N_B - \frac{D-1}{2} N_F + D \quad , \tag{2.5.13}$$

where r is given by

$$r = \frac{D-2}{2} b + \frac{D-1}{2} f + \delta - D \quad . \tag{2.5.14}$$

The above quantity r consists only of b, f and δ which are inherent to the interaction Lagrangian $\mathscr{L}_1$. It is called the *index of divergence* of the interaction $\mathscr{L}_1$. It should be noted here that $-r$ is precisely equal to the mass dimension of the coupling constant. In fact, if we write symbolically

$$\mathscr{L}_1 \sim g(\partial)^\delta (\phi)^b (\psi)^f \quad , \tag{2.5.15}$$

where ϕ and ψ are the boson and fermion fields, respectively, we have, by dimension counting, (see, e.g., Eq. (2.4.58))

$$D = \dim [g] + \frac{D-2}{2} b + \frac{D-1}{2} f + \delta \quad . \tag{2.5.16}$$

Hence we obtain $\dim [g] = -r$ [Sak 51, 52, Ume 56].

The generalization of our result (2.5.13) to the case where more than one interaction term exist is straightforward. Let us replace the interaction term $\mathscr{L}_1$ by the sum of terms, $\sum_i \mathscr{L}_i$, and attach the index i to the quantities peculiar to the interaction term $\mathscr{L}_i$, i.e., g_i, b_i, f_i, δ_i, n_i. We then have instead of Eq. (2.5.13)

$$d = \sum_i r_i n_i - \frac{D-2}{2} N_B - \frac{D-1}{2} N_F + D \quad , \tag{2.5.17}$$

with

$$r_i = \frac{D-2}{2} b_i + \frac{D-1}{2} f_i + \delta_i - D \quad . \tag{2.5.18}$$

We are now ready to classify all interactions into renormalizable and nonrenormalizable ones by applying the notion of the superficial degree of divergence. As we have seen in the last subsection, the Feynman integral I_G for diagram G is divergent if at least one of d_H is nonnegative for $H \subseteq G$.[28] When we discuss the renormalizability of a theory, we first eliminate divergences of all lower order subdiagrams in a given diagram and see whether any more divergences are left. Hence, without loss of generality, we may confine

[28] Here we consider only the case where no reduction of the degree of divergence occurs by symmetry reasons. If such reduction of the degree of divergence occurs, our argument in the following will be slightly modified.

ourselves to Feynman diagrams with only an *overall divergence*, where by overall divergence we mean the divergence coming from the whole diagram G, i.e., the divergence which remains after eliminating divergences of subdiagrams by a suitable renormalization. Overall divergence occurs when momenta on all internal lines of the whole diagram G become simultaneously large.

Consider a theory in which at least one of the r_i is positive, $r_i > 0$. For higher orders when n_i is large, the superficial degree of divergence d grows indefinitely and hence an unlimited number of new types of divergences show up for higher orders. This means that the divergences cannot be removed by a finite number of renormalization constants and interaction parameters. Thus theories of this category are *nonrenormalizable*.

In a theory for which all r_i satisfy $r_i \leq 0$, the superficial degree of divergence d has an upper bound for varying orders n_i. Hence the number of types of divergences is finite, i.e., the number of external lines for Feynman amplitudes with overall divergences is bounded from above. Thus there is a possibility that the divergences may be removed by a finite number of renormalization constants and interaction parameters. Therefore theories of this category are candidates for renormalizable theories. In particular, if $r_i < 0$ for all i, the theory is called *super renormalizable* since in this case not only the number of the types of divergent diagrams but also the number of divergent diagrams becomes finite. The theory in which $r_i = 0$ for all i is called *renormalizable* in a narrow sense.

Hereafter we shall use the term "renormalizable" in the narrow sense. Since $\dim [g_i] = -r_i$, renormalizable theories have dimensionless coupling constants and the interaction Lagrangian consists of terms with dimension D. If at least one of coupling constants in the theory has an inverse mass dimension, the theory is nonrenormalizable.

For example, consider the ϕ^3 theory for which

$$r = (D-6)/2 \quad . \tag{2.5.19}$$

Hence the theory is renormalizable for $D = 6$ and the superficial degree of divergence d reads

$$d = 2(3-N_B) \quad \text{for } D = 6. \tag{2.5.20}$$

We thus realize that ϕ^3 theory is presumably renormalizable in 6 dimensions and Feynman amplitudes with overall divergences have at most three external lines ($N_B \leq 3$). A more rigorous discussion of renormalizability for ϕ^3 theory in 6 dimensions will be left for the next subsection.[29] Another example is quantum chromodynamics (QCD) in which four interaction terms are present as in Eq. (2.3.109). It is easy to show that for all of these four terms $r_i \propto D-4$. Hence QCD is renormalizable in 4 dimensions. It is an easy exercise to show that four-fermion interaction $(\bar{\psi}\psi)^2$ is renormalizable in 2 dimensions.

Exercise

1. Calculate the index of divergence r for Lagrangians of the form $\bar{\psi}\psi\phi^2$ and ϕ^6, and find the space-time dimensions for which these interactions become renormalizable.

2.5.3. Renormalization in ϕ^3 theory

In this subsection we shall present a sketch of the proof of renormalizability. Since the proof itself is rather complicated, we had better choose a theory of the simplest possible form. We take ϕ_6^3 *theory* (ϕ^3 theory in 6 dimensions) here as such an example. The Lagrangian for ϕ^3 theory reads

$$\mathscr{L} = \mathscr{L}_0 + \mathscr{L}_1 \quad ,$$

$$\mathscr{L}_0 = \frac{1}{2}(\partial^\mu\phi\partial_\mu\phi - m^2\phi^2) \quad , \tag{2.5.21}$$

$$\mathscr{L}_1 = \frac{g}{3!}\phi^3 \quad .$$

Note that in the notation of Eq. (2.3.55) $\mathscr{L}_1 = -V(\phi)$.

We start by repeating the definition of renormalizability of the theory. The theory is said to be renormalizable if all the divergences appearing in the truncated connected Green function $\tilde{G}_n^{tc}$ (including the S-matrix elements as a special case) may be absorbed, at each order of perturbation series, into the change of the normalization of the Green functions and the redefinition of the mass m and coupling constant g. This definition of renormalizability is rather restrictive in the sense that it does not allow for an introduction of any terms other than the ones appearing in Eq. (2.5.21). One may relax the definition so as to allow for an addition of a finite number of new terms to the Lagrangian (2.5.21) in order to remove the divergences of the Green functions. In the following argument, however, we adopt the above restrictive definition of renormalizability.

[29] In the more restrictive definition of renormalizability which will be given later in Sec. 2.5.3, the ϕ^3 theory is not renormalizable since, for the renormalization of the theory, a term linear in ϕ is required in the Lagrangian although it does not exist in the original Lagrangian.

Renormalizability in the above sense may be restated by the use of the generating functional $W[J, g, m]$ defined in Eq. (2.3.26). (Here we have explicitly shown the dependence of W on g and m.) If the theory is renormalizable, the divergence of the generating functional $W[J, g, m]$ is removed, order by order, by the redefinition of J, g and m in the following manner:

$$J = Z_3^{-1/2} J_r \quad , \quad g = Z_g g_r \quad , \quad m^2 = Z_m m_r^2 \quad , \tag{2.5.22}$$

where Z_3, Z_g and Z_m are renormalization constants to be determined by suitable renormalization conditions. Note here that the redefinition of J corresponds to the field redefinition $\phi = Z_3^{1/2}\phi_r$. By the above redefinition (2.5.22) the same generating functional $W[J, g, m]$ which appeared to be divergent is rewritten as

$$W[J, g, m] = W_r[J_r, g_r, m_r] \quad , \tag{2.5.23}$$

which has a finite expression as a functional of J_r, g_r and m_r. The renormalized connected Green functions are obtained through the generating functional W_r given by Eq. (2.5.23).

As we saw in the last subsection, the superficial degree of divergence for connected Green functions with N_B external lines is given by Eq. (2.5.20) for ϕ_6^3 theory. Hence, in this theory, only Feynman amplitudes with $N_B \leq 3$ can develop overall divergences. Thus the number of types of Feynman diagrams responsible for the overall divergence is restricted to four and the diagrams have at most three external lines as shown in Fig. 2.5.4. As we have seen in Sec. 2.3.2, the vacuum diagram of type (a) in Fig. 2.5.4 is eliminated by normalizing the generating functional $Z[J]$ in terms of $Z[0]$ and hence we need not consider the contribution of the diagram (a) to the Feynman amplitude. The diagram (b) in Fig. 2.5.4 is often called a tadpole diagram and corresponds to the amplitude

$$G_1^c = \left.\frac{\delta W[J]}{\delta J}\right|_{J=0} = \left.\frac{-i}{Z[0]}\frac{\delta Z[J]}{\delta J}\right|_{J=0} = \langle 0|\hat{\phi}(x)|0\rangle \quad . \tag{2.5.24}$$

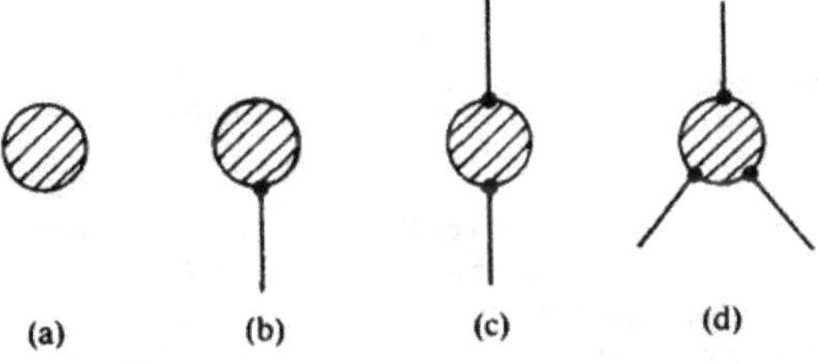

Fig. 2.5.4. Four types of diagrams in ϕ_6^3 theory which may have overall divergence.

This contribution of the diagram (b) is also irrelevant if we adopt the renormalization condition at each order[30]

$$\langle 0|\hat{\phi}(x)|0\rangle = 0 \quad . \tag{2.5.25}$$

Thus we are only left with the diagrams of the types (c) and (d) in Fig. 2.5.4. They are 2- and 3-point truncated connected Green functions which correspond to the inverse propagator and vertex function, respectively. We shall first show that, at the one-loop level, these two Green functions which have overall divergences (only these Green functions are divergent in one-loop order) can be made finite by the redefinition of ϕ, g and m.

(1) *One-loop renormalizability*

Following the argument given in Sec. 2.3.2 we easily derive the formulas for the truncated connected Green functions $\tilde{G}_2^{\mathrm{tc}}$ and $\tilde{G}_3^{\mathrm{tc}}$ to one-loop order:

$$\begin{aligned} -i\tilde{G}_2^{\mathrm{tc}}(p) &= m^2 - p^2 \\ &\quad -\frac{g^2}{2}\int\frac{d^Dk}{(2\pi)^D i}\frac{1}{m^2-k^2}\frac{1}{m^2-(k+p)^2} + O(g^4) \quad , \end{aligned} \tag{2.5.26}$$

$$\begin{aligned} -i\tilde{G}_3^{\mathrm{tc}}(p,q) &= g \\ &\quad +g^3\int\frac{d^Dk}{(2\pi)^D i}\frac{1}{m^2-k^2}\frac{1}{m^2-(k+p)^2}\frac{1}{m^2-(k-q)^2} + O(g^5) \quad , \end{aligned} \tag{2.5.27}$$

where we use dimensional regularization and dropped the tadpole contribution on account of the condition (2.5.25). The Feynman diagrams corresponding to Eq. (2.5.26) and (2.5.27) are pesented in Fig. 2.5.5. As before we define the inverse propagator by $\Delta^{-1}(p^2) = -i\tilde{G}_2^{\mathrm{tc}}(p)$ and hence the propagator to one-loop order takes the following form,

$$\Delta(p^2) = 1/[m^2 - p^2 - \Pi(p^2)] \quad , \tag{2.5.28}$$

where $\Pi(p^2)$ is the self-energy part given by

$$\Pi(p^2) = \frac{g^2}{2}\int\frac{d^Dk}{(2\pi)^D i}\frac{1}{m^2-k^2}\frac{1}{m^2-(k+p)^2} + O(g^4) \quad . \tag{2.5.29}$$

[30] In order to fulfill this renormalization condition, it is necessary to prepare a term linear in ϕ, i.e., $f\phi$ with f a parameter, in the original Lagrangian. Hence the Lagrangian (2.5.21) is nonrenormalizable in the stringent sense. Here we shall not take this problem too seriously and assume that the theory is renormalizable with the renormalization condition (2.5.25). The problem is similar to that of the photon mass term in QED.

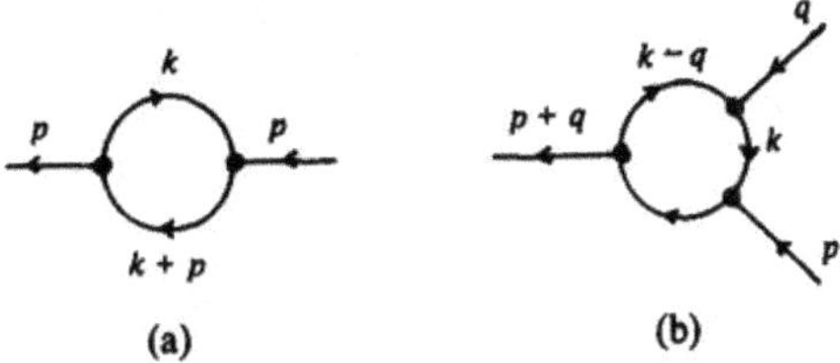

Fig. 2.5.5. The one-loop diagrams for the two- and three-point Green functions.

The vertex function $\Gamma(p, q)$ is defined by $\Gamma(p, q) = -i\tilde{G}_3^{\text{tc}}(p, q)$ and has the form,

$$\Gamma(p, q) = g + gA(p, q) \quad , \tag{2.5.30}$$

with $A(p, q)$ given by

$$A(p, q) = g^2 \int \frac{d^D k}{(2\pi)^D i} \frac{1}{m^2 - k^2} \frac{1}{m^2 - (k+p)^2} \frac{1}{m^2 - (k-q)^2} + O(g^4) \quad . \tag{2.5.31}$$

For $D = 6$ the self-energy part $\Pi(p^2)$ to one-loop order (2.5.29) is quadratically divergent and has the following divergence structure

$$\Pi(p^2) \sim \lim_{K \to \infty} K^2 + \lim_{K \to \infty} \ln K + \text{finite terms.} \tag{2.5.32}$$

The vertex function $A(p, q)$ to one-loop order (2.5.31) is logarithmically divergent for $D = 6$. These divergence properties of $\Pi(p^2)$ and $A(p, q)$ are in conformity with observations based on power counting: $d = 2$ for $\Pi(p^2)$ and $d = 0$ for $A(p, q)$.

We shall now give explicit expressions of $\Pi(p^2)$ and $A(p, q)$ to one-loop order by using dimensional regularization and then confirm that the divergences in $\Pi(p^2)$ and $A(p, q)$ can be absorbed into the redefinition of $\phi(x)$, g and m. We first discuss the self-energy part $\Pi(p^2)$. Applying the Feynman parametrization (2.4.32) and using the formula (2.4.46), we derive the following expression for the self-energy part (2.5.29):

$$\Pi(p^2) = \frac{g^2}{2(4\pi)^{D/2}} \Gamma\left(2 - \frac{D}{2}\right) \int_0^1 dx [m^2 - x(1-x)p^2]^{D/2-2} \quad . \tag{2.5.33}$$

We would like to see the behavior of $\Pi(p^2)$ when the space-time dimension D is slightly smaller than 6. For this purpose we define the quantity

$$\varepsilon = \frac{6-D}{2} \quad . \tag{2.5.34}$$

Using the formula

$$\Gamma\left(2-\frac{D}{2}\right) = \Gamma(\varepsilon - 1) = -\frac{1}{\varepsilon} + \gamma - 1 + O(\varepsilon) \quad , \tag{2.5.35}$$

where γ is the Euler constant and expanding the integrand in Eq. (2.5.33) in powers of ε, we obtain

$$\Pi(p^2) = \frac{g_0^2}{2(4\pi)^3}\left[\left(-m^2 + \frac{p^2}{6}\right)\left(\frac{1}{\varepsilon} - \gamma + \ln 4\pi + 1\right) + \int_0^1 dx\{m^2 - x(1-x)p^2\} \ln \frac{m^2 - x(1-x)p^2}{\mu^2}\right] , \tag{2.5.36}$$

where we set $\varepsilon = 0$ in all the terms of $O(\varepsilon)$, the external momentum p is kept in the space-like region $p^2 < 0$ and μ is the parameter with mass dimension,

$$g = g_0 \mu^{\varepsilon} \quad . \tag{2.5.37}$$

By substituting Eq. (2.5.36) into Eq. (2.5.28) we find

$$\Delta^{-1}(p^2) = m^2\left[1 + \frac{g_0^2}{2(4\pi)^3}\left(\frac{1}{\varepsilon} - \gamma + \ln 4\pi + 1\right)\right] - p^2\left[1 + \frac{g_0^2}{12(4\pi)^3}\left(\frac{1}{\varepsilon} - \gamma + \ln 4\pi + 1\right)\right] - \frac{g_0^2}{2(4\pi)^3}\int_0^1 dx[m^2 - x(1-x)p^2] \ln \frac{m^2 - x(1-x)p^2}{\mu^2} \quad . \tag{2.5.38}$$

It is important to note that, in the above expression, the divergences in $D = 6$ exist only in the coefficients of m^2 and p^2. They are not present in any other places. In other words the divergences can be disposed of through the modification of the free propagator

$$\Delta_0(p^2) = \frac{1}{m^2 - p^2} \quad . \tag{2.5.39}$$

This manipulation amounts to a redefinition of the terms $m^2\phi^2$ and $\partial^\mu\phi\partial_\mu\phi$ in the free Lagrangian $\mathscr{L}_0$ given by Eq. (2.5.21). In fact, noting that

$$\int d^4x(\partial^\mu\phi\partial_\mu\phi - m^2\phi^2) = -\int d^4x\phi(\partial^2 + m^2)\phi \quad ,$$

and $\partial^2 + m^2$ is the inverse free propagator in configuration space, we realize that the m^2 and p^2 terms in the free momentum space propagator $\Delta_0(p^2)$ correspond to $m^2\phi^2$ and $\partial^\mu\phi\partial_\mu\phi$, respectively, in the free Lagrangian.

Now the strategy for removing divergences is self-evident: Let us redefine m^2 and ϕ through the relations

$$m_r^2\phi_r^2 = m^2\phi^2(1-A), \quad A = \frac{-g_0^2}{2(4\pi)^3}\frac{1}{\varepsilon} + \text{const.}, \tag{2.5.40}$$

$$\partial^\mu\phi_r\partial_\mu\phi_r = \partial^\mu\phi\partial_\mu\phi(1-B), \quad B = \frac{-g_0^2}{12(4\pi)^3}\frac{1}{\varepsilon} + \text{const.}, \tag{2.5.41}$$

and regard m_r and ϕ_r as finite quantities. It should be noted here that there is an ambiguity in defining A and B in Eqs. (2.5.40) and (2.5.41), i.e., we always have the freedom of adding arbitrary finite constants to the infinite part (the pole term $1/\varepsilon$) in A and B. This ambiguity is nothing but the arbitrariness of choosing the renormalization scheme which was discussed previously in Sec. 2.4.3. The choice of the quantities A and B without finite constants in Eqs. (2.5.40) and (2.5.41) corresponds to the MS scheme [tHo 73] introduced in Sec. 2.4.3.

Inserting Eqs. (2.5.40) and (2.5.41) into the free Lagrangian $\mathcal{L}_0$ and keeping all terms to order g_0^2, we find

$$\mathcal{L}_0 = \frac{1}{2}(\partial^\mu\phi_r\partial_\mu\phi_r - m_r^2\phi_r^2) + \frac{1}{2}(B\partial^\mu\phi_r\partial_\mu\phi_r - m_r^2A\phi_r^2) \quad . \tag{2.5.42}$$

The terms with the infinite constants A and B in Eq. (2.5.42) serve to cancel the divergences which occur at one-loop level and are called *counter terms*. The above method of eliminating the divergences in terms of the counter terms is essentially the same as the multiplicative renormalization given previously in Eq. (2.5.22). In fact we easily see that

$$Z_3 = 1 + B + O(g_0^4), \quad Z_m = 1 + A - B + O(g_0^4) \quad . \tag{2.5.43}$$

Let us now perform the multiplicative renormalization of the propagator $\Delta(p^2)$. For this purpose we expand $\Pi(p^2)$ in a Taylor series around $p^2 = m_r^2$,

$$\Pi(p) = \Pi(m_r^2) + (p^2 - m_r^2)\,\Pi'(m_r^2) - (p^2 - m_r^2)^2\,\Pi_1(p^2) \quad , \tag{2.5.44}$$

where $\Pi_1(p^2)$ designates an aggregate of all the terms beyond the second term in the Taylor series. Substituting Eq. (2.5.44) into Eq. (2.5.28) we obtain

$$\Delta(p^2) = 1/[\kappa + (m_r^2 - p^2)\{1 + \Pi'(m_r^2) + (m_r^2 - p^2)\,\Pi_1(p^2)\}] \quad , \qquad (2.5.45)$$

where κ is defined by

$$\kappa = m^2 - \Pi(m_r^2) - m_r^2 \quad . \qquad (2.5.46)$$

If we adopt the on-shell renormalization scheme, m_r has to be equal to the physical mass of the field ϕ which is a pole of the propagator $\Delta(p^2)$. Hence it satisfies

$$m^2 - m_r^2 - \Pi(m_r^2) = 0 \quad . \qquad (2.5.47)$$

According to Eq. (2.5.47), $\kappa = 0$ in the on-shell renormalization scheme and, through the definition (2.5.22), we find

$$Z_m = 1 + \Pi(m_r^2)/m_r^2 \quad . \qquad (2.5.48)$$

For simplicity we remain in this renormalization scheme for the moment. Define

$$Z_3 = 1/[1 + \Pi'(m_r^2)] \quad , \qquad (2.5.49)$$

and

$$\Pi_r(p^2) = Z_3 \Pi_1(p^2) \quad , \qquad (2.5.50)$$

where $\Pi'(m_r^2) = d\Pi(p^2)/dp^2|_{p^2=m_r^2}$. We then have

$$\Delta(p^2) = \frac{Z_3}{m_r^2 - p^2}\,\frac{1}{1 + (m_r^2 - p^2)\,\Pi_r(p^2)} \quad . \qquad (2.5.51)$$

Thus we recognize that the divergence of $\Delta(p^2)$ is multiplicatively removed by the factor Z_3. The factor Z_3 defined by Eq. (2.5.49) is nothing but that appearing in Eq. (2.5.22) since $\Delta(p^2)$ is a Fourier transform of $-Z[0]^{-1}\delta^2 Z[J]/\delta J(x_1)\delta J(x_2)|_{J=0}$ and so the renormalization factor in $\Delta(p^2)$ is related to that of $J(x)$ (or $\phi(x)$).

We next turn our attention to the calculation of the vertex function given in Eq. (2.5.31). Using the Feynman parametrization (2.4.105) with $n = 3$ and $\alpha_1 = \alpha_2 = \alpha_3 = 1$, we find

$$\Lambda(p, q) = 2g^2 \int_0^1 dx \int_0^{1-x} dy \int \frac{d^D k'}{(2\pi)^D i}\,\frac{1}{(-k'^2 + K)^3} \quad , \qquad (2.5.52)$$

where $k' = k + yp - xq$ and

$$K = m^2 - x(1-x)q^2 - y(1-y)p^2 - 2xyp{\cdot}q. \qquad (2.5.53)$$

We perform the k'-integration in Eq. (2.5.52) by applying the formula (2.4.76) to obtain

$$\Lambda(p,q) = \frac{g_0^2}{(4\pi)^3}\Gamma(\varepsilon)\int_0^1 dx\int_0^{1-x} dy\left(\frac{K}{4\pi\mu^2}\right)^{-\varepsilon} , \qquad (2.5.54)$$

with ε given by Eq. (2.5.34). Making a Laurent expansion of $\Lambda(p,q)$ in ε we have

$$\Lambda(p,q) = \frac{g_0^2}{(4\pi)^3}\left[\frac{1}{2}\left(\frac{1}{\varepsilon} - \gamma + \ln 4\pi\right) - \int_0^1 dx\int_0^{1-x} dy \ln\frac{K}{\mu^2} + O(\varepsilon)\right] \qquad (2.5.55)$$

Inserting Eq. (2.5.55) into Eq. (2.5.30) we finally obtain

$$\Gamma(p,q) = g_0\left[1 + \frac{g_0^2}{2(4\pi)^3}\left(\frac{1}{\varepsilon} - \gamma + \ln 4\pi\right)\right] - \frac{g_0^3}{(4\pi)^3}\int_0^1 dx\int_0^{1-x} dy \ln\frac{K}{\mu^2} , \qquad (2.5.56)$$

where we have neglected the mass dimension factor μ^ε coming from the extra g. We immediately see that the divergence in the vertex function $\Gamma(p,q)$ to one-loop order can be absorbed into the redefinition of the coupling constant g_0. Note here that the pole term $1/\varepsilon$ is accompanied by the finite constant $-\gamma + \ln 4\pi$ which is an artefact of dimensional regularization. This was also the case in Eq. (2.5.37) suggesting the advantage of the $\overline{\text{MS}}$ scheme [Bar 78].

The divergence in the vertex function may be removed by the following redefinition of the coupling constant g_0,

$$g_r\phi_r^3 = g_0\phi^3(1-C), \quad C = \frac{-g_0^2}{2(4\pi)^3}\frac{1}{\varepsilon} + \text{const.} \qquad (2.5.57)$$

The relation between C and Z_g defined in Eq. (2.5.22) is given by[31]

$$Z_g = 1 + C - \frac{3}{2}B + O(g_0^4) . \qquad (2.5.58)$$

The full Lagrangian is now rewritten as

$$\mathscr{L} = \frac{1}{2}(\partial^\mu\phi_r\partial_\mu\phi_r - m_r^2\phi_r^2) + \frac{g_r}{3!}\phi_r^3 + \frac{1}{2}(B\partial^\mu\phi_r\partial_\mu\phi_r - m_r^2 A\phi_r^2) + \frac{g_r}{3!}C\phi_r^3 . \qquad (2.5.59)$$

[31] Traditionally the symbol Z_1 is used to represent vertex renormalization. In our case $Z_1 = 1 + C + O(g_0^4)$. It is related to Z_g by $Z_g = Z_1 Z_3^{-3/2}$.

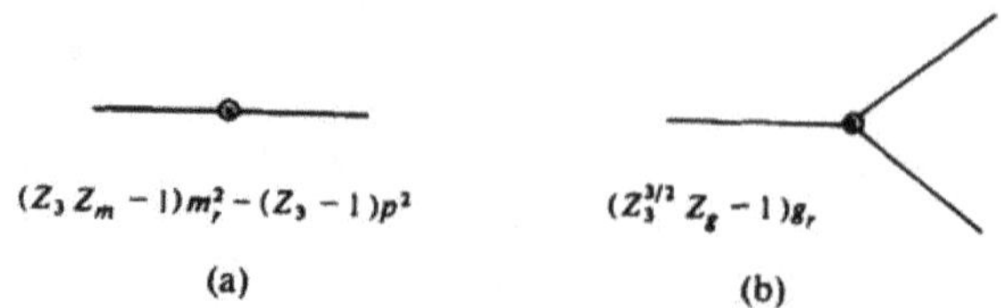

Fig. 2.5.6. Diagrammatic representation of counter terms for the overall divergences of the two- and three-point Green functions.

In terms of the multiplicative renormalization constants Z_3, Z_g and Z_m, the above counter terms are reexpressed in the following way

$$\mathscr{L} = \frac{1}{2}(\partial^\mu \phi_r \partial_\mu \phi_r - m_r^2 \phi_r^2) + \frac{g_r}{3!}\phi_r^3$$
$$+ \frac{1}{2}[(Z_3 - 1)\partial^\mu \phi_r \partial_\mu \phi_r - (Z_3 Z_m - 1)m_r^2 \phi_r^2] + (Z_3^{3/2} Z_g - 1)\frac{g_r}{3!}\phi_r^3 \tag{2.5.60}$$

The counter term method consists of the following procedure. First we rearrange the original full Lagrangian in the form of Eq. (2.5.60) and regard the first term as the free Lagrangian. We then make perturbative expansions with respect to this free Lagrangian where the counter terms are represented by Feynman diagrams shown in Fig. 2.5.6. Note that the calculation is essentially the same as before if we replace the parameters m and g by m_r and g_r respectively. The counter terms are determined in each other so as to cancel the divergences of Green functions under suitable renormalization conditions.

We have seen that the divergent truncated connected Green functions in the one-loop level are made finite by a finite number of subtractions corresponding to the redefinition of the field, coupling constant and mass, and this renormalization process is equivalent to a rearrangement of the Lagrangian into the renormalized free Lagrangian and the interaction part including the counter terms. We would like to show that the above renormalization process works to all orders of perturbation series. Before doing this, however, it is quite useful for us to deal with an example of explicit calculations of the two-loop self-energy part since it displays the overlapping divergence which gives a clue to the all-order proof of renormalizability.

(2) *Two-loop example*

As one of the simplest nontrivial examples of the two-loop renormalization, we present an explicit calculation of the self-energy part $\Pi(p^2)$. The Feynman

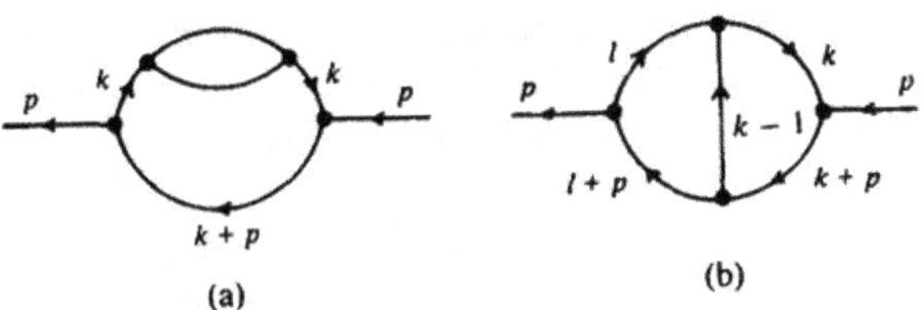

Fig. 2.5.7. Two-loop diagrams for the self-energy part in ϕ_6^3 theory.

diagrams contributing to the two-loop self-energy part are shown in Fig. 2.5.7. In this two-loop example we have a new feature which we have not encountered in the one-loop example discussed previously. The new feature is the appearance of divergences inherent to the subdiagrams of these two-loop diagrams. In the following the divergence emerging from subdiagrams will be called the subdivergence. In Fig. 2.5.7 we have two types of subdivergences which come from the one-loop self-energy part Fig. 2.5.5(a) and the one-loop vertex Fig. 2.5.5(b), respectively. The diagram in Fig. 2.5.7(a) provides us with an example of nested subdiagrams while that in Fig. 2.5.7(b) illustrates subdiagrams overlapping with each other. Here the terms "nested" and "overlapping" are defined as follows: Consider diagrams H_1 and H_2. If H_1 is completely included in H_2 ($H_1 \subset H_2$) as a subdiagram, we say that H_1 is nested in H_2. The diagram H_2 in this case is called the *nested diagram*. If they are not included in each other but have common internal lines and vertices, they are said to overlap. The union of H_1 and H_2 ($H_1 \cup H_2$) is called the *overlapping diagram*. If they are neither nested in each other nor overlapping, they are said to be *disjoint* ($H_1 \cap H_2 = \phi$, where ϕ represents the empty set). Thus nonoverlapping diagrams are either nested or disjoint. In Fig. 2.5.8 we explain the above terminology in a diagrammatic way. The divergence arising from the overlapping subdiagram is called the *overlapping divergence*.

In the following calculations we are only interested in the divergence structure of the amplitude. Since the ultraviolet divergence is caused by the high-momentum region of the Feynman integral, the mass m of the field ϕ

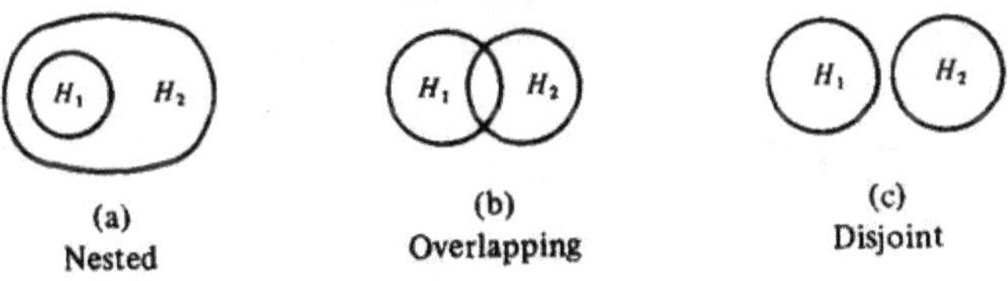

Fig. 2.5.8. Graphical explanation of the concepts of (a) "nested," (b) "overlapping" and (c) "disjoint" graphs.

may be safely neglected for our purposes. We shall consider the massless ϕ_6^3 theory in the following [Mac 74]. Let us first calculate the contribution $\Pi^{(a)}(p^2)$ of Fig. 2.5.7(a) to the self-energy part $\Pi(p^2)$:

$$\Pi^{(a)}(p^2) = \int \frac{d^Dk}{(2\pi)^D i} \frac{-1}{(k^2)^2 (k+p)^2} g^2 \Pi_0(k^2) \quad , \tag{2.5.61}$$

where $\Pi_0(k^2)$ is the one-loop self-energy part given in Eq. (2.5.29) with $m = 0$. The one-loop self-energy part $\Pi_0(k^2)$ has already been calculated for the massive theory as in Eq. (2.5.33). Setting $m = 0$ in Eq. (2.5.33) we have

$$\Pi_0(k^2) = \frac{g^2}{2(4\pi)^{D/2}} \Gamma(2-\tfrac{D}{2})\, B(\tfrac{D}{2}-1, \tfrac{D}{2}-1)\, (-k^2)^{D/2-2} \quad . \tag{2.5.62}$$

It should be noted here that for the massless theory no mass renormalization is needed. We substitute Eq. (2.5.62) into Eq. (2.5.61) and obtain

$$\Pi^{(a)}(p^2) = \frac{-g^4}{2(4\pi)^{D/2}} \Gamma(2-\tfrac{D}{2})\, B(\tfrac{D}{2}-1, \tfrac{D}{2}-1) \times \int \frac{d^Dk}{(2\pi)^D i} \frac{1}{(-k^2)^{4-D/2} (k+p)^2} \quad . \tag{2.5.63}$$

We apply the Feynman parametrization (2.4.105) to the above loop integral and then perform the integrations. After some calculation we find

$$\Pi^{(a)}(p^2) = \frac{-g_0^4}{2(4\pi)^6} p^2 \left(\frac{-p^2}{4\pi\mu^2}\right)^{D-6} \times \frac{\Gamma(2-D/2)\,\Gamma(5-D)\,\Gamma(D/2-1)^3\,\Gamma(D-4)}{\Gamma(4-D/2)\,\Gamma(D-2)\,\Gamma(3D/2-5)} \quad , \tag{2.5.64}$$

where we have employed the dimensionless coupling constant g_0 defined by Eq. (2.5.37). In the above result (2.5.64) we immediately see that there are two sources of the divergence at $D=6$: one from the factor $\Gamma(2-D/2)$ and the other from the factor $\Gamma(5-D)$. The former divergence corresponds to the subdivergence due to the nested one-loop self-energy part $\Pi_0(k^2)$ and the latter represents the two-loop overall divergence. Expanding Eq. (2.5.64) in a Laurent series with respect to $\varepsilon = (6-D)/2$, we obtain

$$\Pi^{(a)}(p^2) = \frac{g_0^4}{(4\pi)^6} p^2 \frac{1}{2(3!)^2} \left[-\frac{1}{2\varepsilon^2} + \frac{1}{\varepsilon}\left(\ln \frac{-p^2}{4\pi\mu^2} + \gamma - \frac{43}{12} \right) + O(\varepsilon^0) \right] \quad , \tag{2.5.65}$$

where we have shown only the divergent parts.

Next we calculate the contribution $\Pi^{(b)}(p^2)$ of Fig. 2.5.7(b) to the self-energy part:

$$\Pi^{(b)}(p^2) = \frac{g^2}{2}\int \frac{d^D l}{(2\pi)^D i}\frac{1}{l^2(l+p)^2}\Lambda_0(p, l) \quad , \tag{2.5.66}$$

where $\Lambda_0(p, l)$ is the one-loop vertex given in Eq. (2.5.31) with $q = l$. We substitute Eq. (2.5.54) with $m = 0$ into Eq. (2.5.66) to obtain

$$\Pi^{(b)}(p^2) = \frac{g^4}{2(4\pi)^{D/2}}\Gamma\left(3-\frac{D}{2}\right)\int_0^1 dx\int_0^{1-x} dy\int \frac{d^D l}{(2\pi)^D i}\frac{1}{l^2(l+p)^2 K^{3-D/2}} \quad , \tag{2.5.67}$$

where K is given by Eq. (2.5.53) with $q = l$ and $m = 0$,

$$K = -x(1-x)l^2 - y(1-y)p^2 - 2xyp\cdot l \quad . \tag{2.5.68}$$

In Eq. (2.5.67) we use the Feynman parametrization (2.4.105) in the following way:

$$\frac{1}{l^2(l+p)^2(K/x(1-x))^{3-D/2}}$$

$$= \frac{\Gamma(5-D/2)}{\Gamma(3-D/2)}\int_0^1 d\xi\int_0^{1-\xi} d\eta \frac{\xi^{2-D/2}}{[-(1-\xi-\eta)l^2 - \eta(l+p)^2 + \xi K/x(1-x)]^{5-D/2}} \,. \tag{2.5.69}$$

Inserting Eq. (2.5.69) into Eq. (2.5.67) and performing the loop integration, we have

$$\Pi^{(b)}(p^2) = \frac{-g_0^4}{2(4\pi)^6}p^2\left(\frac{-p^2}{4\pi\mu^2}\right)^{D-6}\Gamma(5-D)$$

$$\times \int_0^1 du\int_0^1 dv\int_0^1 dx\, x^{2-D/2}(1-x)^{D/2-2}$$

$$\times \int_0^1 d\xi\, \xi^{2-D/2}(1-\xi)\left[\xi u(1-u) + x(1-\xi)\left(v(1-v) + \xi(u-v)^2\right)\right]^{D-5} , \tag{2.5.70}$$

where we made the change of variables $y = (1-x)u$ and $\eta = (1-\xi)v$. It should be noted here that, while the overall divergence is manifest in the factor $\Gamma(5-D)$ in Eq. (2.5.70), the subdivergence is still hidden in the parametric integrals in x and ξ. This feature is characteristic of overlapping divergences.

The Laurent expansion in ε of the parametric integrals in Eq. (2.5.70) is derived by making partial integrations in x and ξ, and the result reads[32]

$$\Pi^{(b)}(p^2) = \frac{-g_0^4}{2(4\pi)^6}p^2\left(\frac{-p^2}{4\pi\mu^2}\right)^{-2\varepsilon}\Gamma(2\varepsilon-1)\frac{1}{3!}\left(\frac{1}{\varepsilon}+4+O(\varepsilon)\right)$$

$$= \frac{g_0^4}{(4\pi)^6}p^2\frac{1}{2\cdot 3!}\left[\frac{1}{2\varepsilon^2}-\frac{1}{\varepsilon}\left(\ln\frac{-p^2}{4\pi\mu^2}+\gamma-3\right)\right.$$

$$\left.+O(\varepsilon^0)\right] . \tag{2.5.71}$$

The full self-energy part $\Pi(p^2)$ up to two loops is given by

$$\Pi(p^2) = \Pi_0(p^2) + \Pi^{(a)}(p^2) + \Pi^{(b)}(p^2) . \tag{2.5.72}$$

In the expressions (2.5.65) and (2.5.71) we recognize that the double-pole term $(1/\varepsilon^2)$ has the same structure as the counter term which is proportional to p^2 for the massless case while the simple-pole term $(1/\varepsilon)$ carries the coefficient $\ln(-p^2/4\pi\mu^2)+\gamma$ which is foreign to the counter term. Thus the double-pole terms are apparently renormalizable by direct subtraction according to the counter term method. In order to show the renormalizability of the simple-pole terms, we need to consider the subtraction of subdivergences (divergences of subdiagrams).

Let us now see whether the subdivergence in Eq. (2.5.72) is successfully subtracted by means of the counter term method. To determine one-loop counter terms, we isolate the divergences in $\Pi_0(p^2)$ and $\Lambda_0(p,q)$,

$$\Pi_0(p^2) = p^2\frac{g_0^2}{(4\pi)^3}\frac{1}{2\cdot 3!}\left(\frac{1}{\varepsilon}-\ln\frac{-p^2}{4\pi\mu^2}-\gamma+\frac{8}{3}\right) , \tag{2.5.73}$$

$$\Lambda_0(p,q) = \frac{g_0^2}{(4\pi)^3}\frac{1}{2}\left(\frac{1}{\varepsilon}+O(\varepsilon^0)\right) . \tag{2.5.74}$$

For simplicity we adopt the MS scheme ['t Ho 73] to determine the counter terms. According to Eqs. (2.5.73) and (2.5.74) we obtain the following one-loop counter terms in the MS scheme,

[32] The integration in Eq. (2.5.66) can be performed analytically by using a method based on the Gegenbauer polynomial expansion in configuration space [Che 79, 80],

$$\Pi^{(b)}(p^2) = \frac{-g^4}{(4\pi)^D}\frac{(-p^2)^{D-5}}{4}\frac{\Gamma(5-D)\Gamma(D/2-2)^3}{\Gamma(3D/2-5)}$$

$$\times\left[\frac{2}{3-D}+B\left(3-D,\frac{3D}{2}-5\right)-B\left(3-D,3-\frac{D}{2}\right)\right] .$$

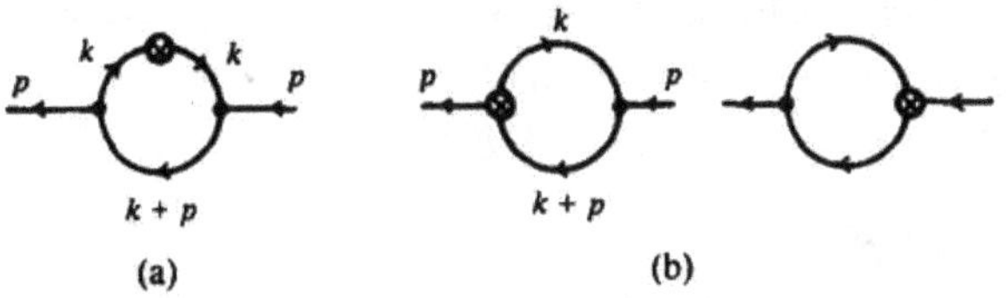

Fig. 2.5.9. Diagrammatic representation of counter terms for subdivergences of (a) the self-energy type and (b) the vertex type.

$$B_0 p^2 = \text{———}\otimes\text{———} = \frac{-g_0^2}{(4\pi)^3}\frac{p^2}{2\cdot 3!}\frac{1}{\varepsilon}\,, \tag{2.5.75}$$

$$C_0 = \text{———}\otimes\!\!< \; = \frac{-g_0^2}{(4\pi)^3}\frac{1}{2\varepsilon}\,. \tag{2.5.76}$$

We first consider the subdivergence in $\Pi^{(a)}(p^2)$. This subdivergence may be subtracted out by taking into account the contribution of Fig. 2.5.9(a) which includes the counter term (2.5.75) as a part. This contribution $\Pi^{(a)}_{CS}(p^2)$ is readily calculated to be

$$\begin{aligned}\Pi^{(a)}_{CS}(p^2) &= g^2\int\frac{d^Dk}{(2\pi)^D i}\frac{-B_0k^2}{(k^2)^2(k+p)^2} = -2B_0\Pi_0(p^2)\\ &= \frac{g_0^4}{(4\pi)^6}p^2\frac{1}{2(3!)^2}\left[\frac{1}{\varepsilon^2}-\frac{1}{\varepsilon}\left(\ln\frac{-p^2}{4\pi\mu^2}+\gamma-\frac{8}{3}\right)\right]\,.\end{aligned} \tag{2.5.77}$$

By adding Eq. (2.5.77) to Eq. (2.5.65) we see that the term $[\ln(-p^2/4\pi\mu^2) + \gamma]/\varepsilon$ cancels out, i.e.,

$$\Pi^{(a)}(p^2) + \Pi^{(a)}_{CS}(p^2) = \frac{g_0^4}{(4\pi)^6}p^2\frac{1}{4(3)!^2}\left[\frac{1}{\varepsilon^2}-\frac{11}{6\varepsilon}+O(\varepsilon^0)\right] \tag{2.5.78}$$

guaranteeing the renormalizability of the subdivergence. We are now left only with the overall divergence in Eq. (2.5.78) and the corresponding counter term of $O(g_0^4)$ is

$$B^{(a)}p^2 = \frac{g_0^4}{(4\pi)^6}p^2\frac{1}{4(3!)^2}\left(-\frac{1}{\varepsilon^2}+\frac{11}{6\varepsilon}\right)\,. \tag{2.5.79}$$

We next discuss the subdivergences in $\Pi^{(b)}(p^2)$. These are the overlapping subdivergences and may be subtracted out by considering the diagrams of Fig. 2.5.9(b) which have the counter term (2.5.76) as a part of them. Note here that we need two subtractions corresponding to the two different vertex parts of the diagram (Fig. 2.5.7(b)). The reason for this will become clear in Sec. 2.5.4. The calculation of the contribution of these diagrams to the self-energy part, $\Pi_{CS}^{(b)}(p^2)$, is straightforward, resulting in

$$\begin{aligned}\Pi_{CS}^{(b)}(p^2) &= 2C_0\Pi_0(p^2)\\ &= \frac{g_0^4}{(4\pi)^6}p^2\frac{1}{2\cdot 3!}\left[-\frac{1}{\varepsilon^2}+\frac{1}{\varepsilon}\left(\ln\frac{-p^2}{4\pi\mu^2}+\gamma-\frac{8}{3}\right)\right] . \end{aligned} \tag{2.5.80}$$

Adding Eq. (2.5.80) to Eq. (2.5.71) we again see that the nonlocal divergence $[\ln(-p^2/4\pi\mu^2)+\gamma]/\varepsilon$ cancels out. Here we employed the term "nonlocal," since $\ln(-p^2)$ is not a polynomial of p^2 and exhibits the nonlocal effect in configuration space. We are again left only with the overall divergence in the following combination,

$$\Pi^{(b)}(p^2)+\Pi_{CS}^{(b)}(p^2)=\frac{g_0^4}{(4\pi)^6}p^2\frac{1}{2\cdot 3!}\left[-\frac{1}{2\varepsilon^2}+\frac{1}{3\varepsilon}+O(\varepsilon^0)\right] . \tag{2.5.81}$$

Thus the counter term for the overall divergence in the diagram Fig. 2.5.7(b) is given by

$$B^{(b)}p^2=\frac{g_0^4}{(4\pi)^6}p^2\frac{1}{2\cdot 3!}\left(\frac{1}{2\varepsilon^2}-\frac{1}{3\varepsilon}\right) . \tag{2.5.82}$$

Summing up all contributions, we obtain the counter term Bp^2 to two-loop order as follows,

$$\begin{aligned}Bp^2 &= (B_0+B^{(a)}+B^{(b)})p^2\\ &= p^2\left[-\frac{g_0^2}{(4\pi)^3}\frac{1}{2\cdot 3!}\frac{1}{\varepsilon}+\frac{g_0^4}{(4\pi)^6}\frac{1}{4(3!)^2}\left(\frac{5}{\varepsilon^2}-\frac{13}{6\varepsilon}\right)\right] .\end{aligned} \tag{2.5.83}$$

(3) *Renormalization to all orders*

We have seen that our renormalization program works in one-loop order and in the self-energy part at the two-loop level. Now we would like to extend the proof of renormalizability to all orders. Although it is well-known that the BPHZ method provides us with one of the most elegant and satisfactory way of proving renormalizability, we start here with explaining Dyson's original method [Dys 49, 49a] since it clarifies the historical reason why overlapping divergences spoiled the proof of renormalizability and the way the difficulty was circumvented. The details of the BPHZ method will be described in the next subsection.

In order to show that our renormalization program works to all orders, it is natural for us to use induction with respect to powers of g of the Feynman amplitudes. We expand the full self-energy part $\Pi(p^2)$ and the full vertex part $\Gamma(p, q)$ in powers of g,

$$\Pi(p^2) = g^2\pi_2(p^2) + g^4\pi_4(p^2) + \dots, \tag{2.5.84}$$

$$\Gamma(p, q) = g[1 + g^2\gamma_2(p, q) + g^4\gamma_4(p, q) + \dots] \quad , \tag{2.5.85}$$

where by the term "full" we mean that all the higher-order corrections are included in the above amplitudes. We have already shown that our renormalization program works for $g^2\pi_2(p^2)$, $g^2\gamma_2(p, q)$ and $g^4\pi_4(p^2)$. Here we assume that the renormalization program works to order g^{2n} and try to show that it also works to order g^{2n+2}. For this purpose it is convenient to use the Dyson-Schwinger equation which was originally derived in quantum electrodynamics [Dys 49a, Sch 51a]. The Dyson-Schwinger equations in ϕ_6^3 theory are given by

$$\Pi(p^2) = \frac{g}{2}\int \frac{d^Dk}{(2\pi)^D i}\Delta(k^2)\,\Delta((k+p)^2)\,\Gamma(p, k) \quad , \tag{2.5.86}$$

$$\Gamma(p, q) = g + \int \frac{d^Dk}{(2\pi)^D i}\Gamma(k+p, k-q)$$
$$\times\, \Delta((k-q)^2)\,\Gamma(q, k-q)\,\Delta(k^2)\,\Gamma(p, k)\,\Delta((k+p)^2) + \dots, \tag{2.5.87}$$

where the full propagator $\Delta(k^2)$ is defined by Eq. (2.5.28). Diagrammatically the above equations are represented as in Figs. 2.5.10(a) and (b) respectively.

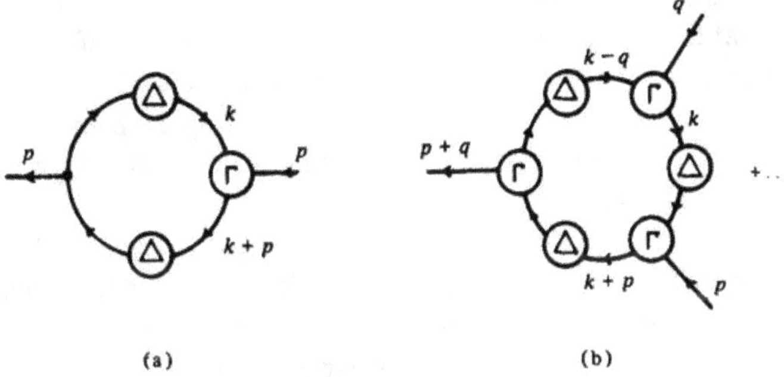

Fig. 2.5.10. Diagrams representing the right-hand sides of the Dyson-Schwinger equations for (a) the self-energy part and (b) the vertex part.

We insert Eqs. (2.5.84) and (2.5.85) into Eqs. (2.5.86) and (2.5.87) through the use of Eq. (2.5.28), and obtain relations expressing $\Pi(p^2)$ and $\Gamma(p, q)$ of the $(n+2)$-th order in terms of $\Pi(p^2)$ and $\Gamma(p, q)$ up to n-th order. We first assume that all the divergences in $\Pi(p^2)$ and $\Gamma(p, q)$ to n-th order are removed by means of our renormalization program. We then use the above relations to show that the divergences in $\Pi(p^2)$ and $\Gamma(p, q)$ of $(n+2)$-th order can be subtracted out by the use of our counter terms. At first sight one may get the impression that the proof seems to be completed along this line. Unfortunately, however, the proof is found to be unsatisfactory in the sense that overlapping divergences are not properly taken care of in the method under consideration. In fact we easily find a simple counter example in the renormalization of the two-loop self-energy part discussed previously. Consider the contribution $\Pi^{(b)}(p^2)$ coming from the diagram of Fig. 2.5.7(b). It was defined by Eq. (2.5.66) and was found to exhibit an overlapping divergence. In order to properly subtract the overlapping subdivergence in $\Pi^{(b)}(p^2)$, we needed two contributions of the vertex counter term as shown in Fig. 2.5.9(b). However, according to the method based on the Dyson-Schwinger equation (2.5.86), we make the subtraction only at one of the corners in the diagram of Fig. 2.5.7(b). Thus, in so far as we rely on the Dyson-Schwinger equation, the overlapping divergence cannot be removed properly, obstructing our all-order proof of renormalizability. This difficulty is known as the overlap problem [Sal 51].

The lesson which we have learned from the above difficulty is that we had better find a systematic way of proving renormalizability dealing only with nonoverlapping (i.e., nested or disjoint) subdiagrams. There are several ways of doing this. In the case of quantum electrodynamics (QED), according to the gauge symmetry, the self-energy part of electrons is related to the vertex part [War 50, Tak 57]. Hence the argument of renormalizing the self-energy part of electrons can be completely replaced by that of the vertex part. It can be shown in QED that the overlapping divergence occurs only in the self-energy part [Bj 65]. (The same is true in ϕ_6^3 theory.) Thus we can get rid of the overlap problem in the electron self-energy part [War 51]. The overlap problem in the photon self-energy part (vacuum polarization) is also circumvented by a similar but slightly different argument.

A more general and rigorous way of solving the overlap problem and of giving the proof of renormalizability for a wide class of quantum field theories is the so-called BPHZ method in which subtractions of subdivergences are performed systematically referring only to nonoverlapping diagrams (without any recourse to overlapping divergences). The method was originally developed by Bogoliubov and Parasiuk [Bog 57, 80] and completed by Hepp [Hep 66]. It was reformulated by Zimmermann in a more transparent way

[Zim 69, 71]. We shall complete the proof of renormalizability of ϕ_6^3 theory in the next subsection.

Digression

POWER DIVERGENCE AND DIMENSIONAL REGULARIZATION: If the divergence in a Feynman integral $F(p^2)$ in one-loop order is of degree n, its divergence structure in terms of the cut-off momentum K is given by

$$F(p^2) \sim \lim_{K\to\infty} (K^n + K^{n-2} + \ldots + \ln K) + \text{finite terms}. \tag{2.5.88}$$

For example the one-loop self-energy part $\Pi_0(p^2)$ in ϕ_6^3 theory is quadratically divergent and has the divergence structure shown in Eq. (2.5.32). It should be noted that in any power-divergent Feynman integral the logarithmic divergence is included as a nonleading term.

The one-loop self-energy part $\Pi_0(p^2)$ for $m = 0$ is given in the dimensional regularization method by Eq. (2.5.62) and its Laurent expansion around $D = 6$ ($\varepsilon = 0$) has the form (2.5.73). In this expression we have only a simple pole $1/\varepsilon$ which corresponds to the logarithmic divergence, and so one might wonder where the quadratic divergence has gone.

To see this point more carefully, we go back to Eq. (2.5.62), which holds for arbitrary dimension D. We find that there is a pole at $D = 4$ in addition to the one at $D = 6$ in the factor $\Gamma(2 - D/2)$. This pole at $D = 4$ is a vestige of the quadratic divergence that existed in an expression like Eq. (2.5.88) in terms of the cut-off parameter K. In general only the logarithmic divergence shows up explicitly as a pole in the space-time dimension for power-divergent Feynman integrals if dimensional regularization is employed. A trace of the power divergence is always found as a lower dimensional pole.

Exercises

1. Derive Eq. (2.5.71) from Eq. (2.5.70). Note that the integral in Eq. (2.5.70) can be written as

$$\int_0^1 du \int_0^1 dv \int_0^1 dx \int_0^1 d\xi [x^{\varepsilon-1}(1-x)^{1-\varepsilon}\xi^{\varepsilon}(1-\xi)u(1-u) + x^{\varepsilon}(1-x)^{1-\varepsilon}\xi^{\varepsilon-1}(1-\xi)^2 v(1-v)$$
$$+ x^{\varepsilon}(1-x)^{1-\varepsilon}\xi^{\varepsilon}(1-\xi)^2(u-v)^2]\,[\xi u(1-u)$$
$$+ x(1-\xi)(v(1-v)+\xi(u-v)^2)]^{-2\varepsilon},$$

and the pole in ε comes from terms with factors $x^{\varepsilon-1}$ and $\xi^{\varepsilon-1}$.

2. Perform the integrations in Eq. (2.5.70) to obtain the exact result shown in footnote 32.

2.5.4. BPHZ method in ϕ^3 theory

The BPHZ method offers a systematic way of subtracting an overall divergence as well as subdivergences in any Feynman integral. In the present subsection we shall explain the BPHZ method in ϕ_6^3 theory and complete the proof of renormalizability to all orders. We shall first show that the Feynman integral after the BPHZ subtractions is finite and then prove inductively that the subtractions are local, i.e., they correspond to the local counter terms which have the same form as those present in the original Lagrangian.

As mentioned in the last subsection, the BPHZ method was developed first by Bogoliubov and Parasiuk [Bog 57, 80] who derived the recursion equations for the subtractions of divergences in Feynman integrals. The method was completed by Hepp [Hep 66]. Zimmermann [Zim 69, 71] gave a solution to the recursion equations and reformulated the method.

Here we shall not follow this historical order in presenting our argument. We rather derive Zimmerman's solution (the forest formula) directly by using the method developed in [Zav 65, App 69, Ani 73, Ber 74][33] and show the finiteness of the subtracted Feynman integral. We then show that the recursion equation holds according to the forest formula. The recursion equation is useful to prove the locality of the subtractions.

(1) *Forest formula*

We start with defining an operation t^H to pick out a divergent part of the Feynman integral F_H corresponding to diagram H. Here the Feynman integral F_H is assumed to have only an overall divergence, i.e., there is no subdivergence in F_H. In ϕ_6^3 theory F_H may be expanded in the following Laurent series with respect to the expansion parameter $\varepsilon = (6-D)/2$, provided that dimensional regularization is employed:

$$F_H = \sum_{n=-N}^{\infty} a_n \varepsilon^n, \tag{2.5.89}$$

where N is the positive integer less than or equal to the number of loops in H. The expansion of the type (2.5.89) is guaranteed by the fact that only singularities in F_H with respect to the variable ε are isolated poles (i.e., there exists no term like $\ln\varepsilon$) and the order of poles is less than or equal to the number of loops ['t Ho 74, Spe 76]. We define operation t^H associated with the diagram H in the following way,

$$t^H F_H = \sum_{n=-N}^{-1} a_n \varepsilon^n \equiv F_H^{\text{div}} \quad . \tag{2.5.90}$$

Here F_H^{div} is a divergent part of F_H as $\varepsilon \to 0$. Accordingly the finite part F_H^{fin} of F_H is given by

$$F_H^{\text{fin}} = F_H - F_H^{\text{div}} = (1-t^H)F_H$$
$$= \sum_{n=0}^{\infty} a_n \varepsilon^n \to a_0 \text{ (as } \varepsilon \to 0). \tag{2.5.91}$$

[33] A nice review is given in [Nak 75].

Thus $1-t^H$ is the operation to extract the finite piece from F_H. Of course there is an ambiguity in removing the divergent part as mentioned earlier in Sec. 2.4.3. The prescription we are using now is nothing but the MS scheme ['t Ho 73]. Reflecting this situation we call $1-t^H$ the minimal subtraction operator. By definition, $t^H F_H = 0$ if F_H is convergent. It should be remarked here that the operation t^H was originally introduced by using the method of the Taylor expansion of F_H with respect to external momenta for the diagram H [Bog 57, 80, Hep 66, Zim 69, 71]. Throughout the present book we would like to use dimensional regularization and adopt the MS (or $\overline{\text{MS}}$) renormalization scheme. Hence we prefer to employ the definition (2.5.90) for t^H [Spe 74, Col 75, Bre 77, Cas 82].

Consider the Feynman diagram G and let H be a subdiagram of G. The diagram H may develop divergences which correspond to subdivergences in the Feynman integral F_G for the diagram G. If H is improper (one-particle reducible), the divergences in F_H are split into those for proper (one-particle irreducible) parts of H. Hence it is sufficient to take only proper subdiagrams as H. A superficially divergent ($d_H \geq 0$) proper subdiagram H of the diagram G is called the *renormalization part* [Zim 69, 71] where d_H is the superficial degree of divergence for the diagram H.

Let us apply operations $1-t^H$ to F_G with respect to all the renormalization parts H in G, i.e.,

$$\bar{R}_G F_G \equiv [\prod_{H \in \Phi} (1-t^H)]\, F_G \quad , \tag{2.5.92}$$

where Φ is a set composed of all the renormalization parts H in G; $\Phi = \{H|H \subset G,\ H = \text{proper},\ d_H \geq 0\}$. In Eq. (2.5.92) the product of operators $1 - t^H$ should be so arranged that $1 - t^H$ comes to the right of $1 - t^{H'}$ if $H \subset H'$ (nested). If H and H' are disjoint or overlapping, their order is irrelevant. We make a further subtraction $1 - t^G$ in Eq. (2.5.92) for the possible overall divergence in F_G,

$$R_G F_G \equiv (1-t^G)\, \bar{R}_G F_G = (1-t^G)\, [\prod_{H \in \Phi} (1-t^H)] F_G \quad . \tag{2.5.93}$$

If there is no overall divergence in F_G, then $t^G \bar{R}_G F_G = 0$. Here the operation

$$R_G = (1 - t^G) \prod_{H \in \Phi} (1 - t^H) \quad , \tag{2.5.94}$$

corresponds to the so-called R-operation in Bogoliubov's terminology [Bog 57, 80]. According to the power counting theorem discussed in Sec. 2.5.1, $R_G F_G$ as given by Eq. (2.5.93) is finite because the operations $1 - t^G$ and $1 - t^H$ render the superficial degrees of divergence d_G and d_H negative. Thus $R_G F_G$ gives the renormalized Feynman integral in the MS scheme.

In Eq. (2.5.93) the set Φ contains all the renormalization parts of G which are divided into two categories, nonoverlapping (nested or disjoint) and overlapping. We shall show that overlapping subdiagrams of G can be discarded in the set Φ so that Φ is composed only of nonoverlapping renormalization parts of G. For this purpose we need the following lemma:

Lemma: If two renormalization parts H_1 and H_2 of G overlap with each other, we have

$$(1-t^{H_{12}})t^{H_1}t^{H_2}=0 \quad , \tag{2.5.95}$$

for the suitable renormalization part H_{12} of G which includes both H_1 and H_2 as subdiagrams. The property (2.5.95) can be generalized for the product of an arbitrary number of t^{H_i} ($i = 1, 2, 3, \ldots$) if the H_i's are overlapping.

The proof of this lemma based on the general grounds may be found, for example, in Appendix III of [Ber 74]. Here we shall be content with giving an illustration in the two-loop self-energy part of Fig. 2.5.7(b). Two overlapping renormalization parts H_1 and H_2 of the whole diagram G are shown in Fig. 2.5.11.

In the present case, H_{12} is taken to be the whole diagram G. According to Eqs. (2.5.66) and (2.5.74) we obtain

$$t^{H_2}F_G=\frac{g_0^2}{2}\int\frac{d^Dl}{(2\pi)^Di}\frac{1}{l^2(l+p)^2}\frac{g_0^2}{(4\pi)^3}\frac{1}{2\varepsilon} \quad . \tag{2.5.96}$$

Hence we have

$$t^{H_1}t^{H_2}F_G=p^2\frac{g_0^4}{(4\pi)^6}\frac{1}{4\cdot 3!}\frac{1}{\varepsilon^2} \quad . \tag{2.5.97}$$

This is exactly equal to the $1/\varepsilon^2$ part of the divergence of F_G as seen in Eq. (2.5.71) and

$$(1-t^G)\,t^{H_1}t^{H_2}F_G=0 \quad . \tag{2.5.98}$$

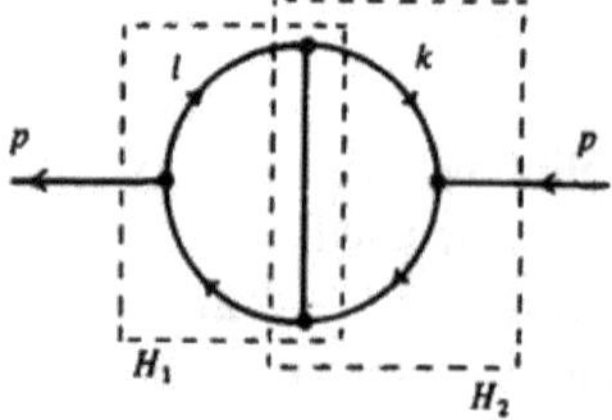

Fig. 2.5.11. An example of a pair of overlapping renormalization parts H_1 and H_2 in the two-loop self-energy part. The portions of the diagram surrounded by dotted lines are the renormalization parts H_1 and H_2, respectively.

Thus we have shown an example of the above lemma.

Now we would like to apply our lemma (2.5.95) to Eq. (2.5.93). Expanding Eq. (2.5.93) we have

$$R_G F_G = (1 - t^G) \sum_i \prod_{H \in \Phi_i} (-t^H) F_G \quad , \tag{2.5.99}$$

where the Φ_is ($i = 0, 1, 2, \ldots$) constitute all possible subsets of Φ and the union of them is equal to Φ. Note that among the Φ_i's we have to include $\Phi_0 = \{\phi\}$ with $\{\phi\}$ an empty set. It can be proven [Ber 74] that for any set of overlapping renormalization parts in G we may always arrange a suitable renormalization part which contains these renormalization parts and corresponds to H_{12} in Eq. (2.5.95). Hence, owing to the lemma, we may drop all t^H in Eq. (2.5.99) if H overlaps with other renormalization parts in Φ_i. Accordingly the sets Φ_i in Eq. (2.5.99) are restricted to all the possible subsets of

$$\Phi \equiv \{H | H \subset G, H = \text{proper, nonoverlapping}, d_H \geq 0\} \quad . \tag{2.5.100}$$

Any set Φ_i consisting of only nonoverlapping renormalization parts of diagram G is called the *forest* [Zim 69, 71]. For example the following three sets are the forest in the case of Fig. 2.5.11,

$$\Phi_0 = \{\phi\}, \quad \Phi_1 = \{H_1\}, \quad \Phi_2 = \{H_2\}, \tag{2.5.101}$$

and the set $\{H_1, H_2\}$ is not the forest since H_1 overlaps with H_2. The formula (2.5.99) with the forests Φ_i was first derived by Zimmermann [Zim 69] and is called the *forest formula*. In the case of Fig. 2.5.11 the forest formula reads

$$R_G F_G = (1 - t^G)(1 - t^{H_1} - t^{H_2}) F_G \quad , \tag{2.5.102}$$

which exactly corresponds to the procedure employed in Sec. 2.5.3 to remove the overlapping divergences in $\Pi^{(b)}(p^2)$ of Fig. 2.5.7(b).

(2) *Recursion equations*

We have seen that the Feynman integral F_G for diagram G may be renormalized in the MS scheme through the use of the forest formula and the renormalized Feynman integral $R_G F_G$ is absolutely convergent thanks to the power counting theorem. We, however, have not shown that the above renormalization process is local, i.e., the process is equivalent to the method based on the subtractions of local counter terms. The forest formula is not convenient for this purpose since the forests are defined without any recourse to the order of perturbation series and so the relation of the formula to the order-by-order subtraction procedure is rather obscure. In the following we shall derive recursion equations for the R-operation starting from the forest formula.

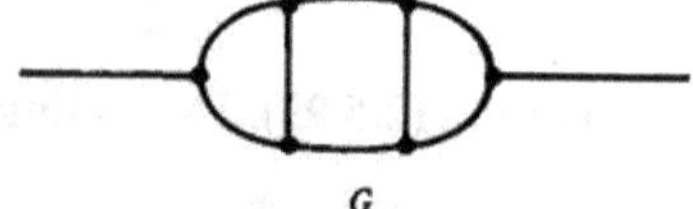

Fig. 2.5.12. A typical diagram contributing to the three-loop self-energy part.

Here again we use a typical example to explain the derivation of the equations. We take the example of the three-loop self-energy part G shown in Fig. 2.5.12. For this diagram G we have 8 forests $\Phi_0, \Phi_1, \ldots, \Phi_7$ with $\Phi_0 = \{\phi\}$ (empty) as defined in Fig. 2.5.13. According to the forest formula (2.5.99), we find

$$R_G F_G = (1 - t^G)\, \bar{R}_G F_G \quad , \tag{2.5.103}$$

$$\begin{aligned} \bar{R}_G F_G &= \sum_{i=0}^{7} \prod_{H \in \Phi_i} (-t^H) F_G \quad , \\ &= [1 + (-t^{H_1}) + (-t^{H_2}) + (-t^{H_3}) + (-t^{H_4}) + (-t^{H_1})(-t^{H_2}) \\ &\quad + (-t^{H_3})(-t^{H_1}) + (-t^{H_4})(-t^{H_2})] F_G , \end{aligned} \tag{2.5.104}$$

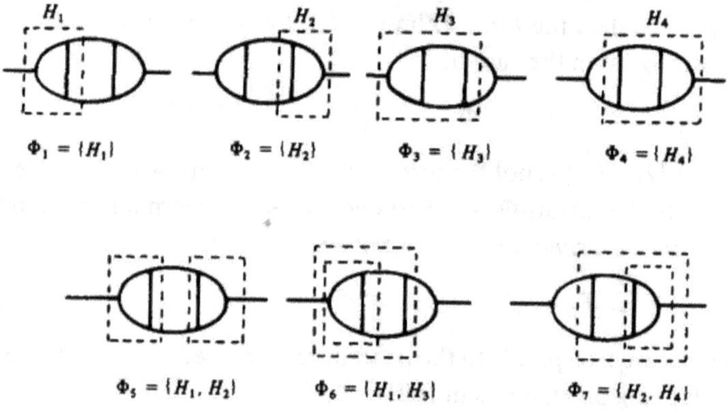

H_1 H_2 H_3 H_4

$\Phi_1 = \{H_1\}$ $\Phi_2 = \{H_2\}$ $\Phi_3 = \{H_3\}$ $\Phi_4 = \{H_4\}$

$\Phi_5 = \{H_1, H_2\}$ $\Phi_6 = \{H_1, H_3\}$ $\Phi_7 = \{H_2, H_4\}$

Fig. 2.5.13. The seven nontrivial forests $\Phi_1, \Phi_2, \ldots, \Phi_7$ in the diagram of Fig. 2.5.12, where the trivial one Φ_0 is omitted.

where $\bar{R}_G$ is the operation to subtract only subdivergences. It should be noted here that the product of operators is arranged in such a way that $(-t^H)$ comes to the right of $(-t^{H'})$ if H is nested in H' ($H \subset H'$). The operation $(-t^H)$ is equivalent to introducing the local counter term corresponding to the diagram H. For example the operation $(-t^{H_1})$ for the subdiagram H_1 defined in Fig. 2.5.13 generates the counter term as shown below,

$$(-t^{H_1})\, F_G \quad \longleftrightarrow \tag{2.5.105}$$

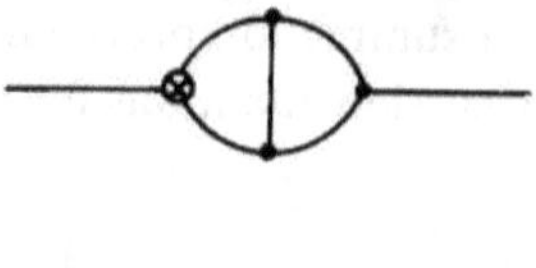

In other words $(-t^{H_1})$ has the effect of crushing the diagram H_1 to a point and multiplying the counter term to the resulting Feynman integral. This process may be expressed in the following way,

$$(-t^{H_1})\,F_G = F_{G/H_1}\,(-t^{H_1}\,F_{H_1}) \quad ; \tag{2.5.106}$$

where G/H_1 represents the diagram obtained from G by reducing H_1 to a point in G. With this notation we rewrite Eq. (2.5.104) in the form

$$\begin{aligned} \bar{R}_G F_G = \; & F_G + F_{G/H_1}(-t^{H_1}F_{H_1}) + F_{G/H_2}(-t^{H_2}F_{H_2}) \\ & + F_{G/H_3}(-t^{H_3}F_{H_3}) + F_{G/H_4}(-t^{H_4}F_{H_4}) \\ & + F_{G/\{H_1,H_2\}}(-t^{H_1}F_{H_1})(-t^{H_2}F_{H_2}) \\ & + F_{G/H_3}(-t^{H_3}(F_{H_3/H_1}(-t^{H_1}F_{H_1}))) \\ & + F_{G/H_4}(-t^{H_4}(F_{H_4/H_2}(-t^{H_2}F_{H_2}))) \quad , \end{aligned} \tag{2.5.107}$$

where $G/\{H_1, H_2\}$ is the diagram obtained from G by reducing both H_1 and H_2 to points. A crucial observation here is that, if the forest formula is applied to the nested diagram H_3 of Fig. 2.5.13, one finds

$$\begin{aligned} \bar{R}_{H_3}F_{H_3} &= F_{H_3} + (-t^{H_1})\,F_{H_3} \quad , \\ &= F_{H_3} + F_{H_3/H_1}(-t^{H_1}F_{H_1}) \quad . \end{aligned} \tag{2.5.108}$$

The same formula as Eq. (2.5.108) holds for the nested diagram H_4 as well. On the other hand, for the diagram H_1 we have

$$\bar{R}_{H_1}F_{H_1} = F_{H_1} \quad , \tag{2.5.109}$$

since there is no more renormalization part in H_1. The same is true for the diagram H_2. Using Eqs. (2.5.108) and (2.5.109) and similar ones for diagrams H_4 and H_2, we re-express Eq. (2.5.107):

$$\begin{aligned} \bar{R}_G F_G = \; & F_G + F_{G/H_1}(-t^{H_1}\bar{R}_{H_1}F_{H_1}) + F_{G/H_2}(-t^{H_2}\bar{R}_{H_2}F_{H_2}) \\ & + F_{G/H_3}(-t^{H_3}\bar{R}_{H_3}F_{H_3}) + F_{G/H_4}(-t^{H_4}\bar{R}_{H_4}F_{H_4}) \\ & + F_{G/\{H_1,H_2\}}(-t^{H_1}\bar{R}_{H_1}F_{H_1})(-t^{H_2}\bar{R}_{H_2}F_{H_2}) \quad . \end{aligned} \tag{2.5.110}$$

Equation (2.5.110) can be written in a more compact form

$$\begin{aligned} \bar{R}_G F_G = \; & (1 - t^{H_1}\bar{R}_{H_1} - t^{H_2}\bar{R}_{H_2} \\ & - t^{H_3}\bar{R}_{H_3} - t^{H_4}\bar{R}_{H_4} + t^{H_1}\bar{R}_{H_1}\,t^{H_2}\bar{R}_{H_2})\,F_G \quad . \end{aligned} \tag{2.5.111}$$

This equation together with Eq. (2.5.108), etc. constitutes a system of the recursive equations for $\bar{R}_H$ operations for diagram G of Fig. 2.5.12. Note here that only disjoint diagrams participate in each term in Eq. (2.5.111).

The recursion equations (2.5.110) and (2.5.111) can be generalized to any Feynman diagram [Bog 57, 80]. (See also [Man 83, Col 84].) The recursion equation for Feynman diagram G reads

$$\bar{R}_G F_G = \sum_{\psi} F_{G/\psi} \prod_{H\in\psi} (-t^H \bar{R}_H F_H) \quad ,$$

$$= \sum_{\psi} \prod_{H\in\psi} (-t^H \bar{R}_H) F_G \quad , \tag{2.5.112}$$

where ψ is a set of disjoint renormalization parts of G (ψ may be empty),

$$\psi = \{H | H \subset G,\ H = \text{proper, disjoint},\ d_H \geq 0\} \quad , \tag{2.5.113}$$

and G/ψ represents the diagram obtained from G by crushing all H in ψ to a point. It is very important to observe that only the set of disjoint renormalization parts is needed in the above recursion equation. This situation should be contrasted with that of the forest formula where it was necessary to consider nested as well as disjoint renormalization parts.

As we have shown in Sec. 2.5.3, ϕ_6^3 theory is renormalizable in one-loop order. Hence in Eq. (2.5.112) the counter terms are properly provided in the one-loop renormalization parts. Thus the locality in the one-loop renormalization part is self-evident. We then consider the two-loop renormalization parts. Using the recursion equations for two-loop subdiagrams $H(2)$,

$$\bar{R}_{H(2)} F_{H(2)} = \sum_{\psi_1} F_{H(2)/\psi_1} \prod_{H(1)\in\psi_1} (-t^{H(1)} F_{H(1)}) \quad , \tag{2.5.114}$$

we can eliminate all the one-loop subdivergences in the two-loop subdiagrams, where ψ_1 is a set of one-loop renormalization parts $H(1)$. Hence the locality of the two-loop renormalization parts is proven. We simply repeat this process until we remove all the subdivergences in F_G, and then we finally apply $1 - t^G$ to subtract the possible overall divergence. The above procedure apparently completes the proof of the equivalence of the BPHZ method to the ordinary counter term method. To be more rigorous we may apply the method of induction to the above proof.

Summing up, the proof of renormalizability in ϕ_6^3 theory is given in the following two steps:

(1) Proof of finiteness. The forest formula provides us with an explicit expression for the renormalized Feynman integral $R_G F_G$ which is shown to be finite according to the power counting theorem.

(2) Proof of locality. The procedure of (1) to obtain renormalized Feynman integrals is proven to be equivalent to the ordinary counter term method. The proof essentially relies on the recursion equations for the R-operation.

Finally it should be stressed that we have not given a rigorous derivation of some formulas and the proof of some theorems as we considered it more important in the present book to get a general idea on the whole structure of the proof of renormalizability in ϕ_6^3 theory. Nevertheless, mathematical rigor is decisively important, in particular, for the proof of the power counting theorem and the lemma (2.5.95) since these play a principal role in the proof of renormalizability. The serious reader is urged to examine original papers on this problem.

2.5.5. Quantum chromodynamics

The basic Lagrangian for *quantum chromodynamics* is given in Eq. (2.3.99). As discussed in Sec. 2.3.5, it is more convenient to write the FP-ghost term in the form of Eq. (2.3.162) and hence we take the following Lagrangian as our starting point,

$$\mathscr{L} = -\frac{1}{4}F^a_{\mu\nu}F^{a\mu\nu} - \frac{1}{2\alpha}(\partial^\mu A^a_\mu)^2 + i(\partial^\mu\chi^a_1)\,D^{ab}_\mu\chi^b_2 + \bar{\psi}^i\,(i\gamma^\mu D^{ij}_\mu - m\delta^{ij})\,\psi^j \quad , \tag{2.5.115}$$

where D^{ab}_μ and D^{ij}_μ refer to the adjoint and fundamental representations of the color SU(3) respectively, and

$$F^a_{\mu\nu} = \partial_\mu A^a_\nu - \partial_\nu A^a_\mu + gf^{abc}A^b_\mu A^c_\nu \quad , \tag{2.5.116}$$

$$D_\mu = \partial_\mu - igT^c A^c_\mu \quad . \tag{2.5.117}$$

The Lagrangian (2.5.115) may be decomposed into a free and interaction part, $\mathscr{L}_0$ and $\mathscr{L}_1$, such that

$$\mathscr{L} = \mathscr{L}_0 + \mathscr{L}_1 \quad , \tag{2.5.118}$$

$$\begin{aligned}\mathscr{L}_0 = &-\frac{1}{4}(\partial_\mu A^a_\nu - \partial_\nu A^a_\mu)(\partial^\mu A^{a\nu} - \partial^\nu A^{a\mu}) - \frac{1}{2\alpha}(\partial^\mu A^a_\mu)^2 \\ &+ i(\partial^\mu\chi^a_1)(\partial_\mu\chi^a_2) + \bar{\psi}^i\,(i\gamma^\mu\partial_\mu - m)\,\psi^i \quad ,\end{aligned} \tag{2.5.119}$$

$$\begin{aligned}\mathscr{L}_1 = &-\frac{g}{2}f^{abc}\,(\partial_\mu A^a_\nu - \partial_\nu A^a_\mu)A^{b\mu}A^{c\nu} - \frac{g^2}{4}f^{abe}f^{cde}A^a_\mu A^b_\nu A^{c\mu}A^{d\nu} \\ &- igf^{abc}\,(\partial^\mu\chi^a_1)\,\chi^b_2 A^c_\mu + g\bar{\psi}^i T^a_{ij}\gamma^\mu\psi^j A^a_\mu .\end{aligned} \tag{2.5.120}$$

We have four interaction terms in $\mathscr{L}_1$. They represent three-gluon, four-gluon, ghost-gluon and quark-gluon interactions, for which the indices of divergence, r_i, ($i = 1, 2, 3, 4$ referring to these terms in the above order) read, following the definition (2.5.17),

$$r_1 = r_3 = r_4 = \frac{D-4}{2} \quad , \qquad r_2 = D-4 \quad . \tag{2.5.121}$$

Hence QCD is renormalizable in 4-dimensional space-time.

Here we could try to generalize our previous argument for ϕ_6^3 theory to the above more complicated case with four interaction terms and could give, without any recourse to gauge symmetry, a rigorous proof of renormalizability of QCD by applying the BPHZ method. This is in principle possible, but the work needed is tremendous so that the proof practically becomes hopeless. Obviously a clever way is to make full use of the gauge symmetry possessed by the theory to reduce the labor. In fact the four interaction terms in Eq. (2.5.120) are interrelated by the gauge symmetry, i.e., the BRS symmetry, and only one coupling constant is left independent. According to the BRS symmetry, we may derive generalized Ward-Takahashi identities (Slavnov-Taylor identities) which relate various Green functions among themselves. These identities help to reduce the number of independent Green functions whose divergences are subject to the renormalization.

In the following, we rewrite the QCD Lagrangian to specify the counter terms, single out Feynman diagrams with overall divergence by power counting, illustrate renormalizability to one-loop order, and explain the proof of renormalizability to all orders based on the generalized Ward-Takahashi identities.

(1) *Renormalized Lagrangian*

We redefine the fields A_μ^a, χ_1^a, χ_2^a and ψ by

$$A_\mu^a = Z_3^{1/2} A_{r\mu}^a \quad , \quad \chi_{1,2}^a = \tilde{Z}_3^{1/2} \chi_{1,2r}^a \quad , \quad \psi = Z_2^{1/2} \psi_r \quad , \tag{2.5.122}$$

and the parameters g, α and m by

$$g = Z_g g_r \quad , \qquad \alpha = Z_3 \alpha_r \quad , \qquad m = Z_m m_r \quad , \tag{2.5.123}$$

where the constants Z_3, $\tilde{Z}_3$ and Z_2 are called the gluon-field, ghost-field and quark-field renormalization constants, respectively, while the constants Z_g and Z_m may be called the coupling-constant and mass renormalization constants. The renormalization constant for the gauge parameter α is chosen to be the same as that for the gluon field A_μ^a so that the gauge fixing term is kept in the same form under the above redefinition. (The reason for this becomes clear later.)

Inserting Eqs. (2.5.122) and (2.5.123) into Eq. (2.5.115), we obtain

$$\mathscr{L} = \mathscr{L}_{r0} + \mathscr{L}_{r1} + \mathscr{L}_{C} \quad , \tag{2.5.124}$$

where $\mathscr{L}_{r0}$ and $\mathscr{L}_{r1}$ are precisely equal to $\mathscr{L}_0$ and $\mathscr{L}_1$ respectively if the quantities, A^a_μ, $\chi^a_{1,2}$, ψ, g, α and m are replaced by the renormalized ones, $A^a_{r\mu}$, $\chi^a_{1,2r}$ ψ_r, g_r, α_r and m_r. The counter-term Lagrangian $\mathscr{L}_C$ is given by

$$\begin{aligned}
\mathscr{L}_C = & -(Z_3-1)\frac{1}{4}(\partial_\mu A^a_{r\nu}-\partial_\nu A^a_{r\mu})(\partial^\mu A^{a\nu}_r-\partial^\nu A^{a\mu}_r) \\
& + (\tilde{Z}_3-1)i(\partial^\mu\chi^a_{1r})(\partial_\mu\chi^a_{2r}) + (Z_2-1)\bar{\psi}^i_r(i\gamma^\mu\partial_\mu-m_r)\psi^i_r \\
& - Z_2(Z_m-1)m_r\bar{\psi}^i_r\psi^i_r \\
& - (Z_g Z_3^{3/2}-1)\frac{1}{2}g_r f^{abc}(\partial_\mu A^a_{r\nu}-\partial_\nu A^a_{r\mu})A^{b\mu}_r A^{c\nu}_r \\
& -(Z_g^2 Z_3^2-1)\frac{1}{4}g_r^2 f^{abe}f^{cde}A^a_{r\mu}A^b_{r\nu}A^{c\mu}_r A^{d\nu}_r \\
& -(Z_g\tilde{Z}_3 Z_3^{1/2}-1)ig_r f^{abc}(\partial^\mu\chi^a_{1r})\chi^b_{2r}A^c_{r\mu} \\
& +(Z_g Z_2 Z_3^{1/2}-1)g_r\bar{\psi}^i_r T^a_{ij}\gamma^\mu\psi^j_r A^a_{r\mu} \quad .
\end{aligned} \tag{2.5.125}$$

In practical calculations with the renormalized Lagrangian we need the Feynman rules for the counter terms. For this purpose we rewrite Eq. (2.5.125) in the following form (up to a total divergence),

$$\begin{aligned}
\mathscr{L}_C = & (Z_3-1)\frac{1}{2}A^{a\mu}_r\delta_{ab}(g_{\mu\nu}\partial^2-\partial_\mu\partial_\nu)A^{b\nu}_r + (\tilde{Z}_3-1)\chi^a_{1r}\delta_{ab}(-i\partial^2)\chi^b_{2r} \\
& +(Z_2-1)\bar{\psi}^i_r(i\gamma^\mu\partial_\mu-m_r)\psi^i_r - Z_2(Z_m-1)m_r\bar{\psi}^i_r\psi^i_r \\
& -(Z_1-1)\frac{1}{2}g_r f^{abc}(\partial_\mu A^a_{r\nu}-\partial_\nu A^a_{r\mu})A^{b\mu}_r A^{c\nu}_r \\
& -(Z_4-1)\frac{1}{4}g_r^2 f^{abe}f^{cde}A^a_{r\mu}A^b_{r\nu}A^{c\mu}_r A^{d\nu}_r -(\tilde{Z}_1-1)ig_r f^{abc}(\partial^\mu\chi^a_{1r})\chi^b_{2r}A^c_{r\mu} \\
& +(Z_{1F}-1)g_r\bar{\psi}^i_r T^a_{ij}\gamma^\mu\psi^j_r A^a_{r\mu} \quad ,
\end{aligned} \tag{2.5.126}$$

where Z_1, Z_4, $\tilde{Z}_1$ and Z_{1F} will be defined in Eq. (2.5.127). From this counter-term Lagrangian we immediately obtain the Feynman rules given below.

Feynman rules for counter terms

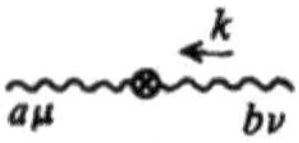

$$(Z_3-1)\delta_{ab}(k_\mu k_\nu - k^2 g_{\mu\nu})$$

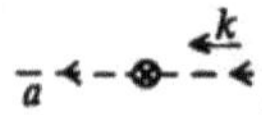

$$(Z_4-1)(-1)g_r^2 W^{a_1\ldots a_4}_{\mu_1\ldots\mu_4}$$

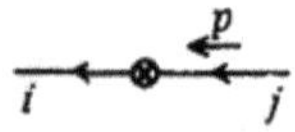

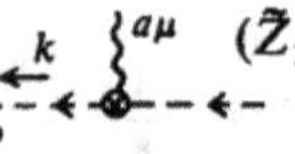

$$(\tilde{Z}_3-1)\delta_{ab}k^2$$

$$(\tilde{Z}_1-1)(-i)g_r f^{abc}k_\mu$$

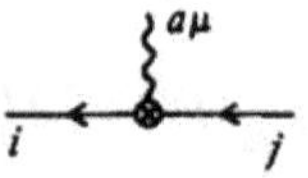

$$[(Z_2-1)\not{p}-(Z_2Z_m-1)m_r]\delta_{ij}$$

$$(Z_{1F}-1)g_r T^a_{ij}\gamma_\mu$$

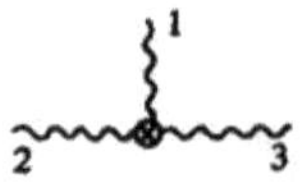

$$(Z_1-1)(-i)g_r f^{a_1a_2a_3}V_{\mu_1\mu_2\mu_3}(k_1,k_2,k_3)$$

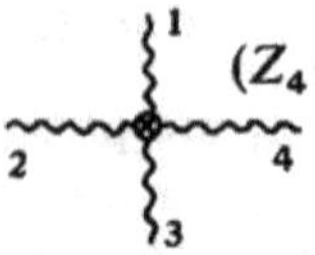

Note here that the above rules are given for ghost fields χ^a and χ^{a*} in conformity with our original Feynman rules in Sec. 2.3.3. The renormalization constants, Z_3, $\tilde{Z}_3$, Z_2, Z_m and Z_g should be determined by adjusting the counter terms so as to cancel overall divergences appearing in higher-order Feynman amplitudes. Of course a suitable renormalization scheme is to be chosen in this process.

It is very important to note that the renormalization constant Z_g can be determined by using any one of the last four counter terms in Eq. (2.5.126). Thus we have four ways of determining Z_g. The problem here is whether the values of Z_g so obtained are the same or not. In considering the gauge symmetry obeyed by the Lagrangian, one might expect that these four values of Z_g should be the same. This, however, is not self-evident and needs a proof. The proof may be given by the use of the generalized Ward-Takahashi identities with the gauge-invariant regularization (such as the dimensional regularization). Let us define four new renormalization constants by

$$Z_1 \equiv Z_g Z_3^{3/2} \ , \qquad Z_4 \equiv Z_g^2 Z_3^2 \ ,$$
$$\tilde{Z}_1 \equiv Z_g \tilde{Z}_3 Z_3^{1/2} \ , \qquad Z_{1F} \equiv Z_g Z_2 Z_3^{1/2} \ . \tag{2.5.127}$$

If the Z_g's in Eq. (2.5.127) are all the same, these four constants are not independent and the following constraints exist,

$$\frac{Z_1}{Z_3} = \frac{\tilde{Z}_1}{\tilde{Z}_3} = \frac{Z_{1F}}{Z_2} = \frac{Z_4}{Z_1} \ . \tag{2.5.128}$$

Equation (2.5.128) is called the *Slavnov-Taylor identity* [Tay 71, Sla 72] in the narrow sense and corresponds to a similar relation $Z_{1F} = Z_2$ in quantum electrodynamics. The Slavnov-Taylor identity guarantees the universality of the renormalized coupling constant g_r. We shall come back to this issue in Sec. 2.5.6.

Our task is now to ensure that the counter terms in Eq. (2.5.126) are sufficient to cancel all the divergences in Feynman integrals to all orders.

(2) *Power counting*

The superficial degree of divergence d for Feynman diagrams with N_B bosonic and N_F fermionic external lines is given by Eq. (2.5.17). In the case of QCD it reads in four dimensions

$$d = 4 - N_B - \frac{3}{2}N_F \quad . \tag{2.5.129}$$

Here in Eq. (2.5.129) we counted the gluons and FP-ghosts as bosonic fields so that N_B represents the total number of the gluon and FP-ghost external lines in the Feynman diagram. It should, however, be noted that in the Feynman rule for the ghost-gluon vertex the momentum factor k_μ appears only on the ghost line that carries the ghost number flowing out of the vertex. According to this special feature of the ghost-gluon interaction, the power counting rule has to be slightly modified. Since half of the FP-ghost internal lines adjacent to FP-ghost external lines in the Feynman diagram do not actually contribute to Eq. (2.5.4) as a factor δ_v, the superficial degree of divergence d given in Eq. (2.5.129) is overestimated by a half of the number of the FP-ghost external lines. Hence we have to replace N_B by

$$N_G + N_{FP} + \frac{1}{2}N_{FP} \quad ,$$

in Eq. (2.5.129) where N_G and N_{FP} represent the numbers of the gluon and FP-ghost external lines respectively. We then obtain

$$d = 4 - N_G - \frac{3}{2}(N_{FP} + N_F) \quad . \tag{2.5.130}$$

By examining the condition $d \geq 0$ for superficially divergent Feynman amplitudes, we find the following eight cases only: $(N_G, N_{FP}, N_F) = (0,0,0)$, $(2,0,0)$, $(0,2,0)$, $(0,0,2)$, $(1,2,0)$, $(1,0,2)$, $(3,0,0)$, $(4,0,0)$. Here, of course, we discarded the case like $(1,0,0)$ because of Lorentz invariance. The case $(0,0,0)$ corresponds to the vacuum diagram and is to be absorbed into the normalization of the generating functional. We are left with seven Feynman amplitudes which possess overall divergences. As shown in Fig. 2.5.14 they

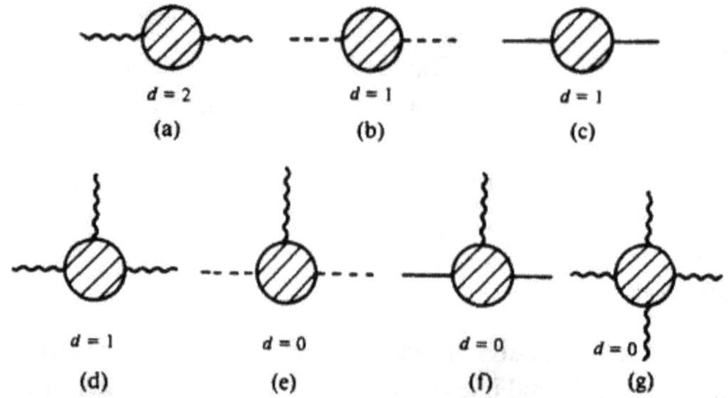

Fig. 2.5.14. Diagrams with the non-negative superficial degree of divergence in QCD: (a) the gluon self-energy part, (b) the FP-ghost self-energy part, (c) the quark self-energy part, (d) the three-gluon vertex part, (e) the ghost-gluon vertex part, (f) the quark-gluon vertex part, and (g) the four-gluon vertex part.

represent gluon, FP-ghost and quark self-energy parts and three-gluon, ghost-gluon, quark-gluon and 4-gluon vertex parts. The overall divergences of all these Feynman amplitudes may be taken care of by the counter terms in Eq. (2.5.126) since these amplitudes are of the same structure as the counter terms. This fact already suggests the renormalizability of QCD.

A remark is in order here that, although the superficial degrees of divergence d for the self-energy part for the gluon (Fig. 2.5.14(a)), FP-ghost (b) and quark (c) and the 3-gluon vertex (d) are 2, 1, 1 and 1, respectively, actual degrees of divergence of these amplitudes are all logarithmic due to the gauge symmetry and Lorentz invariance. This will be seen in an explicit way in the following arguments.

(3) *One-loop example*

In Figs. 2.5.15 to 2.5.21 one-loop contributions to the seven superficially divergent Feynman amplitudes are presented. We calculate the divergent parts of these one-loop contributions. Relegating the details of the calculations to Appendix A we summarize only the resulting expressions in the following.

(i) The *gluon self-energy part* $\Pi^{ab}_{\mu\nu}(k)$ is

$$\Pi^{ab}_{\mu\nu}(k) = \delta_{ab}\,(k_\mu k_\nu - k^2 g_{\mu\nu})\,\Pi(k^2) \quad , \tag{2.5.131}$$

$$\Pi(k^2) = \frac{g_r^2}{(4\pi)^2}\left[\frac{4}{3}T_R N_f - \frac{1}{2}C_G\left(\frac{13}{3} - \alpha_r\right)\right]\frac{1}{\varepsilon} + Z_3 - 1 + \text{finite terms}, \tag{2.5.132}$$

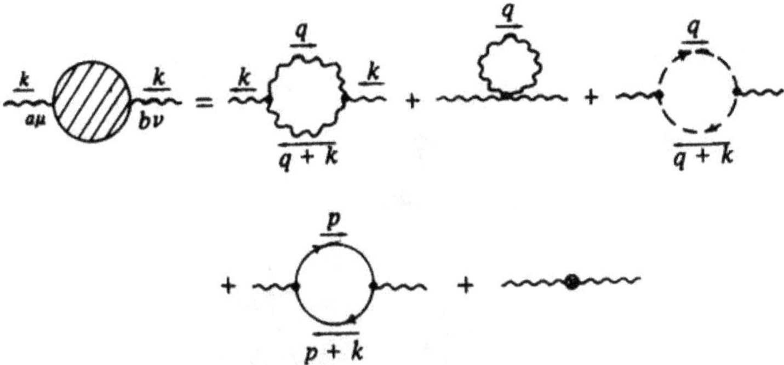

Fig. 2.5.15. The one-loop contribution to the gluon self-energy part. The last diagram on the right-hand side represents the one-loop counter term.

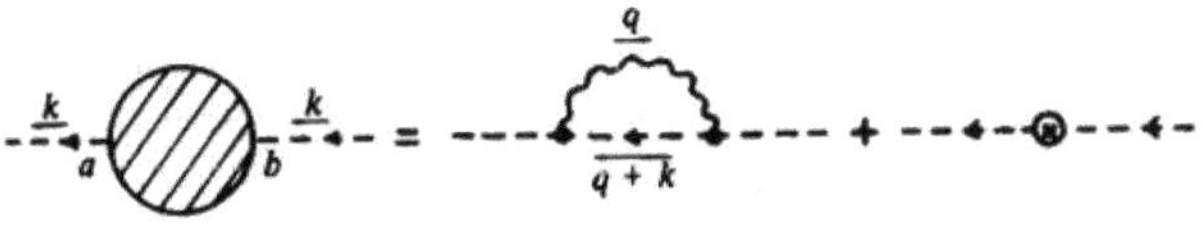

Fig. 2.5.16. One-loop contribution to the FP-ghost self-energy part with one-loop counter term.

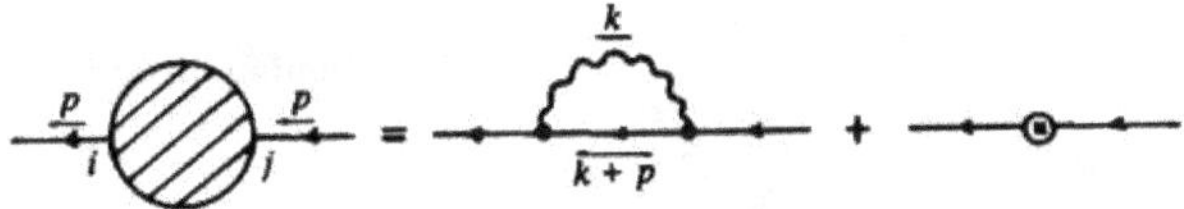

Fig. 2.5.17 One-loop contribution to the quark self-energy part.

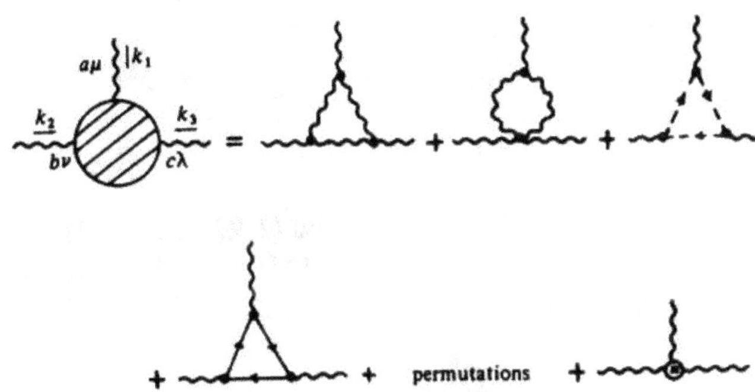

Fig. 2.5.18. One-loop contribution to the three-gluon vertex part.

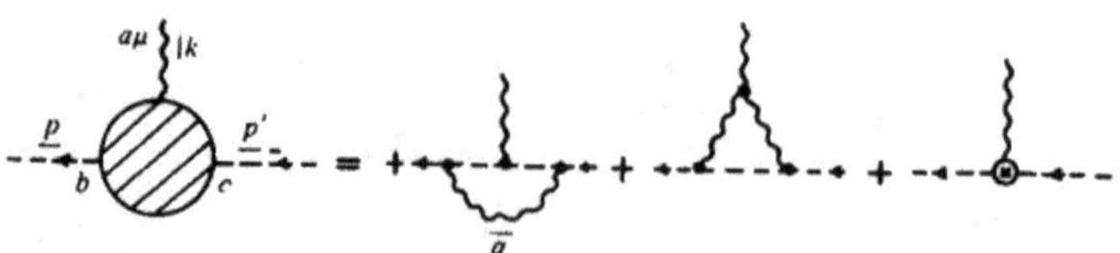

Fig. 2.5.19. One-loop contribution to the ghost-gluon vertex part.

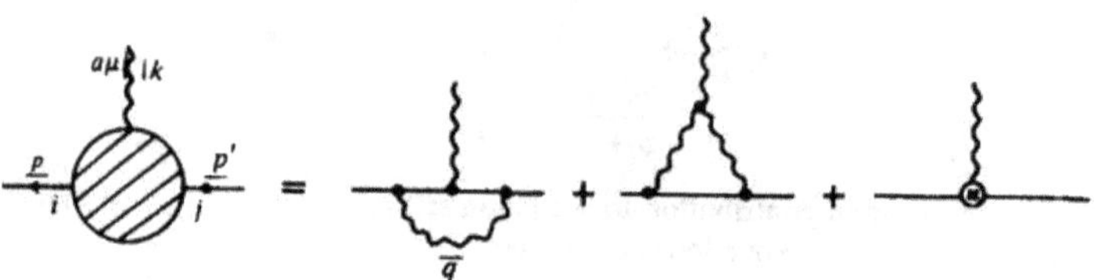

Fig. 2.5.20. One-loop contribution to the quark-gluon vertex part.

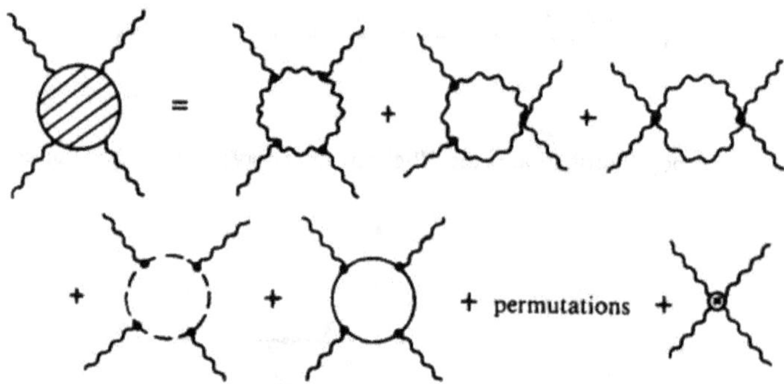

Fig. 2.5.21. One loop contribution to the four-gluon vertex part.

where $\varepsilon = (4-D)/2$. In Eq. (2.5.132) we have taken into account N_f flavors of quarks, and T_R and C_G are the constants defined by

$$T_r[T^a T^b] = \delta_{ab} T_R \quad ,$$

$$f^{acd} f^{bcd} = \delta_{ab} C_G \quad . \tag{2.5.133}$$

As we have already seen in Eqs. (2.4.14) and (2.4.23), $T_R = 1/2$ and $C_G = N$ for SU(N). Just by observing Eq. (2.5.131) we find that the one-loop contribution to the gluon self-energy part satisfies

$$k^\mu \Pi^{ab}_{\mu\nu}(k) = 0 \quad . \tag{2.5.134}$$

We shall see in Sec. 2.5.6 that Eq. (2.5.134) is nothing but one of the generalized Ward-Takahashi identities and is a natural consequence of the gauge symmetry. Because of this constraint the amplitude $\Pi^{ab}_{\mu\nu}(k)$ acquires the factor $k_\mu k_\nu - k^2 g_{\mu\nu}$ as in Eq. (2.5.131) and the degree of divergence for $\Pi^{ab}_{\mu\nu}(k)$ is lowered by 2 units. This structure of $\Pi^{ab}_{\mu\nu}(k)$ does not allow for a mass term and so there is no mass renormalization. Hence the gluon remains massless under the radiative corrections as is required. In the MS scheme the gluon-field renormalization constant Z_3 is determined to be

$$Z_3 = 1 - \frac{g_r^2}{(4\pi)^2}\left[\frac{4}{3}T_R N_f - \frac{1}{2}C_G\left(\frac{13}{3} - \alpha_r\right)\right]\frac{1}{\varepsilon} + O(g_r^4) \quad . \tag{2.5.135}$$

(ii) The *FP-ghost self-energy part* $\tilde{\Pi}^{ab}(k)$ is

$$\tilde{\Pi}^{ab}(k) = \delta_{ab}k^2\left[-\frac{g_r^2}{(4\pi)^2}C_G\frac{3-\alpha_r}{4}\frac{1}{\varepsilon} + \tilde{Z}_3 - 1\right] + \text{finite terms.} \tag{2.5.136}$$

We note that the above divergent part is proportional to k^2 and so there is no mass renormalization. Thus the FP-ghost also remains massless after radiative corrections are made. The ghost-field renormalization constant $\tilde{Z}_3$ in the MS scheme is given by

$$\tilde{Z}_3 = 1 + \frac{g_r^2}{(4\pi)^2}C_G\frac{3-\alpha_r}{4}\frac{1}{\varepsilon} + O(g_r^4) \quad . \tag{2.5.137}$$

(iii) The *quark self-energy part* $\sum^{ij}(p)$ has the form

$$\sum{}^{ij}(p) = \delta_{ij}[(Am_r - B\not{p}) - (Z_2 Z_m - 1)m_r + (Z_2 - 1)\not{p}] + \text{finite terms,}$$

$$A = \frac{-g_r^2}{(4\pi)^2}C_F(3+\alpha_r)\frac{1}{\varepsilon} + O(g_r^4) \quad ,$$

$$B = -\frac{g_r^2}{(4\pi)^2}C_F\alpha_r\frac{1}{\varepsilon} + O(g_r^4) \quad . \tag{2.5.138}$$

We see in Eq. (2.5.138) that the divergence in the quark self-energy part is of two kinds, the mass type Am_r and the kinetic-energy type $-B\not{p}$, just as the case of ϕ_6^3 theory. The mass and quark-field renormalization constants in the MS scheme are

$$Z_m = 1 + A - B + O(g_r^4) \quad ,$$

$$Z_2 = 1 + B + O(g_r^4) \quad . \tag{2.5.139}$$

(iv) The *3-gluon vertex* $\Lambda^{abc}_{\mu\nu\lambda}(k_1, k_2, k_3)$ [Cel 79a] is given by

$$\Lambda^{abc}_{\mu\nu\lambda}(k_1,k_2,k_3) = -ig_r f^{abc} V_{\mu\nu\lambda}(k_1,k_2,k_3)\left[\frac{g_r^2}{(4\pi)^2}\left(C_G\left(-\frac{17}{12}+\frac{3\alpha_r}{4}\right)\right.\right.$$
$$\left.\left.+\frac{4}{3}T_R N_f\right)\frac{1}{\varepsilon}+Z_1-1\right]+\text{finite terms,} \qquad (2.5.140)$$

where $V_{\mu\nu\lambda}(k_1, k_2, k_3)$ is the same as the function defined in Eq. (2.3.139). The three-gluon vertex renormalization constant Z_1 in the MS scheme is found to be

$$Z_1 = 1 - \frac{g_r^2}{(4\pi)^2}\left[C_G\left(-\frac{17}{12}+\frac{3\alpha_r}{4}\right)+\frac{4}{3}T_R N_f\right]\frac{1}{\varepsilon}+O(g_r^4) \quad .(2.5.141)$$

(v) The *ghost-gluon vertex* $\tilde{\Lambda}^{abc}_\mu(k, p, p')$ is

$$\tilde{\Lambda}^{abc}_\mu(k,p,p') = -ig_r f^{abc} p_\mu\left[\frac{g_r^2}{(4\pi)^2}C_G\frac{\alpha_r}{2}\frac{1}{\varepsilon}+\tilde{Z}_1-1\right]+\text{finite terms,} \qquad (2.5.142)$$

where the momentum p_μ refers to the ghost-line which carries the ghost number flowing out of the vertex. The ghost-gluon vertex renormalization constant $\tilde{Z}_1$ in the MS scheme is given by

$$\tilde{Z}_1 = 1 - \frac{g_r^2}{(4\pi)^2}C_G\frac{\alpha_r}{2}\frac{1}{\varepsilon} + O(g_r^4) \quad . \qquad (2.5.143)$$

(vi) The *quark-gluon vertex* $\Lambda^{aij}_{F\mu}(k, p, p')$ has the form

$$\Lambda^{aij}_{F\mu}(k,p,p') = g_r\gamma_\mu T^a_{ij}\left[\frac{g_r^2}{(4\pi)^2}\left(\frac{3+\alpha_r}{4}C_G+\alpha_r C_F\right)\frac{1}{\varepsilon}+Z_{1F}-1\right]$$
$$+\text{ finite terms.} \qquad (2.5.144)$$

In the MS scheme Z_{1F} reads

$$Z_{1F} = 1 - \frac{g_r^2}{(4\pi)^2}\left(\frac{3+\alpha_r}{4}C_G+\alpha_r C_F\right)\frac{1}{\varepsilon}+O(g_r^4) \quad . \qquad (2.5.145)$$

(vii) The *four-gluon vertex* $\Lambda^{a_1\ldots a_4}_{\mu_1\ldots\mu_4}(k_1, \ldots, k_4)$ [Pas 80] reads

$$\Lambda^{a_1\ldots a_4}_{\mu_1\ldots\mu_4}(k_1,\ldots,k_4)$$
$$= -g_r^2 W^{a_1\ldots a_4}_{\mu_1\ldots\mu_4}\left[\frac{g_r^2}{(4\pi)^2}\left(\left(-\frac{2}{3}+\alpha_r\right)C_G+\frac{4}{3}T_R N_f\right)\frac{1}{\varepsilon}\right.$$
$$\left.+Z_4-1\right]+\text{finite terms,} \qquad (2.5.146)$$

where $W^{a_1\cdots a_4}_{\mu_1\cdots\mu_4}$ is the function given previously in Eq. (2.3.140). The four-gluon vertex renormalization constant Z_4 is

$$Z_4 = 1 - \frac{g_r^2}{(4\pi)^2}\left[\left(-\frac{2}{3}+\alpha_r\right)C_G+\frac{4}{3}T_R N_f\right]\frac{1}{\varepsilon} + O(g_r^4) \quad . \quad (2.5.147)$$

We can now see that all the one-loop divergences in the seven superficially divergent Feynman amplitudes can be cancelled by the contributions of the counter terms defined in Eq. (2.5.126). Thus we have proven directly the renormalizability of QCD in one-loop order. Note that we have not taken any account of the gauge symmetry. As was mentioned earlier the gauge symmetry imposes constraints on these seven Feynman amplitudes and, as a special case, the Slavnov-Taylor identity results. In fact by using Eqs. (2.5.135, 137, 139, 141, 143, 145, 147) we can check that the identity is satisfied in one-loop order with the MS scheme:

$$\frac{Z_1}{Z_3} = \frac{\tilde{Z}_1}{\tilde{Z}_3} = \frac{Z_{1F}}{Z_2} = \frac{Z_4}{Z_1} = 1 - \frac{g_r^2}{(4\pi)^2}C_G\frac{3+\alpha_r}{4}\frac{1}{\varepsilon} + O(g_r^4) \quad . \quad (2.5.148)$$

(4) *Renormalization to all orders*

The general proof of renormalizability of non-Abelian gauge field theories (including QCD as a special case) has been pursued by the full use of the constraints imposed by the gauge symmetry [tHo 71, 72b, Sla 72, Tay 71, Lee 72, Klu 75, 75a, Bec 76]. The standard method utilized nowadays for the proof seems to be to first write down identities satisfied by the generating functional under the gauge symmetry (generalized Ward-Takahashi identities) and then use them inductively to prove the renormalizability order by order [Lee 76, Tay 76, Fad 80, Itz 80]. Some other methods for the proof have also been proposed. For example, Baker and Lee [Bak 77] derive the Dyson-Schwinger equations generalized to non-Abelian gauge field theories, which are proven to be free of overlapping divergences, and use them for the proof of renormalizability of the theories. We here employ yet another method introduced by Brandt [Bra 76] and developed by Collins [Col 84]. In this method we directly deal with the relation among Green functions which results from the BRS symmetry and hence the proof carries on in an intuitive way.

As we have seen in Sec. 2.3.5, the quantized version of the gauge symmetry of the classical Lagrangian is the BRS symmetry [Bec 76]. Accordingly Green functions are, as will be shown later in Sec. 2.5.6, invariant under the BRS transformation, i.e., the generalized Ward-Takahashi identities hold for Green functions. Because of these constraints on Green functions the number of independent superficially divergent Feynman amplitudes is less than what we had originally. The superficially divergent Feynman amplitudes are of seven types as shown in Fig. 2.5.14, i.e.,

$\Pi^{ab}_{\mu\nu}$	gluon self-energy part,
$\tilde{\Pi}^{ab}$	ghost self-energy part,
Σ^{ij}	quark self-energy part.
$\Lambda^{abc}_{\mu\nu\lambda}$	3-gluon vertex,
$\tilde{\Lambda}^{abc}_{\mu}$	ghost-gluon vertex,
$\Lambda^{aij}_{F\mu}$	quark-gluon vertex,
$\Lambda^{a_1\ldots a_4}_{\mu_1\ldots\mu_4}$	4-gluon vertex.

In the following we shall show that, according to the BRS symmetry,

(1) the gluon self-energy part is purely transversal and requires only one renormalization constant Z_3, and

(2) only four amplitudes $\Pi^{ab}_{\mu\nu}$, $\tilde{\Pi}^{ab}$, Σ^{ij} and $\tilde{\Lambda}^{abc}_{\mu}$ are independent as far as the renormalization procedure is concerned and the other three are automatically made finite if the above four are renormalized (the other choices of the set of four independent amplitudes are equally allowed).

Once the above statement is proven, we apply the BPH recursion equations to the set of the four amplitudes $\Pi^{ab}_{\mu\nu}$, $\tilde{\Pi}^{ab}$, Σ^{ij} and $\tilde{\Lambda}^{abc}_{\mu}$ and verify the renormalizability order by order inductively. According to our choice of the set of the above four independent amplitudes, independent renormalization constants are restricted to be Z_3, $\tilde{Z}_3$, Z_2, Z_m and $\tilde{Z}_1$.

We shall now present the proof of the statement given above. For this purpose we need the renormalized version of the BRS transformations (2.3.168, 169, 173, 174). Applying the renormalization process (2.5.122 – 123) with $Z_g = \tilde{Z}_1\tilde{Z}_3^{-1}Z_3^{-1/2}$ supplemented by the redefinition of the Grassmann parameter $\delta\lambda$,

$$\delta\lambda = Z_3^{1/2}\tilde{Z}_3^{1/2}\delta\lambda_r, \tag{2.5.149}$$

we have the following renormalized BRS transformations,

$$\begin{aligned} \delta A^a_{r\mu} &\doteq \delta\lambda_r \tilde{Z}_3 D^{ab}_{\mu}\chi^b_{2r} \quad , \\ \delta\psi_r &= \delta\lambda_r i\tilde{Z}_1 g_r T^a \chi^a_2 \psi_r \quad , \\ \delta\chi^a_{1r} &= \delta\lambda_r (i/\alpha_r)\,\partial^\mu A^a_{r\mu} \quad , \\ \delta\chi^a_{2r} &= \delta\lambda_r(-1/2)\tilde{Z}_1 g_r f^{abc}\chi^b_{2r}\chi^c_{2r} \quad , \end{aligned} \tag{2.5.150}$$

where D_μ^{ab} is given in the renormalized form by

$$D_\mu^{ab} = \delta_{ab}\partial_\mu - (\tilde{Z}_1/\tilde{Z}_3)g_r f^{abc} A_{r\mu}^c \quad . \tag{2.5.151}$$

We first show that our statement (1) is indeed true. We consider the following identity (for simplicity we suppress index r for renormalized quantities in all the following expressions),[34]

$$\delta\langle 0|\mathrm{T}[(\partial^\mu \hat{A}_\mu^a(x))\hat{\chi}_1^b(y)]|0\rangle = 0 \quad , \tag{2.5.152}$$

where the symbol $\hat{}$ is attached on the top of the fields A_μ^a and χ_1^b since they are regarded as local operators in the canonical formalism (Sec. 2.3.5). Equation (2.5.152) is rewritten as

$$\langle 0|\mathrm{T}[\delta(\partial^\mu \hat{A}_\mu^a(x))\hat{\chi}_1^b(y)]|0\rangle + \langle 0|\mathrm{T}[(\partial^\mu \hat{A}_\mu^a(x))\delta\hat{\chi}_1^b(y)]|0\rangle = 0. \tag{2.5.153}$$

Here we note that

$$\delta(\partial^\mu \hat{A}_\mu^a) = \partial^\mu(\delta\hat{A}_\mu^a) = \delta\lambda[\tilde{Z}_3\Box\hat{\chi}_2^a - \tilde{Z}_1 g f^{abc}\partial^\mu(\hat{\chi}_2^b\hat{A}_\mu^c)] = 0 \quad , \tag{2.5.154}$$

owing to the field equation for the $\hat{\chi}_2^a$ field:

$$\partial^\mu \frac{\partial\mathscr{L}}{\partial(\partial^\mu\hat{\chi}_1^a)} - \frac{\partial\mathscr{L}}{\partial\hat{\chi}_1^a} = i\tilde{Z}_3\Box\hat{\chi}_2^a - i\tilde{Z}_1 g f^{abc}\partial^\mu(\hat{\chi}_2^b\hat{A}_\mu^c) = 0 \quad . \tag{2.5.155}$$

Hence Eq. (2.5.153) implies that

$$\begin{aligned} 0 &= \frac{i}{\alpha}\langle 0|\mathrm{T}\left[\left(\frac{\partial}{\partial x_\mu}\hat{A}_\mu^a(x)\right)\left(\frac{\partial}{\partial y_\nu}\hat{A}_\nu^b(y)\right)\right]|0\rangle \\ &= \frac{i}{\alpha}\frac{\partial}{\partial x_\mu}\langle 0|\mathrm{T}\left[\hat{A}_\mu^a(x)\frac{\partial}{\partial y_\nu}\hat{A}_\nu^b(y)\right]|0\rangle - \frac{i}{\alpha}\delta(x_0-y_0)\langle 0|[\hat{A}_0^a(x), \frac{\partial}{\partial y_\nu}\hat{A}_\nu^b(y)]|0\rangle \end{aligned} \tag{2.5.156}$$

According to Eq. (2.2.11) we find that

$$\partial^\nu \hat{A}_\nu^a = -Z_3\alpha\hat{\Pi}_0^a \quad . \tag{2.5.157}$$

[34] See Eq. (2.5.186) in Sec. 2.5.6.

Using the equal-time commutation relation,

$$\delta(x_0-y_0)\,[\hat{A}_0^a(x),\hat{\Pi}_0^b(y)] = \frac{i}{Z_3}\delta_{ab}\delta^4(x-y) \quad , \tag{2.5.158}$$

we finally obtain from Eq. (2.5.156)

$$\frac{i}{\alpha}\frac{\partial}{\partial x_\mu}\frac{\partial}{\partial y_\nu}\langle 0|\mathrm{T}[\hat{A}_\mu^a(x)\hat{A}_\nu^b(y)]|0\rangle = \delta_{ab}\delta^4(x-y) \quad . \tag{2.5.159}$$

We define the Fourier transform $\tilde{D}_{\mu\nu}^{ab}(k)$ of the full gluon propagator such that

$$i\langle 0|\mathrm{T}[\hat{A}_\mu^a(x)\hat{A}_\nu^b(y)]|0\rangle = \int\frac{d^4k}{(2\pi)^4}e^{-ik\cdot(x-y)}\,\tilde{D}_{\mu\nu}^{ab}(k) \quad . \tag{2.5.160}$$

Then Eq. (2.5.159) may be cast in the form

$$\frac{1}{\alpha}k^\mu k^\nu\tilde{D}_{\mu\nu}^{ab}(k) = \delta_{ab} \quad . \tag{2.5.161}$$

Note that the gluon self-energy part $\Pi_{\mu\nu}^{ab}(k)$ is related to the full gluon propagator $\tilde{D}_{\mu\nu}^{ab}(k)$ by

$$\tilde{D}_{\mu\nu}^{ab} = \tilde{D}_{0\mu\nu}^{ab} + \tilde{D}_{0\mu\lambda}^{ac}\,\Pi^{cd,\lambda\rho}\,\tilde{D}_{0\rho\nu}^{db} + \dots, \tag{2.5.162}$$

where $\tilde{D}_{0\mu\nu}^{ab}$ is the free gluon propagator,

$$\tilde{D}_{0\mu\nu}^{ab}(k) = \frac{\delta_{ab}}{k^2}\left[g_{\mu\nu} - (1-\alpha)\frac{k_\mu k_\nu}{k^2}\right] \quad .$$

The generalized Ward-Takahashi identity (2.5.161) applied to Eq. (2.5.162) implies that

$$k^\mu k^\nu\Pi_{\mu\nu}^{ab}(k) = 0 \quad . \tag{2.5.163}$$

Equation (2.5.163) is the generalized Ward-Takahashi identity for the gluon self-energy part, which gives

$$\Pi_{\mu\nu}^{ab}(k) = \delta_{ab}(k_\mu k_\nu - k^2 g_{\mu\nu})\,\Pi(k^2) \quad . \tag{2.5.164}$$

Accordingly the full gluon propagator possesses the following form

$$\tilde{D}_{\mu\nu}^{ab}(k) = \frac{\delta_{ab}}{k^2}\left[\frac{g_{\mu\nu} - k_\mu k_\nu/k^2}{1+\Pi(k^2)} + \alpha\frac{k_\mu k_\nu}{k^2}\right] \quad . \tag{2.5.165}$$

Thus we realize that only the transversal part of the gluon propagator is subject to higher-order corrections and the gluon propagator may be renormalized with only one renormalization constant Z_3. The above consequence that the longitudinal part is free from the higher-order corrections is in conformity with our previous choice of the renormalization constant for the gauge parameter α in Eq. (2.5.123): the renormalized gluon propagator is obtained by multiplying the unrenormalized one by Z_3^{-1} so that the gauge parameter should be renormalized as given in Eq. (2.5.123).

We next show that our statement (2) is indeed the case. In order to see that the three-gluon vertex is finite, we examine the generalized Ward-Takahashi identity[35]

$$\delta\langle 0|\mathrm{T}[\hat{A}^a_\mu(x)\hat{A}^b_\nu(y)\hat{\chi}^c_1(z)]|0\rangle = 0 \quad . \tag{2.5.166}$$

From Eq. (2.5.166) we obtain

$$\langle 0|\mathrm{T}[\delta\hat{A}^a_\mu(x)\hat{A}^b_\nu(y)\hat{\chi}^c_1(z)]|0\rangle + \langle 0|\mathrm{T}[\hat{A}^a_\mu(x)\delta\hat{A}^b_\nu(y)\hat{\chi}^c_1(z)]|0\rangle$$
$$+ \langle 0|\mathrm{T}[\hat{A}^a_\mu(x)\hat{A}^b_\nu(y)\delta\hat{\chi}^c_1(z)]|0\rangle = 0 \quad . \tag{2.5.167}$$

In Eq. (2.5.167) the first two terms are finite. This can be seen as follows: note first that

$$\frac{\partial}{\partial x_\mu}\langle 0|\mathrm{T}[\delta\hat{A}^a_\mu(x)\hat{A}^b_\nu(y)\hat{\chi}^c_1(z)]|0\rangle = \langle 0|\mathrm{T}[\delta(\partial^\mu\hat{A}^a_\mu(x))\hat{A}^b_\nu(y)\hat{\chi}^c_1(z)]|0\rangle = 0 \quad , \tag{2.5.168}$$

according to Eq. (2.5.154). Since the operation $\partial/\partial x_\mu$ is equivalent to the multiplication of k^μ in momentum space, Eq. (2.5.168) implies that $\langle 0|\mathrm{T}[\delta\hat{A}^a_\mu(x)\hat{A}^b_\nu(y)\hat{\chi}^c_1(z)]|0\rangle$ is finite. The same argument applies also to $\langle 0|\mathrm{T}[\hat{A}^a_\mu(x)\delta\hat{A}^b_\nu(y)\hat{\chi}^c_1(z)]|0\rangle$. Hence, by using Eq. (2.5.150), we find from Eq. (2.5.167) that

$$\frac{\partial}{\partial z_\lambda}\langle 0|\mathrm{T}[\hat{A}^a_\mu(x)\hat{A}^b_\nu(y)\,\hat{A}^c_\lambda(z)]|0\rangle \text{ is finite.} \tag{2.5.169}$$

The Fourier transform of the Green function $\langle 0|\mathrm{T}[\hat{A}^a_\mu(x)\,\hat{A}^b_\nu(y)\,\hat{A}^c_\lambda(z)]|0\rangle$ is equal to the three-gluon vertex multiplied by gluon propagators. Therefore Eq. (2.5.169) implies that the three-gluon vertex is finite (the gluon-propagator is assumed to be finite).

The finiteness of the four-gluon vertex is deduced in a similar way by observing the identity

$$\delta\langle 0|\mathrm{T}[\hat{A}^a_\mu(x)\hat{A}^b_\nu(y)\hat{A}^c_\lambda(z)\hat{\chi}^d_1(w)]|0\rangle = 0 \quad . \tag{2.5.170}$$

[35] See Eq. (2.5.186).

From Eq. (2.5.170) it follows that

$$\frac{\partial}{\partial w_\rho}\langle 0|\mathrm{T}[\hat{A}^a_\mu(x)\hat{A}^b_\nu(y)\hat{A}^c_\lambda(z)\hat{A}^d_\rho(w)]|0\rangle \text{ is finite.} \tag{2.5.171}$$

Hence the 4-gluon vertex is finite.

To show that the quark-gluon vertex is finite, we consider the identity

$$\delta\langle 0|\mathrm{T}[\hat{\psi}_i(x)\hat{\bar{\psi}}_j(y)\hat{\chi}^a_1(z)]|0\rangle = 0 \quad . \tag{2.5.172}$$

From Eq. (2.5.172) we have

$$\delta\lambda\frac{i}{\alpha}\frac{\partial}{\partial z_\mu}\langle 0|\mathrm{T}[\hat{\psi}_i(x)\hat{\bar{\psi}}_j(y)\hat{A}^a_\mu(z)]|0\rangle = -\langle 0|\mathrm{T}[\hat{\psi}_i(x)\delta\hat{\bar{\psi}}_j(y)\hat{\chi}^a_1(z)]|0\rangle$$
$$-\langle 0|\mathrm{T}[\delta\hat{\psi}_i(x)\hat{\bar{\psi}}_j(y)\hat{\chi}^a_1(z)]|0\rangle \quad . \tag{2.5.173}$$

The right-hand side can be shown to be finite and hence the quark-gluon vertex is finite (the quark propagator is finite by assumption). To show that the right-hand side is finite is rather involved and we shall not go into details here. The complete argument may be found in [Col 84].

Digression

A SPECIFIC FORMULA IN DIMENSIONAL REGULARIZATION: In Appendix A the following formula is employed,

$$\int\frac{d^Dq}{q^2} = 0 \quad .$$

This is actually a special case of a more general formula in dimensional regularization,

$$\int\frac{d^Dq}{(-q^2)^\alpha} = 0 \quad , \qquad \alpha > 0 \quad . \tag{2.5.174}$$

We shall now show that formula (2.5.174) can be justified in the sense of dimensional regularization.

Applying a Wick rotation, the left-hand side of Eq. (2.5.174) may be written as

$$\int\frac{d^Dq}{(-q^2)^\alpha} = i\frac{\pi^{D/2}}{\Gamma(D/2)}\int_0^\infty (Q^2)^{D/2-\alpha-1}dQ^2 \quad , \tag{2.5.175}$$

where $Q^2 = -q^2$. We see that Eq. (2.5.175) develops an ultraviolet divergence for $D > 2\alpha$ while it has an infrared divergence for $D < 2\alpha$. Thus the above integral has no mathematically meaningful region in D. This situation is contrasted with that for the case of the quark self-energy part (2.4.31) discussed previously where the integral is mathematically meaningful if $D < 3$. In order to give mathematical meaning to the integral in Eq. (2.5.175), we split the integration in Q^2 into two parts:[36] the ultraviolet part $Q^2 > \Lambda^2$ and the infrared part $Q^2 < \Lambda^2$,

$$\int\frac{d^Dq}{(-q^2)^\alpha} = i\frac{\pi^{D/2}}{\Gamma(D/2)}\left[\int_0^{\Lambda^2}(Q^2)^{D/2-\alpha-1}dQ^2 + \int_{\Lambda^2}^\infty (Q^2)^{D/2-\alpha-1}dQ^2\right] \tag{2.5.176}$$

On the right-hand side of Eq. (2.5.176) the first integral is convergent for $D > 2\alpha$ while the second one is convergent for $D < 2\alpha$. Here the space-time dimension D acts as a regulator for the infrared as well as ultraviolet divergences. To distinguish the nature of the divergences we designate $D = D_{\mathrm{I}}$ for the first integral and $D = D_{\mathrm{U}}$ for the second. Performing the integrations in Eq. (2.5.176) for $D_{\mathrm{I}} > 2\alpha$ and $D_{\mathrm{U}} < 2\alpha$ we obtain

$$\frac{\Gamma(D/2)}{\pi^{D/2}i}\int\frac{d^Dq}{(-q^2)^\alpha} = \frac{\Lambda^{D_{\mathrm{I}}-2\alpha}}{\frac{1}{2}D_{\mathrm{I}}-\alpha} - \frac{\Lambda^{D_{\mathrm{U}}-2\alpha}}{\frac{1}{2}D_{\mathrm{U}}-\alpha} \tag{2.5.177}$$

We see that the two terms in Eq. (2.5.177) develop poles at $D_{\mathrm{I}} = D_{\mathrm{U}} = 2\alpha$ corresponding to the infrared and ultraviolet divergence, respectively. The right-hand side of Eq. (2.5.177) can be continued analytically to arbitrary values of D_{I} and D_{U} and hence the constraints $D_{\mathrm{I}} > 2\alpha$ and $D_{\mathrm{U}} < 2\alpha$ can be removed. If we identify D_{I} with D_{U} in Eq. (2.5.177), the right-hand side obviously vanishes. In this sense Eq. (2.5.174) is justified. It should be noted that equation (2.5.174) is a prescription in dimensional regularization which we shall employ throughout the book.

Exercise

1. Prove the following formulas by using the Feynman-parametrization method:

$$\int\frac{d^Dq}{(2\pi)^Di}\frac{1}{(-q^2)^\alpha(-(q+k)^2)^\beta} = (4\pi)^{-D/2}(-k^2)^{D/2-\alpha-\beta}\frac{\Gamma(\alpha+\beta-D/2)}{\Gamma(\alpha)\Gamma(\beta)}B\left(\frac{D}{2}-\alpha,\frac{D}{2}-\beta\right) . \tag{2.5.178}$$

$$\int\frac{d^Dq}{(2\pi)^Di}\frac{q_\mu}{(-q^2)^\alpha(-(q+k)^2)^\beta} = -(4\pi)^{-D/2}k_\mu(-k^2)^{D/2-\alpha-\beta}$$
$$\times\frac{\Gamma(\alpha+\beta-D/2)}{\Gamma(\alpha)\Gamma(\beta)}B\left(\frac{D}{2}-\alpha+1,\frac{D}{2}-\beta\right) , \tag{2.5.179}$$

$$\int\frac{d^Dq}{(2\pi)^Di}\frac{q_\mu q_\nu}{(-q^2)^\alpha(-(q+k)^2)^\beta} = (4\pi)^{-D/2}(-k^2)^{D/2-\alpha-\beta}\frac{\Gamma(\alpha+\beta-D/2)}{\Gamma(\alpha)\Gamma(\beta)}$$
$$\times\left[k^2g_{\mu\nu}\frac{B(D/2-\alpha+1,D/2-\beta+1)}{2(\alpha+\beta-1-D/2)}+k_\mu k_\nu B\left(\frac{D}{2}-\alpha+2,\frac{D}{2}-\beta\right)\right] . \tag{2.5.180}$$

2.5.6. Generalized Ward-Takahashi identities

The gauge symmetry present in the classical Lagrangian is inherited in the quantized Lagrangian as the BRS symmetry. Because of the presence of this

[36] Here we regularize the infrared divergence by using dimensional regularization. One could alternatively introduce a gluon mass μ which has the effect of replacing q^2 in Eq. (2.5.174) by $q^2-\mu^2$. In this case Eq. (2.5.174) cannot be made legitimate.

symmetry Green functions are mutually related. In other words the generating functional for Green functions is constrained by the BRS symmetry. The constraints on the generating functional or on the Green functions are called the *generalized Ward-Takahashi identities* (or *Slavnov-Taylor identities*).

The generalized Ward-Takahashi identities were first discussed by 't Hooft [tHo 71] in connection with the renormalization of non-Abelian gauge theories. In his argument, however, only a part of the system of the identities was taken into account. Later Slavnov [Sla 72] and Taylor [Tay 71] presented a full account of the derivation of the generalized Ward-Takahashi identities.

We first show that Green functions are unaltered by the BRS transformation if the gauge-invariant regularization and subtraction scheme are chosen. For this purpose we consider the generating functional (2.3.178),

$$Z[J,\xi_1,\xi_2,\eta,\bar{\eta}] = \int [dA]\,[d\chi_1]\,[d\chi_2]\,[d\psi]\,[d\bar{\psi}]\exp\{i\int d^4x(\mathscr{L}+AJ + \chi_1\xi_1+\chi_2\xi_2+\bar{\psi}\eta+\bar{\eta}\psi) \quad . \tag{2.5.181}$$

It is easy to show that the functional integral measure in Eq. (2.5.181) is invariant under the BRS transformation,

$$[dA]\,[d\chi_1]\,[d\chi_2]\,[d\psi]\,[d\bar{\psi}] = [dA']\,[d\chi_1']\,[d\chi_2']\,[d\psi']\,[d\bar{\psi}'] \quad , \tag{2.5.182}$$

where the fields with prime A', χ_1', χ_2', ψ' and $\bar{\psi}'$ are BRS-transformed fields. We already proved in Sec. 2.3.5 that the quantized Lagrangian is invariant under the BRS transformation. It can be shown that the above statement remains true if we deal with the renormalized fields and parameters.

We now make the change of variables,

$$A\to A', \quad \chi_1\to\chi_1', \quad \chi_2\to\chi_2', \quad \psi\to\psi', \quad \bar{\psi}\to\bar{\psi}' \quad , \tag{2.5.183}$$

representing the BRS transformation. Then Eq. (2.5.181) is rewritten as

$$Z[J,\xi_1,\xi_2,\eta,\bar{\eta}] = \int [dA]\,[d\chi_1]\,[d\chi_2]\,[d\psi]\,[d\bar{\psi}]\exp\{i\int d^4x(\mathscr{L}+A'J + \chi_1'\xi_1+\chi_2'\xi_2+\bar{\psi}'\eta+\bar{\eta}\psi')\} \quad , \tag{2.5.184}$$

where we have taken into account the invariance of the integral measure (2.5.182) and of the Lagrangian. Since Eq. (2.5.183) is merely a change of variables, Eq. (2.5.184) has to be equal to Eq. (2.5.181). This means that Green functions derived from Eq. (2.5.181) have to be the same as those derived from Eq. (2.5.184):

$$\langle 0|\mathrm{T}[\hat{A}^a_\mu(x)\ldots]|0\rangle = \langle 0|\mathrm{T}[\hat{A}'^a_\mu(x)\ldots]|0\rangle \quad . \tag{2.5.185}$$

Thus we obtain the generalized Ward-Takahashi identities for Green functions,

$$\delta\langle 0|T[\hat{A}_\mu^a(x)\ldots]|0\rangle = 0 \quad , \tag{2.5.186}$$

which have been extensively utilized in Sec. 2.5.5. It should be noted here that Eq. (2.5.186) is valid only when the gauge-invariant regularization and subtraction scheme are used [Pas 80, Chi 81].

The generalized Ward-Takahashi identities may also be represented in the form of constraints on the generating functional. Noting that $A' = A + \delta A$, $\chi_1' = \chi_1 + \delta\chi_1$, $\chi_2' = \chi_2 + \delta\chi_2$, $\psi' = \psi + \delta\psi$ and $\bar{\psi}' = \bar{\psi} + \delta\bar{\psi}$, and equating Eq. (2.5.184) to Eq. (2.5.181), we have

$$\int [dA][d\chi_1][d\chi_2][d\psi][d\bar{\psi}] \int d^4x(\delta AJ + \delta\chi_1\xi_1 + \delta\chi_2\xi_2 + \delta\bar{\psi}\eta + \bar{\eta}\delta\psi)$$

$$\times \exp\{i\int d^4x(\mathscr{L} + AJ + \chi_1\xi_1 + \chi_2\xi_2 + \bar{\psi}\eta + \bar{\eta}\psi)\} = 0 \quad . \tag{2.5.187}$$

This identity leads us to the generalized Ward-Takahashi identities for the generating functional. In fact, by inserting Eq. (2.5.150) into the above identity and by rewriting the resulting expression in the form of the functional derivative of the generating functional $Z[J, \xi_1, \xi_2, \eta, \bar{\eta}]$ with respect to the sources J, ξ_1, ξ_2, η and $\bar{\eta}$, we obtain constraint equations for the generating functional. This expression of the constraints, however, is rather inconvenient since it contains functional derivatives in a nonlinear way.

With the purpose of linearizing the above-mentioned constraint equations, we introduce extra source terms for composite fields $D_\mu^{ab}\chi_2^b$, $(-1/2)gf^{abc}\chi_2^b\chi_2^c$ and $igT^a\chi_2^a\psi$, and replace the source terms in Eq. (2.5.181) by [Klu 75]

$$\sum \equiv A_\mu^a J^{a\mu} + \chi_1^a\xi_1^a + \chi_2^a\xi_2^a + \bar{\psi}\eta + \bar{\eta}\psi + K^{a\mu}D_\mu^{ab}\chi_2^b - \frac{1}{2}gf^{abc}\zeta^a\chi_2^b\chi_2^c$$

$$+ ig\chi_2^a\bar{\omega}T^a\psi + ig\chi_2^a\bar{\psi}T^a\omega \quad . \tag{2.5.188}$$

where $K^{a\mu}$, ζ^a and ω are new source functions for the above composite fields ($K^{a\mu}$ is Grassmann while ζ^a and ω are ordinary). Note that the above composite fields correspond to the BRS transformations δA_μ^a, $\delta\chi_2^a$ and $\delta\psi$ respectively. According to the nilpotency of the BRS transformation which will be shown in the "Digression", the BRS transformation on $\sum$ yields

$$\delta\sum = \delta A_\mu^a J^{a\mu} + \delta\chi_1^a\xi_1^a + \delta\chi_2^a\xi_2^a + \delta\bar{\psi}\eta + \bar{\eta}\delta\psi \quad , \tag{2.5.189}$$

where the source functions are assumed to be unchanged under the BRS transformation. Hence we have an identity similar to Eq. (2.5.187) but with the source term $\sum$:

$$\int [dA]\,[d\chi_1]\,[d\chi_2]\,[d\psi]\,[d\bar{\psi}]\int d^4x(\delta AJ+\delta\chi_1\xi_1+\delta\chi_2\xi_2+\delta\bar{\psi}\eta+\bar{\eta}\delta\psi)$$

$$\times\ \exp\{i\int d^4x(\mathscr{L}+\textstyle\sum)\} = 0 \quad . \qquad (2.5.190)$$

Since the generating functional (2.5.181) is now modified so that the source term is given by $\sum$, the following relations follow

$$\delta\lambda\frac{\delta Z}{\delta K^a_\mu} = i\langle\delta A^{a\mu}\rangle \quad ,$$

$$\delta\lambda\frac{\delta Z}{\delta\zeta^a} = i\langle\delta\chi^a_2\rangle \quad , \qquad (2.5.191)$$

$$\delta\lambda\frac{\delta Z}{\delta\omega} = i\langle\delta\bar{\psi}\rangle \quad ,$$

$$\delta\lambda\frac{\delta Z}{\delta\bar{\omega}} = i\langle\delta\psi\rangle \quad ,$$

where an expression like $\langle F\rangle$ is defined by

$$\langle F\rangle = \int [dA][d\chi_1][d\chi_2][d\psi][d\bar{\psi}]F\exp\{i\int d^4x(\mathscr{L}+\textstyle\sum)\} \quad . \qquad (2.5.192)$$

Hence Eq. (2.5.190) may be rewritten in the form

$$\int d^4x\left(\frac{\delta Z}{\delta K^a_\mu}J^a_\mu+\frac{i}{\alpha}\partial_\mu\frac{\delta Z}{\delta J^a_\mu}\xi^a_1+\frac{\delta Z}{\delta\zeta^a}\xi^a_2+\frac{\delta Z}{\delta\omega}\eta-\bar{\eta}\frac{\delta Z}{\delta\bar{\omega}}\right)=0 \quad . \qquad (2.5.193)$$

The generating functional W for connected Green functions is related to Z through $Z = \exp(iW)$ as was shown in Sec. 2.3.1. Hence the functional $W[J, \xi_1, \xi_2, \eta, \bar{\eta}, K, \zeta, \omega, \bar{\omega}]$ satisfies the same identity as Eq. (2.5.193). Equation (2.5.193) is the generalized Ward-Takahashi identity for the generating functionals Z and W. By suitably differentiating Eq. (2.5.193) with respect to the source functions, we may derive the generalized Ward-Takahashi identities for Green functions which are identical to the previously obtained set of identities Eq. (2.5.186).

In order to discuss the renormalizability of gauge theories, it is useful to convert the identity (2.5.193) to the one for the effective action (see Eq. (2.3.41)) which is a generating functional for proper (one-particle irreducible) Green functions. We define the effective action Γ in QCD by

$$\begin{aligned}\Gamma[v_A, v_{\chi_1}, v_{\chi_2}, v_\psi, v_{\bar\psi}; K, \zeta, \omega, \bar\omega] &= W[J, \xi_1, \xi_2, \eta, \bar\eta, K, \zeta, \omega, \bar\omega] \\ &\quad - \int d^4x (v_A^{a\mu} J_\mu^a + v_{\chi_1}^a \xi_1^a + v_{\chi_2}^a \xi_2^a + v_{\bar\psi}\eta + \bar\eta v_\psi) \quad , \end{aligned} \qquad (2.5.194)$$

where new variables $v_A, v_{\chi_1}, v_{\chi_2}, v_\psi$ and $v_{\bar\psi}$ are defined by

$$v_A^{a\mu} = \frac{\delta W}{\delta J_\mu^a}, \quad v_{\chi_{1,2}}^a = \frac{\delta W}{\delta \xi_{1,2}^a}, \quad v_\psi = \frac{\delta W}{\delta \bar\eta}, \quad v_{\bar\psi} = -\frac{\delta W}{\delta \eta} \quad . \qquad (2.5.195)$$

Using Eq. (2.5.194) we rewrite Eq. (2.5.193) with Z replaced by W so that

$$\begin{aligned}\int d^4x \Bigg[\frac{\delta\Gamma}{\delta K_\mu^a}\frac{\delta\Gamma}{\delta v_A^{a\mu}} + \frac{i}{\alpha}(\partial_\mu v_A^{a\mu})\frac{\delta\Gamma}{\delta v_{\chi_1}^a} + \frac{\delta\Gamma}{\delta\zeta^a}\frac{\delta\Gamma}{\delta v_{\chi_2}^a} \\ + \frac{\delta\Gamma}{\delta\omega}\frac{\delta\Gamma}{\delta v_{\bar\psi}} + \frac{\delta\Gamma}{\delta v_\psi}\frac{\delta\Gamma}{\delta\bar\omega}\Bigg] = 0 \quad . \end{aligned} \qquad (2.5.196)$$

Next we use the field equation for the ghost field χ_2^a under the effective Lagrangian $\mathscr{L} + \sum$, i.e.,

$$\langle [i\partial^\mu(D_\mu^{ab}\chi_2^b) - \xi_1^a]\rangle = 0 \quad . \qquad (2.5.197)$$

In terms of the generating functional Z, Eq. (2.5.197) is written as

$$i\,\partial_\mu \frac{\delta Z}{\delta K_\mu^a} = \frac{\delta Z}{\delta \chi_1^a} \quad , \qquad (2.5.198)$$

which, for the effective action Γ, takes the form

$$i\,\partial_\mu \frac{\delta\Gamma}{\delta K_\mu^a} = -\frac{\delta\Gamma}{\delta v_{\chi_1}^a} . \qquad (2.5.199)$$

Inserting Eq. (2.5.199) into Eq. (2.5.196) and making an integration by parts, we have

$$\int d^4x \left[\frac{\delta\Gamma}{\delta K_\mu^a}\left(\frac{\delta\Gamma}{\delta v_A^{a\mu}} - \frac{1}{\alpha}\partial_\mu\partial_\nu v_A^{a\nu}\right) + \frac{\delta\Gamma}{\delta\zeta^a}\frac{\delta\Gamma}{\delta v_{\chi_2}^a} + \frac{\delta\Gamma}{\delta\omega}\frac{\delta\Gamma}{\delta v_{\bar\psi}} + \frac{\delta\Gamma}{\delta v_\psi}\frac{\delta\Gamma}{\delta\bar\omega}\right] = 0 \quad . \qquad (2.5.200)$$

To simplify Eq. (2.5.200) we define a new generating functional $\tilde{\Gamma}$ such that

$$\tilde{\Gamma} = \Gamma + \frac{1}{2\alpha}\int d^4x(\partial_\nu v_A^{a\nu})^2 \quad . \tag{2.5.201}$$

We have finally the generalized Ward-Takahashi identity for the effective action $\tilde{\Gamma}$,

$$\int d^4x\left(\frac{\delta\tilde{\Gamma}}{\delta K_\mu^a}\frac{\delta\tilde{\Gamma}}{\delta v_A^{a\mu}} + \frac{\delta\tilde{\Gamma}}{\delta\zeta^a}\frac{\delta\tilde{\Gamma}}{\delta v_{\chi_1}^a} + \frac{\delta\tilde{\Gamma}}{\delta\omega}\frac{\delta\tilde{\Gamma}}{\delta v_{\bar{\psi}}} + \frac{\delta\tilde{\Gamma}}{\delta v_\psi}\frac{\delta\tilde{\Gamma}}{\delta\bar{\omega}}\right) = 0 \quad . \tag{2.5.202}$$

In the standard approach to the proof of renormalizability [Lee 76, Tay 76, Fad 80, Itz 80] the above form of the generalized Ward-Takahashi identity (2.5.202) is extensively used to give an elegant inductive proof of renormalizability of gauge theories.

It should be stressed here that, although the above derivation of the generalized Ward-Takahashi identity is given in terms of unrenormalized quantities, the same result can be obtained if we replace them by renormalized quantities. In that case, however, we assume the existence of a gauge-invariant regularization and subtraction scheme. A typical example of such a regularization and subtraction scheme is dimensional regularization and minimal subtraction which we have been employing throughout the book.

Finally we remark that the Slavnov-Taylor identity in the narrow sense (2.5.128) can be derived by direct use of the generalized Ward-Takahashi identities for appropriate Green functions. For example, the first equality in Eq. (2.5.128) may be proven by observing the identity (2.5.167) which reads [Sla 72]

$$\begin{aligned}\tilde{Z}_3\{\langle 0|\mathrm{T}[(\hat{D}_\mu^{ad}\hat{\chi}_2^d(x))\hat{A}_\nu^b(y)\hat{\chi}_1^c(z)]|0\rangle + \langle 0|\mathrm{T}[\hat{A}_\mu^a(x)\,(\hat{D}_\nu^{bd}\hat{\chi}_2^d(y))\hat{\chi}_1^c(z)]|0\rangle\}\\ + \frac{i}{\alpha}\frac{\partial}{\partial z_\lambda}\langle 0|\mathrm{T}[\hat{A}_\mu^a(x)\hat{A}_\nu^b(y)\hat{A}_\lambda^c(z)]|0\rangle = 0 \quad ,\end{aligned} \tag{2.5.203}$$

where fields $\hat{A}^a$, $\hat{\chi}_1^a$ and $\hat{\chi}_2^a$ stand for the renormalized ones.

Digression

NILPOTENCY OF THE BRS TRANSFORMATION: In deriving Eq. (2.5.189) we used a remarkable property of the BRS transformation called nilpotency which indicates that the BRS transformation applied twice gives a null result:

$$\delta_B(\delta_B\phi) = 0 \quad , \tag{2.5.204}$$

or more symbolically

$$\delta_B^2 = 0 \quad , \tag{2.5.205}$$

where ϕ represents any one of the fields A_μ^a, χ_1^a, χ_2^a and ψ, and we attached index B to the symbol δ_B to emphasize that it stands for the BRS transformation.

The proof of nilpotency is straightforward. For the field A^a_μ we have (for simplicity we consider the unrenormalized BRS transformations (2.3.168, 169, 173, 174))

$$\begin{aligned}\delta_B(\delta_B A^a_\mu) &= \delta\lambda\delta_B(D^{ab}_\mu \chi^b_2) \\ &= \delta\lambda\{\partial_\mu \delta_B \chi^a_2 - g f^{abc}(\delta_B A^c_\mu \chi^b_2 + A^c_\mu \delta_B \chi^b_2)\} \quad .\end{aligned} \tag{2.5.206}$$

Inserting Eqs. (2.3.168) and (2.3.174) into the above expression, and noting the Jacobi identity (2.1.39), we find

$$\delta_B(\delta_B A^a_\mu) = 0, \quad \text{or} \quad \delta_B(D^{ab}_\mu \chi^b_2) = 0. \tag{2.5.207}$$

For the field χ^a_1 it is straightforward to show that

$$\begin{aligned}\delta_B(\delta_B \chi^a_1) &= \delta\lambda \frac{i}{\alpha} \delta_B(\partial^\mu A^a_\mu) \\ &= \delta\lambda\delta\lambda' \frac{i}{\alpha}\{\Box\chi^a_2 - g f^{abc}\partial^\mu(A^c\chi^b_2)\} = 0 \quad ,\end{aligned} \tag{2.5.208}$$

where the last equality holds on account of the field equation for the ghost field χ^a_2.

For the field χ^a_2 we immediately see

$$\delta_B(\delta_B \chi^a_2) = \delta\lambda\left(-\frac{1}{2}\right) g f^{abc} \delta_B(\chi^b_2 \chi^c_2) = 0 \ , \tag{2.5.209}$$

according to the Jacobi identity. Expressing this result in the form we used in deriving Eq. (2.5.189), we have

$$\delta_B\left(-\frac{1}{2} g f^{abc} \chi^b_2 \chi^c_2\right) = 0 \quad . \tag{2.5.210}$$

For the fermion field ψ nilpotency is also manifest:

$$\delta_B(\delta_B\psi) = \delta\lambda\delta_B(igT^a\chi^a_2\psi) = 0 \quad , \tag{2.5.211}$$

since

$$\begin{aligned}T^a\delta_B(\chi^a_2\psi) &= T^a(\delta_B\chi^a_2\psi + \chi^a_2\delta_B\psi) \\ &= \delta\lambda g\left(-\frac{1}{2} f^{abc}T^c - iT^aT^b\right)\chi^a_2\chi^b_2\psi \\ &= \delta\lambda(i/2)g(if^{abc}T^c - [T^a, T^b])\chi^a_2\chi^b_2\psi = 0 \quad .\end{aligned} \tag{2.5.212}$$

CHAPTER

THREE

RENORMALIZATION GROUP METHOD

The finite renormalization of Green functions constitutes a group called the renormalization group. The physical predictions are invariant under renormalization group transformations. An analytic expression of this property is given by *the renormalization group equation*. We derive the renormalization group equation in the minimal subtraction scheme and compare it with the other renormalization group equations. We present its general solution, and through it we characterize the short-distance behavior of the underlying theory. The method of the renormalization group equation is applied to quantum chromodynamics leading to an understanding of asymptotic freedom. The meaning of the quantities appearing in the renormalization group equation is clarified in terms of the scale transformation.

3.1. RENORMALIZATION GROUP

3.1.1. Finite renormalization

According to the renormalization program we subtract all the divergences from the Green functions systematically order by order in perturbation theory. In the subtraction procedure there exists an arbitrariness of how to define a divergent piece in a Green function, i.e., how much of the finite piece is to be subtracted together with the infinity. This arbitrariness which has already been mentioned in Sec. 2.4.3 is equivalent to the arbitrariness in splitting the Lagrangian into a renormalized Lagrangian and the counter terms and results in a variety of renormalization schemes.

The above arbitrariness persists in the condition for defining the renormalized quantities. For example, in quantum chromodynamics, the renormalized coupling constant g_r may be defined in terms either of the triple-gluon vertex or of the quark-gluon vertex. In general different coupling constants g_r are obtained corresponding to these two different definitions (if one imposes the Slavnov-Taylor identity, these two coupling constants coincide).

In subtracting the divergences we inevitably introduce an arbitrary mass scale μ which we call the *renormalization scale*. For example, in the on-shell subtraction scheme, the renormalization scale μ is taken to be the physical mass of the relevant particle at which the renormalization condition is set up. In the off-shell subtraction scheme, the renormalization scale μ is the off-shell value of the relevant momentum. The renormalization scale in the MS (or $\overline{\text{MS}}$) scheme is rather implicit. In this scheme only the pole in the space-time dimension is subtracted and, at first sight, no mass scale seems to be needed. As a matter of fact, however, the mass dimension of the coupling constant in arbitrary space-time dimensions plays a role of the renormalization scale. The renormalization scale μ is entirely arbitrary (except for the on-shell subtraction scheme) and remains in the finite part of the Green functions thus leaving an arbitrariness for the renormalized Green functions after the subtraction of divergences.

Summing up, the arbitrariness in subtracting divergences is two-fold: (1) arbitrariness of choosing the renormalization condition (arbitrariness in setting up the condition to subtract divergences) and (2) arbitrariness of fixing the renormalization scale μ (arbitrariness of the mass scale at which the subtraction is made). In a broader sense the second arbitrariness may be absorbed in the notion of the first one. It is, however, more elucidating to keep noting the above distinction. Just as the second arbitrariness is characterized by the scale parameter μ, it is also possible to give an explicit parametrization for the first arbitrariness by using suitable renormalization parameters [Ste 81, 81a, Pet 82]. We shall come back to this topics in Sec. 5.3.1.

According to this arbitrariness we have many possible expressions for one physical quantity depending on the choice of the renormalization scheme and scale. These different expressions are connected by a finite renormalization as will be shown in a moment. A natural question here is whether these different expressions for one physical quantity are really equivalent or not. Since they are obtained for one physical quantity starting from the unique Lagrangian, they describe a unique physical phenomenon and hence have to be equivalent. In other words, physical quantities such as S-matrix elements are invariant under a finite renormalization.

Let us explain the above circumstances in a more quantitative way by using an example. As before we employ the ϕ_6^3 theory (ϕ^3 theory in 6 dimensions) whose renormalization was fully described in Sec. 2.5.3. While the generating functional $W[J, g, m]$ for connected Green functions has a divergent expression, the same functional acquires a finite form when it is re-expressed in terms of the renormalized quantities J_r, g_r and m_r as in Eq. (2.5.23). The above renormalized quantities are defined by Eq. (2.5.22), i.e.,

$$J_r = Z_3^{1/2} J, \quad g_r = Z_g^{-1} g, \quad m_r^2 = Z_m^{-1} m^2 \quad . \tag{3.1.1}$$

According to the arbitrariness of the renormalization procedure mentioned before, we may choose a subtraction procedure differing from the one used to obtain Eq. (3.1.1). The renormalized quantities obtained in this new procedure will be distinguished from the ones in Eq. (3.1.1) by attaching a prime:

$$J'_r = Z'^{1/2}_3 J, \quad g'_r = Z'^{-1}_g g, \quad m'^2_r = Z'^{-1}_m m^2 \quad . \tag{3.1.2}$$

Since the physical prediction in the above two renormalization procedures should be the same, we have

$$W'_r[J'_r, g'_r, m'_r, \mu'] = W_r[J_r, g_r, m_r, \mu] \quad , \tag{3.1.3}$$

where we have written down an explicit dependence on the renormalization scale μ. The transformation from the scheme (J_r, g_r, m_r, μ) to (J'_r, g'_r, m'_r, μ') is effected by a finite renormalization obtained from Eqs. (3.1.1) and (3.1.2),

$$J'_r = z_3^{1/2} J_r, \quad g'_r = z_g g_r, \quad m'^2_r = z_m m_r^2, \tag{3.1.4}$$

where z_3, z_g and z_m are defined by

$$z_3 = Z'_3/Z_3, \quad z_g = Z_g/Z'_g, \quad z_m = Z_m/Z'_m \quad . \tag{3.1.5}$$

The constants z_3, z_g and z_m are, in fact, finite since the divergent part in Z'_3 (Z'_g, Z'_m) cancels out by that of Z_3 (Z_g, Z_m) owing to its multiplicative nature. Equation (3.1.3) is the quantitative statement of the fact that the physical prediction is invariant under the finite renormalization.

By differentiating the generating functional $W_r[J_r, g_r, m_r, \mu]$ with respect to the source function J_r, we obtain connected Green functions from which the S-matrix elements $S(p, g_r, m_r, \mu)$ are derived as described in Sec. 2.3.1 where p represents an aggregate of external momenta. The uniqueness of the physical prediction implies that $S(p, g_r, m_r, \mu)$ is invariant under the finite renormalization,

$$S'(p, g'_r, m'_r, \mu') = S(p, g_r, m_r, \mu) \quad , \tag{3.1.6}$$

where g'_r and m'_r are the renormalized coupling constant and mass in the scheme different from the one for g_r and m_r. Here g'_r and m'_r may be obtained from g_r and m_r by finite renormalization and are functions of g_r and m_r:

$$\begin{aligned} g'_r &= g'_r(g_r, m_r) \quad , \\ m'_r &= m'_r(g_r, m_r) \quad . \end{aligned} \tag{3.1.7}$$

Of course the functional form of $S'(p, g'_r, m'_r, \mu')$ as a function of g'_r and m'_r is apparently different from that of $S(p, g_r, m_r, \mu)$, while their values have to be the same. We compute physical quantities such as cross sections on the basis of $S(p, g_r, m_r, \mu)$ and compare the cross sections with experimental data. We then determine the values of g_r and m_r and predict the cross sections as functions of p. The same procedure as above may be applied to the S-matrix element $S'(p, g'_r, m'_r, \mu')$ and then the values of g'_r and m'_r are determined by fitting the cross sections with experimental data. The values of g'_r and m'_r together with g_r and m_r must satisfy Eq. (3.1.7). In this way $S(p, g_r, m_r, \mu)$ and $S'(p, g'_r, m'_r, \mu')$ describe the same physics.

The above statement may be directly confirmed if we knew the exact form of the S-matrix element $S(p, g_r, m_r, \mu)$. Unfortunately, however, we can calculate $S(p, g_r, m_r, \mu)$ only through a perturbation series and we truncate the series after the first few orders. To guarantee the smallness of the neglected higher order terms, it is better to choose the scheme in which the value of g_r is small (when it is determined through the comparison of the physical prediction with experimental data). Since we neglect the higher order terms, $S'(p, g'_r, m'_r, \mu')$ and $S(p, g_r, m_r, \mu)$ differ by

$$S'(p, g'_r, m'_r, \mu') - S(p, g_r, m_r, \mu) = O(g_r^{n+2}) \quad , \tag{3.1.8}$$

where n represents the order in the perturbation series to which the S-matrix element is calculated. This fact creates the problem of the renormalization-scheme dependence of physical predictions which will be discussed in more detail in Sec. 5.3.

In the case of quantum electrodynamics the on-shell renormalization scheme is chosen as a natural scheme since the physical electron mass (i.e., the pole of the electron propagator) is directly measurable and the fine structure constant $\alpha = e^2/4\pi$ is precisely determined in the Thomson limit for Compton scattering [Thi 50, Bj 65]. The coupling constant determined in the on-shell scheme is in fact small ($\alpha = 1/137.036$) and perturbation theory works in this scheme. We are, however, not necessarily forced to use only the on-shell renormalization scheme and, in fact, we can choose a different scheme in which the renormalization scale is taken to be an off-shell photon momentum.

3.1.2. Group of finite renormalizations

The renormalized coupling constant g_r and mass m_r depend on the renormalization scale μ at which the subtraction procedure is defined. Writing this dependence explicitly we have [Eri 63]

$$g_r(\mu) = Z_g(\mu)^{-1} g \quad ,$$
$$m_r(\mu) = Z_m(\mu)^{-1/2} m \quad . \tag{3.1.9}$$

The renormalized coupling constants $g_r(\mu)$ and $g_r(\mu')$ which are obtained through two different subtraction procedures characterized by the renormalization scales μ and μ', respectively, are related to each other as in Eq. (3.1.4) by

$$g_r(\mu') = z_g(\mu', \mu) g_r(\mu) \quad , \tag{3.1.10}$$

where the finite renormalization $z_g(\mu', \mu)$ is given by

$$z_g(\mu', \mu) = Z_g(\mu)/Z_g(\mu') \quad . \tag{3.1.11}$$

In the same way we have

$$m_r(\mu') = z_m(\mu', \mu) m_r(\mu) \quad , \tag{3.1.12}$$

where $z_m(\mu', \mu)$ is defined by

$$z_m(\mu', \mu) = [Z_m(\mu)/Z_m(\mu')]^{\frac{1}{2}} \quad . \tag{3.1.13}$$

Let us focus our attention on Eq. (3.1.10) which defines a set of finite renormalizations $\{z_g(\mu', \mu)\}$ for varying renormalization scales μ' and μ. We regard the finite renormalization (3.1.10) as a transformation. We can then show that this set of transformations possesses group properties [Wil 71]. In fact we can define a product of two elements[1] $z_g(\mu'', \mu')$ and $z_g(\mu', \mu)$,

$$z_g(\mu'', \mu')\, z_g(\mu', \mu) \quad , \tag{3.1.14}$$

which represents the change of $g_r(\mu)$ through the successive changes of the scales $\mu \to \mu' \to \mu''$. According to Eq. (3.1.11) the above product is equal to $z_g(\mu'', \mu)$ which is nothing but the finite renormalization of $g_r(\mu)$ caused by the scale change $\mu \to \mu''$. Thus the product $z_g(\mu'', \mu')\, z_g(\mu', \mu)$ also belongs to the set $\{z_g(\mu', \mu)\}$. Moreover, the inverse of $z_g(\mu', \mu)$ may be defined by

$$z_g^{-1}(\mu', \mu) = z_g(\mu, \mu') \tag{3.1.15}$$

as is easily seen, and the identity

$$z_g(\mu, \mu) = 1 \tag{3.1.16}$$

belongs to the set $\{z_g(\mu', \mu)\}$. Thus this set is a group which is Abelian.

A similar argument applies to the set of finite renormalizations $\{z_m(\mu', \mu)\}$ showing it to be an Abelian group.

[1] Note that, in defining the product, we have to match the renormalization scale μ' for two elements $z_g(\mu'', \mu')$ and $z_g(\mu', \mu)$.

We have yet another kind of the finite renormalization which applies to Green functions. Let us consider the truncated connected Green function $\tilde{G}_n^{\rm tc}(p, g, m)$ which is defined in Eq. (2.3.34), where p stands for an aggregate of external momenta $p_1, p_2, \ldots, p_{n-1}$ and the dependence on g and m is explicitly shown. The Green function $-i\tilde{G}_n^{\rm tc}(p, g, m)$ corresponds to the Feynman amplitude as is shown in Sec. 2.3.2. The definition of $\tilde{G}_n^{\rm tc}(p, g, m)$ reads

$$(2\pi)^4\delta^4(p_1 + \ldots + p_n)\,\tilde{G}_2(p_1)\ldots\tilde{G}_2(p_n)\tilde{G}_n^{\rm tc}(p, g, m)$$

$$= \int d^4x_1 \ldots d^4x_n e^{-i(p_1\cdot x_1 + \ldots + p_n\cdot x_n)}\, G_n^{\rm c}(x_1, \ldots, x_n) \quad , \qquad (3.1.17)$$

where $G_n^{\rm c}(x_1, \ldots, x_n)$ is the connected Green function given by Eq. (2.3.25), i.e.,

$$G_n^{\rm c}(x_1, \ldots, x_n) = (-i)^{n-1} \left.\frac{\delta^n W[J, g, m]}{\delta J(x_1)\ldots\delta J(x_n)}\right|_{J=0} \quad . \qquad (3.1.18)$$

On account of Eq. (2.5.23) and the renormalization (3.1.1), i.e., $J_r = Z_3^{1/2}J$, we find that the renormalized connected Green function $G_{rn}^{\rm c}(x_1, \ldots, x_n)$ is related to $G_n^{\rm c}(x_1, \ldots, x_n)$ by

$$G_{rn}^{\rm c} = Z_3^{-n/2}G_n^{\rm c} \quad . \qquad (3.1.19)$$

Defining the renormalized truncated connected Green function $\tilde{G}_{rn}^{\rm tc}(p, g_r, m_r, \mu)$ (an explicit dependence on the renormalization scale μ is indicated) through the equation

$$(2\pi)^4\delta^4(p_1 + \ldots + p_n)\tilde{G}_{r2}(p_1)\ldots\tilde{G}_{r2}(p_n)\tilde{G}_{rn}^{\rm tc}(p, g_r, m_r, \mu)$$

$$= \int d^4x_1 \ldots d^4x_n e^{-i(p_1\cdot x_1 + \ldots + p_n\cdot x_n)}\, G_{rn}^{\rm c}(x_1, \ldots, x_n) \quad , \qquad (3.1.20)$$

we obtain

$$\tilde{G}_{rn}^{\rm tc}(p, g_r, m_r, \mu) = Z_3(\mu)^{n/2}\tilde{G}_n^{\rm tc}(p, g, m) \quad . \qquad (3.1.21)$$

We introduce the renormalized Feynman amplitude $F_n(p, g_r(\mu), m_r(\mu), \mu)$ with renormalization scale μ such that

$$F_n(p, g_r(\mu), m_r(\mu), \mu) = -i\tilde{G}_{rn}^{\rm tc}(p, g_r(\mu), m_r(\mu), \mu) \quad . \qquad (3.1.22)$$

The finite renormalization for F_n is then given by

$$F_n(p, g_r(\mu'), m_r(\mu'), \mu') = z_n(\mu', \mu)F_n(p, g_r(\mu), m_r(\mu), \mu) \quad , \tag{3.1.23}$$

where the renormalization factor $z_n(\mu', \mu)$ is defined by

$$z_n(\mu', \mu) = [Z_3(\mu')/Z_3(\mu)]^{n/2} \quad . \tag{3.1.24}$$

Thus the change of the renormalization scale $\mu \to \mu'$ generates a finite multiplicative renormalization of the Feynman amplitudes which, as before, gives rise to an Abelian group $\{z_n(\mu', \mu)\}$.

We have gotten three sets of finite renormalizations $\{z_g(\mu', \mu)\}$, $\{z_m(\mu', \mu)\}$ and $\{z_n(\mu', \mu)\}$ which constitute Abelian groups generated by the scale change $\mu \to \mu'$. The group thus obtained is called the *renormalization group* [Stu 53, Gel 54, Ovs 56].

3.2. RENORMALIZATION GROUP EQUATIONS

3.2.1. Renormalization group equation in the MS scheme

The transformations (3.1.10), (3.1.12) and (3.1.23) corresponding to the finite renormalization caused by the change of the renormalization scale μ may be regarded as functional equations for $g_r(\mu)$, $m_r(\mu)$ and $F_n(p, g_r(\mu), m_r(\mu), \mu)$ characteristic to the renormalization group. If we restrict ourselves to an infinitesimal change of the renormalization scale μ, these functional equations reduce to differential equations which correspond to the Lie differential equations in the Lie group [Bog 80]. The differential equations expressing the response of Green functions and parameters (e.g., coupling constants and masses) to the change of the renormalization scale μ are, in general, called the *renormalization group equations* [Stu 53, Gel 54, Ovs 56, Sym 70, Cal 70, tHo 73, Wei 73a].

It is very important to note here that the renormalization group equation is independent of perturbation theory although our previous argument on the renormalization has been based on the perturbative treatment. Because of this important property the renormalization group equation may be employed to supplement perturbation theory for the derivation of the proper asymptotic behavior of Green functions for large external momenta.

In this subsection we shall derive the renormalization group equation in the MS (or $\overline{\text{MS}}$) scheme [tHo 73, Bar 78], i.e., under the minimal subtraction (or modified minimal subtraction) renormalization condition (see Sec. 2.4.3). For simplicity we work in the ϕ_6^3 theory as in the previous section. The case of quantum chromodynamics will be dealt with at the end of this subsection.

We first derive the differential equations corresponding to the functional equations (3.1.10) and (3.1.12). We keep μ fixed in Eqs. (3.1.10) and (3.1.12), and differentiate both sides of these functional equations with respect to μ'. In this way we can derive the differential equations for $g_r(\mu')$ and $m_r(\mu')$. It is, however, more convenient in the later practical calculations to start with Eq. (3.1.9) to obtain the same differential equations. In Eq. (3.1.9) we employ dimensional regularization to make the equation mathematically meaningful. Then g and g_r acquire mass dimensions as we saw before in Sec. 2.5.3. We isolate these mass dimensions explicitly,

$$
\begin{aligned}
g &= g_0 \mu_0^{\varepsilon} \quad , \\
g_r &= g_R \mu^{\varepsilon} \quad ,
\end{aligned}
\tag{3.2.1}
$$

where $\varepsilon = (6-D)/2$ and g_0 and g_R are dimensionless coupling constants. Here the mass scale μ_0 for the bare coupling constant g is a fixed scale while the mass scale μ for the renormalized coupling constant g_r is a variable parameter. The mass scale μ is in fact identified with the renormalization scale in the MS (or $\overline{\text{MS}}$) scheme. Using Eq. (3.2.1) we rewrite Eq. (3.1.9) in the following form

$$
g_R(\mu) = \left(\frac{\mu_0}{\mu}\right)^{\varepsilon} Z_g(\mu)^{-1} g_0 \quad . \tag{3.2.2}
$$

The bare parameters g and m are regarded as fixed constants and are free from the renormalization scale μ. Hence we have

$$
\frac{dg}{d\mu} = 0, \quad \frac{dm}{d\mu} = 0. \tag{3.2.3}
$$

According to Eqs. (3.2.2) we obtain, from Eq. (3.2.3), the differential equations for the renormalized parameters g_R and m_R,

$$
\mu \frac{dg_R}{d\mu} = \beta \quad , \tag{3.2.4}
$$

$$
\mu \frac{dm_R}{d\mu} = -m_R \gamma_m \quad , \tag{3.2.5}
$$

where we have rewritten m_r as m_R, i.e., $m_R = m_r$, just to balance the notation, and β and γ_m are given by

$$
\beta = -\varepsilon g_R - \frac{\mu}{Z_g} \frac{dZ_g}{d\mu} g_R \quad , \tag{3.2.6}
$$

$$
\gamma_m = \frac{\mu}{Z_m^{1/2}} \frac{dZ_m^{1/2}}{d\mu} \quad . \tag{3.2.7}
$$

The quantities β and γ_m defined here are finite functions of μ since the divergences in Z_g and Z_m, as $\varepsilon \to 0$, cancel out in the expressions (3.2.6) and (3.2.7). Their dependence on μ is partly direct and partly indirect through $g_R(\mu)$ and $m_R(\mu)$, i.e., they are, in general, functions of μ, g_R and m_R. In the MS (or $\overline{\text{MS}}$) scheme, however, we can show that they depend only on g_R,

$$\beta = \beta(g_R) \quad , \tag{3.2.8}$$

$$\gamma_m = \gamma_m(g_R) \quad . \tag{3.2.9}$$

In fact, Eqs. (3.2.8) and (3.2.9) are justified since in the MS scheme the renormalization constants Z_g and $Z_m^{1/2}$ take the form,

$$Z_g = 1 + g_R^2\frac{A_{11}}{\varepsilon} + g_R^4\left(\frac{A_{22}}{\varepsilon^2} + \frac{A_{21}}{\varepsilon}\right) + \dots \quad , \tag{3.2.10}$$

$$Z_m^{1/2} = 1 + g_R^2\frac{B_{11}}{\varepsilon} + g_R^4\left(\frac{B_{22}}{\varepsilon^2} + \frac{B_{21}}{\varepsilon}\right) + \dots \quad , \tag{3.2.11}$$

where A_{ij} and B_{ij} $(i, j = 1, 2, 3, \dots)$ are constants. Thus the dependence of Z_g and $Z_m^{1/2}$ on μ is solely through g_R. The reason why A_{ij} and B_{il} do not depend on m_R is the following: As we have seen in Sec. 2.5.3, the renormalization constants in the MS scheme are independent of μ, i.e., the μ-dependent divergence such as $(\ln \mu)/\varepsilon$ is always canceled by the divergence coming from counter terms for subdiagrams, and should not be included in the counter terms for the overall divergence. Hence A_{ij} and B_{ij} do not depend on μ. By dimensional reasoning they cannot depend on m_R.

The equations (3.2.4) and (3.2.5) together with Eqs. (3.2.8) and (3.2.9) are the renormalization group equations for g_R and m_R in the MS scheme [tHo 73]. These equations are convenient to deal with because β and γ_m are independent of m_R and hence Eq. (3.2.4) decouples from Eq. (3.2.5). We may thus solve Eq. (3.2.4) independently of Eq. (3.2.5). In this sense the MS scheme is often referred to as the mass-independent renormalization scheme.

We now turn our attention to the differential equation corresponding to the functional equation (3.1.23). We note that the unrenormalized n-point Green function $\tilde{G}_n^{tc}(p, g, m)$ defined in Eq. (3.1.17) is independent of the renormalization scale μ as far as the bare parameters g and m are fixed, i.e.,

$$\frac{d}{d\mu}\tilde{G}_n^{tc}(p, g, m)\bigg|_{g,m} = 0 \quad , \tag{3.2.12}$$

where the g and m affixed to the above equation indicate that g and m are kept fixed. In terms of the renormalized n-point Green function defined by Eq. (3.1.21), we re-express Eq. (3.2.12) in the following way:

$$\frac{d}{d\mu}[Z_3(\mu,g,m)^{-n/2}F_n(p,g_R(\mu,g,m),\, m_R(\mu,g,m),\, \mu)]\bigg|_{g,m} = 0 \quad , \qquad (3.2.13)$$

where the definition (3.1.22) has been employed and the dependence of Z_3, g_R and m_R on μ, g and m is explicitly shown. Applying the chain rule in differentiation to Eq. (3.2.13) we obtain

$$\frac{\partial Z_3^{-n/2}}{\partial\mu}\bigg|_{g,m} F_n + Z_3^{-n/2}\left(\frac{\partial}{\partial\mu} + \frac{\partial g_R}{\partial\mu}\frac{\partial}{\partial g_R} + \frac{\partial m_R}{\partial\mu}\frac{\partial}{\partial m_R}\right)F_n\bigg|_{g,m} = 0 \quad . \qquad (3.2.14)$$

Rewriting Eq. (3.2.14) we have

$$\left(\mu\frac{\partial}{\partial\mu} + \beta\frac{\partial}{\partial g_R} - \gamma_m m_R\frac{\partial}{\partial m_R} - n\gamma\right)F_n = 0 \quad , \qquad (3.2.15)$$

where β, γ_m and γ are defined by

$$\beta = \mu\frac{\partial g_R}{\partial\mu}\bigg|_{g,m} \quad , \qquad (3.2.16)$$

$$\gamma_m = -\frac{\mu}{m_R}\frac{\partial m_R}{\partial\mu}\bigg|_{g,m} \quad , \qquad (3.2.17)$$

$$\gamma = \frac{\mu}{2Z_3}\frac{\partial Z_3}{\partial\mu}\bigg|_{g,m} \quad . \qquad (3.2.18)$$

Equations (3.2.16) and (3.2.17) are essentially the same as Eqs. (3.2.4) and (3.2.5), respectively, and so β and γ_m are identified with the previous ones (3.2.8) and (3.2.9). Hence β and γ_m appearing in Eq. (3.2.15) depend only on g_R. We can show, essentially in the same way as before, that the coefficient γ defined in Eq. (3.2.18) is also independent of m_R. In fact the field renormalization constant Z_3 in the MS scheme acquires the form

$$Z_3 = 1 + g_R^2\frac{C_{11}}{\varepsilon} + g_R^4\left(\frac{C_{22}}{\varepsilon^2} + \frac{C_{21}}{\varepsilon}\right) + \ldots, \qquad (3.2.19)$$

with C_{ij} ($i,j = 1, 2, 3, \ldots$) being the constant independent of m_R. Inserting Eq. (3.2.19) into Eq. (3.2.18) we find that γ depends only on g_R:

$$\gamma = \gamma(g_R) \qquad (3.2.20)$$

Equation (3.2.15) together with Eqs. (3.2.8), (3.2.9) and (3.2.20) constitutes the renormalization group equation for the Green function F_n in the MS scheme [Wei 73a, Hol 74, Col 74]. We call Eq. (3.2.15) the 't Hooft-Weinberg equation. The functions $\beta(g_R)$, $\gamma_m(g_R)$ and $\gamma(g_R)$ are called the renormalization group functions. In particular $\beta(g_R)$ goes by the name of the β function or the Gell-Mann-Low function while $\gamma(g_R)$ is called the anomalous dimension of the field $\phi(x)$. The reason why $\gamma(g_R)$ is the "anomalous" dimension of $\phi(x)$ will be elucidated in Sec. 3.5.

As we have seen above, the renormalization group equations (3.2.4), (3.2.5) and (3.2.15) in the MS scheme are the mere representation of the fact that the bare parameters g and m are μ-independent unique constants. Thus the renormalization group equations guarantee that our theory is based on a unique Lagrangian although the renormalized theory appears to possess an arbitrariness due to the change of the renormalization scale μ. In other words the renormalized theory is warranted by these equations to give unique physical predictions independent of the renormalization scale μ.

It is now easy for us to generalize our result to the case of quantum chromodynamics. Here our basic Lagrangian $\mathscr{L}$ is given by Eq. (2.3.99) or equivalently by Eq. (2.5.116). The bare parameters present in the Lagrangian $\mathscr{L}$ are the gauge coupling constant g, quark mass m and gauge parameter α. We split the Lagrangian into two parts, the renormalized one,

$$\mathscr{L}_r = \mathscr{L}_{r0} + \mathscr{L}_{r1} \quad , \tag{3.2.21}$$

and the counter term $\mathscr{L}_C$ as in Eq. (2.5.125). The renormalized parameters g_r, m_r and α_r are defined by Eq. (2.5.123) in terms of the renormalization constants Z_g, Z_m and Z_3, respectively. We also redefine the gluon field A^a_μ, ghost field χ^a and quark field ψ through the relation (2.5.122) where we need the field renormalization constants for the gluon Z_3, ghost $\tilde{Z}_3$ and quark Z_2. The generalization of the previous renormalization group equations (3.2.4), (3.2.5) and (3.2.15) to the case of QCD is straightforward and reads

$$\begin{aligned}&[\mu\frac{\partial}{\partial\mu}+\beta(g_R,\alpha_R)\frac{\partial}{\partial g_R}-\gamma_m(g_R,\alpha_R)m_R\frac{\partial}{\partial m_R}\\&\quad+\delta(g_R,\alpha_R)\frac{\partial}{\partial\alpha_R}-n_G\gamma_G(g_R,\alpha_R)-n_F\gamma_F(g_R,\alpha_R)]F_{n_G,n_F}=0 \quad ,\end{aligned} \tag{3.2.22}$$

where F_{n_G,n_F} is the truncated connected Green function (times $-i$) with n_G gluon and n_F quark legs in momentum space (we do not consider the Green function with ghost external lines), g_R is the dimensionless renormalized gauge coupling constant defined as in Eq. (3.2.2),

$$g_R = \left(\frac{\mu_0}{\mu}\right)^{\varepsilon} Z_g^{-1} g_0 \quad , \tag{3.2.23}$$

with $g_R = g_r \mu^{-\varepsilon}$, $g_0 = g\, \mu_0^{-\varepsilon}$, $\varepsilon = (4-D)/2$, $m_R = m_r$ and $\alpha_R = \alpha_r$. The renormalization group functions β, γ_m, δ, γ_G and γ_F are defined by

$$\beta(g_R, \alpha_R) = \mu \frac{\partial g_R}{\partial \mu}\bigg|_{g,m,\alpha} \quad , \tag{3.2.24}$$

$$\gamma_m(g_R, \alpha_R) = -\frac{\mu}{m_R}\frac{\partial m_R}{\partial \mu}\bigg|_{g,m,\alpha} \quad , \tag{3.2.25}$$

$$\delta(g_R, \alpha_R) = \mu \frac{\partial \alpha_R}{\partial \mu}\bigg|_{g,m,\alpha} = -2\alpha_R \gamma_G(g_R, \alpha_R) \quad , \tag{3.2.26}$$

$$\gamma_G(g_R, \alpha_R) = \frac{\mu}{2Z_3}\frac{\partial Z_3}{\partial \mu}\bigg|_{g,m,\alpha} , \tag{3.2.27}$$

$$\gamma_F(g_R, \alpha_R) = \frac{\mu}{2Z_2}\frac{\partial Z_2}{\partial \mu}\bigg|_{g,m,\alpha} \quad . \tag{3.2.28}$$

Here γ_G and γ_F are called the anomalous dimensions of the gluon and quark fields, respectively. Note that in Eq. (3.2.26) we used Eqs. (2.5.123) and (3.2.27).

In what follows, the above renormalization group functions play a crucial role in deriving fundamental properties of QCD. Among them the β function is of primary importance, in particular, for the discussion of asymptotic freedom in QCD which will be given in Sec. 3.4. All of these renormalization group functions are, in general, gauge-dependent, i.e., depend on α_R, and hence they are not the direct physical quantities. It is, however, possible to extract some useful physical consequences by observing the behavior of these functions as will be shown in the subsequent sections.

The minimal subtraction renormalization scheme (MS scheme) has the remarkable property that in this scheme the β function and the function γ_m are gauge-independent. The argument for this goes as follows [Cas 74, Gro 76]: In the MS scheme the renormalization constant Z_g defined through the relation (3.2.23) takes the form

$$Z_g = 1 + \frac{a_1(g_R, \alpha_R)}{\varepsilon} + \frac{a_2(g_R, \alpha_R)}{\varepsilon^2} + \dots, \tag{3.2.29}$$

where $a_i(g_R, \alpha_R)$ $(i = 1, 2, 3, \ldots)$ are suitable functions of g_R and α_R. Since the bare coupling constant g_0 is independent of the gauge parameter, we have from Eq. (3.2.23) for fixed μ, μ_0 and ε,

$$\frac{d}{d\alpha_R}(Z_g\, g_R) = 0 \quad . \tag{3.2.30}$$

Inserting Eq. (3.2.29) into Eq. (3.2.30) and rearranging the terms in powers of $1/\varepsilon$, we obtain

$$\frac{dg_R}{d\alpha_R} + \frac{1}{\varepsilon}\left(a_1\frac{dg_R}{d\alpha_R} + \frac{da_1}{d\alpha_R}g_R\right) + \frac{1}{\varepsilon^2}\left(a_2\frac{dg_R}{d\alpha_R} + \frac{da_2}{d\alpha_R}g_R\right) + \ldots = 0 \quad . \tag{3.2.31}$$

Since the terms of different powers of $1/\varepsilon$ in Eq. (3.2.31) are independent (the coefficients of the powers of $1/\varepsilon$ are finite as $\varepsilon \to 0$), each term has to vanish separately, i.e.,

$$\frac{dg_R}{d\alpha_R} = 0, \quad \frac{da_1}{d\alpha_R} = 0, \quad \frac{da_2}{d\alpha_R} = 0 \quad ,\ldots \tag{3.2.32}$$

Hence we find that g_R, a_1, a_2, ... are actually independent of the gauge parameter α_R. This provides us with the immediate consequence that the β function is gauge independent in the MS scheme since it is directly related by definition to g_R, i.e., by Eq. (3.2.24). An argument that is essentially the same as above also applies to the function γ_m and leads to the proof of the gauge independence of γ_m.

It is important to note that in the other renormalization schemes the renormalized coupling constant g_R is, in general, gauge dependent. This is because the expression of Z_g in the renormalization scheme other than the MS (or $\overline{\text{MS}}$) scheme has a term of the form $a_0(g_R, \alpha_R)$ in addition to the terms at the right-hand side of Eq. (3.2.29),

$$Z_g = 1 + a_0(g_R, \alpha_R) + \frac{a_1(g_R,\alpha_R)}{\varepsilon} + \ldots \tag{3.2.33}$$

It is easy to see that the existence of the term $a_0(g_R, \alpha_R)$ prevents us from proving the gauge independence of g_R.

3.2.2. The renormalization group equation in other schemes

In the previous subsection we derived the renormalization group equations on the basis of the MS (or $\overline{\text{MS}}$) scheme and observed that they possessed the expression characteristic to the MS (or $\overline{\text{MS}}$) scheme. The renormalization group equation, in general, takes the form specific to the renormalization scheme employed. If we would have adopted a renormalization scheme other than the MS scheme, we would have obtained a renormalization group equation different from the one in Eq. (3.2.15). In the present subsection we deal with a variety of renormalization group equations in the schemes employed by Gell-Mann and Low [Gel 54], Callan and Symanzik [Cal 70, Sym 70] and Georgi and Politzer [Geo 76], respectively. We make comparison of these renormalization group equations with the one in the MS scheme. For this discussion we again use the ϕ_6^3 theory as a simple example.

In the renormalization scheme adopted by Gell-Mann and Low [Gel 54], the renormalized mass m_R is taken to be the physical mass, i.e., the position of the pole in the propagator for the field ϕ while the renormalized coupling constant g_R and the renormalized field ϕ_R are defined at $p^2 = -\mu^2$ by the off-shell subtraction procedure where μ corresponds to the renormalization scale.[2] Hence m_R is independent of μ while Z_g and Z_3 depend on μ. The renormalized coupling constant in this scheme satisfies the following renormalization group equation

$$\mu\frac{dg_R}{d\mu} = \beta(g_R, \frac{m_R}{\mu}) \quad . \tag{3.2.34}$$

Here, unlike the case of the MS scheme, the β function depends on μ and its dependence is through the ratio m_R/μ because of dimensional reasons. Since the renormalized mass m_R is a fixed constant in this scheme, we have no renormalization group function corresponding to γ_m in the MS scheme. The anomalous dimension γ of the field ϕ is defined analogously to Eq. (3.2.18) and depends on μ through the ratio m_R/μ. The renormalization group equation for the truncated connected n-point Green function (multiplied by $-i$) $F_n(p, g_R, m_R, \mu)$ reads

$$\left[\mu\frac{\partial}{\partial\mu} + \beta\left(g_R, \frac{m_R}{\mu}\right)\frac{\partial}{\partial g_R} - n\gamma\left(g_R, \frac{m_R}{\mu}\right)\right]F_n = 0 \quad . \tag{3.2.35}$$

Equation (3.2.35) together with Eq. (3.2.34) is called the *Gell-Mann-Low equation*. The Gell-Mann-Low equation is simpler in its appearance than Eq. (3.2.15) since the term involving mass differentiation is missing. It is, however, impossible to give a general expression for the solution of Eq. (3.2.35), while the general solution of Eq. (3.2.15) with Eqs. (3.2.8) and (3.2.9) is easily

[2] Here we choose the renormalization point to be space-like in order to avoid a possible singularity in the time-like region.

obtained (see Sec. 3.3). An asymptotic solution of Eq. (3.2.35) for $p, \mu \gg m_R$ may be written down in an explicit way and coincides with that of Eq. (3.2.15) in the MS scheme[3] because $\beta(g_R, 0)$ and $\gamma(g_R, 0)$ are essentially the same as $\beta(g_R)$ and $\gamma(g_R)$ in the MS scheme.

Symanzik [Sym 70] and Callan [Cal 70] adopted the on-shell renormalization scheme to define all the renormalized quantities. Hence the renormalized mass m_R is the physical mass, i.e., the pole position in the propagator. This physical mass m_R is taken to be the renormalization scale and so the renormalized quantities g_R and F_n are defined by subtracting the infinity at $p^2 = m_R^2$.

In order to derive the renormalization group equation in the on-shell renormalization scheme, we start with an unrenormalized Feynman amplitude F_{0n} which is related to the renormalized one F_n by

$$F_n = Z_3^{n/2} F_{0n} \quad , \tag{3.2.36}$$

with Z_3 defined on mass shell. Here we assume that a suitable regularization procedure such as dimensional regularization is applied. We first note that the differential operation $m\partial/\partial m$ with respect to the bare mass m on the unrenormalized Feynman amplitude F_{0n} generates a mass-insertion vertex on an internal line. In fact the following relation apparently proves this statement

$$m\frac{\partial}{\partial m}\frac{1}{m^2-p^2} = -2\frac{1}{m^2-p^2}m^2\frac{1}{m^2-p^2} \quad . \tag{3.2.37}$$

The bare mass insertion on an internal line in F_{0n} may be performed formally by inserting the composite operator (the mass term in $\mathscr{L}$)

$$\hat{\theta} = \frac{1}{2}m^2\hat{\phi}^2 \quad , \tag{3.2.38}$$

among n fields when we define the n-point Green functions, i.e.,

$$\langle 0|\mathrm{T}[\hat{\phi}(x_1)\ldots\hat{\phi}(x_n)\hat{\theta}(x)]|0\rangle_{\mathrm{conn}} . \tag{3.2.39}$$

Performing a truncation (amputation) and Fourier transformation we obtain a new Feynman amplitude F_{0n}^{θ} which corresponds to the Feynman amplitude with the bare mass insertion (3.2.37). Hence we find

$$m\frac{\partial}{\partial m}F_{0n} = -2F_{0n}^{\theta} \quad . \tag{3.2.40}$$

[3] The statement is true as long as $m_R(\mu) \to 0$ as $\mu \to \infty$ in the MS scheme. This condition is satisfied if $\gamma_m(\mu) > -1$. See Sec 3.3.

The Feynman amplitude F^{θ}_{0n} is renormalized in the following way,

$$F^{\theta}_{n} = Z_{\theta}^{-1} Z_3^{n/2} F^{\theta}_{0n} \quad , \tag{3.2.41}$$

where F^{θ}_{n} is the renormalized Feynman amplitude with the insertion of the renormalized composite operator $\hat{\theta}_R$ defined by

$$\hat{\theta}_R = Z_{\theta}^{-1}\,\hat{\theta} \quad , \tag{3.2.42}$$

and Z_{θ} is the renormalization constant, the renormalization condition of which will be given later. Inserting Eqs. (3.2.36) and (3.2.41) into Eq. (3.2.40) and using the relation

$$\frac{\partial F_n}{\partial m} = \frac{\partial m_R}{\partial m}\frac{\partial F_n}{\partial m_R} + \frac{\partial g_R}{\partial m}\frac{\partial F_n}{\partial g_R} \quad , \tag{3.2.43}$$

we find

$$Z_{\theta}^{-1}\left(m\frac{\partial m_R}{\partial m}\frac{\partial}{\partial m_R} + m\frac{\partial g_R}{\partial m}\frac{\partial}{\partial g_R} - n\frac{m}{2Z_3}\frac{\partial Z_3}{\partial m}\right)F_n = -\,2F^{\theta}_{n} \quad . \tag{3.2.44}$$

We choose the renormalization condition to determine Z_{θ} to be

$$Z_{\theta}^{-1}\frac{m}{m_R}\frac{\partial m_R}{\partial m} = 1 \quad . \tag{3.2.45}$$

We define the β function and the anomalous dimension of the field ϕ as follows:

$$\beta(g_R) = Z_{\theta}^{-1} m\frac{\partial g_R}{\partial m} = m_R\frac{\partial g_R}{\partial m_R} \quad , \tag{3.2.46}$$

$$\gamma(g_R) = Z_{\theta}^{-1}\frac{m}{2Z_3}\frac{\partial Z_3}{\partial m} = \frac{m_R}{2Z_3}\frac{\partial Z_3}{\partial m_R} \quad . \tag{3.2.47}$$

Here β and γ depend only on g_R. The reason for this is that the only available parameter with mass dimension is the physical mass m_R in the renormalized theory while β and γ are dimensionless. Thus β and γ cannot depend on m_R. We finally obtain

$$\left[m_R\frac{\partial}{\partial m_R} + \beta(g_R)\frac{\partial}{\partial g_R} - n\gamma(g_R)\right]F_n = -2F^{\theta}_{n} \quad . \tag{3.2.48}$$

The above renormalization group equation is called the *Callan-Symanzik equation* [Cal 70, Sym 70]. (For reviews, see, e.g., [Col 72, 73b]. An attempt to homogenize the Callan-Symanzik equation is made in [Nis 77].)

If it were not for F_n^θ on the right-hand side of Eq. (3.2.48), we would have been able to write down explicitly a general solution of this differential equation. The presence of F_n^θ prevents us from doing so. However, for the asymptotic limit $\lambda \to \infty$ where $p_i = \lambda k_i$ with k_i fixed momenta ($i = 1, 2, \ldots, n$), F_n^θ may be safely neglected in comparison with the left-hand side of Eq. (3.2.48). This is guaranteed by the Weinberg theorem [Wei 60] (see the end of this section) on the asymptotic behavior of the renormalized Feynman integrals for the large Euclidean momenta. An intuitive argument for the above statement is the following. For $\lambda \to \infty$ all internal momenta in the Feynman amplitude are large. On the other hand, by adding the mass-insertion operator, one creates an extra propagator on an internal line in the Feynman amplitude. Hence F_n^θ is less than F_n for each power of λ. Following this argument we can neglect F_n^θ on the right-hand side of Eq. (3.2.48) and then we can write down an explicit solution of the equation. Actually it is possible to show that in the asymptotic regime the 't Hooft-Weinberg equation, the Gell-Mann-Low equation and the Callan-Symanzik equation all provide us with the same solution.

Yet another form of the renormalization group equation was proposed by Georgi and Politzer [Geo 76] on the basis of the fully off-shell renormalization scheme where the renormalized mass m_R as well as other renormalized quantities is defined at an off-shell position of momentum. In particular the off-shell position is taken to be space-like, i.e., $p^2 = -\mu^2$, and the renormalized mass m_R is defined in such a way that the renormalized propagator $\Delta_R(p)$ agrees with the free propagator for a space-like point, i.e.,

$$\Delta_R(p) = 1/(m_R^2 - p^2) \quad \text{at} \quad p^2 = -\mu^2 \quad . \tag{3.2.49}$$

This renormalization scheme is, in a sense, useful in defining quark masses in quantum chromodynamics because the quarks in QCD are most likely not to be observed in isolation due to the confinement mechanism and it is meaningless to talk about quarks on their mass shell.

The renormalization group equation in the off-shell scheme may be obtained by taking into account the fact that the total derivative of the unrenormalized Feynman amplitude F_{0n} with respect to μ vanishes so that physical predictions extracted from F_{0n} are independent of μ:

$$\frac{dF_{0n}}{d\mu} = 0 \quad . \tag{3.2.50}$$

By using the chain rule for differentiation we have, from Eq. (3.2.50), for the renormalized Feynman amplitude F_n,

$$\left[\mu\frac{\partial}{\partial\mu}+\beta\left(g_R,\frac{m_R}{\mu}\right)\frac{\partial}{\partial g_R}-\gamma_m\left(g_R,\frac{m_R}{\mu}\right)m_R\frac{\partial}{\partial m_R}-n\gamma\left(g_R,\frac{m_R}{\mu}\right)\right]F_n=0 \quad , \tag{3.2.51}$$

where both of g_R and m_R depend on μ and

$$\beta\left(g_R,\frac{m_R}{\mu}\right)=\mu\frac{dg_R}{d\mu} \quad , \tag{3.2.52}$$

$$\gamma_m\left(g_R,\frac{m_R}{\mu}\right)=-\frac{\mu}{m_R}\frac{dm_R}{d\mu} \quad , \tag{3.2.53}$$

$$\gamma\left(g_R,\frac{m_R}{\mu}\right)=\frac{\mu}{2Z_3}\frac{dZ_3}{d\mu} \tag{3.2.54}$$

We call the renormalization group equation (3.2.51) the *Georgi-Politzer equation.*

Digression

WEINBERG THEOREM: In Sec. 2.5.1 we explained the power counting theorem as the basis of the renormalization theory. The power counting theorem provides us with the convergence condition for a Feynman amplitude through the observation of the degree of divergence for all the subdiagrams. In connection with the power counting theorem we can derive a theorem on the asymptotic behavior of the Feynman amplitude for large external momenta. This theorem is called the *Weinberg theorem* [Wei 60].

We consider renormalizable theories and make all the relevant momenta Euclidean, i.e., $p_0 \to ip_0$ through the Wick rotation. The Weinberg theorem states the following:

Weinberg theorem: Consider a subset $(p_{i_1}, \ldots, p_{i_m})$ of external momenta $p_1, \ldots, p_n$ for the renormalized Feynman amplitude $F_n(p_1, \ldots, p_n)$ corresponding to a Feynman diagram G. We set

$$p_{i_l} = \lambda k_{i_l}; \quad l = 1, 2, \ldots, m, \tag{3.2.55}$$

and let $\lambda \to \infty$ keeping k_{i_l} fixed.[4] Then

$$F_n(p_1, \ldots, p_n) \to \lambda^a \ln^b\lambda \quad , \tag{3.2.56}$$

with b undetermined and a given by

$$a = \operatorname*{Max}_{H\subseteq G} d(H) \quad , \tag{3.2.57}$$

[4] The momenta $p_{i_1}, \ldots, p_{i_m}$ are assumed to be nonexceptional. (The momentum configuration for which no partial sum of the momenta vanishes is called nonexceptional.)

where $d(H)$ is the superficial degree of divergence for subdiagram H. Here by subdiagram we mean that part of G consisting of the continuous path of lines which is connected to the external lines with momenta $p_{i_1}, \ldots, p_{i_m}$.

In particular, if the subset $(p_{i_1}, \ldots, p_{i_m})$ is taken to be the whole set of the external momenta $p_1, \ldots, p_n$, the theorem states that

$$F_n(p_1, \ldots, p_n) \to \lambda^{d(G)} \ln^c \lambda \quad , \tag{3.2.58}$$

where c is undetermined and $d(G)$ is the superficial degree of divergence for diagram G. The constant $d(G)$ is given by Eq. (2.5.13) with $r = 0$ since we are concerned with renormalizable theories. For example the constant $d(G)$ reads for the ϕ_6^3 theory ($D = 6$)

$$d(G) = 2(3 - n) \quad . \tag{3.2.59}$$

Note that Eq. (3.2.59) is just equal to the mass dimension of the Feynman amplitude $F_n(p_1, \ldots, p_n)$ as is seen by the naive dimension counting in the definition (2.3.34). Hence the Weinberg theorem tells us that the asymptotic limit of the Feynman amplitude in the deep Euclidean region $\lambda \to \infty$ is given by naive power counting up to a logarithmic factor.

As an example of the Weinberg theorem we consider the four-point Feynman amplitude in the ϕ_6^3 theory which corresponds to the diagram given in Fig. 3.2.1. We let the momenta q and q' be large in the Euclidean region, i.e., $\lambda \to \infty$ for $q = \lambda k$ and $q' = \lambda k'$ with k and k' fixed and Euclidean ($k^2 < 0$ and $k'^2 < 0$). The subdiagrams to be considered are shown in Figs. 3.2.2(a), (b) and (c). (The portion of the diagram with boldfaced lines represents the subdiagram.) For subdiagrams (a), (b) and (c) the superficial degree of divergence reads $d = -2, -6$ and -2, respectively. Hence the diagrams (a) and (c) dominantly contribute to the asymptotic behavior of the Feynman amplitude F_4 of Fig. 3.2.1,

$$F_4(q, q', p, p') \to \lambda^{-2} \ln^b \lambda \quad , \tag{3.2.60}$$

where b is undetermined.

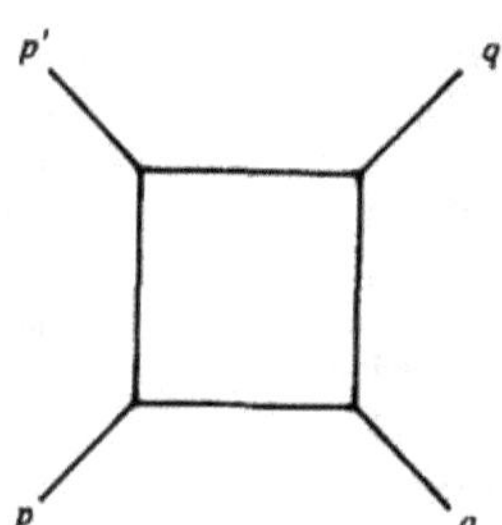

Fig. 3.2.1. The Feynman diagram corresponding to the one-loop four-point amplitude in the ϕ_6^3 theory to be considered as an example of the Weinberg theorem.

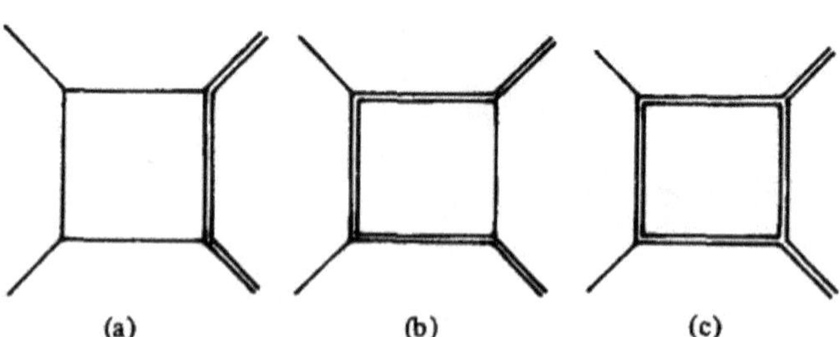

Fig. 3.2.2. Three different subdiagrams (a), (b) and (c) of the diagram of Fig. 3.2.1. They are indicated by double lines.

3.3. SOLUTION OF THE RENORMALIZATION GROUP EQUATIONS

3.3.1. Solution of the 't Hooft-Weinberg equation

We continue to use the ϕ_6^3 theory for our discussion in providing an explicit solution of the renormalization group equation. We choose the 't Hooft-Weinberg equation as our basic renormalization group equation since we intend to remain in the MS scheme and this equation allows us to give an explicit presentation of the solution. The 't Hooft-Weinberg equation for the n-point Feynman amplitude $F_n(\lambda p, g, m, \mu)$ in the ϕ_6^3 theory reads

$$\left[\mu\frac{\partial}{\partial\mu}+\beta(g)\frac{\partial}{\partial g}-\gamma_m(g)m\frac{\partial}{\partial m}-n\gamma(g)\right]F_n(\lambda p,g,m,\mu)=0 \quad , \tag{3.3.1}$$

with $p = (p_1, p_2, \ldots, p_n)$ where we introduced the dimensionless scale parameter λ for later convenience and suppressed suffix R in the renormalized quantities g and m as no confusion will be expected to occur. The renormalization group functions β, γ_m and γ are defined in Eqs. (3.2.16–18).

Based on naive dimension counting in the definition of the truncated connected Green function, Eq. (2.3.34), we find that the Feynman amplitude F_n has the following form in six dimensions

$$F_n(\lambda p, g, m, \mu) = \mu^{6-nd_0}\Phi_n\left(\frac{\lambda p}{\mu}, g, \frac{m}{\mu}\right) \quad , \tag{3.3.2}$$

where d_0 is the naive mass dimension $(D-2)/2$ of the field $\phi(x)$ with $D=6$ and Φ_n is a dimensionless function. Taking into account the above general form (3.3.2) we obtain an identity,

$$\left(\mu\frac{\partial}{\partial\mu}+m\frac{\partial}{\partial m}+\lambda\frac{\partial}{\partial\lambda}-6+nd_0\right)F_n(\lambda p, g, m, \mu)=0 \quad . \tag{3.3.3}$$

We combine Eq. (3.3.3) with Eq. (3.3.1) to derive

$$\left[-\lambda\frac{\partial}{\partial\lambda}+\beta(g)\frac{\partial}{\partial g}-(1+\gamma_m(g))m\frac{\partial}{\partial m}+6-nd_0-n\gamma\right]F_n(\lambda p,g,m,\mu)=0 \quad . \tag{3.3.4}$$

By setting

$$t = -\ln\lambda, \qquad \omega_n(g) = 6 - nd_0 - n\gamma(g) \quad , \tag{3.3.5}$$

we rewrite Eq. (3.3.4) in the following form,

$$\left[\frac{\partial}{\partial t}+\beta(g)\frac{\partial}{\partial g}-(1+\gamma_m(g))m\frac{\partial}{\partial m}+\omega_n(g)\right]F_n(e^{-t}p,g,m,\mu)=0 \quad . \quad (3.3.6)$$

This is a linear homogeneous partial differential equation for F_n of order one and may be solved following the standard method. We introduce the running coupling constant $\bar{g}(t)$ and running mass $\bar{m}(t)$ which are the renormalized coupling constant and mass, respectively, defined at the renormalization scale μ/λ instead of μ, i.e.,

$$\frac{\mu}{\lambda}\frac{d\bar{g}}{d(\mu/\lambda)}=\beta(\bar{g}) \quad , \quad \frac{\mu}{\lambda}\frac{1}{\bar{m}}\frac{d\bar{m}}{d(\mu/\lambda)}=-1-\gamma_m(\bar{g}) \quad .$$

If we regard λ as a variable instead of μ, we obtain

$$\frac{d\bar{g}}{dt}=\beta(\bar{g}), \quad \frac{1}{\bar{m}}\frac{d\bar{m}}{dt}=-1-\gamma_m(\bar{g}) \quad . \quad (3.3.7)$$

If we replace g and m in Eq. (3.3.6) by $\bar{g}$ and $\bar{m}$, respectively, we find, with the help of Eq. (3.3.7), that Eq. (3.3.6) is transformed into a total derivative

$$\left[\frac{d}{dt}+\omega_n(\bar{g})\right]F_n(e^{-t}p,\bar{g},\bar{m},\mu)=0 \quad . \quad (3.3.8)$$

This equation can be easily integrated to give

$$F_n(e^{-t}p,\bar{g}(t),\bar{m}(t),\mu)=F_n(p,g,m,\mu)\exp\left[-\int_0^t dt'\omega_n(\bar{g}(t'))\right] \quad , \quad (3.3.9)$$

where we took into account the condition,

$$\bar{g}(0)=g, \quad \bar{m}(0)=m \quad . \quad (3.3.10)$$

Equation (3.3.9) is the general solution to the 't Hooft-Weinberg equation and is the basis of our analysis of the asymptotic behavior of the Feynman amplitude for large momenta.

For our purposes in discussing the behavior of the Feynman amplitude, it is more convenient to rewrite Eq. (3.3.9) in a slightly different form. We replace $e^{-t}p$ by p and rearrange Eq. (3.3.9) to obtain

$$F_n(e^t p,g,m,\mu)=F_n(p,\bar{g}(t),\bar{m}(t),\mu)\exp\left[(6-nd_0)t-n\int_0^t dt'\gamma(\bar{g}(t'))\right] \quad . \quad (3.3.11)$$

Here the running coupling constant $\bar{g}(t)$ and running mass $\bar{m}(t)$ are solutions of Eq. (3.3.7) with the initial condition (3.3.10). The integrated form of Eq. (3.3.7) with Eq. (3.3.10) incorporated is given by

$$t = \int_g^{\bar{g}(t)} \frac{dg'}{\beta(g')} \quad , \tag{3.3.12}$$

$$\bar{m}(t) = m \exp\left[-t - \int_0^t dt' \gamma_m(\bar{g}(t')) \right] \quad . \tag{3.3.13}$$

The expression (3.3.11) is the most convenient for studying the behavior of $F_n(e^t p, g, m, \mu)$ as a function of t: on the right-hand side the t dependence comes through $\bar{g}(t)$, $\bar{m}(t)$ and $\gamma(\bar{g}(t))$ and is, in principle, calculable although the functional form of F_n is unknown.

Exercises

1. Write down the general solution of the 't Hooft-Weinberg equation in the case of quantum chromodynamics, Eq. (3.2.22).
2. Find the general solution for the Georgi-Politzer equation.

3.3.2. Behavior of the solution

We are now in a position to discuss the behavior of the Feynman amplitudes, making full use of the general solution (3.3.11) of the renormalization group equation. In this subsection we discuss the behavior of the high as well as low momentum limit of Feynman amplitudes.

If the theory under investigation were a finite theory, i.e., had no divergence, there would be no renormalization and so $\beta = \gamma_m = \gamma = 0$. In this case the running coupling constant $\bar{g}(t)$ and running mass $\bar{m}(t)$ would be given by

$$\bar{g}(t) = g, \quad \bar{m}(t) = me^{-t} \quad . \tag{3.3.14}$$

Hence Eq. (3.3.11) would read

$$F_n(e^t p, g, m) = (e^t)^{6 - nd_0} F_n(p; g, me^{-t}) \quad , \tag{3.3.15}$$

where we accounted for the fact that the Feynman amplitudes would not depend on the renormalization scale μ. Equation (3.3.15) clearly shows that the behavior of the Feynman amplitude for $t \to \infty$ coincides with what would be expected by naive power counting (we assume the existence of the massless limit).

In practice, however, the divergences are present and are to be disposed of in renormalizable theories. The renormalization effects modify the prediction (3.3.15) of the naive power counting due to the existence of the functions β, γ_m and γ. In the following argument we simplify our model by setting $m = 0$, i.e., we work with the massless ϕ_6^3 theory.[5] The effect of the mass will be considered later. Discarding the mass dependence we have

$$F_n(e^t p, g, \mu) = F_n(p, \bar{g}(t), \mu) \exp\left[(6 - nd_0)t - n\int_0^t dt' \gamma(\bar{g}(t'))\right] , \quad (3.3.16)$$

$$t = \int_g^{\bar{g}(t)} \frac{dg'}{\beta(g')} . \quad (3.3.17)$$

It should be noted here that in the massless case all the renormalization group equations introduced in the previous section (i.e., the 't Hooft-Weinberg, Gell-Mann-Low, Callan-Symanzik and Georgi-Politzer equations) reduce to the same equation and give an identical solution.

The deviation of the behavior of the Feynman amplitude from what is expected by naive power counting comes, on the right-hand side of Eq. (3.3.16), through the running coupling constant $\bar{g}(t)$ in the amplitude $F_n(p, \bar{g}(t), \mu)$ and in the exponent with the anomalous dimension $\gamma(\bar{g})$. Hence we first study the behavior of the running coupling constant $\bar{g}(t)$. Scrutiny of Eq. (3.3.17) elucidates the following general behavior of $\bar{g}(t)$ as $t \to \infty$. Note here that $\beta(0) = 0$ because $\beta(g)$ is obtained through a loop calculation and has the form $\beta(g) = bg^n + O(g^{n+1})$ with $n \geq 1$.

(1) If $\beta(\bar{g})$ has no zero except at $\bar{g} = 0$ and is positive as shown in Fig.3.3.1(a), $\bar{g}(t)$ behaves as

$$\bar{g}(t) \to \begin{cases} \infty \\ 0 \end{cases} \quad \text{for } t \to \pm\infty . \quad (3.3.18)$$

For the negative semi-definite $\beta(\bar{g})$ as in Fig. 3.3.1(b) we have

$$\bar{g}(t) \to \begin{cases} 0 \\ \infty \end{cases} \quad \text{for } t \to \pm\infty . \quad (3.3.19)$$

[5] According to the Kinoshita-Poggio-Quinn theorem [Kin 62, 76, Pog 76] which will be discussed in Sec. 6.3.2, Green functions with off-shell external lines are free from infrared divergence and hence we can safely deal with Green functions in massless theories as far as the Green functions are off the mass shell.

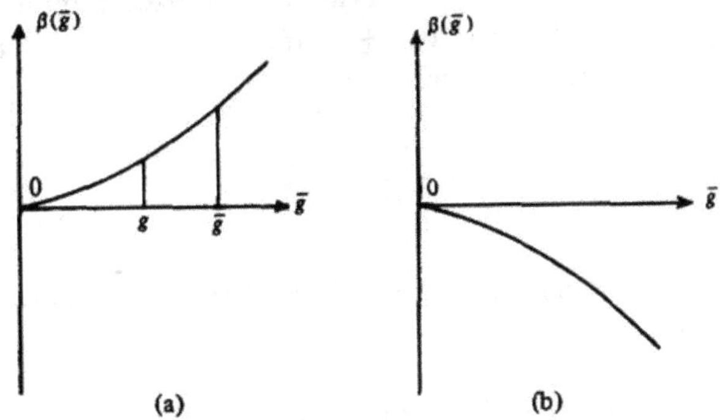

Fig. 3.3.1. The behavior of the β function as a function of the running coupling constant $\bar{g}$: (a) the case in which $\beta(\bar{g})$ is a monotonically increasing function of $\bar{g}$ and (b) the case with $\beta(\bar{g})$ a monotonically decreasing function of $\bar{g}$.

(2) If $\beta(\bar{g})$ has a zero at $\bar{g} = g_c$ as shown in Fig. 3.3.2, we find that, if $\beta'(g_c) < 0$ with $\beta'(\bar{g}) = d\beta/d\bar{g}$,

$$\bar{g}(t) \to g_c \quad \text{as} \quad t \to +\infty \quad , \tag{3.3.20}$$

and, if $\beta'(g_c) > 0$,

$$\bar{g}(t) \to g_c \quad \text{as} \quad t \to -\infty \quad . \tag{3.3.21}$$

The case (1) above may be regarded as the special case of (2) for $g_c = 0$ and so we shall consider only the case (2) in the following. Since large t implies a large momentum scale, the limit $t \to \infty$ is called the ultraviolet (UV) limit. Correspondingly the limit $t \to -\infty$ is called the infrared (IR) limit. The limiting value g_c of $\bar{g}(t)$ for the UV limit (IR limit) is known as the ultraviolet (infrared) fixed point.

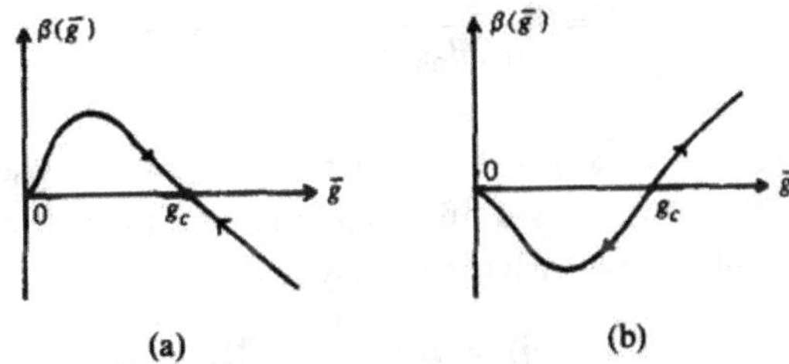

Fig. 3.3.2. The behavior of the β function as a function of the running coupling constang $\bar{g}$: the case in which $\beta(\bar{g})$ has a nontrivial zero at $\bar{g} = g_c$ with (a) $\beta'(g_c) < 0$ and (b) $\beta'(g_c) > 0$. The arrow attached to the curve indicates the direction to which $\bar{g}(t)$ moves as $t \to \infty$.

We shall now see what is the implication of the fixed point for the behavior of the Feynman amplitude as $t \to \pm\infty$. If we have a fixed point g_c, i.e., $\bar{g}(t) \to g_c$ for $t \to +\infty$ (or $-\infty$), we have

$$\int_0^t dt' \gamma(\bar{g}(t')) \to \gamma(g_c)t \quad \text{as} \quad t \to \pm\infty \ . \tag{3.3.22}$$

Hence, as can be seen from Eq. (3.3.16), the Feynman amplitude F_n behaves, for $t \to \infty$, as

$$F_n(e^t p, g, \mu) \to F_n(p, g_c, \mu)\,(e^t)^{6 - nd_0 - n\gamma(g_c)} \ . \tag{3.3.23}$$

The asymptotic behavior (3.3.23) clearly shows the deviation from the canonical behavior (3.3.15) given by naive power counting. The difference comes from the anomalous dimension $\gamma(g_c)$ at the fixed point g_c. Thus the presence of the fixed point has an important significance for the asymptotic behavior of the Feynman amplitude.

In the case that $g_c = 0$ the behavior of the Feynman amplitude is roughly the same as the canonical one (3.3.15) since $\gamma(0) = 0$. Hence a theory of this type is almost equivalent to a free field theory and we call this property *asymptotic freedom* if $g_c = 0$ is the UV fixed point. A theory with asymptotic freedom is called an asymptotically free theory. In order to obtain the precise asymptotic behavior of the Feynman amplitude F_n for the case $g_c = 0$, we need to specify the behavior of $\beta(\bar{g})$ near $\bar{g} = 0$. We assume that[6]

$$\beta(\bar{g}) = -\beta_0 \bar{g}^3 + O(\bar{g}^5) \ , \tag{3.3.24}$$

where β_0 is the positive constant. Inserting Eq. (3.3.24) into Eq. (3.3.17) we have

$$\bar{g}(t)^2 = \frac{g^2}{1 + 2\beta_0 g^2 t} \quad \text{as} \quad t \to +\infty \ . \tag{3.3.25}$$

Since $\bar{g}(t) \to 0$ as $t \to +\infty$ for the β function (3.3.24), the use of perturbation theory is legitimate for $t \to +\infty$ in this theory, and the anomalous dimension $\gamma(\bar{g})$ may be calculated perturbatively, say

$$\gamma(\bar{g}) = \gamma_0 \bar{g}^2 + O(\bar{g}^4) \ , \tag{3.3.26}$$

[6] As will be seen later this behavior is characteristic to ϕ_6^3 theory and quantum chromodynamics.

where γ_0 is a constant to be calculated in the one-loop order. Substituting Eqs. (3.3.25) and (3.3.26) into Eq. (3.3.16) we obtain for $t \to +\infty$

$$F_n(e^t p, g, \mu) \to F_n(p,0,\mu)\,(e^t)^{6-nd_0}(2\beta_0 g^2 t)^{-n\gamma_0/2\beta_0} \quad . \tag{3.3.27}$$

Equation (3.3.27) shows that the theory under investigation is close to a free field theory but not quite. The deviation from the canonical behavior (3.3.15) is rather minor but still significant. We shall see later that this property will lead to scaling violations in deep inelastic scattering.

In the case of quantum electrodynamics the β function is of the form[7]

$$\beta(\bar{g}) = \beta_0 \bar{g}^3 + O(\bar{g}^5) \quad , \tag{3.3.28}$$

where $\beta_0 > 0$. Hence $\bar{g} = 0$ is the infrared fixed point. The running coupling constant $\bar{g}(t)$ behaves, for $t \to -\infty$, as

$$\bar{g}(t)^2 = \frac{g^2}{1-2\beta_0 g^2 t} \quad \text{as} \quad t \to -\infty \quad . \tag{3.3.29}$$

Although Eq. (3.3.29) is valid only for $t \to -\infty$, we simply extrapolate it to the region $t > 0$. We then observe that, if the momentum scale e^t approaches $\exp(1/2\beta_0 g^2)$, the running coupling constant $\bar{g}(t)$ grows without bound and perturbation theory breaks down. This momentum scale corresponds to a pole in t in Eq. (3.3.29) which is often called the *Landau pole* [Lan 54, 56].

So far we disregarded the effect of the mass by setting $m = 0$. For $m \neq 0$ the t dependence of the Feynman amplitude through the running mass $\bar{m}(t)$ should be taken into account. As can be seen in Eq. (3.3.13), the running mass $\bar{m}(t)$ vanishes in the UV limit as far as $\gamma_m(\bar{g}) > -1$. In ordinary cases where the perturbative expansion is legitimate, $\gamma_m(\bar{g})$ is of the form

$$\gamma_m(\bar{g}) = \gamma_{m0}\bar{g}^2 + O(\bar{g}^4) \quad , \tag{3.3.30}$$

and is expected to satisfy $|\gamma_m(\bar{g})| \ll 1$. Thus we find that the running mass $\bar{m}(t)$ vanishes sufficiently fast in the UV limit $t \to +\infty$. The behavior of $\bar{m}(t)$ in the case with the UV fixed point $\bar{g} = g_c$ is given by

$$\bar{m}(t) \to m e^{-(1+\gamma_m(g_c))t} \quad \text{as} \quad t \to \infty \quad . \tag{3.3.31}$$

For asymptotically free theories we employ the expressions (3.3.24) and (3.3.30) to obtain

$$\bar{m}(t) \to m e^{-t}(2\beta_0 g^2 t)^{-\gamma_{m0}/2\beta_0} \quad . \tag{3.3.32}$$

In any case $\bar{m}(t) \to 0$ as $t \to \infty$ and so the effect of the mass may be safely neglected in the UV limit $t \to \infty$.

[7] See Sec. 3.4.3.

Exercises

1. Derive the ultraviolet behavior of $\bar{g}(t)$ and $F_n(e^t p, g, \mu)$ in the case that the β function and anomalous dimension are given by
$$\beta(\bar{g}) = \bar{g}(a - \bar{g}^2), \quad \gamma(\bar{g}) = b\bar{g}^2 \quad ,$$
where $a > 0$ and $b > 0$ are known constants.

2. In quantum electrodynamics the β function to two-loop order is given by [deR 74]
$$\beta(e) = \frac{e^3}{12\pi^2} + \frac{e^5}{64\pi^4} + O(e^7) \quad .$$
Find the behavior of the running coupling constant $\bar{e}(t)$ in the infrared region $t \to -\infty$.

3.4. ASYMPTOTIC FREEDOM

3.4.1. Renormalization group functions in QCD

In this subsection we calculate the renormalization group functions $\beta(g)$, $\gamma_m(g)$, $\delta(g, \alpha)$, $\gamma_G(g, \alpha)$ and $\gamma_F(g, \alpha)$ in quantum chromodynamics. We first show the details of the calculation in one-loop order and exhibit the results for two- and three-loop orders. The calculation, of course, is performed in the MS (or $\overline{\text{MS}}$) scheme. In the following calculation we restore the suffix R for the renormalized quantities just to avoid possible confusion.

In order to calculate the β function to one-loop order [Gro 73, 73a, Pol 73, Geo 74] we need to know the renormalized coupling constant g_R in one-loop order with the renormalization scale μ. There are four different ways of doing this because Z_g can be calculated by using the four different definitions (2.5.127). These four methods are, however, equivalent owing to the Slavnov-Taylor identity (2.5.128). Perhaps the easiest way of calculating Z_g is to use the definition

$$Z_g = \tilde{Z}_1/(\tilde{Z}_3 Z_3^{1/2}) \quad . \tag{3.4.1}$$

Here we need to calculate, in one-loop order, the ghost-gluon vertex, the ghost self-energy part and the gluon self-energy part in order to obtain $\tilde{Z}_1$, $\tilde{Z}_3$ and Z_3, respectively. The calculation has already been done and the result can be found in Eqs. (2.5.135), (2.5.137) and (2.5.143):

$$\tilde{Z}_1 = 1 - \frac{g_R^2}{(4\pi)^2} C_G \frac{\alpha_R}{2} \frac{1}{\varepsilon} + O(g_R^4) \quad , \tag{3.4.2}$$

$$\tilde{Z}_3 = 1 + \frac{g_R^2}{(4\pi)^2} C_G \frac{3 - \alpha_R}{4} \frac{1}{\varepsilon} + O(g_R^4) \quad , \tag{3.4.3}$$

$$Z_3 = 1 + \frac{g_R^2}{(4\pi)^2}\left[\frac{1}{2}C_G\left(\frac{13}{3} - \alpha_R\right) - \frac{4}{3}T_R N_f\right]\frac{1}{\varepsilon} + O(g_R^4) \quad . \tag{3.4.4}$$

Substituting Eqs. (3.4.2–4) into Eq. (3.4.1) we have

$$Z_g = 1 - \frac{g_R^2}{(4\pi)^2}\frac{1}{\varepsilon}\frac{1}{6}(11C_G - 4T_R N_f) + O(g_R^4) \quad . \tag{3.4.5}$$

It should be noted here that, as is expected by the argument in Sec. 3.2.1, the dependence on the gauge parameter α_R cancels out in Z_g. Now we insert Eq. (3.4.5) into Eq. (3.2.23) and perform a differentiation with respect to μ, keeping g fixed. We find, according to Eq. (3.2.6),

$$\begin{aligned}
\beta(g_R) &= -\varepsilon g_R - \frac{\mu}{Z_g}\frac{dZ_g}{d\mu}g_R \quad , \\
&= -\varepsilon g_R + \frac{11C_G - 4T_R N_f}{3}\frac{g_R^2}{(4\pi)^2}\frac{1}{\varepsilon}\beta(g_R) + O(g_R^5) \quad , \\
&= -\frac{1}{(4\pi)^2}\frac{11C_G - 4T_R N_f}{3}g_R^3 + O(g_R^5, \varepsilon) \quad .
\end{aligned} \tag{3.4.6}$$

Thus we find that the coefficient β_0 defined in Eq. (3.3.24) is given by

$$\beta_0 = \frac{1}{(4\pi)^2}\frac{11C_G - 4T_R N_f}{3} \quad . \tag{3.4.7}$$

Asymptotic freedom occurs if $\beta_0 > 0$, i.e., $11C_G - 4T_R N_f > 0$. Since $C_G = 3$ and $T_R = 1/2$ for SU(3), the condition for the asymptotic freedom reads

$$N_f < 33/2 \quad . \tag{3.4.8}$$

Hence quantum chromodynamics enjoys the property of asymptotic freedom in so far as the number of quark flavors is less than 16. It should be noted that, for $N_f = 0$, i.e., for a world made up only of gluons, the coefficient β_0 is positive definite. It is the presence of quarks that can spoil asymptotic freedom. The fundamental origin of asymptotic freedom may be traced back to the existence of the three-gluon coupling term in the Lagrangian. As this term is peculiar to the Yang-Mills theory, we realize that asymptotic freedom is inherent in the nature of a Yang-Mills theory.

We turn our attention to the calculation of the other renormalization group functions. The function $\gamma_m(g)$ is calculated by using Eqs. (3.2.25) and (2.5.139):

$$\gamma_m(g_R) = \frac{\mu}{Z_m}\frac{dZ_m}{d\mu} = -\frac{6g_R}{(4\pi)^2}C_F\frac{1}{\varepsilon}\beta(g_R)$$

$$= \frac{6g_R^2}{(4\pi)^2}C_F + O(g_R^4, \varepsilon) \quad . \tag{3.4.9}$$

Hence the constant γ_{m0} defined in Eq. (3.3.30) reads

$$\gamma_{m0} = \frac{6C_F}{(4\pi)^2} \quad , \tag{3.4.10}$$

where $C_F = 4/3$ for SU(3). Remembering Eq. (3.3.32) we obtain the UV behavior of the running mass $\bar{m}(t)$ such that

$$\bar{m}(t) \rightarrow me^{-t}\left[\frac{2g^2}{(4\pi)^2}\left(11 - \frac{2}{3}N_f\right)t\right]^{-4/(11-2N_f/3)} \tag{3.4.11}$$

Here and in the following the suffix R is neglected. The anomalous dimensions $\gamma_G(g, \alpha)$ and $\gamma_F(g, \alpha)$ are straightforwardly calculated through the definitions (3.2.27) and (3.2.28),

$$\gamma_G(g,\alpha) = -\frac{g^2}{(4\pi)^2}\left[\frac{1}{2}C_G\left(\frac{13}{3} - \alpha\right) - \frac{4}{3}T_R N_f\right] + O(g^4,\varepsilon) \quad , \tag{3.4.12}$$

$$\gamma_F(g,\alpha) = \frac{g^2}{(4\pi)^2}C_F\alpha + O(g^4,\varepsilon) \quad . \tag{3.4.13}$$

The remaining renormalization group function $\delta(g, \alpha)$ is readily obtained by noting the relation (3.2.26),

$$\delta(g, \alpha) = \frac{g^2}{(4\pi)^2}\alpha\left[C_G\left(\frac{13}{3} - \alpha\right) - \frac{8}{3}T_R N_f\right] + O(g^4, \varepsilon) \quad . \tag{3.4.14}$$

From the point of view of the simplicity of the renormalization group equation, the Landau gauge $\alpha = 0$ seems to be the most convenient since $\gamma_F = 0$ and $\delta = 0$ in this gauge. Moreover, by studying the equation for the running gauge parameter $\bar{\alpha}(t)$,

$$\frac{d\bar{\alpha}}{dt} = \delta(\bar{g}, \bar{\alpha}) \quad , \tag{3.4.15}$$

one finds that the Landau gauge corresponds to the UV fixed point for $\bar{\alpha}(t)$ if

$$13C_G - 8T_R N_f < 0 \quad . \tag{3.4.16}$$

In fact, inserting Eq. (3.3.25) with Eq. (3.4.7) into Eq. (3.4.15) and solving the differential equation, we obtain

$$\bar{\alpha}(t) = \frac{\alpha a}{\alpha + (a-\alpha)\,(1+2\beta_0 g^2 t)^b} \quad , \tag{3.4.17}$$

where $\alpha = \bar{\alpha}(0)$ and

$$a = \frac{13}{3} - \frac{8T_R N_f}{3C_G} \quad ,$$

$$b = -\frac{13C_G - 8T_R N_f}{22C_G - 8T_R N_f} \quad . \tag{3.4.18}$$

From Eq. (3.4.17) we see that $\bar{\alpha}(t) \to 0$ as $t \to \infty$, if $b > 0$. For SU(3) color Eq. (3.4.16) implies that

$$10 \le N_f \le 16 \quad , \tag{3.4.19}$$

where the upper bound comes from the condition of asymptotic freedom. If the number of flavors satisfies the condition (3.4.19), the Landau gauge $\bar{\alpha} = 0$ corresponds to the UV fixed point. Hence we always reach the Landau gauge if we start with an arbitrary gauge in the renormalization process. In the case that $N_f < 10$, the Landau gauge $\bar{\alpha} = 0$ turns out to be the IR fixed point and we have no convenient gauge in the UV limit.[8]

Beyond one-loop order the calculation of the renormalization group functions becomes much harder. For example, in order to calculate the β function to two loops, we have to take into account all the diagrams shown in Fig. 3.4.1. The calculation of the β function has been performed up to four loops in the MS scheme. A successful two-loop calculation was first made by Jones [Jon 74] and Caswell [Cas 74a]. (See also [Bel 74, Vla 77, Ego 78, Tar 81].) The formidable three-loop and four-loop calculations have been performed by Tarasov, Vladimirov and Zharkov [Tar 80] and van Ritbergen, Vermaseren and Larin [Rit 97]. As will be shown later[9] the β function in a perturbative expansion is renormalization-scheme independent up to two loops and beyond two loops it depends on the renormalization scheme employed. The β function up to two loops follows straightforwardly by the use of Eqs. (C.5), (C.6) and (C.8) and the formula (3.4.45) given at the end of this subsection. The expression for the β function up to three loops reads

[8] When $N_f < 10$ (i.e, $b < 0$), we see in Eq. (3.4.17) that $\bar{\alpha} \to a$ as $t \to \infty$ and hence $\bar{\alpha} = a$ appears to be the UV fixed point. This, however, is uncertain because, for the nonvanishing fixed point, the higher order effect is important and the conclusion may be drastically affected by higher order corrections.

[9] See Sec. 5.3.2.

$$\beta(g) = -\beta_0 g^3 - \beta_1 g^5 - \beta_2 g^7 + O(g^9) \quad , \tag{3.4.20}$$

where β_0 is given in Eq. (3.4.7) and

$$\beta_1 = \frac{1}{(4\pi)^4}\left[\frac{34}{3}C_G^2 - 4\left(\frac{5}{3}G_G + C_F\right)T_R N_f\right] \quad , \tag{3.4.21}$$

$$\beta_2 = \frac{1}{(4\pi)^6}\left[\frac{2857}{54}C_G^3 - \left(\frac{1415}{27}C_G^2 + \frac{205}{9}C_G C_F - 2C_F^2\right)T_R N_f \right.$$
$$\left. + \left(\frac{158}{27}C_G + \frac{44}{9}C_F\right)T_R^2 N_f^2\right] \quad . \tag{3.4.22}$$

For SU(3) color the coefficients β_0, β_1 and β_2 take the following form,

$$\beta_0 = \frac{1}{(4\pi)^2}\left(11 - \frac{2}{3}N_f\right) \quad , \tag{3.4.23}$$

$$\beta_1 = \frac{1}{(4\pi)^4}\left(102 - \frac{38}{3}N_f\right) \quad , \tag{3.4.24}$$

$$\beta_2 = \frac{1}{(4\pi)^6}\left(\frac{2857}{2} - \frac{5033}{18}N_f + \frac{325}{54}N_f^2\right) \quad . \tag{3.4.25}$$

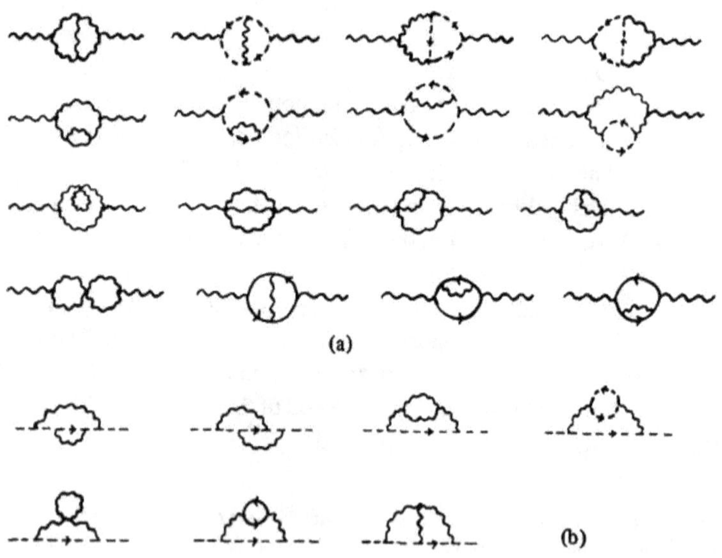

Fig. 3.4.1.

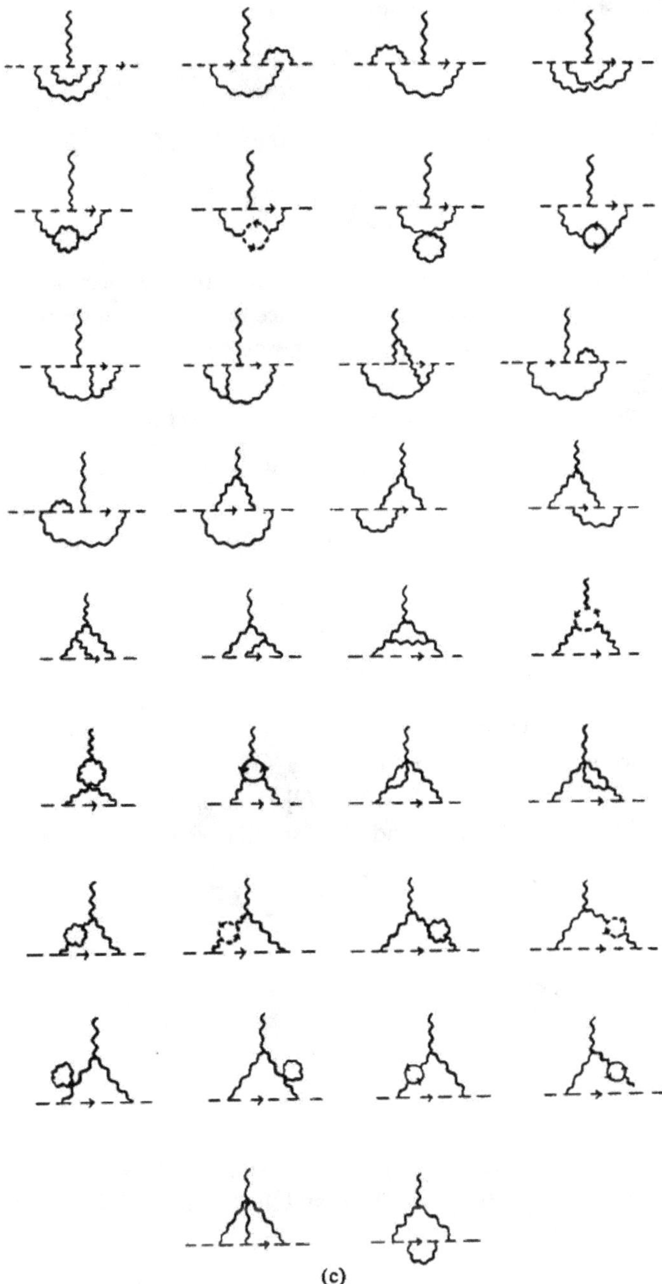

(c)

Fig. 3.4.1. The diagrams to be considered in the computation of the β function to two-loop order: (a) the gluon self-energy part to obtain Z_3, (b) the ghost self-energy part to obtain $\tilde{Z}_3$ and (c) the ghost-gluon vertex part to obtain $\tilde{Z}_1$.

In the case that the number of flavors N_f falls in the region

$$9 \leq N_f \leq 16 , \tag{3.4.26}$$

the coefficient β_1 has the opposite sign to that of β_0:

$$\beta_0 > 0, \quad \beta_1 < 0 . \tag{3.4.27}$$

If the higher order terms are negligible, it is tempting to expect that there exists a zero of $\beta(g)$ near $g^2 = \beta_0/(-\beta_1)$ and hence the nonvanishing IR fixed point emerges. This anticipation, however, is not necessarily justified since the higher orders, when summed up, may drastically modify the fixed-point property of $\beta(g)$. For example, assume that the exact form of $\beta(g)$ is

$$\beta(g) = -\beta_0 g^3/(1 + \alpha g^2) , \tag{3.4.28}$$

where $\beta_0 > 0$ and $\alpha > 0$ are the numerical constants. If we expand Eq. (3.4.28) in powers of g, we have

$$\beta(g) = -\beta_0 g^3 + \alpha\beta_0 g^5 + O(g^7) . \tag{3.4.29}$$

Truncating the perturbative series (3.4.29) at order g^5 one may expect that a fixed point at $g^2 = 1/\alpha$ develops. This, however, is obviously erroneous.

The renormalization group function $\gamma_m(g)$ is calculated up to two-loop order [Tar 81, Nac 81]. (See also [Nan 79].) The expression for $\gamma_m(g)$ may be obtained by using Eq. (C.4) and the formula which will be given by Eq. (3.4.47),

$$\gamma_m(g) = \gamma_{m0} g^2 + \gamma_{m1} g^4 + O(g^6) , \tag{3.4.30}$$

where γ_{m0} is given by Eq. (3.4.10) and

$$\gamma_{m1} = \frac{C_F}{(4\pi)^4}\left(3C_F + \frac{97}{3}C_G - \frac{20}{3}T_R N_f\right) . \tag{3.4.31}$$

The anomalous dimensions of the gluon and quark fields are calculated through the use of Eqs. (C.5) and (C.7) and the formulas (3.4.48) and (3.4.49),

$$\gamma_G(g, \alpha) = \gamma_{G0} g^2 + \gamma_{G1} g^4 + O(g^6) , \tag{3.4.32}$$

$$\gamma_F(g, \alpha) = \gamma_{F0} g^2 + \gamma_{F1} g^4 + O(g^6) , \tag{3.4.33}$$

where γ_{G0} and γ_{F0} have already appeared in Eqs. (3.4.12) and (3.4.13),

$$\gamma_{G0} = -\frac{1}{(4\pi)^2}\left[\frac{1}{2}C_G\left(\frac{13}{3}-\alpha\right)-\frac{4}{3}T_R N_f\right] , \tag{3.4.34}$$

$$\gamma_{F0} = \frac{1}{(4\pi)^2}C_F\alpha , \tag{3.4.35}$$

and γ_{G1} and γ_{F1} are given by

$$\gamma_{G1} = -\frac{1}{(4\pi)^4}\left[\frac{1}{4}C_G\left(\frac{59}{2}-\frac{11}{2}\alpha-\alpha^2\right)-(4C_F+5C_G)T_R N_f\right] , \tag{3.4.36}$$

$$\gamma_{F1} = \frac{1}{(4\pi)^4}C_F\left[\frac{1}{4}C_G(25+8\alpha+\alpha^2)-2T_R N_f-\frac{3}{2}C_F\right] . \tag{3.4.37}$$

Finally it should be mentioned that in this subsection we have dealt with the renormalization group functions calculated only in the MS (or $\overline{\text{MS}}$) scheme. The renormalization group functions in the MOM scheme have also been calculated by many authors. For references see, e.g. [Hag 82, Mat 80].

Digression

FORMULAS FOR RENORMALIZATION GROUP FUNCTIONS: Once an explicit expression of the renormalization constants Z_g, Z_m, Z_3 and Z_2 is given, it is straightforward to obtain the renormalization group functions $\beta(g)$, $\gamma_m(g)$, $\gamma_G(g,\alpha)$ and $\gamma_F(g,\alpha)$ through the definitions (3.2.24–28). Here we present explicit formulas for these functions in the MS scheme.

Let us write the renormalization constants in the MS scheme in the following form,

$$Z_g = 1 + \frac{A_g}{\varepsilon} + \frac{B_g}{\varepsilon^2} + \dots, \tag{3.4.38}$$

$$Z_m = 1 + \frac{A_m}{\varepsilon} + \frac{B_m}{\varepsilon^2} + \dots, \tag{3.4.39}$$

$$Z_3 = 1 + \frac{A_3}{\varepsilon} + \frac{B_3}{\varepsilon^2} + \dots, \tag{3.4.40}$$

$$Z_2 = 1 + \frac{A_2}{\varepsilon} + \frac{B_2}{\varepsilon^2} + \dots, \tag{3.4.41}$$

where A_g, A_m, A_3, A_2, B_g, B_m, ... are constants depending only on g the renormalized coupling constant. The β function is given by Eq. (3.2.6), i.e.,

$$\beta(g) = -\varepsilon g - g f(g), \quad f(g) = \frac{\mu}{Z_g}\frac{dZ_g}{d\mu} . \tag{3.4.42}$$

Using Eq. (3.4.38) and noting that A_g depends only on g, we have

$$\left(1+\frac{A_g}{\varepsilon}+\frac{B_g}{\varepsilon^2}+\ldots\right)f(g)=\frac{1}{\varepsilon}\beta(g)\left(\frac{dA_g}{dg}+\frac{1}{\varepsilon}\frac{dB_g}{dg}+\ldots\right) . \tag{3.4.43}$$

We substitute Eq. (3.4.42) into Eq. (3.4.43). Since $f(g)$ is finite, the above equality should hold for each coefficient of the power of $1/\varepsilon$. Hence we have

$$f(g)=-g\frac{dA_g}{dg} . \tag{3.4.44}$$

Inserting Eq. (3.4.44) into Eq. (3.4.42) and letting $\varepsilon \to 0$, we obtain

$$\beta(g)=g^2\frac{dA_g}{dg} . \tag{3.4.45}$$

This formula provides us with the expression of the β function once Z_g is calculated.

The function $\gamma_m(g)$ is expressed in terms of A_m in a similar way. We use the definition (3.2.25) and obtain

$$\gamma_m(g)=\frac{\beta(g)}{Z_m}\frac{dZ_m}{dg} . \tag{3.4.46}$$

Through the same argument as before applied to Eq. (3.4.39) we find

$$\gamma_m(g)=-g\frac{dA_m}{dg} . \tag{3.4.47}$$

The anomalous dimensions γ_G and γ_F defined in Eqs. (3.2.27) and (3.2.28) are also calculated in similar manner,

$$\gamma_G(g,\alpha)=-\frac{g}{2}\frac{dA_3}{dg} , \tag{3.4.48}$$

$$\gamma_F(g,\alpha)=-\frac{g}{2}\frac{dA_2}{dg} . \tag{3.4.49}$$

3.4.2. Asymptotic freedom in QCD

In quantum electrodynamics we know that the coupling constant, i.e., the expansion parameter, defined on the mass shell is small enough to guarantee the perturbative expansion. Hence perturbation theory can be safely used in most of the phenomenological applications of QED.

In quantum chromodynamics, however, a method independent of perturbation theory is not known to determine experimentally the size of the coupling constant. We are ignorant about the validity of perturbation theory in QCD until we perform practical perturbative calculations: we first tentatively disregard the question of validity of perturbation theory and compute the β function in the lowest order of perturbation theory. We then find that the renormalized coupling constant tends to be small as the relevant momentum scale grows. According to this property of asymptotic freedom we realize that our perturbative calculation is justified for the large momentum scale. Thus in QCD perturbation theory is perfectly legitimate in the large momentum region.

Among the known renormalizable quantum field theories in four dimensions only quantum chromodynamics, or Yang-Mills theories in general, enjoys the property of asymptotic freedom [Zee 73, Col 73a]. This property is attributed to the gluon self-interactions as was already mentioned in Sec. 3.4.1.

The β function up to three loops is given in Eq. (3.4.20). The running coupling constant $\bar{g}(t)$ at the momentum scale e^t is given by Eq. (3.3.7) or by its integrated form (3.3.12). We choose the momentum scale to be

$$e^t = \sqrt{-q^2}/\mu \quad , \tag{3.4.50}$$

where q is the typical momentum under consideration which is taken to be space-like and μ the fixed momentum scale (we choose it to be the renormalization scale for g: $\bar{g}(0) = g$). We insert Eq. (3.4.20) into Eq. (3.3.12) to obtain

$$t = -\frac{1}{2}\int_{g^2}^{\bar{g}^2} \frac{d\lambda}{\lambda^2} \frac{1}{\beta_0 + \beta_1\lambda + \beta_2\lambda^2 + O(\lambda^3)} \quad . \tag{3.4.51}$$

If we choose g sufficiently small, λ is also kept small since $\bar{g} < g$. Hence to this approximation we may safely truncate the perturbative series for the β function. Keeping only the one-loop order we have from Eq. (3.4.51)

$$t = \frac{1}{2\beta_0}\left(\frac{1}{\bar{g}^2} - \frac{1}{g^2}\right) \quad . \tag{3.4.52}$$

Hence the running coupling constant $\bar{g}$ is given by

$$\bar{g}^2 = \frac{g^2}{1 + 2\beta_0 g^2 t} = \frac{1}{\beta_0 \ln(-q^2/\Lambda^2)} \quad , \tag{3.4.53}$$

where the new momentum scale Λ is defined by

$$\Lambda = \mu \exp\left[-1/(2\beta_0 g^2)\right] \quad . \tag{3.4.54}$$

The momentum scale Λ is often referred to as the QCD *scale parameter* and is the only adjustable parameter in QCD except for the quark masses. In fact, the free parameter g present in the original Lagrangian is replaced by the scale parameter Λ through Eq. (3.4.54) where the renormalization scale μ is introduced by hand. This is an example of the so-called dimensional transmutation [Col 73] in which a dimensionless parameter in the theory transforms itself into a dimensional parameter. The scale parameter Λ should be determined by comparing the QCD predictions with experimental data.

The formula (3.4.53) may be improved by taking into account the two-loop term, i.e. the term with the coefficient β_1 in Eq. (3.4.51). Performing the integration we obtain

$$t = \frac{1}{2\beta_0}\left[\frac{1}{\bar{g}^2} - \frac{1}{g^2} + \frac{\beta_1}{\beta_0}\ln\frac{\bar{g}^2(\beta_0+\beta_1 g^2)}{g^2(\beta_0+\beta_1\bar{g}^2)}\right] . \tag{3.4.55}$$

We rewrite Eq. (3.4.55) in the following form,

$$\frac{1}{\bar{g}^2} + \frac{\beta_1}{\beta_0}\ln\frac{\beta_0\bar{g}^2}{1+\beta_1\bar{g}^2/\beta_0} = \beta_0\ln(-q^2/\Lambda^2) \quad , \tag{3.4.56}$$

with the scale parameter Λ defined by

$$\Lambda = \mu e^{-1/(2\beta_0 g^2)}\left(\frac{1+\beta_1 g^2/\beta_0}{\beta_0 g^2}\right)^{\beta_1/(2\beta_0^2)} . \tag{3.4.57}$$

Note here that Eq. (3.4.57) reduces to Eq. (3.4.54) if we set $\beta_1 = 0$. Equation (3.4.56) may be solved for $\bar{g}^2$ iteratively provided that $-q^2 \gg \Lambda^2$,

$$\bar{g}^2 = \frac{1}{\beta_0\ln(-q^2/\Lambda^2)}\left[1 - \frac{\beta_1}{\beta_0^2}\frac{\ln\ln(-q^2/\Lambda^2)}{\ln(-q^2/\Lambda^2)} + \ldots\right] . \tag{3.4.58}$$

The second term in the parentheses in Eq. (3.4.58) represents the next-to-leading order which corresponds to the two-loop effect.

3.4.3. Renormalization group functions in other theories

In four-dimensional space-time it has been found that only Yang-Mills theory exhibits the property of asymptotic freedom [Col 73a]. In fact all the other known quantum field theories have been shown not to enjoy asymptotic freedom [Zee 73]. We shall show some typical examples in the following.

The Lagrangian for quantum electrodynamics is given in Eq. (2.3.146). The β function for this Lagrangian has been calculated by de Rafael and Rosner [deR 74] up to three loops. For $Q = -1$ in Eq. (2.3.146) the β function reads

$$\beta(e) = \frac{e^3}{12\pi^2} + \frac{e^5}{64\pi^4} - \frac{121e^7}{18432\pi^6} + O(e^9) . \tag{3.4.59}$$

We see that $e = 0$ is obviously not the ultraviolet stable fixed point but the infrared one. The theory is not asymptotically free in the ultraviolet limit.

The ϕ^4 theory in which the interaction Lagrangian $\mathscr{L}_1$ is given by

$$\mathscr{L}_1 = -\lambda\phi^4/4! \quad , \tag{3.4.60}$$

has a stable vacuum only if $\lambda > 0$. The β function in the one-loop order is given by

$$\beta(\lambda) = \frac{3}{16\pi^2}\lambda^2 + O(\lambda^3) \quad . \tag{3.4.61}$$

Hence the theory is not asymptotically free. The β function in ϕ^4 theory has been calculated to five-loop order in [Che 83a]. (To four-loop order, see [Bre 74, Vla 78, 79].) As was suggested by Symanzik [Sym 73], the ϕ^4 theory turns out to be asymptotically free if λ is negative. This is because $\beta(\lambda) = -3\lambda^2/(16\pi^2) + O(\lambda^3)$ for $\lambda < 0$. The theory, however, is unstable in this case.

The β function for the Yukawa theory,

$$\mathscr{L}_1 = g\,\bar{\psi}\gamma_5\psi\phi \quad , \tag{3.4.62}$$

has been calculated to two-loop oder [Vla 75]. (See also [Mac 84].) To one-loop order it reads [Zee 73]

$$\beta(g) = \frac{5}{16\pi^2}g^3 + O(g^5) \quad , \tag{3.4.63}$$

and so the theory is not asymptotically free.

Except for Yang-Mills theories we find no consistent theory which is asymptotically free in four dimensions [Zee 73, Col 73a]. In other space-time dimensions, however, we can find some simple examples of asymptotically free theories. The ϕ_6^3 theory is one of these theories. The one-loop β function in the ϕ_6^3 theory can be easily obtained by using Eqs. (2.5.58) and (3.4.45),

$$\beta(g) = -\frac{3}{4(4\pi)^3}g^3 + O(g^5) \quad . \tag{3.4.64}$$

Obviously Eq. (3.4.64) exhibits the property of asymptotic freedom. The β function up to two loops may be found in [Mac 74].

3.5. ANOMALOUS DIMENSIONS

In Sec. 3.2.1 we introduced the idea of anomalous dimension through the definition (3.2.18). We saw that it played an important role in governing the asymptotic behavior of Feynman amplitudes as discussed in Sec. 3.3.2. We, however, have not explained why it is called the "anomalous dimension." Here in this section we shall clarify its relevance to the dimension of renormalized fields.

Let us start with scale transformations (or dilatations) in space-time (see, e.g., [Col 71, 72, Cal 73]),

$$x_\mu \to e^{-t}x_\mu \quad , \tag{3.5.1}$$

where t is a parameter representing the dilatation. The scale transformation (3.5.1) generates the corresponding transformation on the field operator. For simplicity, we restrict ourselves to the neutral scalar theory with field $\phi(x)$. The scale transformation on the field operator $\hat{\phi}(x)$ reads

$$\hat{\phi}(x) \to U^\dagger \hat{\phi}(x) U = e^{-dt}\hat{\phi}(e^{-t}x) \quad , \tag{3.5.2}$$

where d is a constant characteristic to the field $\phi(x)$ and U is the unitary operator which gives rise to the scale transformation of $\hat{\phi}(x)$.

The constant d is considered to be the mass dimension of the field operator $\hat{\phi}(x)$. For example, if $\hat{\phi}(x)$ is a free field, we find that

$$d = \frac{D-2}{2} = 1 \quad \text{for } D = 4 \quad , \tag{3.5.3}$$

where D is the space-time dimension. Equation (3.5.3) is simply obtained by applying the scale transformation (3.5.2) to the canonical commutation relation,

$$\left[\hat{\phi}(x), \frac{\partial \hat{\phi}(y)}{\partial y_0}\right]\delta(x_0 - y_0) = i\delta^D(x-y) \quad . \tag{3.5.4}$$

The dimension d in Eq. (3.5.3) is just the naive mass dimension of $\hat{\phi}(x)$ obtained by simple power counting and is called *canonical dimension* which we denote by d_0:

$$d_0 = \frac{D-2}{2} \quad . \tag{3.5.5}$$

If $\hat{\phi}(x)$ is the renormalized interacting field, the constant d does not necessarily coincide with the naive dimension d_0 of $\hat{\phi}(x)$. Thus the constant d is, in general, called the *scale dimension* of the field $\hat{\phi}(x)$.

If we consider a massless theory, the action is scale-invariant, i.e. invariant under the scale transformation (3.5.2). For example, the action for the massless ϕ_6^3 theory,

$$A = \int d^6x \left[\frac{1}{2}(\partial^\mu \phi)(\partial_\mu \phi) + \frac{g}{3!}\phi^3 \right] , \tag{3.5.6}$$

is obviously scale invariant for $d = d_0$. (Note that $D = 6$.) Due to the presence of the interaction term, however, it is not evident whether the resulting renormalized Green functions reflect the nature of the scale invariance. In fact we have seen in Sec. 3.3.2 that the behavior of the Feynman amplitude in general differs from the one expected by naive power counting although the latter behavior would be required by the scale invariance of the theory. If the theory possesses an ultraviolet fixed point at $g = g_c$, the β function vanishes there, $\beta(g_c) = 0$, and scale invariance is recovered since the renormalization group equation (3.3.6) for $g = g_c$ reduces to

$$\left(\frac{\partial}{\partial t} - 6 + nd_0 + n\gamma \right) F_n(e^t p, g, \mu) = 0 , \tag{3.5.7}$$

where $d_0 = 2$ for the ϕ_6^3 theory. The solution to Eq. (3.5.7) is readily given by Eq. (3.3.23), i.e.,

$$F_n(e^t p, g_c, \mu) = F_n(p, g_c, \mu)\,(e^t)^{6 - n(d_0 + \gamma)} . \tag{3.5.8}$$

On the other hand, since the theory is scale invariant for $g = g_c$, by applying the scale transformation (3.5.2) to the Feynman amplitude $F_n(p, g_c, \mu)$ defined by Eqs. (3.1.20) and (3.1.22) we find that

$$F_n(e^t p, g_c, \mu) = F_n(p, g_c, \mu)\,(e^t)^{6 - nd} . \tag{3.5.9}$$

Comparing Eq. (3.5.8) with Eq. (3.5.9) we have

$$d = d_0 + \gamma . \tag{3.5.10}$$

The above relation clearly shows that the quantity γ defined by Eq. (3.2.18) is the difference between the scale and canonical dimension of the field $\phi(x)$. Since γ represents the deviation of the scale dimension from the canonical one, it is quite natural to call γ the *anomalous dimension*.

The anomalous dimension γ of the field $\phi(x)$ in the ϕ_6^3 theory can be calculated in the one-loop order by using the expression of Z_3 in Eq. (2.5.43),

$$\gamma = \frac{g^2}{12(4\pi)^3} + O(g^4) . \tag{3.5.11}$$

In some model field theories which are exactly solvable, one may calculate the anomalous dimension of the relevant field without recourse to perturbation theory. For example, in the Thirring model which is a two-dimensional self-coupled fermion theory with the interaction Lagrangian

$$\mathscr{L}_1 = \lambda\bar{\psi}\gamma_\mu\psi\bar{\psi}\gamma^\mu\psi \quad , \tag{3.5.12}$$

the anomalous dimension γ of the fermion field ψ can be calculated exactly and is given by [Wil 70, Low 70]

$$\gamma = \left(\frac{\lambda}{4\pi}\right)^2 \Bigg/ \left[1 - \left(\frac{\lambda}{4\pi}\right)^2\right] . \tag{3.5.13}$$

In this model the β function is known to vanish identically for any λ and hence any value of λ corresponds to the fixed point.

The notion of the anomalous dimension introduced here for the field operators may be generalized for any operator composed of the field operators. Typical examples of such operators are the electromagnetic and color currents[10]

$$J_\mu(x) = \bar{\psi}(x)\gamma_\mu\psi(x) \quad , \tag{3.5.14}$$

$$J_\mu^a(x) = \bar{\psi}(x)T^a\gamma_\mu\psi(x) \quad , \tag{3.5.15}$$

and the composite operators which have appeared in the BRS transformations (Sec. 2.3.5). In the next chapter we shall discuss the operator-product expansion. There new types of composite operators participate and the anomalous dimension of these composite operators plays an important role in deriving the asymptotic behavior of the nucleon structure functions.

As is well known, the conserved currents such as the electromagnetic and color currents are not subjected to renormalization [Boy 67] owing to the Ward-Takahashi identity. Accordingly the anomalous dimension of the conserved currents vanishes [Gro 76, Ynd 83] so that the scale dimension of the conserved currents coincides with their canonical dimension. More generally the currents, even if they are not conserved, acquire no anomalous dimensions if they are generators of a certain symmetry algebra: define canonical operators corresponding to the generators,

$$Q^a(t) = \int d^3x J_0^a(x); \tag{3.5.16}$$

[10] For simplicity we neglect the caret ^ on canonical operators in the following.

then $Q^a(t)$ generate the symmetry algebra

$$[Q^a(t), Q^b(t)] = if^{abc}Q^c(t) \quad . \tag{3.5.17}$$

The above follows from the fact that $Q^a(t)$ should be dimensionless in order that Eq. (3.5.17) holds and so J_0^a must have the canonical dimension 3. Hence by covariance $J_\mu^a(x)$ should not have anomalous dimension.

CHAPTER

FOUR

OPERATOR-PRODUCT EXPANSION

The product of two electromagnetic (or weak) currents relevant to deep inelastic lepton-hadron scatterings is written as a series expansion called the operator-product expansion which enables us to extract a short distance piece in the scattering cross sections. This piece is under our control in the sense that it is calculable through the QCD Lagrangian by the use of the renormalization group method. We prove the operator-product expansion to all orders of perturbation theory by employing the BPHZ renormalization procedure. The renormalization group equation for the expansion coefficients is derived for later use.

4.1. OPERATOR PRODUCTS

4.1.1. Composite operators

In the present section we would like to expose the theoretical and phenomenological motivations for requiring the operator-product expansion. We first concentrate on the theoretical side.

We have already seen some examples of products of fields at the same space-time point in the BRS transformation and in the definition of currents. The product of fields at the same space-time point is called the *composite field* (or *composite operator*). Strictly speaking the composite field is not well defined if one takes a product of fields in a naive way. To make the argument simpler we shall confine ourselves to the case of the neutral scalar field $\phi(x)$.

Even for free fields we can show that the composite operator is not well defined. Let us consider the time-ordered product of two fields $\mathrm{T}[\phi(x)\phi(y)]$. Its vacuum expectation value is the propagator of the free field $\phi(x)$ (times $-i$),

$$\langle 0|\mathrm{T}[\hat{\phi}(x)\hat{\phi}(y)]|0\rangle = -i\Delta(x-y) = -i\int\frac{d^4p}{(2\pi)^4}\frac{e^{-ip\cdot(x-y)}}{m^2-p^2-i\varepsilon}. \tag{4.1.1}$$

As we let $y\to x$ in Eq. (4.1.1), we see that the momentum integral at the right-hand side diverges. Furthermore, not only for the vacuum expectation value (4.1.1) but also for general Green functions $\langle 0|\mathrm{T}[\hat{\phi}(x)\hat{\phi}(y)\hat{\phi}(x_1)\ldots\hat{\phi}(x_n)]|0\rangle$, one may show that the divergence occurs as $y \to x$. Thus the composite operator $\lim_{y\to x} \mathrm{T}[\hat{\phi}(x)\hat{\phi}(y)]$ for free fields is not well defined. To see the situation more clearly, we perform the momentum integration in Eq. (4.1.1) explicitly [Bog 80],

$$\Delta(x-y) = \frac{1}{4\pi}\delta((x-y)^2) + i\frac{m}{4\pi^2}\frac{K_1(m\sqrt{-(x-y)^2+i\varepsilon})}{\sqrt{-(x-y)^2+i\varepsilon}}, \tag{4.1.2}$$

where $K_1(z)$ is the modified Bessel function of the second kind [Erd 54a, Mor 60]. The right-hand side of Eq. (4.1.2) is obviously divergent for $x=y$. In the case of free fields we may find a meaningful definition of the composite field $\phi(x)^2$ by subtracting the singularity of its vacuum expectation value from the naive product of the field operators,

$$:\hat{\phi}(x)^2: = \lim_{y\to x}\{\hat{\phi}(x)\hat{\phi}(y) - \langle 0|\hat{\phi}(x)\hat{\phi}(y)|0\rangle\}. \tag{4.1.3}$$

The composite operator $:\hat{\phi}(x)^2:$ defined in this way is nothing but the normal product of free fields appearing in any standard textbook of quantum field theory (e.g. [Sch 61, Bj 65, Bog 80, Itz 80]). For the interacting fields, this simple manipulation cannot be generalized in a straightforward manner.

For interacting fields we also have a simple argument showing that the composite operator $\phi(x)^2$ is ill-defined. We take the vacuum expectation value of the composite operator $\hat{\phi}(x)^2$ and find that [Leh 54]

$$\langle 0|\hat{\phi}(x)^2|0\rangle = \int\frac{d^3p}{(2\pi)^3 2p_0}\sum_n|\langle 0|\hat{\phi}(0)|p,n\rangle|^2. \tag{4.1.4}$$

Here we have inserted between the two $\hat{\phi}(x)$'s the complete set of eigenstates $|p,n\rangle$ of the four-momentum operator $\hat{P}_\mu$,

$$\hat{P}_\mu|p,n\rangle = p_\mu|p,n\rangle, \tag{4.1.5}$$

where n is the quantum number other than momentum p, labelling the eigenstates, and we have also used the translation invariance of the theory,

$$\hat{\phi}(x) = e^{i\hat{P}\cdot x}\,\hat{\phi}(0)\,e^{-i\hat{P}\cdot x}. \tag{4.1.6}$$

The complete set of states $|p,n\rangle$ includes a one-particle state $|p,1\rangle$ as a subset and hence

$$\sum_n|\langle 0|\hat{\phi}(0)|p,n\rangle|^2 \geq |\langle 0|\hat{\phi}(0)|p,1\rangle|^2. \tag{4.1.7}$$

Here the matrix element in Eq. (4.1.7) depends only on p^2 by covariance. In particular the right-hand side of Eq. (4.1.7) is independent of p_μ since $p^2 = m^2$ for the one-particle state where m is the mass of field $\phi(x)$. Therefore we have, by combining Eq. (4.1.4) with Eq. (4.1.7),

$$\langle 0|\hat{\phi}(x)^2|0\rangle \geq N\int \frac{d^3p}{(2\pi)^4 2p_0}, \quad p_0 = \sqrt{\mathbf{p}^2 + m^2}, \tag{4.1.8}$$

the right-hand side of which is divergent, and $N = |\langle 0|\hat{\phi}(0)|p, 1\rangle|^2$. Thus the composite operator $\hat{\phi}(x)^2$ in general gives rise to divergent matrix elements and is not a mathematically well-defined object.

It is the operator-product expansion proposed by Wilson [Wil 64, 69] that may serve to give a meaningful definition of the composite operator. By the operator-product expansion we mean that the product of operators, say $\hat{A}(x)$ and $\hat{B}(y)$, is expanded in a series of well-defined local operators $\hat{O}_i(x)$ with singular c-number coefficients $C_i(x)$ ($i=0, 1, 2, 3, \ldots$),

$$\hat{A}(x)\hat{B}(y) = \sum_{i=0}^{\infty} C_i(x-y)\, \hat{O}_i\left(\frac{x+y}{2}\right), \tag{4.1.9}$$

where $\hat{A}(x)$ and $\hat{B}(y)$ may be the field $\hat{\phi}(x)$ or any other local operators. The local operator $\hat{O}_i(x)$ is regular in the sense that the singularity of the product $\hat{A}(x)\hat{B}(y)$ for $y=x$ is fully contained in the coefficient functions $C_i(x-y)$. In Eq. (4.1.9) we arranged each term in the order of decreasing singularity. Hence $C_0(x-y)$ is the most singular as $y\to x$ and the next most singular one is $C_1(x-y)$ and so on. The operator $\hat{O}_0(x)$ is usually an identity operator. We now use Eq. (4.1.9) in order to define a composite operator in a regularized form. Let us consider the case $\hat{A}(x) = \hat{B}(x) = \hat{\phi}(x)$ in the ϕ^4 theory. As a regularized composite operator denoted by $[\hat{\phi}(x)^2]$ we take

$$[\hat{\phi}(x)^2] = \lim_{y\to x} \frac{\hat{\phi}(x)\hat{\phi}(y) - C_0(x-y)}{C_1(x-y)} = \hat{O}_1(x). \tag{4.1.10}$$

Thus the operator-product expansion serves as a means of defining the composite operator. Later we shall consider Green functions which are extended to include the insertions of composite operators and discuss the renormalization of such Green functions. It is worth noting here that the problem of defining the composite operators dates back to the works on bound states in quantum field theory [Nis 58, Haa 58, Zim 58].

The operator-product expansion (4.1.9) was proven by Zimmermann [Zim 71, 73] within the framework of perturbation theory through the application of the BPHZ method. The details of the proof will be described in Sec. 4.2.4

4.1.2. Product of currents in deep inelastic scattering

In Sec. 4.1.1 we explained the theoretical motivation for the need of the operator-product expansion and we now describe the phenomenological motivation in deep inelastic lepton-hadron scatterings. To do this we start with the kinematics of lepton-hadron scattering processes.

Consider the process in which a lepton (electron, muon or neutrino) beam bombards a target hadron (usually the nucleon N) and scatters off the target inelastically,

$$l + \mathrm{N} \to l + \mathrm{X}, \tag{4.1.11}$$

where l represents the lepton, N the nucleon and X the system of hadrons produced through inelastic processes. The corresponding Feynman diagram for the process is depicted in Fig. 4.1.1 where the lowest-order approximation is assumed for the electroweak interactions. For definiteness we restrict ourselves to the case of an electron beam hitting a proton target. The process takes place mainly through the electromagnetic interaction. In this process high energy electrons knock on the target proton producing many hadrons (denoted by X in Eq. (4.1.11)) in the final state. Experimentally we do not observe the produced hadrons X and measure only the electron momentum in the final state. Experiments of this type are called *inclusive experiments* [Fey 69a].

There are three independent kinematical variables in this inclusive experiment for which we may choose

$$s=(p+k)^2, \quad q^2=(k-k')^2, \quad W^2=(p+q)^2, \tag{4.1.12}$$

where k and k' are momenta of the initial and final electrons respectively, p the proton momentum and

$$q=k-k'. \tag{4.1.13}$$

The variable s corresponds to the total energy squared of the electron-proton system in the center-of-mass frame, q^2 is the momentum transfer squared of

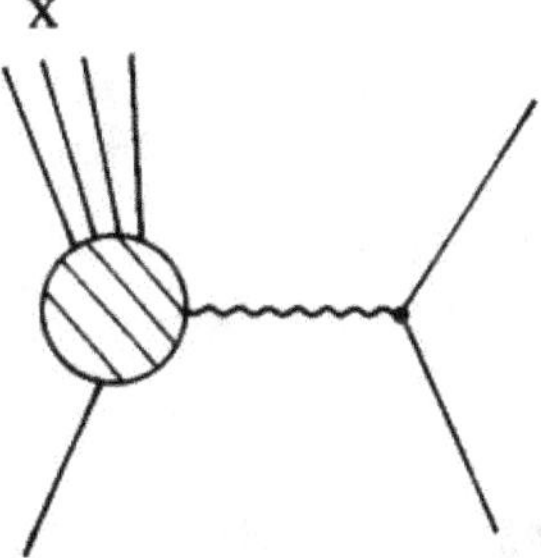

Fig. 4.1.1. The Feynman diagram for the inelastic lepton-hadron scattering in one-photon (or one-weak-boson) approximation where X represents the system of unobserved hadrons.

the electron and W^2 is the invariant mass squared of the final hadron system X. In the laboratory frame (the target rest frame) $p = (M, 0, 0, 0)$ where M is the proton mass, and the variables s, q^2 and W^2 are rewritten as

$$s = M(2E + M),$$

$$q^2 = -4EE' \sin^2 \frac{\theta}{2},$$

$$W^2 = M^2 + 2M(E - E') + q^2, \tag{4.1.14}$$

where we neglected the electron mass, and $E = k_0$, $E' = k'_0$ and θ is the scattering angle of the electron in the laboratory frame. The physical region of the process is given by

$$s \geq M^2, \quad q^2 \leq 0, \quad W^2 \geq (M+m_\pi)^2, \tag{4.1.15}$$

where m_π is the pion mass and again the electron mass is neglected. In place of W^2 we frequently use the variable ν defined by

$$\nu = p{\cdot}q/M. \tag{4.1.16}$$

In terms of ν the physical region condition on W^2 reads

$$2M\nu + q^2 \geq m_\pi(2M + m_\pi). \tag{4.1.17}$$

The transition matrix element defined in Eq. (2.3.9) is given for this process in the lowest order of the electromagnetic interaction by

$$\langle \mathrm{eX}|T|\mathrm{eN}\rangle = \bar{u}_{\sigma'}(k')e\gamma^\mu u_\sigma(k)\frac{1}{q^2}\langle \mathrm{X}|(-e)j_\mu(0)|\mathrm{p}\lambda\rangle, \tag{4.1.18}$$

where σ', σ and λ are spin components of the scattered electron, initial electron and target proton respectively, $|\mathrm{p}\lambda\rangle$ is the proton state, $|\mathrm{X}\rangle$ the state representing the final hadron system, and $j_\mu(x)$ the quark part of the electromagnetic current. (Here and in the following we shall omit the caret ^ on operators.) The interaction Lagrangian for the electromagnetic interactions is given by

$$\mathscr{L}_1 = (-e\bar{\psi}_e\gamma_\mu\psi_e + ej_\mu)A^\mu, \tag{4.1.19}$$

in conformity with Eq. (2.3.146) in Sec. 2.3.3, where ψ_e is the electron field.

The unpolarized total cross section σ for the process (4.1.11) is written down by applying the formula (2.3.10),

$$\sigma = \frac{1}{16k{\cdot}p}\sum_{\sigma'\sigma,\lambda}\int\frac{d^3k'}{(2\pi)^3 2k'_0}\sum_{\mathrm{X}}(2\pi)^4\delta^4(p_{\mathrm{X}}+k'-k-p)|\langle \mathrm{eX}|T|\mathrm{eN}\rangle|^2, \tag{4.1.20}$$

where the summation Σ_x extends over all the variables of the system X, and p_X is the total momentum of the system X. In the inclusive experiment like (4.1.11) we measure only the momentum distribution of electrons in the final state and the corresponding invariant cross section $k'_0 d\sigma/d^3k'$ is called the inclusive cross section. It is given by

$$k'_0 \frac{d\sigma}{d^3k'} = \frac{1}{32(2\pi)^3 k\cdot p} \sum_{\sigma',\sigma,\lambda} \sum_{X} (2\pi)^4 \delta^4(p_X + k' - k - p) |\langle eX|T|eN\rangle|^2. \quad (4.1.21)$$

Substituting Eq. (4.1.18) into Eq. (4.1.21) we obtain

$$k'_0 \frac{d\sigma}{d^3k'} = \frac{1}{k\cdot p}\left(\frac{\alpha}{q^2}\right)^2 L^{\mu\nu} W_{\mu\nu}, \quad (4.1.22)$$

where α is the fine structure constant $\alpha = e^2/(4\pi)$, $L^{\mu\nu}$ and $W_{\mu\nu}$ are the leptonic and hadronic tensors, respectively, which are given by

$$L^{\mu\nu} = \frac{1}{4} \sum_{\sigma'\sigma} \left(\bar{u}_{\sigma'}(k')\gamma^\mu u_\sigma(k)\right)^* \left(\bar{u}_{\sigma'}(k')\,\gamma^\nu u_\sigma(k)\right), \quad (4.1.23)$$

$$W_{\mu\nu} = \frac{1}{2\pi} \sum_{X} (2\pi)^4 \delta^4\,(p_X - p - q) \frac{1}{2} \sum_{\lambda} \langle p\lambda| j_\mu(0)|X\rangle\langle X| j_\nu(0)|p\lambda\rangle. \quad (4.1.24)$$

We use Eq. (1.4.11) to rewrite Eq. (4.1.23),

$$\begin{aligned} L^{\mu\nu} &= \frac{1}{4} \mathrm{Tr}[(\not{k} + m_e)\,\gamma^\mu(\not{k}' + m_e)\gamma^\nu], \\ &= k'^\mu k^\nu + k'^\nu k^\mu + (q^2/2)\,g^{\mu\nu}. \end{aligned} \quad (4.1.25)$$

It is an easy exercise to show that the hadronic tensor $W_{\mu\nu}$ may be expressed in a more compact form,

$$W_{\mu\nu} = \frac{1}{2\pi} \int d^4x\, e^{iq\cdot x} \frac{1}{2} \sum_{\lambda} \langle p\lambda| j_\mu(x) j_\nu(0)|p\lambda\rangle. \quad (4.1.26)$$

Combining Eqs. (4.1.22), (4.1.25) and (4.1.26) we find that the inclusive cross section for inelastic electron-proton scatterings is expressed in terms of the Fourier transform of the expectation value of the current product $j_\mu(x)j_\nu(0)$ in the proton state. Thus we realize that the product of two electromagnetic currents is relevant in inelastic electron-proton scatterings. Furthermore in the following we shall show that the product of currents $j_\mu(x)j_\nu(0)$ near the light cone $x^2 = 0$ plays an important role in *deep* inelastic electron-proton scattering.

In the following we shall use the simplifying notation

$$\langle p|O|p\rangle \equiv \frac{1}{2} \sum_{\lambda} \langle p\lambda|O|p\lambda\rangle, \quad (4.1.27)$$

where O is an arbitrary operator. (Remember that we have been omitting the caret ^ on operators.) We then have

$$W_{\mu\nu} = \frac{1}{2\pi}\int d^4x\, e^{iq\cdot x}\langle p|j_\mu(x)j_\nu(0)|p\rangle. \tag{4.1.28}$$

For physical processes $E > E'$ and so $q_0 > 0$. It can be shown that for $q_0 > 0$

$$\int d^4x\, e^{iq\cdot x}\langle \mathrm{p}|j_\nu(0)j_\mu(x)|\mathrm{p}\rangle = 0. \tag{4.1.29}$$

In fact, by sandwiching the completeness condition $\sum_X |X\rangle\langle X| = 1$ between two currents in Eq. (4.1.29) and using translation invariance, we obtain

$$\int d^4x\, e^{iq\cdot x}\langle \mathrm{p}|j_\mu(0)j_\nu(x)|\mathrm{p}\rangle$$

$$= \sum_X (2\pi)^4\delta^4(q-p+p_X)\langle \mathrm{p}|j_\nu(0)|X\rangle\langle X|j_\mu(0)|\mathrm{p}\rangle. \tag{4.1.30}$$

In Eq. (4.1.30) we can show that $q-p+p_X \neq 0$, and hence Eq. (4.1.29) results. The reason why $q-p+p_X \neq 0$ is the following: Assume that $q-p+p_X=0$ holds. Then we have $(p-q)^2 = p_X^2 = W^2$ which leads to $q_0 = (M^2+q^2-W^2)/(2M)$ in the laboratory frame. According to the physical region condition $W^2 \geq M^2$ and $q^2 \leq 0$, we find that $q_0 \leq 0$. This contradicts our original assumption $q_0 > 0$.

We use the property (4.1.29) to rewrite Eq. (4.1.28). We have

$$W_{\mu\nu} = \frac{1}{2\pi}\int d^4x\, e^{iq\cdot x}\langle \mathrm{p}|[j_\mu(x), j_\nu(0)]|\mathrm{p}\rangle, \tag{4.1.31}$$

where $[j_\mu(x), j_\nu(0)]$ represents the commutator of the currents. According to the causality requirement this commutator has to vanish for space-like x,

$$[j_\mu(x), j_\nu(0)] = 0 \quad \text{for} \quad x^2 < 0, \tag{4.1.32}$$

and the integrand in Eq. (4.1.31) has a support only for $x^2 \geq 0$. We shall now prove the following statement.

Light-cone dominance: For deep inelastic scatterings in which $-q^2 \to \infty$ with $-q^2/\nu$ fixed, the dominant contribution to $W_{\mu\nu}$ comes from the region $0 \leq x^2 \lesssim \text{const.}/(-q^2)$ of the integration in Eq. (4.1.31).

According to the above statement we realize that in deep inelastic scattering, the inclusive cross section (4.1.22) is dominated by the light-cone region $x^2 \sim 0$

of the space-time integration in Eq. (4.1.31) [Iof 69, Fri 70]. We shall now give a proof to this statement. Let us focus our attention on the integration in Eq. (4.1.31) and see which region of the integration gives rise to the dominant contribution in the deep inelastic limit $-q^2 \to \infty$ with $-q^2/\nu$ fixed. First of all we recognize (according to the Riemann-Lebesgue theorem [Tit 37, Lig 58]) that the region of x for which $|q\cdot x| \to \infty$ in the deep inelastic limit makes vanishing contribution to the integral since $\exp(iq\cdot x)$ in the integrand oscillates without bound. Hence we only deal with the region of x, where $|q\cdot x|$ is finite in the deep inelastic limit. Now on the other hand we observe, in the laboratory frame where $\nu = q_0$, that

$$q\cdot x = \nu(x_0 - \sqrt{1 - q^2/\nu^2}\, r), \tag{4.1.33}$$

where r is the component of $\mathbf{x}$ in the direction of $\mathbf{q}(r = \mathbf{q}\cdot\mathbf{x}/|\mathbf{q}|)$. In the deep inelastic limit, i.e., $\nu \to \infty$ with $\xi = -q^2/(2M\nu)$ fixed, Eq. (4.1.33) reads

$$q\cdot x = \nu(x_0 - r) - M\xi r + O(1/\nu). \tag{4.1.34}$$

In order to keep $|q\cdot x|$ finite in the deep inelastic limit we must have

$$|x_0 - r| \lesssim \text{const.}/\nu, \quad |r| \lesssim \text{const.}/\xi. \tag{4.1.35}$$

Hence we obtain

$$x_0^2 \lesssim \left(r + \frac{\text{const.}}{\nu}\right)^2 \simeq r^2 + \text{const.}\frac{r}{\nu} < \mathbf{x}^2 + \frac{\text{const.}}{\xi\nu}, \tag{4.1.36}$$

which implies that

$$x^2 \lesssim \frac{\text{const.}}{-q^2}. \tag{4.1.37}$$

Since we already know that the integrand of Eq. (4.1.31) has a support for $x^2 \geq 0$, our previous statement immediately follows from Eq. (4.1.37).

From the statement proven here we realize that the inclusive cross section for the deep inelastic scattering is governed by the product of currents $j_\mu(x)j_\nu(0)$ near the light cone $x^2 \sim 0$.

We next show that the amplitude $W_{\mu\nu}$ defined by Eq. (4.1.28) or (4.1.31) is related to the absorptive part of the forward virtual Compton amplitude,[1]

[1] The amplitude (4.1.38) corresponds to the Compton scattering off protons in the forword direction since $\epsilon^\mu(q)T_{\mu\nu}\epsilon^\nu(q)$ with $\epsilon^\mu(q)$ the polarization vector of the photon of momentum q is the transition amplitude for the process $\gamma + N \to \gamma + N$ (γ represents the photon) in the forward direction as can be shown by the use of the reduction formula explained in Sec.2.3.1. The amplitude is called "virtual" since the photon is not on mass shell ($q^2 \neq 0$).

$$T_{\mu\nu} = i\int d^4x\, e^{iq\cdot x}\langle p|\mathrm{T}[j_\mu(x)j_\nu(0)]|p\rangle, \tag{4.1.38}$$

in the following way,

$$W_{\mu\nu} = \frac{1}{\pi}\,\mathrm{Abs}\,T_{\mu\nu}. \tag{4.1.39}$$

Here by the absorptive part (i.e. Abs $T_{\mu\nu}$) we mean the discontinuity of $T_{\mu\nu}$ across the cut along the line $q_0 > 0$ in the complex q_0 plane,

$$\mathrm{Abs}\,T_{\mu\nu} = \frac{1}{2i}[T_{\mu\nu}(q_0+i\varepsilon) - T_{\mu\nu}(q_0-i\varepsilon)], \tag{4.1.40}$$

with ε an infinitesimal positive constant.

In order to prove the relation (4.1.39) we remember the definition of the time-ordered product of two (bosonic) operators $A(x)$ and $B(y)$,

$$\mathrm{T}[A(x)B(y)] = \theta(x_0-y_0)A(x)B(y)+\theta(y_0-x_0)B(y)A(x), \tag{4.1.41}$$

where $\theta(x_0-y_0)$ is the step function: $\theta(x_0-y_0)=1$ for $x_0 > y_0$ and $\theta(x_0-y_0) = 0$ for $x_0 < y_0$. Using Eq. (4.1.41) in Eq. (4.1.38) and sandwiching the completeness condition $\sum_X |X\rangle\langle X| = 1$ between two currents we obtain

$$T_{\mu\nu} = \frac{(2\pi)^3\delta^3(\mathbf{p}_X-\mathbf{q}-\mathbf{p})}{p_{X0}-q_0-p_0-i\varepsilon}\sum_X\langle p|j_\mu(0)|X\rangle\langle X|j_\nu(0)|p\rangle$$

$$+\frac{(2\pi)^3\delta^3(\mathbf{p}_X+\mathbf{q}-\mathbf{p})}{p_{X0}+q_0-p_0-i\varepsilon}\sum_X\langle p|j_\nu(0)|X\rangle\langle X|j_\mu(0)|p\rangle. \tag{4.1.42}$$

Since we can argue that $p_{X0}+q_0-p_0\neq 0$ for $q_0>0$ in the same way as in the previous argument, the second term of the right-hand side of Eq. (4.1.42) has no discontinuity and only the first term contributes to the discontinuity of $T_{\mu\nu}$ in the complex q_0. Taking the absorptive part of Eq. (4.1.42) we find that Eq. (4.1.39) holds.

It is customary to rewrite the cross section (4.1.22) in terms of the so-called structure functions W_1 and W_2 which are defined by referring to the general form of the hadronic tensor $W_{\mu\nu}$. The most general form of $W_{\mu\nu}$ is derived under the requirement of the Lorentz, space-inversion and time-reversal invariance, hermiticity and current conservation,

$$W_{\mu\nu} = W_1(\nu, -q^2)\left(\frac{q_\mu q_\nu}{q^2} - g_{\mu\nu}\right)$$

$$+ W_2(\nu, -q^2)\frac{1}{M^2}\left(p_\mu - \frac{p\cdot q}{q^2}q_\mu\right)\left(p_\nu - \frac{p\cdot q}{q^2}q_\nu\right), \qquad (4.1.43)$$

where the invariant functions W_1 and W_2 called the *structure functions* of nucleons are functions of two independent invariant variables ν and $-q^2$. (It is assumed that q^2 is space-like so that $q^2 < 0$.) Substituting Eq. (4.1.43) into Eq. (4.1.22) and using Eq. (4.1.25) we have

$$k_0' \frac{d\sigma}{d^3k'} = \frac{\alpha^2}{2p\cdot k(-q^2)}\left[2W_1 + \left(\frac{4p\cdot k\, p\cdot k'}{-M^2q^2} - 1\right)W_2\right]. \qquad (4.1.44)$$

In the laboratory frame where $p = (M, 0, 0, 0)$, Eq. (4.1.44) is rewritten as

$$\frac{d\sigma}{d\Omega dE'} = \left(\frac{d\sigma}{d\Omega}\right)_M\left(2W_1\tan^2\frac{\theta}{2} + W_2\right)\Big/(2M), \qquad (4.1.45)$$

with $(d\sigma/d\Omega)_M$ the Mott cross section defined in Eq. (1.3.1).

Exercises

1. Show that Eq. (4.1.26) reduces to Eq. (4.1.24) if one applies the completeness condition,

$$\sum_X |X\rangle\langle X| = 1,$$

and translation invariance,

$$j_\mu(x) = e^{iP\cdot x}j_\mu(0)e^{-iP\cdot x},$$

where P is the energy-momentum operator which satisfies the eigenvalue equation,

$$P^\mu|X\rangle = p_X^\mu|X\rangle.$$

2. Obtain the physical region for the Bjorken scaling variable $\xi = -q^2/(2M\nu)$.

4.1.3. Product of currents in e^+e^- annihilations

Another phenomenological motivation for considering the operator-product expansion is found in e^+e^- annihilation total cross sections. Let us consider the process in which a positron e^+ and an electron e^- annihilate through the electromagnetic interaction producing a number of hadrons. Here, for simplicity, we do not take into account the weak interaction effect which may become significant in the energy region of the mass of the weak neutral boson Z^0.

Fig. 4.1.2. The Feynman diagram for the e$^+$ e$^-$ annihilation in one-photon approximation with X denoting the system of hadrons.

The process of our interest is written as

$$e^+ + e^- \to X, \tag{4.1.46}$$

where X represents the final hadron system. The Feynman diagram for the process (4.1.46) in the lowest order of the electromagnetic interaction is drawn in Fig. 4.1.2. The corresponding Feynman amplitude is given by

$$\langle X|T|e^+e^-\rangle = \bar{v}_{\lambda_2}(p_2)e\gamma^\mu u_{\lambda_1}(p_1)\frac{1}{q^2}\langle X|(-e)j_\mu(0)|0\rangle, \tag{4.1.47}$$

where $p_1(p_2)$ and $\lambda_1(\lambda_2)$ are the momentum and spin component of the electron (positron) respectively with q the total momentum $q=p_1+p_2$, $u_\lambda(p)$ $(\bar{v}_\lambda(p))$ is the Dirac spinor of the electron (positron) and $j_\mu(x)$ is the quark part of the electromagnetic current.

The total cross section for the annihilation process (4.1.46) is easily written down by the use of Eq. (2.3.10),

$$\sigma = \frac{1}{2s}\frac{1}{4}\sum_{\lambda_1,\lambda_2}\sum_X(2\pi)^4\delta^4(p_X-q)|\langle X|T|e^+e^-\rangle|^2, \tag{4.1.48}$$

where the electron mass is neglected and

$$s=q^2 = (p_1+p_2)^2. \tag{4.1.49}$$

Inserting Eq. (4.1.47) in Eq. (4.1.48) we obtain

$$\sigma = \frac{e^4}{2s^3}l^{\mu\nu}w_{\mu\nu}, \tag{4.1.50}$$

where

$$l^{\mu\nu} = p_1^\mu p_2^\nu + p_1^\nu p_2^\mu - (q^2/2)g^{\mu\nu}, \tag{4.1.51}$$

$$w_{\mu\nu} = \sum_X (2\pi)^4 \delta^4(p_X - q)\langle 0|j_\mu(0)|X\rangle\langle X|j_\nu(0)|0\rangle. \tag{4.1.52}$$

In the same way as in Sec 4.1.2 we can show that $w_{\mu\nu}$ in Eq. (4.1.52) is rewritten as

$$w_{\mu\nu} = \int d^4x\, e^{iq\cdot x}\langle 0|j_\mu(x)j_\nu(0)|0\rangle. \tag{4.1.53}$$

For the process $e^+ + e^- \to X$ to be physical, the total energy q_0 of the initial state should be positive and then Eq. (4.1.53) can be written in the following form just as before,

$$w_{\mu\nu} = \int d^4x\, e^{iq\cdot x}\langle 0|[j_\mu(x), j_\nu(0)]|0\rangle. \tag{4.1.54}$$

Thus the total cross section for e^+e^- annihilations is expressed in terms of the current commutator. In the center-of-mass system, we have $q = (q_0, 0, 0, 0)$. Bearing in mind high energy annihilations we let $q_0 \to \infty$ and we find that only the region $x_0 \sim 0$ makes a major contribution to the integral (4.1.54) according to the Riemann-Lebesgue theorem [Tit 37, Lig 58]. On the other hand the integrand in Eq. (4.1.54) has a support only when $x^2 \geq 0$ due to the causality requirement so that $x_0 \sim 0$ implies $x \sim 0$. Hence we conclude that the total cross section for high energy e^+e^- annihilations is governed by the current commutator at short distances.

For later convenience we further rewrite Eq. (4.1.50). The general tensor structure of $w_{\mu\nu}$ may be easily deduced following the requirements of Lorentz invariance and current conservation. We find that $w_{\mu\nu}$ is expressed in terms of only one invariant amplitude $w(q^2)$,

$$w_{\mu\nu} = (q_\mu q_\nu - q^2 g_{\mu\nu})\frac{1}{6\pi}\, w(q^2), \tag{4.1.55}$$

where the extra factor $1/(6\pi)$ is attached for later convenience. Substituting Eq. (4.1.55) into Eq. (4.1.50) we have

$$\sigma = \frac{4\pi\alpha^2}{3s}\, w(s), \tag{4.1.56}$$

where $\alpha = e^2/(4\pi)$. On the other hand it is easy to show that the total cross section for the process,

$$e^+ + e^- \to \mu^+ + \mu^-, \tag{4.1.57}$$

in the lowest order of the electromagnetic interaction is equal to

$$\sigma_{\mu\mu} = \frac{4\pi\alpha^2}{3s}, \tag{4.1.58}$$

where the electron and muon masses are neglected. (The derivation of Eq. (4.1.58) proceeds in parallel with the argument given in Sec. 2.3.4.) It is customary to define the ratio called the *R-ratio* (or *Drell ratio*),

$$R = \frac{\sigma}{\sigma_{\mu\mu}} = w(s), \tag{4.1.59}$$

to discuss the high energy e^+e^- annihilation process. The R-ratio has been already defined in Eq. (1.2.11). Here in Eq. (4.1.59) we have a direct relationship between the R-ratio and $w(s)$ defined in Eq. (4.1.55). Thus the R-ratio is closely related to the current commutator at short distances.

4.2. OPERATOR-PRODUCT EXPANSION IN PERTURBATION THEORY

4.2.1. Free field theory

Our goal in the present section is to derive the operator-product expansion of the form (4.1.9). The expansion was conjectured by Wilson [Wil 64, 69, Bra 67] and was shown to hold in the Thirring model [Wil 70, Low 70]. It was later proven in a rigorous manner by using the BPHZ method within the framework of perturbation theory [Zim 71, 73, Bra 69]. Although there exists no nonperturbative proof of the operator-product expansion except for illustrations in some model field theories [Wil 70, Low 70, Shi 79, Nov 80, 84, 85, Dav 82, 84], it is generally expected that it holds beyond the limitation of perturbation theory (see, e.g., [Shif 79, Nov 80, 84]).

In the present subsection we shall exemplify the operator-product expansion in free field theories. One of the simplest examples of the operator-product expansion in free field theories may be, as mentioned already in Sec. 4.1.1, the Wick theorem applied to the time-ordered product of two free neutral scalar fields

$$\mathrm{T}[\phi(x)\phi(y)] = :\phi(x)\phi(y): + \langle 0|\mathrm{T}[\phi(x)\phi(y)]|0\rangle, \tag{4.2.1}$$

where $:\phi(x)\phi(y):$ represents the normal product (see, e.g., [Sch 61, Bj 65, Bog 80, Itz 80]). Also for free fermions we have

$$T[\psi(x)\bar{\psi}(y)] = :\psi(x)\bar{\psi}(y): + \langle 0|T[\psi(x)\bar{\psi}(y)]|0\rangle. \tag{4.2.2}$$

As remarked before in Sec. 4.1.1, the normal product of free fields may be used to define a composite operator as it is regular even in the limit $x \to y$. Define the electromagnetic current $j_\mu(x)$ by the normal product for the quark fields,

$$j_\mu(x) = :\bar{\psi}(x)\gamma_\mu\psi(x):\,; \tag{4.2.3}$$

we derive the expansion of the product of two currents by applying the Wick theorem [Sch 61, Bj 65, Bog 80, Itz 80],

$$\begin{aligned} T[j_\mu(x)j_\nu(0)] &= -\mathrm{Tr}[\langle 0|T\,[\psi(0)\bar{\psi}(x)]|0\rangle\gamma_\mu\langle 0|T[\psi(x)\bar{\psi}(0)]|0\rangle\gamma_\nu] \\ &\quad + :\bar{\psi}(x)\gamma_\mu\langle 0|T[\psi(x)\bar{\psi}(0)]|0\rangle\gamma_\nu\psi(0): \\ &\quad + :\bar{\psi}(0)\gamma_\nu\langle 0|T[\psi(0)\bar{\psi}(x)]|0\rangle\gamma_\mu\psi(x): \\ &\quad + :\bar{\psi}(x)\gamma_\mu\psi(x)\bar{\psi}(0)\gamma_\nu\psi(0):. \end{aligned} \tag{4.2.4}$$

Note here that for free quark field $\psi(x)$

$$i\langle 0|T[\psi(x)\bar{\psi}(0)]|0\rangle = S(x) = \int\frac{d^4p}{(2\pi)^4}\frac{1}{m-\not{p}-i\varepsilon}e^{-ip\cdot x}, \tag{4.2.5}$$

where $S(x)$ is the free quark propagator defined in Eq. (2.3.123). As is easily seen in Eq. (4.2.5), $S(x)$ is singular for $x \to 0$. Since on the right-hand side of Eq. (4.2.4) we have two $S(x)$'s in the first term, one $S(x)$ in the second and third, and none in the fourth, we realize that each term at the right-hand side of Eq. (4.2.4) is arranged in the order of decreasing singularity for $x \sim 0$. Equation (4.2.4) is clearly an example of the operator-product expansion (4.1.9).

The free quark propagator $S(x)$ is related to the free neutral scalar propagator $\Delta(x)$ defined previously in Eq. (4.1.1), i.e.,

$$S(x) = (i\not{\partial} + m)\Delta(x). \tag{4.2.6}$$

The explicit form of $\Delta(x)$ is given by Eq. (4.1.2). The leading singularity of $\Delta(x)$ may be extracted from Eq. (4.1.2) and is found to be independent of the quark mass [Bog 80],

$$\Delta(x) = \frac{1}{4\pi^2 i}\frac{1}{x^2 - i\varepsilon} + \text{less singular terms}. \tag{4.2.7}$$

The singularity lies on the light cone and so is called the light-cone singularity. It is important to note here that, the more singular the behavior of $\Delta(x)$ near the light cone, the larger the power of q^2 in the Fourier transform $\tilde{\Delta}(q)$ of $\Delta(x)$. The following formula is a typical example of this property in the one-dimensional Fourier transformation [Gel 64a],

$$\int_{-\infty}^{\infty} dx \frac{e^{iqx}}{(x-i\varepsilon)^{\alpha}} = \frac{2\pi e^{i\alpha\pi/2}}{\Gamma(\alpha)} \theta(q) q^{\alpha-1}. \tag{4.2.8}$$

Hence it is enough for us to examine the most singular part of the c-number coefficients in Eq. (4.2.4) in order to see the dominant contribution of the current product to the hadronic tensor $W_{\mu\nu}$ and the tensor $w_{\mu\nu}$.

We extract the most singular part of Eq. (4.2.4) near the light cone by using Eqs. (4.2.6) and (4.2.7) (or by setting $(m = 0)$ [Fri 70, Fri 71] (see also [Fri 74, Ell 77, Itz 80, Che 84]),

$$\begin{aligned} \mathrm{T}[j_\mu(x)j_\nu(0)] &= \frac{x^2 g_{\mu\nu} - 2x_\mu x_\nu}{\pi^4 (x^2 - i\varepsilon)^4} + \frac{i x^\lambda}{2\pi^2 (x^2 - i\varepsilon)^2} \sigma_{\mu\lambda\nu\rho} O_V^\rho(x,0) \\ &\quad + \frac{x^\lambda}{2\pi^2 (x^2 - i\varepsilon)^2} \varepsilon_{\mu\lambda\nu\rho} O_A^\rho(x,0) + O_{\mu\nu}(x,0), \end{aligned} \tag{4.2.9}$$

where $O_V^\rho(x,0)$, $O_A^\rho(x,0)$ and $O_{\mu\nu}(x,0)$ are regular bilocal operators defined by

$$O_V^\mu(x,y) = :\bar{\psi}(x)\gamma^\mu\psi(y) - \bar{\psi}(y)\gamma^\mu\psi(x):, \tag{4.2.10}$$

$$O_A^\mu(x,y) = :\bar{\psi}(x)\gamma^\mu\gamma_5\psi(y) + \bar{\psi}(y)\gamma^\mu\gamma_5\psi(x):, \tag{4.2.11}$$

$$O_{\mu\nu}(x,y) = :\bar{\psi}(x)\gamma_\mu\psi(x)\bar{\psi}(y)\gamma_\nu\psi(y):, \tag{4.2.12}$$

and $\sigma_{\mu\lambda\nu\rho}$ is given by

$$\sigma_{\mu\lambda\nu\rho} = g_{\mu\lambda}g_{\nu\rho} + g_{\mu\rho}g_{\nu\lambda} - g_{\mu\nu}g_{\lambda\rho}. \tag{4.2.13}$$

In deriving Eq. (4.2.9) we have used the formula,

$$\gamma_\mu\gamma_\lambda\gamma_\nu = (\sigma_{\mu\lambda\nu\rho} + i\varepsilon_{\mu\lambda\nu\rho}\gamma_5)\gamma^\rho, \tag{4.2.14}$$

where $\varepsilon_{\mu\lambda\nu\rho}$ is the totally antisymmetric tensor of rank 4 defined in Eq. (2.2.108). The derivation of Eq. (4.2.14) is straightforward. As a basis of the Clifford algebra in four dimensions we may choose the set $(1, \gamma_5, \gamma_\mu, \gamma_5\gamma_\mu, \sigma_{\mu\nu})$ where

$$\sigma_{\mu\nu} = \frac{i}{2}[\gamma_\mu, \gamma_\nu], \tag{4.2.15}$$

and hence $\gamma_\mu\gamma_\lambda\gamma_\nu$ is expressed as a linear combination of these matrices,

$$\gamma_\mu\gamma_\lambda\gamma_\nu = a^{(1)}_{\mu\lambda\nu} + a^{(2)}_{\mu\lambda\nu}\gamma_5 + a^{(3)}_{\mu\lambda\nu\rho}\gamma^\rho + a^{(4)}_{\mu\lambda\nu\rho}\gamma_5\gamma^\rho + a^{(5)}_{\mu\lambda\nu\rho\kappa}\sigma^{\rho\kappa}. \tag{4.2.16}$$

Taking suitable traces in Eq. (4.2.16) we find

$$a^{(1)}_{\mu\lambda\nu} = a^{(2)}_{\mu\lambda\nu} = a^{(5)}_{\mu\lambda\nu\rho\kappa} = 0$$

$$a^{(3)}_{\mu\lambda\nu\rho} = \sigma_{\mu\lambda\nu\rho}, \quad a^{(4)}_{\mu\lambda\nu\rho} = i\varepsilon_{\mu\lambda\nu\rho}, \tag{4.2.17}$$

and Eq. (4.2.14) follows.

It is important to note here that the operator-product expansion (4.2.9) provides us with a clear separation of the short distance effects from the long distance effects. In fact the singular c-number coefficients in the expansion characterize the short distance behavior of the product of currents while the regular bilocal operators include full information on the long distance properties of the theory and are unimportant in the short distance region.

We may transform Eq. (4.2.9) into the formula for the current commutator $[j_\mu(x), j_\nu(0)]$. For this purpose we first note that

$$\mathrm{T}[j_\mu(x)\, j_\nu(0)] - \mathrm{T}[j_\mu(x)\, j_\nu(0)]^\dagger = \varepsilon(x_0)\,[j_\mu(x), j_\nu(0)], \tag{4.2.18}$$

where we took into account that the current $j_\mu(x)$ is hermitean and $\varepsilon(x_0)$ is the sign function,

$$\varepsilon(x_0) = \frac{x_0}{|x_0|}. \tag{4.2.19}$$

We then use the fundamental relation

$$\frac{1}{x^2 - i\varepsilon} = \frac{P}{x^2} + i\pi\delta(x^2), \tag{4.2.20}$$

where P denotes the principal part. Differentiating Eq. (4.2.20) $n-1$ times with respect to x^2 we have

$$\frac{1}{(x^2 - i\varepsilon)^n} = \frac{P}{(x^2)^n} + i\pi\frac{(-1)^{n-1}}{(n-1)!}\,\delta^{(n-1)}(x^2), \tag{4.2.21}$$

where

$$\delta^{(n)}(x^2) = \frac{d^n}{d(x^2)^n}\,\delta(x^2). \tag{4.2.22}$$

From Eq. (4.2.21) we immediately obtain

$$\frac{1}{(x^2 - i\varepsilon)^n} - \frac{1}{(x^2 + i\varepsilon)^n} = 2\pi i\frac{(-1)^{n-1}}{(n-1)!}\,\delta^{(n-1)}(x^2). \tag{4.2.23}$$

Using Eqs. (4.2.9), (4.2.18) and (4.2.23), we finally obtain the desired formula,

$$\begin{aligned}\varepsilon(x_0)\,[j_\mu(x), j_\nu(0)] &= \frac{i}{3\pi^3}(2x_\mu x_\nu - x^2 g_{\mu\nu})\delta^{(3)}(x^2)\\ &+ \frac{1}{\pi} x^\lambda \delta^{(1)}(x^2)\sigma_{\mu\lambda\nu\rho} O_V^\rho(x,0)\\ &- \frac{i}{\pi} x^\lambda \delta^{(1)}(x^2)\varepsilon_{\mu\lambda\nu\rho} O_A^\rho(x,0)\\ &+ O_{\mu\nu}(x,0) - O_{\nu\mu}(0,x).\end{aligned} \tag{4.2.24}$$

4.2.2. The parton model

As we have mentioned in Sec. 1.3, the parton model consists of the assumption that at short distances the constituents of hadrons behave as if they are free point-like particles [Fey 69, 69a]. This means that in the parton model we pick up only the lowest-order electromagnetic (or weak) contributions to physical quantities like cross sections and neglect all the QCD corrections. We shall now show that the above assumption is in conformity with the previously derived free field operator-product expansion at short distances, i.e., Eq. (4.2.24).

We start with an application of Eq. (4.2.24) to the e^+e^- annihilation cross section. We consider the tensor $w_{\mu\nu}$ given in Eq. (4.1.54) which is the Fourier transform of the vacuum expectation value of the current commutator $[j_\mu(x), j_\nu(0)]$. We insert Eq. (4.2.24) in Eq. (4.1.54). Since O_V, O_A and $O_{\mu\nu}$ are of the form of a normal product, we realize that only the first term on the right-hand side of Eq. (4.2.24) contributes to $w_{\mu\nu}$. Hence we have

$$w_{\mu\nu} = \frac{i}{3\pi^3}\left(g_{\mu\nu}\frac{\partial}{\partial q}\cdot\frac{\partial}{\partial q} - 2\frac{\partial}{\partial q^\mu}\frac{\partial}{\partial q^\nu}\right)I_3, \tag{4.2.25}$$

where nonleading contributions are neglected and I_n is given by

$$I_n = \int d^4x\, e^{iq\cdot x}\varepsilon(x_0)\delta^{(n)}(x^2). \tag{4.2.26}$$

After some calculation [El1 77], I_n is found to be

$$I_n = \frac{i\pi^2}{4^{n-1}(n-1)!}(q^2)^{n-1}\varepsilon(q_0)\theta(q^2), \tag{4.2.27}$$

where $\theta(q^2)$ is the step function. We then obtain

$$w_{\mu\nu} = \frac{1}{6\pi}(q_\mu q_\nu - q^2 g_{\mu\nu})\,\varepsilon(q_0)\,\theta(q^2). \tag{4.2.28}$$

Substituting Eq. (4.2.28) into Eq. (4.1.50) we immediately find for the e^+e^- annihilation total cross section

$$\sigma = \frac{4\pi\alpha^2}{3s}, \tag{4.2.29}$$

where we took into account that $q_0>0$ and $q^2>0$ for the physical process $e^+ + e^- \to X$. Here in the above argument we started from the electromagnetic current given by Eq. (4.2.3) and so the quark was assumed to have unit charge $Q=1$. If we would have started from the electromagnetic current of the form

$$j_\mu(x) = \sum_{i=1}^{N_f} Q_i \sum_{j=1}^{N_c} :\bar{\psi}_{ij}(x)\gamma_\mu\psi_{ij}(x):, \tag{4.2.30}$$

with N_f quark flavors and N_c colors, we would have obtained

$$\sigma = \frac{4\pi\alpha^2}{3s} N_c \sum_{i=1}^{N_f} Q_i^2, \tag{4.2.31}$$

which is nothing but the parton model result (1.2.10). Thus the free quark operator-product expansion (4.2.24) at short distances is essentially equivalent to the parton model. This is in a sense quite reasonable because the first term of the right-hand side of Eq. (4.2.24) comes from the first term of Eq. (4.2.4) which corresponds to the Feynman diagram depicted in Fig. 4.2.1.

Fig. 4.2.1. The Feynman diagram corresponding to the first term of the expansion of the current product (Eq. (4.2.4.)).

Taking the imaginary part of the Feynman amplitude corresponding to Fig. 4.2.1, we obtain the lowest order contribution of the electromagnetic interaction to the e^+e^- annihilation total cross section just as we have seen in Sec. 2.3.4. Hence we have the parton model prediction to the cross section.

We next turn our attention to the deep inelastic electron-proton scattering. Substituting Eq. (4.2.24) into Eq. (4.1.31) we have

$$\begin{aligned} W_{\mu\nu} &= (\text{First term of Eq. (4.2.24)})\,\langle p|p\rangle \\ &- \frac{1}{2\pi^2}\int d^4x\, e^{iq\cdot x} x^\lambda \varepsilon(x_0)\delta^{(1)}(x^2)\,\langle p|(-\sigma_{\mu\lambda\nu\rho}O_V^\rho(x,0) + i\varepsilon_{\mu\lambda\nu\rho}O_A^\rho(x,0))|p\rangle \\ &+ \frac{1}{2\pi}\int d^4x\, e^{iq\cdot x}\varepsilon(x_0)\,\langle p|[O_{\mu\nu}(x,0) - O_{\nu\mu}(0,x)]|p\rangle. \end{aligned} \tag{4.2.32}$$

The diagrammatical implication of Eq. (4.2.32) is self-evident if we remember that the light-cone singularities in Eq. (4.2.32) come from the free quark propagator. The right-hand side of Eq. (4.2.32) is nothing but the imaginary part of the amplitude represented by the diagram shown in Fig. 4.2.2. The first term of the right-hand side of Eq. (4.2.32) corresponds to the disconnected diagram and hence will be discarded in the following argument. The last term is less singular compared with other terms so that we also neglect this term. Thus we are left with two terms corresponding to two operators O_V and O_A. The diagrams representing these two terms are the lowest-order diagrams as seen in Fig. 4.2.2 and are in accord with the parton model assumption.

The operators $O_V^\rho(x,0)$ and $O_A^\rho(x,0)$ are regular bilocal operators which are finite in the limit $x\to 0$. We can make the Taylor expansion of these operators around the point $x=0$ and express them as an infinite series of local operators. For this purpose we expand the quark field $\psi(x)$ around $x=0$,

$$\psi(x) = \psi(0) + x^\mu[\partial_\mu\psi(x)]_{x=0} + \frac{1}{2!}x^{\mu_1}x^{\mu_2}[\partial_{\mu_1}\partial_{\mu_2}\psi(x)]_{x=0} + \dots \tag{4.2.33}$$

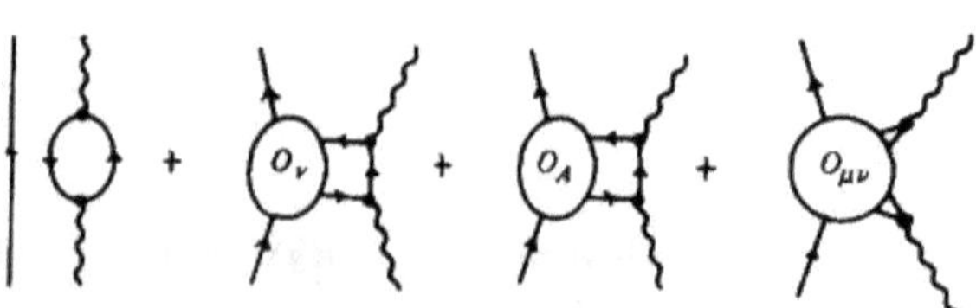

Fig. 4.2.2. The diagrammatic representation of Eq. (4.2.32): the right-hand side of Eq. (4.2.32) is the imaginary part of the amplitude corresponding to these diagrams.

Substituting Eq. (4.2.33) into Eqs. (4.2.10) and (4.2.11) we obtain

$$O^{\rho}_{V}(x,0) = \sum_{n=0}^{\infty} \frac{1}{n!} x^{\mu_1} \dots x^{\mu_n} O^{\rho}_{V\mu_1 \dots \mu_n}(0),$$

$$O^{\rho}_{A}(x,0) = \sum_{n=0}^{\infty} \frac{1}{n!} x^{\mu_1} \dots x^{\mu_n} O^{\rho}_{A\mu_1 \dots \mu_n}(0), \tag{4.2.34}$$

where the local operators $O^{\rho}_{V\mu_1 \dots \mu_n}(x)$ and $O^{\rho}_{A\mu_1 \dots \mu_n}(x)$ are given by

$$O^{\rho}_{V\mu_1 \dots \mu_n}(x) = :(\partial_{\mu_1} \dots \partial_{\mu_n} \bar{\psi}(x))\gamma^{\rho}\psi(x) - \bar{\psi}(x)\gamma^{\rho}\partial_{\mu_1} \dots \partial_{\mu_n}\psi(x):,$$

$$O^{\rho}_{A\mu_1 \dots \mu_n}(x) = :(\partial_{\mu_1} \dots \partial_{\mu_n} \bar{\psi}(x))\gamma^{\rho}\gamma_5\psi(x) + \bar{\psi}(x)\gamma^{\rho}\gamma_5\partial_{\mu_1} \dots \partial_{\mu_n}\psi(x):. \tag{4.2.35}$$

By the use of Eq. (4.2.34) we rewrite Eq. (4.2.32) in the following form,

$$W_{\mu\nu} = -\frac{1}{2\pi^2} \sum_{n=0}^{\infty} \frac{1}{n!} \int d^4x \, e^{iq\cdot x} x^{\lambda} \varepsilon(x_0) \delta^{(1)}(x^2) x^{\mu_1} \dots x^{\mu_n}$$

$$\times (-\sigma_{\mu\lambda\nu\rho} \langle p | O^{\rho}_{V\mu_1 \dots \mu_n}(0) | p \rangle + i\varepsilon_{\mu\lambda\nu\rho} \langle p | O^{\rho}_{A\mu_1 \dots \mu_n}(0) | p \rangle). \tag{4.2.36}$$

We note here that the term with the matrix element $\langle p | O^{\rho}_{A\mu_1 \dots \mu_n}(0) | p \rangle$ in Eq. (4.2.36) does not contribute to the unpolarized cross section (4.1.22) since this term carries the totally antisymmetric tensor which gives a vanishing contribution to Eq. (4.1.22) when it is contracted with the symmetric tensor $L_{\mu\nu}$ of Eq. (4.1.25). It makes a contribution to the hadronic tensor only when we deal with polarized deep-inelastic electron-proton scatterings (for references, see, e.g., [Kod 80]), in which the leptonic tensor $L_{\mu\nu}$ contains antisymmetric pieces. In order to see the contribution of $O^{\rho}_{V\mu_1 \dots \mu_n}$ to $W_{\mu\nu}$, we write down a general form of the matrix element of $O^{\rho}_{V\mu_1 \dots \mu_n}$,

$$\langle p | O^{\rho}_{V\mu_1 \dots \mu_n}(0) | p \rangle = a_n p^{\rho} p_{\mu_1} \dots p_{\mu_n} + \text{ terms containing } g_{\mu\nu}, \tag{4.2.37}$$

where a_n is the Lorentz invariant constant which depends on $p^2 = M^2$ and possibly on quark masses. In deriving Eq. (4.2.37) we used the fact that the matrix element depends only on p_μ and is symmetric in the indices $\mu_1, \mu_2, \dots, \mu_n$ as can be found by Eq. (4.2.35). The terms containing $g_{\mu\nu}$ in Eq. (4.2.37) are of the form, e.g., $p^{\rho}p^2 g_{\mu_1\mu_2} p_{\mu_3} \dots p_{\mu_n}$ and give less singular terms in Eq. (4.2.36) because $g_{\mu_1\mu_2}$ gives rise to a factor x^2. Hence we neglect these terms in the following.[2] We insert Eq. (4.2.37) into Eq. (4.2.36) and find

[2] If we choose the operator $O^{\rho}_{V\mu_1 \dots \mu_n}$ to be of spin $n+1$, its matrix element is a symmetric traceless tensor. In this case the terms containing $g_{\mu\nu}$ in Eq. (4.2.37) serve to make the matrix element traceless.

$$W_{\mu\nu} = \frac{1}{2\pi^2}\sigma_{\mu\lambda\nu\rho}p^\rho \int d^4x\, e^{iq\cdot x}x^\lambda \varepsilon(x_0)\delta^{(1)}(x^2)f(p\cdot x), \tag{4.2.38}$$

where the function $f(z)$ is defined by

$$f(z) = \sum_{n=0}^{\infty} a_n \frac{z^n}{n!}. \tag{4.2.39}$$

Note that a_n vanishes for even n in Eq. (4.2.37) due to the form of $O^q_{V\mu_1\dots\mu_n}$ given in Eq. (4.2.35) and the summation in Eq. (4.2.39) runs only over odd n. Making the Fourier transformation,

$$f(z) = \int_{-\infty}^{\infty} d\zeta\, e^{iz\zeta}\tilde{f}(\zeta), \tag{4.2.40}$$

Eq. (4.2.38) is rewritten as

$$W_{\mu\nu} = \frac{-i}{2\pi^2}\sigma_{\mu\lambda\nu\rho}p^\rho \frac{\partial}{\partial q_\lambda}\int_{-\infty}^{\infty} d\zeta \tilde{f}(\zeta)\int d^4x\, e^{i(q+\zeta p)\cdot x}\varepsilon(x_0)\delta^{(1)}(x^2). \tag{4.2.41}$$

By the use of the formula (4.2.27) we obtain

$$\begin{aligned} W_{\mu\nu} = &\int_{-\infty}^{\infty} d\zeta[-(p\cdot q+\zeta M^2)g_{\mu\nu}+2\zeta p_\mu p_\nu + p_\mu q_\nu + p_\nu q_\mu] \\ &\times \varepsilon(q_0+\zeta p_0)\,\delta(q^2+2\zeta p\cdot q+\zeta^2M^2)\tilde{f}(\zeta). \end{aligned} \tag{4.2.42}$$

We work in the Bjorken limit: $p\cdot q\to\infty$ and $-q^2\to\infty$ with $\xi = -q^2/(2p\cdot q)$ fixed and hence M^2 may be neglected in Eq. (4.2.42). We then have [Fri 70, Fri 71]

$$W_{\mu\nu} = \frac{1}{2}\tilde{f}(\xi)\left(-g_{\mu\nu} - \frac{q^2}{(p\cdot q)^2}p_\mu p_\nu + \frac{p_\mu q_\nu + p_\nu q_\mu}{p\cdot q}\right). \tag{4.2.43}$$

We note that Eq. (4.2.43) manifestly respects the current conservation condition $q^\mu W_{\mu\nu} = 0$. We may rewrite Eq. (4.2.43) in the form of Eq. (4.1.43),

$$W_{\mu\nu} = \frac{1}{2}\tilde{f}(\xi)\left(\frac{q_\mu q_\nu}{q^2} - g_{\mu\nu}\right) + \frac{1}{p\cdot q}\xi\tilde{f}(\xi)\left(p_\mu - \frac{p\cdot q}{q^2}q_\mu\right)\left(p_\nu - \frac{p\cdot q}{q^2}q_\nu\right). \tag{4.2.44}$$

Using the structure functions W_1 and W_2 defined in Eq. (4.1.43) we have

$$\begin{aligned} W_1(\nu, Q^2) &= \frac{1}{2}\tilde{f}(\xi) \equiv F_1(\xi), \\ \nu W_2(\nu, Q^2)/2M &= \frac{1}{2}\xi\tilde{f}(\xi) \equiv F_2(\xi). \end{aligned} \tag{4.2.45}$$

Equation (4.2.45) is nothing but the Bjorken scaling with scaling functions $F_1(\xi)$ and $F_2(\xi)$. Thus, in the framework of the free quark operator-product expansion near the light cone, we recovered the parton model prediction.

It should be remarked finally that the converse to the above statement is also true, i.e., Bjorken scaling leads us to the free field light-cone singularity of the product of currents in the hadronic tensor $W_{\mu\nu}$ [Bra 69, Jac 70, Leu 70]. This can be proven by taking the Fourier transform of $W_{\mu\nu}$ of the scaling form (4.2.44) with two independent structure functions $F_1(\xi)$ and $F_2(\xi)$.

4.2.3. Renormalization of composite operators

According to the operator-product expansion a product of operators is expressed as a series of local composite operators which are regular. The regular composite operator $O(x)$ may be defined if the set of all Green functions of the form,

$$\langle 0|\mathrm{T}[O(x)\phi(x_1)\ldots\phi(x_n)]|0\rangle, \tag{4.2.46}$$

is given, where we have in mind the neutral scalar theory for simplicity. As an example, let us assume that $O(x)$ is a regularized version of composite operator $\phi(x)^2$. In this case the Green functions (4.2.46) are defined as a finite part of the Green functions,

$$\langle 0|\mathrm{T}[\phi(x)^2\phi(x_1)\ldots\phi(x_n)]|0\rangle. \tag{4.2.47}$$

It will be seen later that the regularization of composite operators in the above sense amounts to the renormalization of composite operators. Given the definition (4.2.46) of the Green functions for a composite operator, we may express the operator-product expansion (4.1.9) as a relation among Green functions,

$$\begin{aligned}&\langle 0|\mathrm{T}[A(x)B(y)\phi(x_1)\ldots\phi(x_n)]|0\rangle\\ &= \sum_i C_i(x-y)\langle 0|\mathrm{T}\left[O_i\left(\frac{x+y}{2}\right)\phi(x_1)\ldots\phi(x_n)\right]|0\rangle,\end{aligned} \tag{4.2.48}$$

where the summation on i is understood to include the Lorentz indices.

The question we are now confronted with is whether the Green function with the insertion of the composite operator $O_i(x)$ needs any extra renormalization in addition to the ordinary renormalization of the mass, coupling constant and field operators. This extra renormalization is attributed to the renormalization of the composite operator. This problem has already been discussed in Sec. 3.2.2 in connection with the insertion of the mass term in deriving the Callan-Symanzik equation. Here in the present subsection we shall give a more general discussion on the renormalization of Green functions with insertion of composite operators.

Before going into a discussion of the Green function with the insertion of composite operators, let us recall the notion of superficial degree of divergence d_G for Feynman diagram G without the insertion of composite operators,

$$d_G = \sum_i r_i n_i - d_{BO} N_B - d_{FO} N_F + D, \tag{4.2.49}$$

where d_{BO} and d_{FO} are the canonical dimensions of the boson and fermion fields, respectively,

$$d_{BO} = \frac{D-2}{2}, \qquad d_{FO} = \frac{D-1}{2}. \tag{4.2.50}$$

The Feynman diagram G is assumed to have N_B boson and N_F fermion external lines, and is generated by the interaction terms of the form,

$$\mathscr{L}_i \sim g_i(\partial)^{\delta_i}(\phi)^{b_i}(\psi)^{f_i}. \tag{4.2.51}$$

Here the index of divergence r_i for the interaction Lagrangian $\mathscr{L}_i$ is given by Eq. (2.5.18) and n_i is the number of vertices of the type i (i.e., vertices generated by $\mathscr{L}_i$). As we have shown before, the index of divergence r_i is equal to minus the mass dimension of g_i:

$$r_i = -\dim[g_i]. \tag{4.2.52}$$

Therefore r_i is related to the dimension d_i of the vertex operator $(\partial)^{\delta_i}(\phi)^{b_i}(\psi)^{f_i}$ by

$$r_i = d_i - D. \tag{4.2.53}$$

The definition of the superficial degree of divergence (4.2.49) may be made more elaborate by taking into account the general structure of the Feynman amplitude. Occasionally the external-momentum variables factor out of the Feynman amplitude by symmetry. This factorization reduces the effective degree of divergence for the Feynman amplitude. For example the gluon self-energy part $\Pi^{ab}_{\mu\nu}(k)$ has an overall momentum factor $k_\mu k_\nu - k^2 g_{\mu\nu}$ which serves to reduce the degree of divergence by two. If we regard the Feynman amplitude on the whole as a vertex, this momentum factor corresponds to the derivative ∂ in the effective vertex operator for this amplitude. Hence we denote by δ the power of the external momenta factored out of the Feynman amplitude. It is then straightforward to see how to modify the definition (4.2.49) by taking into account the above argument. We have

$$d_G = \sum_i r_i n_i - d_{BO} N_B - d_{FO} N_F - \delta + D. \tag{4.2.54}$$

We define

$$r_G = d_{BO}N_B + d_{FO}N_F + \delta - D, \tag{4.2.55}$$

which is interpreted as the naive mass dimension of the Feynman amplitude minus the space-time dimension D. The superficial degree of divergence d_G for Feynman diagram G is then given by

$$d_G = \sum_i r_i n_i - r_G, \tag{4.2.56}$$

which is equal to

$$\begin{bmatrix}\text{Total number of the dimension}\\ \text{of vertex operators (minus } D)\end{bmatrix} - \begin{bmatrix}\text{Dimension of the Feynman}\\ \text{amplitude (minus } D)\end{bmatrix}.$$

Let us then consider the insertion of composite operator $O(x)$ to the Green function corresponding to the Feynman diagram G, i.e.,

$$\mathrm{FT}\langle 0|\mathrm{T}[O(x)\phi(x_1)\ldots\phi(x_n)]|0\rangle_{tc}, \tag{4.2.57}$$

where FT denotes the Fourier transform and suffixes t and c refer to "truncated" and "connected", respectively. Here again we have in mind the neutral scalar theory for simplicity. The insertion of the composite operator $O(x)$ is nothing but adding an extra vertex corresponding to $O(x)$ in addition to vertices $\mathscr{L}_i$ in the Feynman diagram G as shown in Fig. 4.2.3. Here a small solid circle in the diagram G' of Fig. 4.2.3 represents an insertion of the composite operator $O(x)$. On account of Eq. (4.2.56) we find that the superficial degree of divergence d'_G for the Feynman diagram with insertion of $O(x)$ is obtained by adding

$$r_O = d_O - D, \tag{4.2.58}$$

to d_G where d_O is the canonical mass dimension of $O(x)$, i.e.

$$d'_G = d_G + r_O. \tag{4.2.59}$$

Fig. 4.2.3. The Feynman diagrams G and G' without and with the insertion of composite operator $O(x)$ respectively.

Now we are ready to discuss the divergence structure of the Feynman amplitude with an operator insertion. If the canonical mass dimension d_O of operator $O(x)$ is smaller (larger) than D, the superficial degree of divergence d'_G for the diagram G' with operator insertion is smaller (larger) than the one d_G for the diagram G without operator insertion. Thus the insertion of a composite operator with $d_O < D$ improves power counting while the insertion of a composite operator with $d_O > D$ makes power counting worse. The insertion of a composite operator with $d_O = D$ does not affect the original power counting.

We now look into the details of renormalization of the Green function with operator insertion. For simplicity we shall confine ourselves to the ϕ_6^3 theory. Since $r_i = 0$, $N_F = 0$ and $\delta = 0$ in this theory, we have

$$d'_G = d_O - 2N_B. \tag{4.2.60}$$

Let us first consider the simplest composite operator

$$O(x) = \frac{1}{2!}\phi(x)^2. \tag{4.2.61}$$

The canonical mass dimension of $O(x)$ in this case is

$$d_O = D - 2 = 4. \tag{4.2.62}$$

The superficial degree of divergence is given by

$$d'_G = 2(2 - N_B). \tag{4.2.63}$$

Hence only two kinds of diagrams with $N_B = 1$ and 2 have overall divergence, i.e., $d'_G \geq 0$ for $N_B = 1, 2$. Disregarding the tadpole diagram ($N_B = 1$) by the

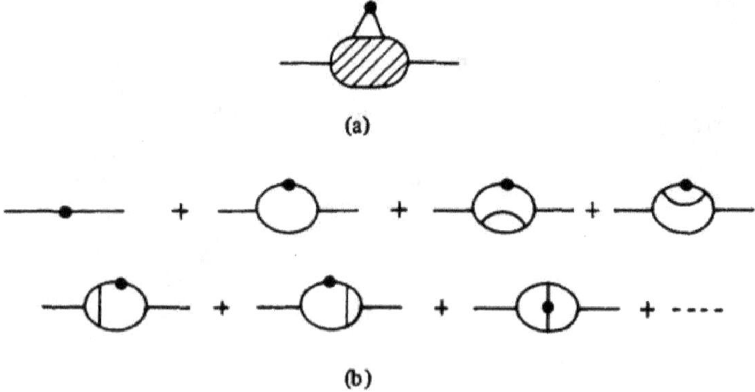

Fig. 4.2.4. The two-point diagram with the insertion of composite operator $O(x)$ in ϕ_6^3 theory which exhibits overall divergence: (a) the full amplitude and (b) its perturbative expansion.

same reasoning as in Sec. 2.5.3, we are left only with the diagram possessing two external lines as shown in Fig. 4.2.4(a). The corresponding Green function $\tilde{G}(p)$ with the zero-momentum insertion of the composite operator $O(x)$ is defined by

$$\tilde{G}(p)\,(2\pi)^6\delta^6(p'+p) = \int d^6x\,d^6x_1\,d^6x_2\,e^{i(p'\cdot x_1+p\cdot x_2)}\,\langle 0|\mathrm{T}[O(x)\phi(x_1)\phi(x_2)]|0\rangle. \tag{4.2.64}$$

Note that external lines of the Green function $\tilde{G}(p)$ are not truncated. The Feynman rule to obtain $\tilde{G}(p)$ is essentially the same as that given in Sec. 2.3.2 except that we need an extra rule for the operator-insertion vertex:

$$\text{—•—} \quad \longleftrightarrow \quad 1 \tag{4.2.65}$$

By the use of rule (4.2.65) we can calculate the contribution to $\tilde{G}(p)$ of all the Feynman diagrams drawn in Fig. 4.2.4(b). These Feynman diagrams give divergent contributions as is easily seen by power counting. A large number of the divergent subdiagrams in Fig. 4.2.4 already appeared in Sec. 2.5.3 (i.e., the self-energy and vertex parts) and may be taken care of by the renormalization constants Z_m, Z_g and Z_3. After performing the renormalization of these subdivergences we are still left with the divergence of the type given, e.g., by the diagram in Fig. 4.2.5. It can be shown that the divergence of this type is disposed of by the renormalization of the composite operator $O(x)$,

$$O_R(x) = Z_O^{-1}\,O(x). \tag{4.2.66}$$

In fact the Green function $\tilde{G}(p)$ (up to one-loop order) is calculated to be

$$\tilde{G}(p) = \frac{Z_3^{1/2}}{m^2-p^2}[1+\Lambda(p)]\frac{Z_3^{1/2}}{m^2-p^2}, \tag{4.2.67}$$

where $\Lambda(p)$ represents the contribution of the one-loop diagram in Fig. 4.2.5 and is given by[3]

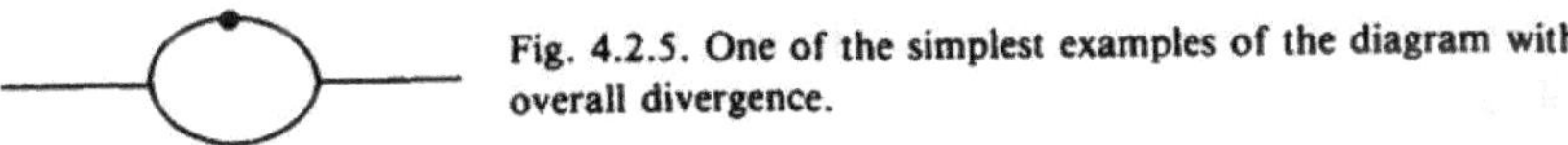

Fig. 4.2.5. One of the simplest examples of the diagram with overall divergence.

[3] Note here that $\Lambda(p)$ is equal to the one-loop vertex function with $q = 0$, i.e., $\Lambda(p) = \Lambda(p, 0)$. See Sec. 2.5.3.

$$\Lambda(p) = \frac{g_0^2}{(4\pi)^3}\left[\frac{1}{2}\left(\frac{1}{\varepsilon} - \gamma\right) - \int_0^1 dx(1-x)\ln\frac{m^2 - x(1-x)p^2}{4\pi\mu^2}\right], \tag{4.2.68}$$

where $\varepsilon = (6-D)/2$. From the above equation we find that, in the MS scheme,

$$Z_O = 1 + \frac{g_0^2}{2(4\pi)^3}\frac{1}{\varepsilon} + O(g_0^4). \tag{4.2.69}$$

The renormalized Green function $\tilde{G}_R(p)$ defined by

$$\tilde{G}_R(p) = Z_O^{-1} Z_3^{-1} \tilde{G}(p), \tag{4.2.70}$$

is then made finite.

Often there exist some composite operators $O_i(i=1, 2, ...)$ which have the same quantum numbers and canonical dimension. In this case the composite operators mix with each other in the renormalization procedure so that the renormalized composite operators O_{Ri} are defined by

$$O_{Ri} = \sum_j Z_{ij}^{-1} O_j, \tag{4.2.71}$$

where Z_{ij} $(i, j = 1, 2, 3, ...)$ are the renormalization constants. The phenomenorr in which operators with the same quantum numbers and canonical dimension mix with each other is called *operator mixing*. The concrete example of this phenomenon will be dealt with in Sec. 5.2.

In physical applications of quantum chromodynamics we are interested in the operator-product expansion of the product of two electromagnetic (or weak) currents,

$$j_\mu(x)j_\nu(0) = \sum_n C_{\mu\nu n}(x)\, O_n(x). \tag{4.2.72}$$

Since the left-hand side of Eq. (4.2.72) is obviously color gauge invariant, the right-hand side as a whole should also be gauge invariant. It is quite natural to expect the set of operators O_n to consists only of gauge invariant composite operators. This naive expectation, however, is invalidated by the participation of gauge-noninvariant composite operators due to the presence of the FP-ghost fields. These gauge-noninvariant composite operators do mix with the gauge-invariant composite operators in the renormalization process. For discussions, see [Gro 73a, 74, Geo 74, Dix 74, Kai 74, Sar 74, 75, Klu 75, 75a, 75b, Jog 76, Lee 76a, Dea 78]. Fortunately it has been found that, in spite of the presence of the mixing among the gauge-invariant and gauge-noninvariant composite operators, the set of the gauge-invariant composite operators which are composed only of gluon and quark fields is renormalized independently of the gauge-noninvariant composite operators and their renormalization constants are gauge independent. On the other hand, for the renormalization of the gauge-noninvariant composite operators which include the FP-ghost fields, both of the gauge-invariant and gauge-noninvariant composite operators have to be taken into account.

The above statement may be summarized in the following formula [Klu 75b, Jog 76, Lee 76a, Dea 78]

$$O^i = Z^{ij} O^j_R, \tag{4.2.73}$$

$$O'^i = Z'^{ij} O^j_R + Z''^{ij} O'^j_R, \tag{4.2.74}$$

where $O^i(O'^i)$ is the gauge-invariant (gauge-noninvariant) composite operators with index i to distinguish operators with the same quantum number and canonical dimension (we neglect the suffix n for simplicity), $O^i_R(O'^i_R)$ is the renormalized composite operator and Z^{ij}, Z'^{ij} and Z''^{ij} are renormalization constants. Here Z^{ij} are gauge independent as mentioned before. The formula (4.2.73) will be found to greatly simplify the calculation of anomalous dimensions of the guage-invariant composite operators.

4.2.4. Proof of the operator-product expansion

We shall now give a general survey on the proof of the operator-product expansion within the framework of perturbation theory. We confine ourselves to ϕ^3_6 theory with mass m in order to explain the proof in the simplest manner. The generalization to gauge theories should be straightforward. In the following we omit the caret ˆ on the top of operators and all the fields are understood to be renormalized.

We consider the "current" $j(x)$ defined by

$$j(x) = \phi(x)^2, \tag{4.2.75}$$

where the product of fields $\phi(x)$ at the same point (i.e., the composite operator $j(x)$) is understood to be renormalized in the manner described in Sec. 4.2.3. We are interested in the proof of the operator-product expansion of the form,

$$j(x)j(0) = \sum_k C_k(x) O_k(0), \tag{4.2.76}$$

where the summation over index k in general includes that on the Lorentz indices. In order to perform the proof of Eq. (4.2.76), it is more convenient to consider its matrix element rather than the operator form (4.2.76) by itself:

$$\langle 0|\mathrm{T}[j(x)j(0)\phi(x_1)\ldots\phi(x_n)]|0\rangle$$
$$= \sum_k C_k(x)\langle 0|T[O_k(0)\phi(x_1)\ldots\phi(x_n)]|0\rangle. \qquad (4.2.77)$$

Equation (4.2.77) may be rewritten in momentum space as a relation among truncated n-point Green functions with insertions of the current product $j(x)j(0)$ and composite operator $O_k(0)$,

$$F(q,p_1,\ldots,p_n) = \sum_k \tilde{C}_k(q)E_k(p_1,\ldots,p_n), \qquad (4.2.78)$$

where

$$F(q,p_1,\ldots,p_n) = \int d^6x\, d^6x_1\ldots d^6x_n\, e^{(iq\cdot x + ip_1\cdot x_1 + \ldots + ip_n\cdot x_n)}$$
$$\times \langle 0|\mathrm{T}[j(x)j(0)\phi(x_1)\ldots\phi(x_n)]|0\rangle\, \Delta^{-1}(p_1)\ldots\Delta^{-1}(p_n), \qquad (4.2.79)$$

$$\tilde{C}_k(q) = \int d^6x\, e^{iq\cdot x}\, C_k(x), \qquad (4.2.80)$$

$$E_k(p_1,\ldots,p_n) = \int d^6x_1\ldots d^6x_n\, e^{(ip\cdot x_1 + \ldots + ip_n\cdot x_n)}$$
$$\times \langle 0|\mathrm{T}[O_k(0)\phi(x_1)\ldots\phi(x_n)]|0\rangle\, \Delta^{-1}(p_1)\ldots\Delta^{-1}(p_n), \qquad (4.2.81)$$

with $\Delta(p)$ the propagator of field $\phi(x)$ defined by $\Delta(p) = -i\tilde{G}_{r2}(p)$ ($\tilde{G}_{r2}$ is the renormalized two-point Green function). Diagrammatically Eq. (4.2.78) is represented as in Fig. 4.2.6. It should be noted here that the operator-product expansion, if represented in the form of Green functions in momentum space

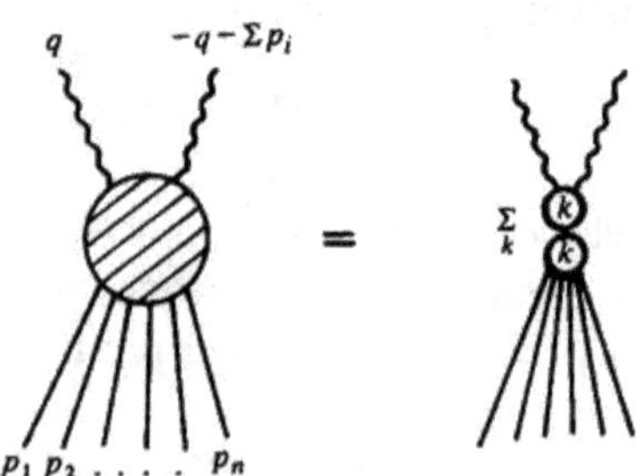

Fig. 4.2.6. Diagrammatic expression of Eq. (4.2.78) where the upper blob carrying index k represents the coefficient function $\tilde{C}_k(q)$ and the lower one corresponds to the amplitude $E_k(p_1, \ldots, p_n)$ with operator insertion.

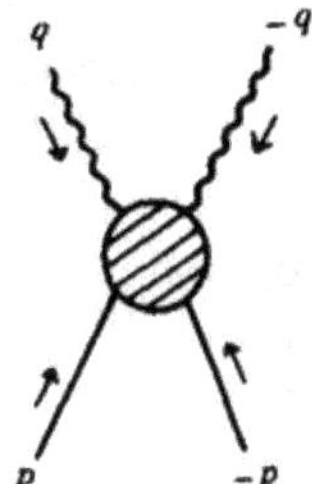

Fig. 4.2.7. The Feynman amplitude for forward scatterings of the current $j(x)$ off ϕ fields.

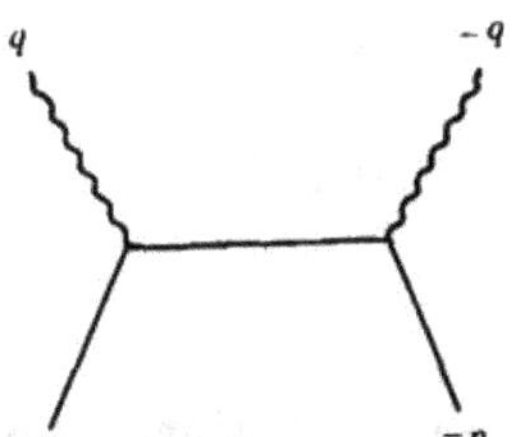

Fig. 4.2.8. The lowest-order contribution to the amplitude $F(p, q)$ of Fig. 4.2.7.

such as Eq. (4.2.78), provides us with the factorization of amplitude $F(q, p_1, \ldots, p_n)$ into the part depending only on q and the part depending only on $p_1, p_2, \ldots, p_n$. It has been shown [Wil 72, Zim 71, 73] that the operator-product expansion in the form of Eq. (4.2.76) is an exact relation which holds for arbitrary values of x if the summation on the right-hand side extends over a series of the infinite number of local composite operators. It is, however, enough for our present purpose to prove the above-mentioned factorization property only at short distances, i.e., for large current momentum q. At short distances only the finite number of the composite operators contributes to the right-hand side of Eq. (4.2.76) if we restrict ourselves to terms with dominant short-distance singularities. In the following we shall prove Eq. (4.2.78) only in the large-q limit

Examples: In order to explain the basic idea for the proof of Eq. (4.2.78), let us first examine a simple example in the lower-order Feynman diagrams. We shall make our argument simpler by considering only the case of forward scatterings with $n=2$ as depicted in Fig. 4.2.7. The amplitude $F(q, p, -p)$ will be denoted by $F(q, p)$.

The lowest-order contribution $F_0(q, p)$ to the Feynman amplitude $F(q, p)$ comes from the tree diagram of Fig. 4.2.8 and is given by

$$F_0(q, p) = \frac{1}{m^2-(q+p)^2} \cdot \tag{4.2.82}$$

We shall restrict ourselves to the short-distance region $x \sim 0$ in the configuration space which corresponds to the region $q \to \infty$ in momentum space. In Minkowski space, q^2 may remain finite even if all the components of q are rendered large. We shall, however, consider only the limit in which $|q^2| \to \infty$ as $q_\mu \to \infty$. The limit corresponds to the Euclidean region and may be expressed as follows,

$$|q^2| \gg |p \cdot q|, \quad |p^2|, \quad m^2. \tag{4.2.83}$$

We decompose the tree amplitude (4.2.82) into a term of order $1/q^2$ and the terms of higher powers of $1/q^2$,

$$\begin{aligned} F_0(q,p) &= \frac{1}{m^2 - q^2} + \left[\frac{1}{m^2 - (q+p)^2} - \frac{1}{m^2 - q^2}\right] \\ &= \frac{1}{-q^2} + O(1/q^4). \end{aligned} \tag{4.2.84}$$

Thus we can pick out the leading term for $|q^2| \to \infty$ in a trivial manner. If we define an operator t_0^p that sets $p = 0$ in an amplitude on which it operates, we can re-express Eq. (4.2.84) in the following form,

$$F_0(q, p) = t_0^p F_0(q, p) + (1 - t_0^p) F_0(q, p). \tag{4.2.85}$$

Hence $1 - t_0^p$ in Eq. (4.2.85) is the operator which increases the power of $1/q^2$. The first term of the right-hand side of Eq. (4.2.84) or (4.2.85) gives the simplest example of the operator-product expansion (4.2.78) at short distances.

At the one-loop level three diagrams shown in Fig. 4.2.9. contribute to the amplitude $F(q, p)$. We take the first diagram (a) of Fig. 4.2.9 as a typical

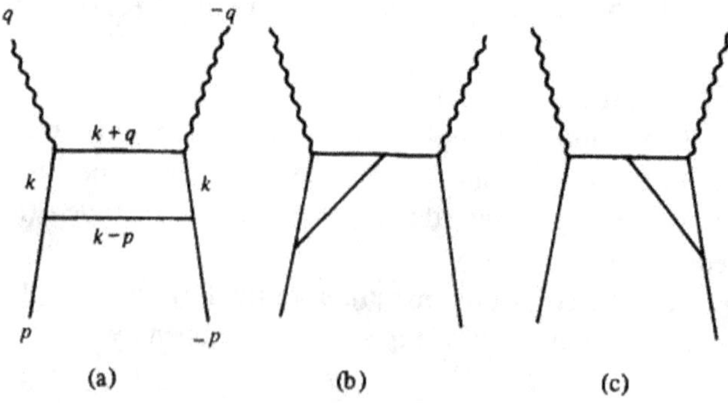

Fig. 4.2.9. The one-loop contribution to the amplitude $F(q, p)$.

example and explain the factorization property (4.2.78) in terms of the corresponding amplitude. The Feynman amplitude $F_1^a(q,p)$ corresponding to the diagram (a) is given by

$$F_1^a(q,p) = \int \frac{d^6k}{(2\pi)^6 i} g^2 I_a(q,p,k), \tag{4.2.86}$$

where

$$I_a(q,p,k) = \frac{1}{m^2-(k+q)^2} \frac{1}{(m^2-k^2)^2} \frac{1}{m^2-(k-p)^2}. \tag{4.2.87}$$

It is easy to observe that the integral over k in Eq. (4.2.86) would diverge logarithmically if we simple-mindedly let $|q^2| \to \infty$ in the integrand (4.2.87) to pick out the leading term:

$$F_1^a(q,p) \to \frac{g^2}{-q^2} \int \frac{d^6k}{(2\pi)^6 i} \frac{1}{(m^2-k^2)^2} \frac{1}{m^2-(k-p)^2}. \tag{4.2.88}$$

This behavior may be represented diagrammatically as in Fig. 4.2.10: the original box-type diagram is effectively reduced, in the limit $|q^2| \to \infty$, to the triangle-type diagram which corresponds to the logarithmically divergent amplitude. The above divergence in Eq. (4.2.88) is a manifestation of the large $|q^2|$ behavior of the original integral (4.2.86) which is of the form $(\ln|q^2|)/q^2$ for $|q^2| \to \infty$. Although the Feynman integral (4.2.86) is convergent, we anticipate the above divergence and rearrange the integrand I_a so that a subtraction at $p=0$ is made,

$$I_a = t_0^p I_a + (1-t_0^p) I_a. \tag{4.2.89}$$

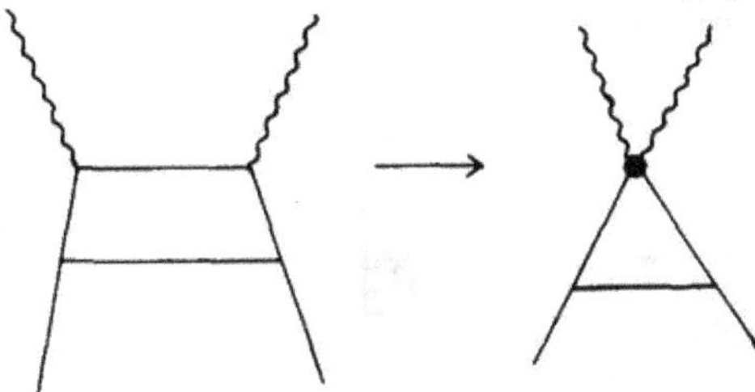

Fig. 4.2.10. The reduction of the box diagram to the triangle-type diagram in the limit $|q^2| \to \infty$. The solid circle in the triangle-type diagram represents an effective $\phi - \phi - j - j$ vertex.

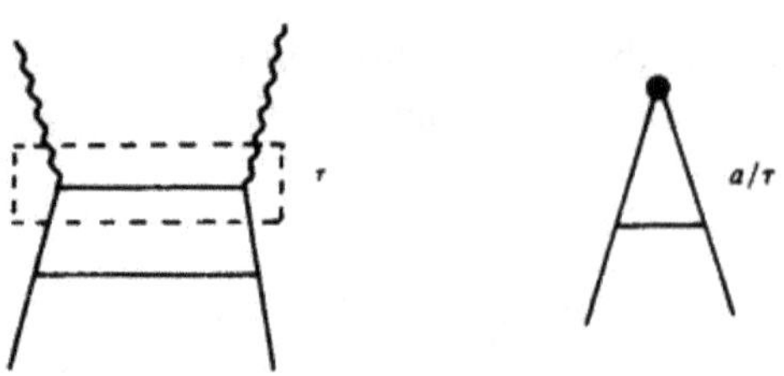

Fig. 4.2.11. The diagrams τ and a/τ.

This subtraction is a typical example of the oversubtraction whose definition will be given later. The second term of the right-hand side of Eq. (4.2.89) provides us with the finite expression even if the limit $|q^2|\to\infty$ is taken, i.e.,

$$\int\frac{d^6k}{(2\pi)^6 i}g^2(1-t_0^p)\,I_a \longrightarrow \frac{g^2}{-q^2}\int\frac{d^6k}{(2\pi)^6 i}\frac{1}{(m^2-k^2)^2}\times\left[\frac{1}{m^2-(k-p)^2}-\frac{1}{m^2-k^2}\right]. \tag{4.2.90}$$

Thus in Eq. (4.2.89) we rearranged the integrand I_a so that the first and second terms behave as $(\ln|q^2|)/q^2$ and $1/q^2$ for $|q^2|\to\infty$, respectively. Furthermore let us pick out terms of order $1/q^4$ by applying t_0^k to the integrand I_a,

$$I_a = t_0^p I_a + (1-t_0^p)\,I_{a/\tau}t_0^k I_\tau + (1-t_0^p)\,I_{a/\tau}(1-t_0^k)I_\tau, \tag{4.2.91}$$

Here in Eq. (4.2.91) the diagram τ corresponding to the Feynman integrand I_τ represents an upper half of the diagram 4.2.9 (a) and the diagram a/τ is obtained by reducing τ to a point in the diagram 4.2.9(a). See Fig. 4.2.11. Equation (4.2.91) may be rewritten in an explicit manner as follows,

$$I_a = \frac{1}{m^2-(k+q)^2}\frac{1}{(m^2-k^2)^3}+\frac{1}{(m^2-k^2)^2}\left[\frac{1}{m^2-(k-p)^2}-\frac{1}{m^2-k^2}\right]\times\frac{1}{m^2-q^2}+\frac{1}{(m^2-k^2)^2}\left[\frac{1}{m^2-(k-p)^2}-\frac{1}{m^2-k^2}\right]\times\left[\frac{1}{m^2-(k+q)^2}-\frac{1}{m^2-q^2}\right]. \tag{4.2.92}$$

The last term of Eq. (4.2.92) is the nonleading term behaving as $1/q^4$ for $|q^2|\to\infty$ and may be neglected. The decomposition (4.2.92) of the integrand I_a can be expressed diagrammatically as shown in Fig. 4.2.12. Substituting Eq.

(4.2.92) into Eq. (4.2.86) and adding the lowest-order Feynman amplitude F_0 to F_1^a we find

$$F_0 + F_1^a = \left[\frac{1}{m^2-q^2} + g^2\int\frac{d^6k}{(2\pi)^6 i}\frac{1}{m^2-(k+q)^2}\frac{1}{(m^2-k^2)^3}\right]$$
$$\times\left[1+g^2\int\frac{d^6k}{(2\pi)^6 i}\frac{1}{(m^2-k^2)^2}\left(\frac{1}{m^2-(k-p)^2}-\frac{1}{m^2-k^2}\right)\right] + O\left(\frac{1}{q^4}\right), \tag{4.2.93}$$

where the term of order g^4 is irrelevant. The diagrammatic expression of Eq. (4.2.93) is obvious as is found in Fig. 4.2.13. The logarithmically divergent constant

$$g^2\int\frac{d^6k}{(2\pi)^6 i}\frac{1}{(m^2-k^2)^3}$$

appearing in the last factor in Eq. (4.2.93) is absorbed as a part of the renormalization constant for the composite operator O:

$$Z_O^{-1} = 1-g^2\int\frac{d^6k}{(2\pi)^6 i}\frac{1}{(m^2-k^2)^3} + \text{other terms.} \tag{4.2.94}$$

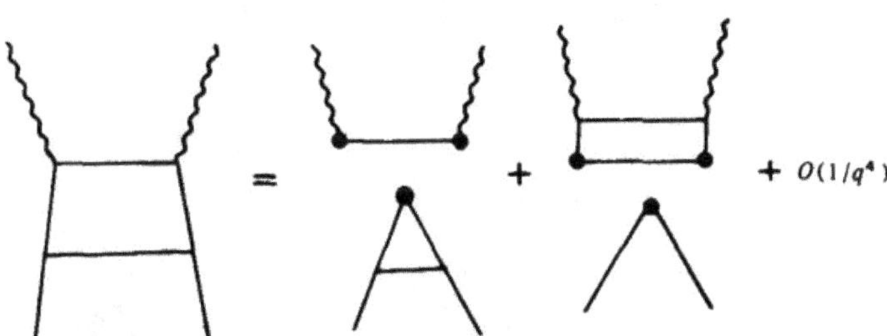

Fig. 4.2.12. The diagrammatic expression of Eq. (4.2.92). The solid circles in the diagrams on the right-hand side indicate that no momentum flows in at these vertices.

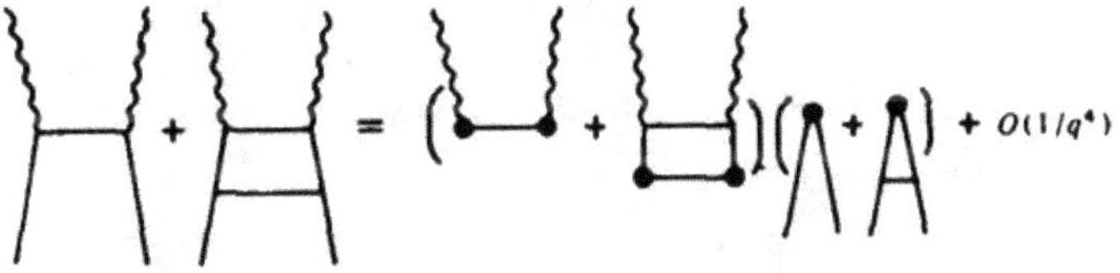

Fig. 4.2.13. The diagrammatic expression of Eq. (4.2.93).

Here it should be noted that we are adopting the renormalization scheme of the zero-momentum subtraction, i.e., the subtraction at $p = 0$. Thus the last factor in Eq. (4.2.93) is nothing but the matrix element of the renormalized composite operator which corresponds to the function $E_k(p_1, ..., p_n)$ in Eq. (4.2.81). Accordingly the first factor in Eq. (4.2.93) is interpreted as the singular c-number function $\tilde{C}_k(q)$. Hence Eq. (4.2.93) is found to be an example of the operator-product expansion (4.2.78). The contribution of the other diagrams (b) and (c) in Fig. 4.2.9 can be evaluated in a similar manner and the operator-product expansion of the form (4.2.78) is obtained for the full one-loop amplitude $F_0 + F_1^a + F_1^b + F_1^c$.

Proof: We are now in a position to elaborate on the above heuristic argument and to complete the proof of the operator-product expansion rigorously. For this purpose the BPHZ method described in Sec. 2.5.4 turns out to be the most useful and hence we shall adopt this method in the following argument. We consider the renormalized truncated n-point Green function $F(q, p_1, ..., p_n)$ with the insertion of the operator $j(x)j(0)$ which is defined by Eq. (4.2.79). We would like to show that the Green function F has the factorized form of the right-hand side of Eq. (4.2.78). Taking into account the perturbative expansion of $\tilde{C}_k(q)$ and $E_k(p_1, ..., p_n)$, we find that the right-hand side of Eq. (4.2.78) may be written as

$$\sum_k \tilde{C}_k E_k = \sum_k [\tilde{C}_k^0 E_k^0 + g^2(\tilde{C}_k^0 E_k^1 + \tilde{C}_k^1 E_k^0) + \\ \dots + g^{2N} \sum_{n=0}^{N} C_k^n E_k^{N-n} + \dots], \tag{4.2.95}$$

where

$$\tilde{C}_k = \sum_{n=0}^{\infty} g^{2n} \tilde{C}_k^{\,n}, \tag{4.2.96}$$

$$E_k = \sum_{n=0}^{\infty} g^{2n} E_k^{\,n}. \tag{4.2.97}$$

The terms of orders g^0 and g^2 in Eq. (4.2.95) were already investigated in the previous example. We now wish to show that the general N-th order term in the expansion of the Green function F has the form of the g^{2N} term in Eq. (4.2.95).

The contribution to F coming from Feynman diagram G with L loops may be written in the form,

$$F_G(q, p_1, ..., p_n) = \int d^6k_1 \dots d^6k_L \, I_G(k_1, ..., k_L; q, p_1, ..., p_n), \tag{4.2.98}$$

where I_G is called the Feynman integrand. Since F_G is a renormalized Feynman amplitude, suitable subtractions are already made in the Feynman integrand I_G so as to eliminate all the ultraviolet divergences. In the BPHZ method such subtractions are made systematically by the Zimmermann forest formula (2.5.99).[4] Here we use the formula expressed in terms of the Feynman integrand with the application of the zero-momentum subtraction which will be explained shortly,

$$I_G = \sum_{\Phi \in U_G} \prod_{H \in \Phi} (-t^H)\, I_G^0, \tag{4.2.99}$$

where I_G^0 is the bare Feynman integrand, i.e. the integrand of the unrenormalized Feynman amplitude, H is a renormalization part of G, Φ is a forest which consists of nonoverlapping renormalization parts H, U_G is a set of all the forests Φ for the diagram G. Note that the overall subtraction $1 - t^G$ present in Eq. (2.5.99) is absorbed in the sum over the forests in Eq. (4.2.99) by allowing the case of $H = G$. Here we employ the zero-momentum subtraction scheme in which the subtraction is made at the vanishing momenta because it is much easier in this subtraction scheme to understand the concept of the oversubtraction (see [Gup 80] for the minimal subtraction method). The zero-momentum subtraction scheme has been employed in the original papers on the BPHZ method [Bog 57, Hep 66, Zim 69, 71]. In the zero-momentum subtraction scheme the operator t^H is defined to be an operation which picks up the first d_H terms in the Taylor expansion of $I_H{}^0$ with respect to external momenta $p_i (i = 1, 2, ..., n_H)$ of diagram H around $p_i = 0$,

$$t^H I_H^0 = \sum_{n=0}^{d_H} \frac{1}{n!} \left(\frac{\partial}{\partial \lambda}\right)^n I_H^0 (\lambda p_1, ..., \lambda p_{n_H})|_{\lambda=0} \tag{4.2.100}$$

where d_H is the superficial degree of divergence for diagram H which is assumed to have only overall divergences and I_H^0 is the bare Feynman integrand for diagram H with external momenta $p_1, p_2, ..., p_{n_H}$. In the above sense t^H is often called the Taylor operator. The operation $t^H I_G^0$ $(H \subseteq G)$ is defined by the procedure

$$t^H I_G^0 = I_{G/H}^0\, t^H I_H^0. \tag{4.2.101}$$

[4] In Sec. 2.5.4. we adopted the minimal subtraction (MS) renormalization scheme and hence the R operation was applied to the Feynman amplitude instead of the Feynman integrand. Here in the present subsection we use the zero-momentum subtraction as in the preceding one-loop example. Hence the Zimmermann forest formula is written down for the Feynman integrand. It should be noted that the original proof of the formula (Zim 69, 71) was made in this subtraction scheme.

Summing up, t^H is the operation that singles out terms in I_H^0 which give rise to divergent integrals (from the logarithmic to the d_H-th-order divergence) and the operator $1-t^H$ plays the role of eliminating all the divergences in the Feynman integration for I_H^0.

With the zero-momentum subtraction performed in the manner of Eq. (4.2.99), we obtain the renormalized Feynman amplitude (4.2.98). We then let $q\to\infty$ in $F_G(q, p_1, ..., p_n)$ and meet with a new divergence as in the previous one-loop example. In order to elimate this divergence we have to make an extra subtraction which we call the *oversubtraction.* In general we need to make extra subtractions to eliminate terms of lower power in $1/q^2$ until we reach the remaining amplitude of the desired power in $1/q^2$. We call all these extra subtractions the oversubtractions.

To make the situation clear we tentatively go back to the previous example. Since the amplitude $F_1{}^a(q, p)$ in Eq. (4.2.86) is finite, no subtraction of the ordinary ultraviolet divergence is required and hence the renormalized integrand I_a is the same as the bare one I_a^0. We made two oversubtractions $1-t_0^p$ and $1-t_0^k$ in Eq. (4.2.91) which correspond to the subtractions for diagrams a/τ and τ, respectively. We may rewrite Eq. (4.2.91) in the following form,

$$I_a = -[(-t_0^p)I_a + (-t_0^k)I_a + (-t_0^p)(-t_0^k)I_a] + (1-t_0^p - t_0^k + t_0^p t_0^k)I_a. \tag{4.2.102}$$

Equation (4.2.102) is easily recognized to be a special case of the general identity,

$$\sum_{\Phi\in U_a}\prod_{H\in\Phi}(-t^H)\,I_G^0 = -\sum_{\Phi\in U}\prod_{H\in\Phi}(-t^H)\,I_G^0 + \sum_{\Phi\in U_{\bar G}}\prod_{H\in\Phi}(-t^H)\,I_G^0, \tag{4.2.103}$$

where U_G is the set of forests Φ corresponding to diagram G (Fig. 4.2.14 (a)), $U_{\bar G}$ the set of forests Φ corresponding to the diagram $\bar G$ which is generated

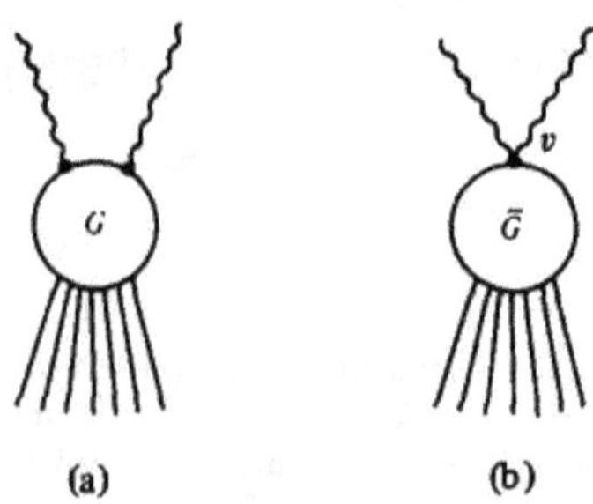

Fig. 4.2.14. (a) The diagram G. (b) The diagram $\bar G$.

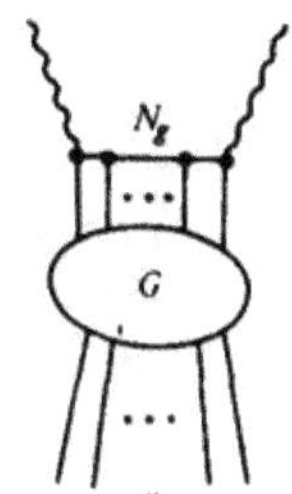

Fig. 4.2.15. The number (N_g) of the ϕ^3-interaction vertices in the diagram G hooking on the line between two currents.

from G by merging two external currents as shown in Fig. 4.2.14 (b), and U the part of $U_{\bar{G}}$ obtained by removing U_G, i.e.,

$$U = U_{\bar{G}} - U_G. \tag{4.2.104}$$

Note here that the forest of $U_{\bar{G}}$ includes tree vertices of composite operators although they are not the renormalization parts. The left-hand side of Eq. (4.2.103) is nothing but the Feynman integrand for the Feynman amplitude (4.2.98) as has been given in Eq. (4.2.99). The second term on the right-hand side of Eq. (4.2.103) is the Green function with oversubtraction and is suppressed by powers of $1/q^2$ for $q \to \infty$. It represents the Feynman integrand for the Green function with the insertion of the renormalized composite operator $\phi(x)^N$, where the power N is determined by the number of ϕ^3-interaction vertices N_g between two current vertices in the diagram G as shown in Fig. 4.2.15: $N = N_g + 2$. The composite operator $\phi(x)^N$ emerges if these two current vertices are pinched to a point. The first term of the right-hand side of Eq. (4.2.103) is of relevance to our present purpose. It gives a leading contribution to the Feynman amplitude for $q \to \infty$. What we would like to show is that the first term of the right-hand side of Eq. (4.2.103) is factorized in the manner shown in Eq. (4.2.95).

In order to prove the above statement we rewrite the first term of the right-hand side of Eq. (4.2.103) so that its factorization is manifest. For this purpose we first look into the structure of the set of forests, U, given in Eq. (4.2.104). We recognize that U is the set of all the forests made of the nonoverlapping renormalization parts, at least one of which contains the vertex v of operator $\phi(x)^N$. This vertex is formed by contracting two current vertices as mentioned before. See Fig. 4.2.14 (b). The renormalization parts containing the vertex v cannot be disjoint and so they are all nested. Therefore we conclude that there exists the smallest renormalization part which we call τ. We then find that the forest Φ in U is expressed as a union of three forests Φ_1, Φ_2 and $\{\tau\}$,

$$\Phi = \Phi_1 \cup \{\tau\} \cup \Phi_2, \tag{4.2.105}$$

where Φ_1 consists of renormalization parts containing τ or disconnected from τ, and Φ_2 is made of renormalization parts contained in τ. Hence we obtain

$$\prod_{H\in\Phi}(-t^H) = \prod_{H\in\Phi_1}(-t^H)(-t^\tau)\prod_{H\in\Phi_2}(-t^H). \tag{4.2.106}$$

Applying the formula (4.2.106) to the first term of the right-hand side of Eq. (4.2.103) we obtain

$$\sum_{\Phi\in U}\prod_{H\in\Phi}(-t^H)I_G^0 = \sum_\tau\sum_{\Phi_1\in U_1}\sum_{\Phi_2\in U_2}\prod_{H\in\Phi_1}(-t^H)(-t^\tau)\prod_{H\in\Phi_2}(-t^H)I_G^0, \tag{4.2.107}$$

where U_1 and U_2 are made of Φ_1 and Φ_2, respectively, i.e., U_1 is the set of the forests consisting of renormalization parts containing τ or disconnected from τ, and U_2 is the set of the forests made of renormalization parts smaller than τ. Equation (4.2.107) is rewritten as

$$\sum_{\Phi\in U}\prod_{H\in\Phi}(-t^H)I_G^0 = \sum_\tau\left[\sum_{\Phi_1\in U_1}\prod_{H\in\Phi_1}(-t^H)I_{G/\tau}^0\right](-t^\tau)\left[\sum_{\Phi_2\in U_2}\prod_{H\in\Phi_2}(-t^H)I_\tau^0\right]. \tag{4.2.108}$$

The right-hand side of Eq. (4.2.108) is obviously of the factorized form, i.e., for the Feynman amplitude F_G it gives a sum of terms which factorize into the part depending only on q and the part depending only on $p_1, \ldots, p_n$. Thus Eq. (4.2.108) completes our proof of the factorization of the first term of the right-hand side of Eq. (4.2.103). Note that the first factor on the right-hand side of Eq. (4.2.108) corresponds to the Green function E_k^n with insertion of the composite operator O_k defined in Eq. (4.2.97), while the last factor corresponds to the singular coefficient function $\tilde{C}_k^n$ in Eq. (4.2.96).

Finally we would like to make several remarks concerning the operator-product expansion just proven above.

(1) The operator-product expansion is proven here in the form of an asymptotic expression for large momentum, $q\to\infty$. The expansion, however, actually holds for arbitrary value of q, since Eqs. (4.2.103) and (4.2.108) are not asymptotic relations but identities.

(2) Although we have used the zero-momentum subtraction scheme in the above proof of the operator-product expansion, it is of course possible to perform the proof in other renormalization schemes such as the MS scheme introduced in Sec. 2.4.3 [Gup 79, 80, Che 82, 83, 84, Tka 83]. In this subtraction scheme the oversubtraction at vanishing momenta is replaced by a

subtraction of poles at a space-time dimension higher than four. Although such poles do not correspond to divergences of Green functions in four dimensions, they represent potential divergences in the limit $q \to \infty$.

(3) In our proof we heavily relied on Zimmermann's forest formula to derive the factorization property of Green functions. The same result, however, can be reached in an entirely different method where the factorization of mass singularities in Green functions is extensively used [El1 78, 79, Ama 78, 78a, Lib 78, Hum 82]. The mass singularity is relevant here for the argument of the operator-product expansion because the limit of vanishing mass $m \to 0$ gives rise to the same effect as that for the limit of large momentum $q \to \infty$ by dimensional reasoning.

(4) The operator-product expansion in the form proven here indicates that any Green function may be expanded in the power series of $1/q^2$ if the large mass scale $\sqrt{|q^2|}$ is present in the Green function. It is straightforward to extend this expansion to the case of large particle masses. In fact the method employed to prove the operator-product expansion has been applied [Kaz 79, 80, Wei 80, Cha 82] to the derivation of the decoupling theorem [App 75a]. The theorem states that heavy particles in the theory decouple for $M \to \infty$ with M the mass of heavy particles, i.e., the contribution of heavy particles to Green functions is suppressed by the power of $1/M$.

(5) The operator-product expansion of the form (4.2.76) is of essential use in deep inelastic scattering to extract the short-distance piece in the cross section. In some other reactions such as the Drell-Yan process we are unable to express the cross section in the form of the imaginary part of the matrix element of the current product and so it is not clear whether the operator-product expansion is applicable to this process. However, by studying the proof of the operator-product expansion closely, we recognize that we have not directly dealt with the operator form (4.2.76) but worked with the relation among Green functions. In our proof we have only used the Zimmermann forest formula in perturbation theory and derived the factorization of the short-distance piece in Green functions. This factorization property is quite general so that it holds for a large class of Green functions and is referred to as the *factorization theorem*. The generalization of the factorization property to other processes such as the Drell-Yan process was first suggested by Politzer [Pol 77, 77a] using the factorization property of the mass singularity. The complete proof of the factorization theorem was given both in the BPHZ method [Mue 78, 81 Hum 82] and in the mass-singularity method [El1 78, 79, Ama 78, 78a, Lib 78, Hum 82] (see also [Lin 83]).

4.2.5. Light-cone expansion

In the preceeding subsection we restricted ourselves to the short-distance region $x \sim 0$ in deriving the operator-product expansion though the validity of the expansion is more general as mentioned earlier. The operator-product expansion in the asymptotic expression valid for $x \sim 0$ is called the *short-distance expansion.* In the short-distance expansion only a limited number of composite operators contribute to terms with the dominant short-distance singularity. Here the summation on k in Eq. (4.2.76) runs only over different types of the composite operators and does not include the summation on Lorentz indices. The short-distance region $x \sim 0$ corresponds to the large q region characterized by Eq. (4.2.83) and so we can neglect terms of order $|p{\cdot}q|$ in comparison with those of order $|q^2|$. For example, in the case of the tree amplitude (4.2.82) we can safely neglect all the terms but $1/(-q^2)$ as shown in Eq. (4.2.84). Thus we recognize that only one composite operator contributes to the dominant term of the tree amplitude.

In our physical applications, however, we need the operator product expansion near the light cone $x^2 \sim 0$ rather than the one in the short-distance region $x \sim 0$. In deep inelastic scatterings the light-cone region $x^2 \sim 0$ corresponds to the region $-q^2 \to \infty$ with $\xi = -q^2/(2p{\cdot}q)$ fixed in momentum space. Thus $|p{\cdot}q|$ is of the same order as $|q^2|$. The region in momentum space corresponding to the light-cone region reads

$$|q^2| \sim |p{\cdot}q| \gg |p^2|, m^2. \tag{4.2.109}$$

Condition (4.2.83) for the Feynman amplitude $F(q, p)$ should now be replaced by Eq. (4.2.109). Then, in the example of the tree amplitude (4.2.82), the correct expansion for the above region (4.2.109) is not the one given in Eq. (4.2.84) but

$$F_0(q, p) = \frac{1}{-q^2}\frac{1}{1-2p{\cdot}q/(-q^2)} + O(1/q^4). \tag{4.2.110}$$

Expanding the first term in Eq. (4.2.110) in powers of $2p{\cdot}q/(-q^2)$ we obtain

$$F_0(q, p) = \frac{1}{-q^2} + \frac{2p{\cdot}q}{q^4} + \frac{(2p{\cdot}q)^2}{-q^6} + \ldots + O(1/q^4). \tag{4.2.111}$$

Hence we find that to the dominant term of $F_0(q, p)$ in the limit (4.2.109) contributes an infinite number of composite operators. To see this situation more clearly we evaluate Fourier transformation of Eq. (4.2.111) and work in configuration space [Gel 64a],

$$\int d^6q\, e^{-iq\cdot x} F_0(q,p) \propto \frac{1}{x^4} + i\frac{p\cdot x}{x^4} - \frac{(p\cdot x)^2}{8x^4} + \ldots, \tag{4.2.112}$$

The n-th term of the right-hand side of Eq. (4.2.112) may be written in the following form,

$$\frac{1}{(x^2)^2} x^{\mu_1}\ldots x^{\mu_n} \langle p|O_{\mu_1\ldots\mu_n}|p\rangle, \quad (n = 0, 1, 2, \ldots), \tag{4.2.113}$$

with $\langle p|O_{\mu_1\ldots\mu_n}|p\rangle$ the matrix element of the composite operator $O_{\mu_1\ldots\mu_n}$.

In general the operator-product expansion near the light cone takes the form,

$$j(x)j(0) = \sum_{i,n} C_n^{(i)}(x^2) x^{\mu_1}\ldots x^{\mu_n} O^{(i)}_{\mu_1\ldots\mu_n}(0), \tag{4.2.114}$$

where index i distinguishes the different types of the composite operators. Denoting by k the indices i and $\mu_1, \ldots, \mu_n$ in the lump and setting

$$C_k(x) \equiv C_n^{(i)}(x^2) x^{\mu_1}\ldots x^{\mu_n}, \tag{4.2.115}$$

we recover Eq. (4.2.76). We call the expansion (4.2.114) the *light-cone expansion*.

4.3. COEFFICIENT FUNCTIONS

4.3.1. Singularity in coefficient functions

The coefficients $C_k(x)$ in the operator product expansion (4.2.76) are often called *coefficient functions*. We have seen in Sec. 4.1 that the singularity structure of the coefficient functions plays a crucial role in making physical predictions in deep inelastic scatterings and e^+e^- annihilations. The operator-product expansion itself does not give us any information on their singularity structure and so we must appeal to some other means to determine it.

In the free field theory we simply count the canonical dimension of operators appearing in the operator-product expansion (4.2.114). In order to match both sides of Eq. (4.2.114) in their mass dimension, the light-cone singularity of the coefficient function $C_n^{(i)}(x^2)$ has to be of the following form,

$$C_n^{(i)}(x^2) \sim (x^2)^{-d_{jo} - n/2 + d_0^i(n)/2}, \tag{4.3.1}$$

where d_{jo} and $d_O^i(n)$ are canonical dimensions of the current $j(x)$ and the composite operator $O^{(i)}_{\mu_1\ldots\mu_n}$, respectively. The above simple power-counting argument is allowed because no other mass scale is present in the operator product expansion for the free field theory. Note here that the strength of the light-cone singularity of $C_n^{(i)}(x^2)$ is governed by the number $d_O^i(n) - n$ in the exponent of Eq. (4.3.1). The smaller is the value of $d_O^i(n) - n$, the stronger the singularity of $C_n^{(i)}(x^2)$. The index

$$\tau_n^i = d_O^i(n) - n, \tag{4.3.2}$$

is called the *twist* [Gro 71] of the composite operator $O^{(i)}_{\mu_1 \ldots \mu_n}$.

For interacting fields the naive dimension counting does not work in determining the light-cone singularity of the coefficient functions. In the case when an ultraviolet fixed point exists in the theory under consideration, the scale invariance is recovered with anomalous dimensions at the fixed point and so the dimension counting with the scale dimension is applicable. Here the canonical dimensions d_{jo} and $d_O^i(x)$ in the previous argument are replaced by the scale dimensions d_j and $d^i(n)$ for the current $j(x)$ and the composite operator $O^{(i)}_{\mu_1 \ldots \mu_n}$ respectively. The light-cone singularity of $C_n^{(i)}(x^2)$ now reads [Wil 69]

$$C_n^{(i)}(x^2) \sim (x^2)^{-d_j - n/2 + d^i(n)/2}. \tag{4.3.3}$$

If the coupling constant of the theory is kept off the fixed point or the theory has no nontrivial fixed point, the above simple argument of counting the scale dimensions does not hold and in general logarithmic corrections to Eq. (4.3.3) develop. The full singularity structure of the coefficient functions on the light cone may be obtained by using the renormalization group equation for the coefficient functions which will be discussed in the next subsection.

4.3.2. Renormalization group equations

The renormalization group method is of essential use in extracting information on the singularity structure of the coefficient function on the light cone.

We consider the operator-product expansion of the form (4.2.78). It is easy to show that the Green function $F(q, p_1, \ldots, p_n)$ satisfies the following renormalization group equation,

$$[\mathscr{D} + 2\gamma_j(g) - n\gamma(g)]F = 0, \tag{4.3.4}$$

where $\gamma_j(g)$ and $\gamma(g)$ are anomalous dimensions of the current $j(x)$ and the field $\phi(x)$ respectively,[5] and $\mathscr{D}$ is the differential operator defined by

[5] The anomalous dimension $\gamma_j(g)$ of the composite operator $j(x)$ is given by $(\mu/Z_j)(\partial Z_j/\partial\mu)$ for the fixed bare mass and coupling constant with Z_j the renormalization constant for the current $j(x)$. If the current is conserved, $\gamma_j(g) = 0$.

$$\mathscr{D} = \mu\frac{\partial}{\partial\mu} + \beta(g)\frac{\partial}{\partial g} - \gamma_m(g)m\frac{\partial}{\partial m}\,, \tag{4.3.5}$$

with $\beta(g)$ and $\gamma_m(g)$ given by Eqs. (3.2.16) and (3.2.17) respectively. On the other hand the Green function $E_k(p_1, \ldots, p_n)$ is shown to obey the renormalization group equation,

$$[\mathscr{D} + \gamma_{O_k}(g) - n\gamma(g)]E_k = 0, \tag{4.3.6}$$

where $\gamma_{O_k}(g)$ is the anomalous dimension of the composite operator $O_k(x)$ defined by

$$\gamma_{O_k}(g) = \frac{\mu}{Z_{O_k}}\frac{\partial Z_{O_k}}{\partial\mu}\bigg|_{g_0, m_0}, \tag{4.3.7}$$

with g_0 and m_0 the bare coupling constant and mass, respectively, and Z_{O_k} is the renormalization constant of the composite operator $O_k(x)$,

$$O_k = Z_{O_k}^{-1}\, O_k^0 \quad (O_k^0, \text{ unrenormalized}). \tag{4.3.8}$$

Here we are of course working in the ϕ_6^3 theory with mass m and neglecting possible operator mixing. We have assumed that the minimal subtraction (MS) scheme is adopted for the renormalization of composite operators $O_k(x)$ as well as for the renormalization of the mass m, coupling constant g and field $\phi(x)$. Hence the operator-product expansion (4.2.78) is understood here to be given in the MS scheme.

We apply the differential operator $\mathscr{D}$ defined in Eq. (4.3.5) to both sides of Eq. (4.2.78) and use Eqs. (4.3.4) and (4.3.6). We then find

$$[\mathscr{D} + 2\gamma_j(g) - \gamma_{O_k}(g)]\, \tilde{C}_k(q) = 0. \tag{4.3.9}$$

This is the renormalization group equation which serves to determine the behavior of the coefficient function $\tilde{C}_k(q)$ for large Euclidean momenta $-q^2 \to \infty$.

In the case of the light-cone expansion, the basic formula is given by Eq. (4.2.114) and is rewritten in momentum space as follows,

$$F(q, p_1, \ldots, p_n) = \sum_{i,N} \tilde{C}_N^{(i)}(q^2)\, q^{\mu_1} \ldots q^{\mu_N} E^{(i)}_{\mu_1 \ldots \mu_N}(p_1, \ldots, p_n), \tag{4.3.10}$$

where $F(q, p_1, \ldots, p_n)$ is given in Eq. (4.2.79) and

$$q^{\mu_1} \ldots q^{\mu_n} \tilde{C}_N^{(i)}(q^2) = \int d^6x\, e^{iq\cdot x} x^{\mu_1} \ldots x^{\mu_N} C_N^{(i)}(x^2), \tag{4.3.11}$$

$$E^{(i)}_{\mu_1\ldots\mu_n}(p_1,\ldots,p_n) = \int d^6x_1\ldots d^6x_n \exp(ip_1\cdot x_1 + \ldots + ip_n\cdot x_n)$$
$$\times \langle 0|T[O^{(i)}_{\mu_1\ldots\mu_N}(0)\phi(x_1)\ldots\phi(x_n)]|0\rangle \Delta^{-1}(p_1)\ldots\Delta^{-1}(p_n). \tag{4.3.12}$$

As before we apply the differential operator $\mathscr{D}$ to both sides of Eq. (4.3.10) to obtain

$$[\mathscr{D} + 2\gamma_j(g) - \gamma_N^{(i)}(g)]\,\tilde{C}_N^{(i)}(q^2) = 0, \tag{4.3.13}$$

where $\gamma_N^{(i)}(g)$ is the anomalous dimension of operator $O^{(i)}_{\mu_1\ldots\mu_N}(x)$. The solution of Eq. (4.3.13) can be obtained in the same way as in Sec. 3.3.1. By setting

$$\tilde{C}_N^{(i)}(q^2) = \tilde{C}_N^{(i)}\left(\frac{-q^2}{\mu^2}, g, m\right), \tag{4.3.14}$$

with μ the renormalization scale, we write down the solution in the following form,

$$\tilde{C}_N^{(i)}\left(\frac{-q^2}{\mu^2}, g, m\right) = \tilde{C}_N^{(i)}(1, \bar{g}(t), \bar{m}(t)) \exp \int_0^t dt' \, [2\gamma_j(\bar{g}(t')) - \gamma_N^{(i)}(\bar{g}(t'))], \tag{4.3.15}$$

where the running coupling constant $\bar{g}(t)$ and the running mass $\bar{m}(t)$ are given by Eq. (3.3.7).

If operator mixing is present in the renormalization of the set of composite operators $\{O^{(i)}_{\mu_1\ldots\mu_n}\}$ with index $i = 1, 2, \ldots$ distinguishing different operators with the same twist and quantum number, Eq. (4.3.13) turns out to be simultaneous differential equations of the following form,

$$[(\mathscr{D} + 2\gamma_j)\delta_{ij} - (\gamma_N)_{ij}]\,\tilde{C}_N^{(j)}(q^2) = 0, \tag{4.3.16}$$

where the summation on $j\,(= 1, 2, \ldots)$ is understood (the suffix j of γ_j designates the current j) and the matrix of the anomalous dimension γ_N is defined by

$$\gamma_N = \mu \frac{\partial Z_N}{\partial \mu} Z_N^{-1}\bigg|_{g_0, m_0} \tag{4.3.17}$$

in terms of the renormalization constant matrix Z_N for the renormalization of the composite operators,

$$O^{(j)}_{\mu_1\ldots\mu_N} = O^{0(i)}_{\mu_1\ldots\mu_N}(Z_N^{-1})_{ij}. \tag{4.3.18}$$

The solution of Eq. (4.3.16) may be obtained essentially in the same way as in the previous case where the operator mixing was absent. One should, however, note that in general $\gamma_N(g)$ does not commute with $\gamma_N(g')$ as it is a matrix. In writing down the solution some care has to be given to this noncommutativity of $\gamma_N(g)$. This problem is solved most elegantly by introducing the notion of "t-ordering" T[Gro 74, Gey 79] which is the analogue of the time ordering employed in expressing the S-matrix element in perturbation theory. Using the t-ordering T we may write down the solution of Eq. (4.3.16) as follows,

$$\mathcal{C}_N^{(i)}\left(\frac{-q^2}{\mu^2}, g, m\right) = \left[T\exp\int_0^t dt'\{2\gamma_j(g(t')) - \gamma_N(g(t'))\}\right]_{ij} \mathcal{C}_N^{(j)}(1, \bar{g}(t), \bar{m}(t)), \tag{4.3.19}$$

where $\gamma_j(g(t'))$ is understood to be accompanied by the identity matrix. For matrix $A(t)$ the t-ordering T is defined by

$$\begin{aligned} T\exp\int_0^t dt' A(t') &= 1 + \int_0^t dt' A(t') + \frac{1}{2!}\int_0^t dt' \int_0^t dt'' T[A(t')A(t'')] + \ldots, \\ &= 1 + \int_0^t dt' A(t') + \int_0^t dt' \int_0^{t'} dt''\, A(t')A(t'') + \ldots \end{aligned} \tag{4.3.20}$$

CHAPTER

FIVE

PHYSICAL APPLICATIONS

Equipped with the operator-product expansion and renormalization group method we are ready to apply perturbative quantum chromodynamics (QCD) to physical processes dominated by short-distance effects. As typical examples we deal with the total cross section of e^+e^- annihilations, deep inelastic structure functions and hadron jet distributions in e^+e^- annihilations. The dependence of the physical predictions on the renormalization scheme chosen in perturbative calculations is explained and its phenomenological relevance is emphasized. The factorization theorem which is a generalization of the operator product expansion is then briefly discussed and applied to the Drell-Yan process. In the course of our discussions of these physical applications there appear infrared (soft and collinear) divergences due to the presence of the massless gluons and light quarks. These infrared divergences are disposed of by a standard method which will be explained in Chapter 6.

5.1. TOTAL CROSS SECTION FOR e^+e^- ANNIHILATIONS

5.1.1. Renormalization group equation

One of the simplest examples of the application of perturbative QCD is found in total cross sections for e^+e^- annihilation processes. The general kinematics for e^+e^- annihilation processes has already been presented in Sec. 4.1.3 and the calculation of the cross sections in the tree appproximation appeared in Sec. 2.3.4.

For e^+e^- annihilations the total cross section is given by Eq.(4.1.56) in terms of the R-ratio (or Drell ratio),

$$R = -\frac{2\pi}{q^2}\int d^4x e^{iq\cdot x}\,\langle 0|j_\mu(x)j^\mu(0)|0\rangle\ , \qquad (5.1.1)$$

where we used Eqs.(4.1.54), (4.1.55) and (4.1.59). We consider the e^+e^- annihilation at very high center-of-mass energies (large $\sqrt{q^2}$) so that all the relevant quark masses are negligible compared with $\sqrt{q^2}$. Then R is a function only of the center-of-mass energy squared, $s = q^2$, the renormalized coupling constant g and the renormalization scale μ,

$$R = R(s/\mu^2, g)\ . \qquad (5.1.2)$$

In Eq.(5.1.2) the dependence of R on s and μ is given by the ratio s/μ^2 for dimensional reasons.

Unlike the case of deep inelastic lepton-hadron scatterings no hadronic state participates in the expression of the R-ratio (5.1.1) and the quantity R consists only of the short-distance piece for large s. Hence we practically need not appeal to the operator product expansion in the present argument.

The operator $j_\mu(x)$ in Eq.(5.1.1) is the electromagnetic current which is conserved and hence its anomalous dimension vanishes as shown in Sec. 3.5. Accordingly the renormalization group equation for the R-ratio reads

$$\left[\mu\frac{\partial}{\partial\mu} + \beta(g)\frac{\partial}{\partial g}\right] R\left(\frac{s}{\mu^2}, g\right) = 0. \tag{5.1.3}$$

The general solution of Eq.(5.1.3) is easily found to be

$$R\left(\frac{s}{\mu^2}, g\right) = R(1, \bar{g}(s)), \tag{5.1.4}$$

where the running coupling constant $\bar{g}(s)$ is defined in terms of the β function such that

$$\frac{d\bar{g}}{dt} = \beta(\bar{g}), \qquad \bar{g}(\mu^2) = g, \tag{5.1.5}$$

with $t = (1/2)\ln(s/\mu^2)$.

The meaning of Eq.(5.1.4) is obvious: the explicit s-dependence of the R-ratio computed by using the coupling constant g can be completely absorbed into the s-dependence of the running coupling constant $\bar{g}(s)$. In asymptotically free field theories, the running coupling constant $\bar{g}(s)$ for large s is found to be small; thus the validity of the perturbative calculation of R is guaranteed.

The R-ratio expressed in terms of the coupling constant g renormalized at scale μ contains, in general, large logs, $\ln(s/\mu^2)$, for large s in each term of the perturbative expansion and hence the effectiveness of the perturbative calculation is spoiled. According to Eq.(5.1.4), however, the calculation is drastically improved if we employ the coupling constant renormalized at the scale of the relevant energy $\sqrt{s}$.

Let us look into the details of the above statement. The R-ratio $R(s/\mu^2, g)$ is given by the perturbative calculation in the following form,

$$R\left(\frac{s}{\mu^2}, g\right) = \sum_i Q_i^2\,[1 + a(s/\mu^2)g^2 + b(s/\mu^2)g^4 + \ldots]\,, \tag{5.1.6}$$

where the first term on the right-hand side has already been calculated in Eq.(2.3.158) (the index i runs over colors and flavors of quarks), and the second, third, ... terms will be discussed in the following subsections. The coefficients a, b, ..., in general, include large logs, $\ln(s/\mu^2)$. Equation (5.1.4) indicates that, if g is replaced by $\bar{g}(s)$, these large logs in the coefficient disappear, i.e.,

$$R\left(\frac{s}{\mu^2}, g\right) = \sum_i Q_i^2 [1 + a(1)\bar{g}(s)^2 + b(1)\bar{g}(s)^4 + \ldots] . \tag{5.1.7}$$

The expansion (5.1.7) is much better than Eq.(5.1.6) in two respects: its expansion coefficients are smaller than those in Eq.(5.1.6) and the expansion parameter $\bar{g}(s)$ is smaller than g for large s ($s \gg \mu^2$) according to the property of asymptotic freedom.

5.1.2. Order-g^2 correction

We shall show how to calculate $a(1)$ in Eq.(5.1.7) in perturbative QCD. The strategy for computing $a(1)$ is first to calculate the R-ratio to order g^2 by using the coupling constant g renormalized at the scale μ and then set $\mu^2 = s$ to obtain $a(1)$.

The Feynman diagrams contributing to the total cross section of the e^+e^- annihilation are shown in Fig. 5.1.1. The total cross section σ receives separate contributions from the final states $q\bar{q}$, $q\bar{q}G$, $q\bar{q}q\bar{q}$, $q\bar{q}GG$, ... with q, $\bar{q}$ and G denoting the quark, antiquark and gluon, respectively. The contribution up to order g^2 may be represented as in Fig. 5.1.2 and is split into three parts: the Born cross section σ_B [Fig. 5.1.2(a)], the virtual (one-loop) gluon contribution σ_V [Fig. 5.1.2(b), (c)] and the real-gluon-emission cross section σ_R[Fig. 5.1.2(d)]. Denoting the full cross section to order g^2 by σ, we have

$$\begin{aligned} \sigma &= Z_2^2 \, \sigma_B + \sigma_V + \sigma_R \ , \\ &= \sigma_B + \tilde{\sigma}_V + \sigma_R \ , \end{aligned} \tag{5.1.8}$$

where $\tilde{\sigma}_V = \sigma_V + (Z_2^2 - 1)\sigma_B$ with Z_2 the field renormalization constant associated with the quark external lines. The factor Z_2^2 in Eq. (5.1.8) is necessary since the field renormalization constant $Z_2^{1/2}$ for each quark external line should be included in the expression of the renormalized S-matrix element as a renormalized truncated Green function. (See Eqs. (2.3.40) and (3.1.21).)

The Born cross section σ_B has already been calculated in Sec. 2.3.4, which, by neglecting electron and quark masses, reads

$$\sigma_B = \frac{4\pi\alpha^2}{3s} \sum_i Q_i^2 \ . \tag{5.1.9}$$

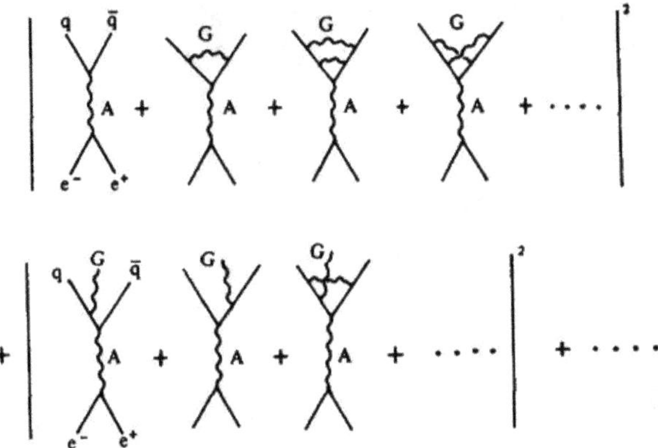

Fig. 5.1.1. The diagrams contributing to the total cross section of the e^+e^- annihilation. The wavy line accompanied by the letter A represents the photon while the one accompanied by the letter G indicates the gluon.

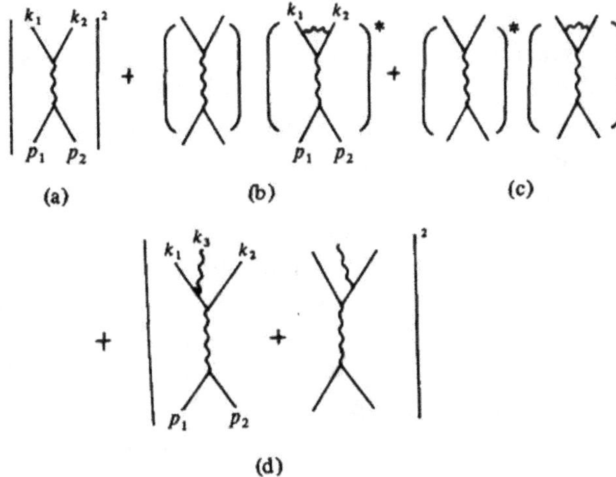

Fig. 5.1.2. The diagrammatic expression of the terms up to order g^2 in the total cross section of the e^+e^- annihilation: (a) the lowest order (Born) term, (b) and (c) the order-g^2 cross term between the lowest and one-loop amplitudes, and (d) the order-g^2 real-gluon emission term.

The one-loop contribution σ_V to the $e^+e^- \to q\bar{q}$ cross section of Figs. 5.1.2 (b) and (c) is calculated in the following way. We consider the e^+e^- annihilations at very high energies so that quark masses are practically negligible. In the following calculations all the quarks are regarded as massless. The contribution of Figs. 5.1.2 (b) and (c) can be written in such a way that

$$\sigma_V = \frac{1}{8s}\int \frac{d^3k_1}{(2\pi)^3 2k_{10}} \frac{d^3k_2}{(2\pi)^3 2k_{20}} (2\pi)^4\, \delta^4(k_1 + k_2 - p_1 - p_2)\, F_V\ , \tag{5.1.10}$$

where F_V is given by

$$F_V = \left(\sum_i Q_i^2\right)\frac{e^4}{q^4}\,\mathrm{Tr}\,[\not p_2\gamma^\mu \not p_1\gamma^\nu]\,\mathrm{Tr}\,[\not k_1\,\Lambda_\mu \not k_2\gamma_\nu] + \text{c.c.}\ , \tag{5.1.11}$$

with c.c. representing the complex conjugate of the first term and Λ_μ the one-loop vertex part corresponding to Fig. 5.1.3,

$$\Lambda_\mu = g^2\, C_F \int \frac{d^Dk}{(2\pi)^D i}\frac{1}{k^2}\gamma_\lambda \frac{1}{\not k - \not k_1}\gamma_\mu \frac{1}{\not k + \not k_2}\gamma^\lambda\ . \tag{5.1.12}$$

Note also that we use the Feynman gauge in the pesent calculation. In Sec. 2.5.5 the constant Z_2 was calculated in the MS scheme. Here in the present case, we would like to calculate it on the mass shell of quarks where quarks are massless. Here naturally we meet with the infrared divergence (mass singularity) arising from the vanishing quark mass. We shall regularize the mass singularity by means of dimensional regularization, a systematic explanation of which will be given in the next chapter. The unrenormalized one-loop self-energy part of the massless quark in the Feynman gauge is given in Eq.(2.4.41). After performing the integration it is found to have the form (2.4.63). It exhibits the mass singularity at $p^2 = 0$. In order to circumvent this mass singularity we employ the method of dimensional regularization. For $p^2 = 0$ the quark self-energy part $\sum(p)$ in the Feynman gauge reads [as may be seen in Eq.(2.4.36)]

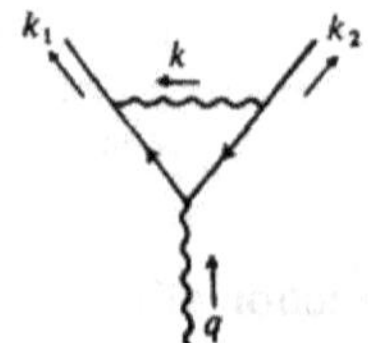

Fig. 5.1.3. The one-loop vertex part appearing in Fig.5.1.2 (b) and (c). The external wavy line with momentum q represents the photon and the internal one with momentum k denotes the gluon.

$$\begin{aligned}\Sigma(p)\,|_{p^2=0} &= g^2 C_F (D-2)\not{p}\int_0^1 dx(1-x)\int\frac{d^Dk}{(2\pi)^D i}\frac{1}{k^4}\,,\\ &= \frac{g^2}{(4\pi)^2}C_F\not{p}\left(\frac{1}{\varepsilon'}-\frac{1}{\varepsilon}\right),\end{aligned} \tag{5.1.13}$$

where we used the method explained in Eq.(2.5.177), and ε' and ε are equal to the parameter $(4-D)/2$ and serve to regularize the ultraviolet and infrared divergences, respectively. In this calculation we do not set $\varepsilon' = \varepsilon$ so that we can distinguish two kinds of divergences. From Eq.(5.1.13) we obtain the renormalization constant Z_2 to one-loop order,

$$Z_2 = 1 + \frac{g^2}{(4\pi)^2}C_F\left(\frac{1}{\varepsilon}-\frac{1}{\varepsilon'}\right). \tag{5.1.14}$$

The integration in Eq.(5.1.12) is performed similarly and results in

$$\Lambda_\mu = \gamma_\mu\frac{g^2}{8\pi^2}C_F\left(\frac{4\pi\mu^2}{-q^2}\right)^\varepsilon\Gamma(1+\varepsilon)\,B(1-\varepsilon,2-\varepsilon)\left(\frac{1}{\varepsilon'}-\frac{2}{\varepsilon^2}-2\right), \tag{5.1.15}$$

where μ is the mass scale introduced to make the coupling constant g dimensionless and

$$q = p_1 + p_2\,. \tag{5.1.16}$$

As expected, the ultraviolet divergences present in Eq.(5.1.15) cancel out in $\tilde{\sigma}_V$ on account of Eqs.(5.1.14). Inserting Eq.(5.1.15) into Eq.(5.1.11) and taking account of Eqs. (5.1.9) and (5.1.14) we find

$$\tilde{\sigma}_V = A_V\sigma_B\,, \tag{5.1.17}$$

where A_V is given by

$$A_V = \frac{\alpha_S}{\pi}C_F\left(\frac{4\pi\mu^2}{s}\right)^\varepsilon\frac{\cos\pi\varepsilon}{\Gamma(1-\varepsilon)}\left(-\frac{1}{\varepsilon^2}-\frac{3}{2\varepsilon}-4+O(\varepsilon)\right), \tag{5.1.18}$$

where α_S is the QCD coupling constant defined by

$$\alpha_s = \frac{g^2}{4\pi}\,. \tag{5.1.19}$$

The calculation of σ_R goes as follows. The cross section σ_R corresponding to Fig. 5.1.2 (d) is written in the form,

$$\sigma_R = \frac{1}{8s}\int \prod_{i=1}^{3} \frac{d^{D-1}k_i}{(2\pi)^{D-1}2k_{i0}} (2\pi)^D \delta^D\left(\sum_{i=1}^{3} k_i - p_1 - p_2\right) F_R\,, \tag{5.1.20}$$

where

$$F_R = -\left(\sum_i Q_i^2\right)\frac{e^4}{q^4} g^2 C_F \,\mathrm{Tr}\,[\not p_2 \gamma^\mu \not p_1 \gamma^\nu]\,\mathrm{Tr}[\not k_1 S_{\lambda\mu} \not k_2 S^\lambda_\nu]\,, \tag{5.1.21}$$

$$S_{\mu\nu} = \gamma_\mu \frac{-1}{\not k_1 + \not k_3}\gamma_\nu + \gamma_\nu \frac{1}{\not k_2 + \not k_3}\gamma_\mu\,. \tag{5.1.22}$$

In Eq.(5.1.20) we had worked in D dimensions rather than in four dimensions because we anticipate possible infrared divergences. We introduce tensors $G_{\mu\nu}$, $L^{\mu\nu}$ and $I_{\mu\nu}$,

$$G_{\mu\nu} = \mathrm{Tr}\,[\not k_1 S_{\lambda\mu} \not k_2 S^\lambda_\nu]\,, \tag{5.1.23}$$

$$L^{\mu\nu} = \mathrm{Tr}\,[\not p_2 \gamma^\mu \not p_1 \gamma^\nu] = 4\left(p_1^\mu p_2^\nu + p_1^\nu p_2^\mu - \frac{q^2}{2} g^{\mu\nu}\right)\,, \tag{5.1.24}$$

$$I_{\mu\nu} = \int \prod_{i=1}^{3} \frac{d^{D-1}k_i}{2k_{i0}} \delta^D\left(\sum_{i=1}^{3} k_i - q\right) G_{\mu\nu}\,. \tag{5.1.25}$$

The cross section σ_R of Eq.(5.1.20) is then expressed in the following form,

$$\sigma_R = \frac{-e^4 g^2}{8s(2\pi)^{2D-3} q^4}\left(\sum_i Q_i^2\right) C_F L^{\mu\nu} I_{\mu\nu}\,. \tag{5.1.26}$$

As can be seen in Eq.(5.1.25), $I_{\mu\nu}$ depends only on q_μ and satisfies the following condition corresponding to the conservation of the electromagnetic current,

$$q^\mu I_{\mu\nu} = 0\,. \tag{5.1.27}$$

Hence the general form of $I_{\mu\nu}$ is given by

$$I_{\mu\nu} = I(q^2)\left(\frac{q_\mu q_\nu}{q^2} - g_{\mu\nu}\right)\,, \tag{5.1.28}$$

with $I(q^2) = -g^{\mu\nu} I_{\mu\nu}/(D-1)$. By means of Eq.(5.1.28) we obtain

$$L^{\mu\nu} I_{\mu\nu} = \frac{D-2}{D-1} q^2 g^{\mu\nu} I_{\mu\nu}\,. \tag{5.1.29}$$

On the other hand we find after some calculation

$$g^{\mu\nu}G_{\mu\nu} = -8(1-\varepsilon)\frac{x_1^2 + x_2^2 - \varepsilon x_3^2}{(1-x_1)(1-x_2)}\ , \tag{5.1.30}$$

where $\varepsilon = (4-D)/2$ and

$$x_i = 2k_i{\cdot}q/q^2\ , \qquad (i = 1, 2, 3)\ . \tag{5.1.31}$$

The three-body phase volume for $g^{\mu\nu}I_{\mu\nu}$ in D dimensions may be rewritten in terms of the variables x_i so that

$$g^{\mu\nu}I_{\mu\nu} = \frac{\pi(\pi s)^{1-2\varepsilon}}{4\Gamma(2-2\varepsilon)}\int_0^1 \prod_{i=1}^{3}(1-x_i)^{-\varepsilon}dx_i\delta(2 - \sum_{i=1}^{3} x_i)\, g^{\mu\nu}G_{\mu\nu}\ . \tag{5.1.32}$$

Combining Eqs.(5.1.26), (5.1.29), (5.1.30) and (5.1.32) we arrive at

$$\sigma_R = \left(\sum_i Q_i^2\right)\alpha^2\alpha_S C_F \frac{2}{s}\left(\frac{4\pi\mu}{s}\right)^{2\varepsilon}\frac{(1-\varepsilon)^2}{(3-2\varepsilon)\,\Gamma(2-2\varepsilon)}K\ , \tag{5.1.33}$$

where μ comes from the mass dimension of the coupling constant g in D dimensions and

$$K = \int_0^1 \prod_{i=1}^{3}(1-x_i)^{-\varepsilon}dx_i\delta(2 - \sum_{i=1}^{3} x_i)\frac{x_1^2 + x_2^2 - \varepsilon x_3^2}{(1-x_1)(1-x_2)}\ . \tag{5.1.34}$$

The constant K can be calculated analytically to order ε^0, i.e.,

$$K = \left(\frac{4}{\varepsilon^2} - \frac{12}{\varepsilon} + 10 - 4\varepsilon\right)B(1-\varepsilon, 2-2\varepsilon)\,B(1-\varepsilon, 1-\varepsilon) + O(\varepsilon). \tag{5.1.35}$$

We wish to express Eq.(5.1.33) in the form of

$$\sigma_R = A_R\,\sigma_B\ , \tag{5.1.36}$$

with A_R to be determined. For this purpose we need to find the expression of the Born cross section σ_B in D dimensions. We repeat the previous calculation in Sec. 2.3.4 in D-dimensional space-time. The calculation is straightforward and results in

$$\sigma_B = \frac{4\pi\alpha^2}{3s}\left(\sum_i Q_i^2\right)\left(\frac{4\pi}{s}\right)^{\varepsilon}\frac{3(1-\varepsilon)\,\Gamma(2-\varepsilon)}{(3-2\varepsilon)\,\Gamma(2-2\varepsilon)}\ . \tag{5.1.37}$$

Comparing Eq.(5.1.33) with Eq.(5.1.37) we find for A_R of Eq.(5.1.36),

$$A_R = \frac{\alpha_s}{\pi} C_F \left(\frac{4\pi\mu^2}{s}\right)^\varepsilon \frac{\cos\pi\varepsilon}{\Gamma(1-\varepsilon)} \left(\frac{1}{\varepsilon^2} + \frac{3}{2\varepsilon} + \frac{19}{4} + O(\varepsilon)\right) . \tag{5.1.38}$$

Substituting Eqs.(5.1.17) and (5.1.36) into Eq.(5.1.8) we finally obtain

$$\sigma = (1 + A_V + A_R)\,\sigma_B = \left(1 + \frac{3}{4} C_F \frac{\alpha_s}{\pi}\right) \sigma_B . \tag{5.1.39}$$

In this result we clearly see that the infrared divergences present in A_V and A_R just cancel out leaving a finite one-loop effect. This result is a special example of an application of the Kinoshita-Poggio-Quinn theorem which will be discussed in Chap. 6. We thus finally obtain $a(s/\mu^2)$ which was defined in Eq.(5.1.6).

It should be noted here that up to this order $a(s/u^2)$ is independent of large logs, $\ln(s/\mu^2)$, and so

$$a(s/\mu^2) = a(1) .$$

Under this circumstance we may, according to Eq.(5.1.4), simply replace α_S in Eq.(5.1.39) by $\bar{\alpha}_S$, the running coupling constant,

$$\bar{\alpha}_S = \frac{\bar{g}^2}{4\pi} , \tag{5.1.40}$$

and obtain [App 73, Zee 73a]

$$R\left(\frac{s}{\mu^2}, g\right) = \sum_i Q_i^2 \left(1 + \frac{3}{4} C_F \frac{\bar{\alpha}_s}{\pi}\right) . \tag{5.1.41}$$

As we had seen in Sec. 3.4.2, owing to asymptotic freedom the running coupling constant of QCD decreases logarithmically as the relevant mass scale grows. Rewriting Eq.(3.4.53) in terms of $\bar{\alpha}_S$ we have

$$\bar{\alpha}_S = \frac{1}{b_0 \ln(s/\Lambda^2)} , \tag{5.1.42}$$

where

$$b_0 = 4\pi\beta_0 = \frac{11 C_G - 4 T_R N_f}{12\pi} . \tag{5.1.43}$$

Accordingly Eq.(5.1.41) tells us that the R-ratio in QCD approaches the parton-model prediction from above as $s \to \infty$. Though this behavior of the R-ratio is characteristic of QCD and is to be tested experimentally, the data on R so far has not attained enough statistics to perform accurate tests (see Fig. 1.2.3).

5.1.3. Higher-order corrections

If we wish to calculate terms to order g^4, we have to take into account diagrams with the $q\bar{q}$ final state up to two loops, diagrams with the $q\bar{q}G$ final state up to one loop, and diagrams with the $q\bar{q}GG$ and $q\bar{q}q\bar{q}$ final states at the tree level. As we have seen in Sec. 5.1.2 the coefficient of g^2, $a(s/\mu)$, does not depend on s so that it is free from large logs. In other words the coefficient $a(s/\mu^2)$ does not contain the renormalization scale μ and hence is independent of the renormalization scheme chosen. However, the coefficient of g^4, $b(s/\mu^2)$, does depend on the renormalization scheme employed to calculate it. Accordingly $b(s/\mu^2)$, in general, includes large logs.

The calculation of $b(s/\mu^2)$ was performed by three groups [Din 79, Che 79, 80, Cel 80, 80a] in the $\overline{\text{MS}}$ [Bar 78] as well as MS [tHo 73] scheme. Applying the renormalization-group improvement described in Sec. 5.1.1, we replace $b(s/\mu^2)$ by $b(1)$ as well as α_S by $\bar{\alpha}_S$ and obtain an improved perturbation series. The result of these calculations may be expressed in such a way that

$$R = \sum_i Q_i^2 \left[1 + \frac{3}{4} C_F \frac{\bar{\alpha}_s}{\pi} + A \left(\frac{\bar{\alpha}_s}{\pi} \right)^2 + \ldots \right] , \tag{5.1.44}$$

where the constant A for SU(N) color with N_f flavors is given in the MS scheme by

$$A = \frac{N^2 - 1}{32N} \left[\frac{3}{4N} + \frac{243}{4} N - 11 N_f \right.$$
$$\left. + (11N - 2N_f)(-4\zeta(3) + \ln 4\pi - \gamma) \right] . \tag{5.1.45}$$

where $\zeta(z)$ is the zeta function [Erd 54, Mor 60] with $\zeta(3) = 1.202 \ldots$ and γ the Euler constant 0.5772... The constant A in the $\overline{\text{MS}}$ scheme can be obtained from Eq.(5.1.45) by eliminating $\ln 4\pi - \gamma$.

It is interesting to see whether the convergence of the perturbative series (5.1.44) is guaranteed or not. Of course, it is impossible to examine the full perturbation series to all orders but one can estimate the size of the first several terms in the series and get an idea of the validity of the perturbative expansion. The size of the QCD running coupling constant $\bar{\alpha}_S(s)$ at around $\sqrt{s} = 30$ GeV has been determined by comparing experimental data with the theoretical predictions on jets in e^+e^- annihilations [Ell 80, 81, Fab 80, 82, Ver 81, Ali 82], (see [Duk 85] for a thorough review on the experimental determination of $\bar{\alpha}_S$.) The running coupling constant $\bar{\alpha}_S$ determined in this way is

$$\bar{\alpha}_S \simeq 0.2 \quad \text{for } \sqrt{s} = 30 \text{ GeV}. \tag{5.1.46}$$

On the other hand the coefficient of $\bar{\alpha}_S$ in Eq.(5.1.44) takes the following value for the SU(3) color ($N = 3$),

$$\frac{3}{4} C_F = 1 \quad . \tag{5.1.47}$$

Hence the size of the second-order term reads

$$\frac{3}{4} C_F \frac{\bar{\alpha}_s}{\pi} \simeq 0.06 \ll 1 \quad , \tag{5.1.48}$$

and so the perturbative expansion is meaningful up to this order. As mentioned before, the fourth-order term is renormalization-scheme dependent. For the MS and $\overline{\text{MS}}$ scheme the constant A amounts to

$$A = \begin{cases} 7.36 - 0.44 N_f & (\text{MS}) \quad , \\ 1.99 - 0.12 N_f & (\overline{\text{MS}}) \quad . \end{cases} \tag{5.1.49}$$

Taking N_f to be 5 (up to the bottom quark), we have

$$A \left(\frac{\bar{\alpha}_s}{\pi} \right)^2 = \begin{cases} 0.021 & (\text{MS}) \quad , \\ 0.006 & (\overline{\text{MS}}) \quad . \end{cases} \tag{5.1.50}$$

The value 0.021 for the MS scheme is not so small compared with that of the second-order term (0.06) and is not enough to guarantee the fast convergence of the series. On the other hand the value 0.006 for the $\overline{\text{MS}}$ scheme is smaller by one order than the second-order term and better suited to assuring the convergence of the series. Thus from the practical point of view the $\overline{\text{MS}}$ scheme is more appropriate than the MS scheme for computation of the R-ratio.

5.2. DEEP INELASTIC LEPTON-HADRON SCATTERINGS

5.2.1. Moment sum rules for structure functions

Deep inelastic lepton-hadron scattering is one of the best-investigated short-distance processes, and is a testing ground for perturbative QCD. We shall now explain how precise predictions for this process can be made within the framework of perturbative QCD.

We first discuss a consequence derived from the operator product expansion for the deep-inelastic structure functions. Our concern here is the product of two electromagnetic currents[1] $j_\mu(x) j_\nu(x')$. Its general form may be

[1] In this section we deal only with electromagnetic processes. The weak processes such as neutrino-nucleon scatterings may be treated in a similar manner (see, e.g., [Bur 80, Rey 81]).

determined by the requirements of Lorentz invariance and current conservation,

$$
\begin{aligned}
j_\mu(x)j_\nu(x') = {} & (\partial_\mu\partial'_\nu - g_{\mu\nu}\partial\cdot\partial')\, O_L(x, x') \\
& + (g_{\mu\lambda}\partial_\rho\partial'_\nu + g_{\rho\nu}\partial_\mu\partial'_\lambda - g_{\mu\lambda}g_{\rho\nu}\partial\cdot\partial' - g_{\mu\nu}\partial_\lambda\partial'_\rho)\, O_2^{\lambda\rho}(x, x') \\
& + i\varepsilon_{\mu\nu\lambda\rho}\,\partial^\lambda O_3^\rho(x, x') \\
& + i(\varepsilon_{\mu\nu\lambda\rho}\,\partial\cdot\partial' - \varepsilon_{\mu\sigma\lambda\rho}\,\partial_\nu\partial'^\sigma + \varepsilon_{\nu\sigma\lambda\rho}\,\partial_\mu\partial'^\sigma)\, O_4^{\lambda\rho}(x, x') ,
\end{aligned}
\tag{5.2.1}
$$

where $\partial'_\mu = \partial/\partial x'^\mu$, and O_L, O_2, O_3 and O_4 are suitable bilocal operators[2]. The terms carrying O_3 and O_4 are antisymmetric in the indices μ and ν, and are irrelevant to the spin-averaged structure functions W_1 and W_2 as was pointed out in Sec. 4.2.2. These antisymmetric terms are shown to contribute to cross sections for polarized deep-inelastic lepton-hadron scatterings [Hey 72]. As in the case of the free field theory in Sec. 4.2.1, we shall, in the following, deal with the time-ordered product $\mathrm{T}[j_\mu(x)j_\nu(x')]$ instead of the simple product of Eq.(5.2.1). The corresponding general expression for $\mathrm{T}[j_\mu(x)j_\nu(x')]$ reads

$$
\begin{aligned}
\mathrm{T}[j_\mu(x)j_\nu(x')] = {} & (\partial_\mu\partial'_\nu - g_{\mu\nu}\partial\cdot\partial')\, O_L(x, x') \\
& + (g_{\mu\lambda}\partial_\rho\partial'_\nu + g_{\rho\nu}\partial_\mu\partial'_\lambda - g_{\mu\lambda}g_{\rho\nu}\partial\cdot\partial' - g_{\mu\nu}\,\partial_\lambda\partial'_\rho)\, O_2^{\lambda\rho}(x, x') \\
& + \text{antisymmetric part} ,
\end{aligned}
\tag{5.2.2}
$$

where we used the same notation O_L and O_2 as before for the bilocal operators since no confusion is expected.

Following the argument given in Sec. 4.2.5 we have the light-cone expansion for O_L and O_2,

$$
\begin{aligned}
O_L(x, x') &= \sum_{i,n} C_{L,n}^{(i)}(y^2)\, y^{\mu_1}\cdots y^{\mu_n}\, O_{L\mu_1\cdots\mu_n}^{(i)}\left(\frac{x + x'}{2}\right) , \\
O_2^{\lambda\rho}(x, x') &= \sum_{i,n} C_{2,n}^{(i)}(y^2)\, y^{\mu_1}\cdots y^{\mu_n}\, O_{2\mu_1\cdots\mu_n}^{(i)\lambda\rho}\left(\frac{x + x'}{2}\right) ,
\end{aligned}
\tag{5.2.3}
$$

where $y = x - x'$. Here the choice of the argument of the composite operators on the right-hand side of Eq.(5.2.3) is rather arbitrary; it could be x'.

[2] We use the subscript L for operator O_L in order to make clear its connection to the longitudinal structure function.

We take the spin-averaged matrix element of Eq.(5.2.2) between the proton states with momentum p [see the definition (4.1.27)], and apply a Fourier transformation on both sides of the equation to obtain

$$\begin{aligned}
&\int d^4x\, e^{iq\cdot x} \langle p|\mathrm{T}[j_\mu(x) j_\nu(0)]|p\rangle \\
&= (q_\mu q_\nu - q^2 g_{\mu\nu}) \sum_{i,n} \langle p|O^{(i)}_{L\mu_1\ldots\mu_n}(0)|p\rangle \int d^4x\, x^{\mu_1}\ldots x^{\mu_n}\, C^{(i)}_{L,n}(x^2)\, e^{iq\cdot x} \\
&\quad + (g_{\mu\lambda} q_\rho q_\nu + g_{\rho\nu} q_\mu q_\lambda - q^2 g_{\mu\lambda} g_{\rho\nu} - g_{\mu\nu} q_\lambda q_\rho) \\
&\quad \times \sum_{i,n} \langle p|O^{(i)\lambda\rho}_{2\mu_1\ldots\mu_n}(0)|p\rangle \int d^4x\, x^{\mu_1}\ldots x^{\mu_n}\, C^{(i)}_{2,n}(x^2)\, e^{iq\cdot x} .
\end{aligned} \tag{5.2.4}$$

We define Fourier transforms of the coefficient functions as follows:

$$\begin{aligned}
&\tilde{C}^{(i)}_{L,n}(-q^2)(-i)(-q^2/2)^{-n-1} q^{\mu_1}\ldots q^{\mu_n} \\
&\qquad = \int d^4x\, e^{iq\cdot x} x^{\mu_1}\ldots x^{\mu_n}\, C^{(i)}_{L,n}(x^2) , \\
&\tilde{C}^{(i)}_{2,n+2}(-q^2)(-2i)(-q^2/2)^{-n-2} q^{\mu_1}\ldots q^{\mu_n} \\
&\qquad = \int d^4x\, e^{iq\cdot x} x^{\mu_1}\ldots x^{\mu_n}\, C^{(i)}_{2,n}(x^2) .
\end{aligned} \tag{5.2.5}$$

The matrix elements of the *composite operators* appearing in Eq.(5.2.4) have the following tensor structures,

$$\begin{aligned}
\langle p|O^{(i)}_{L\mu_1\ldots\mu_n}(0)|p\rangle &= A^{(i)}_{L,n} p_{\mu_1}\cdots p_{\mu_n} + \text{terms containing } g_{\mu_i\mu_j} , \\
\langle p|O^{(i)\lambda\rho}_{2\mu_1\ldots\mu_n}(0)|p\rangle &= A^{(i)}_{2,n+2} p^\lambda p^\rho p_{\mu_1}\cdots p_{\mu_n} + \text{terms containing } g_{\mu_i\mu_j} .
\end{aligned} \tag{5.2.6}$$

We substitute Eqs.(5.2.5) and (5.2.6) into Eq.(5.2.4) and obtain

$$T_{\mu\nu} = 2\sum_{i,n} [e_{\mu\nu} A^{(i)}_{L,n} \tilde{C}^{(i)}_{L,n}(-q^2) + d_{\mu\nu} A^{(i)}_{2,n} \tilde{C}^{(i)}_{2,n}(-q^2)]\omega^n , \tag{5.2.7}$$

where $T_{\mu\nu}$ is the forward virtual Compton amplitude defined by Eq.(4.1.38) and

$$\begin{aligned}
\omega &= 2p\cdot q/(-q^2) = 1/\xi , \\
e_{\mu\nu} &= g_{\mu\nu} - q_\mu q_\nu/q^2 , \\
d_{\mu\nu} &= -g_{\mu\nu} - p_\mu p_\nu q^2/(p\cdot q)^2 + (p_\mu q_\nu + p_\nu q_\mu)/p\cdot q .
\end{aligned} \tag{5.2.8}$$

In Eq.(5.2.7) terms less dominant in the limit $-q^2 \to \infty$ with ω fixed are neglected. It should be noted here that the summation on n in Eq.(5.2.7) runs only over even n in conformity with crossing symmetry,

$$T_{\mu\nu}(-\omega) = T_{\mu\nu}(\omega) ,$$

which may be derived from the definition of $T_{\mu\nu}$, Eq.(4.1.38). Equation (5.2.7) is a power series in ω and the condition $\omega < 1$ is to be fulfilled to warrant the convergence of the series. Unfortunately, however, the kinematical region with $0 < \omega < 1$ is unphysical as can be seen in Eq.(4.1.17), thus Eq.(5.2.7) is not of direct use in our physical argument.

For the purpose of converting the unphysical relation (5.2.7) to a physical one for which $\omega \geq 1$, we apply a Cauchy integration to both sides of Eq.(5.2.7). Since the forward virtual Compton amplitude $T_{\mu\nu}$ is an analytic function of complex ω with branch cuts along the real axis for $\omega \leq -1$ and $\omega \geq 1$, the contour C for the Cauchy integration is taken to be that shown in Fig. 5.2.1. Applying the Cauchy integration with contour C to the left-hand side of Eq.(5.2.7), i.e., $T_{\mu\nu}$, we obtain

$$\frac{1}{2\pi i}\oint_C \frac{T_{\mu\nu}}{\omega^n} = \frac{2}{\pi}\int_1^\infty \frac{d\omega}{\omega^n}\,\mathrm{Abs}\,T_{\mu\nu} = 2\int_0^1 d\xi\,\xi^{n-2}\,W_{\mu\nu} , \tag{5.2.9}$$

where Abs $T_{\mu\nu}$ is defined in Eq.(4.1.40) and use has been made of Eq.(4.1.39) and crossing symmetry. On the other hand, by noting the formula

$$\frac{1}{2\pi i}\oint_C d\omega\,\omega^{m-n} = \delta_{m,n-1} ,$$

we have for the right-hand side of Eq.(5.2.7)

$$2\sum_i [e_{\mu\nu} A^{(i)}_{L,n-1} \tilde{C}^{(i)}_{L,n-1}(-q^2) + d_{\mu\nu} A^{(i)}_{2,n-1}(-q^2)\tilde{C}^{(i)}_{2,n-1}(-q^2)] . \tag{5.2.10}$$

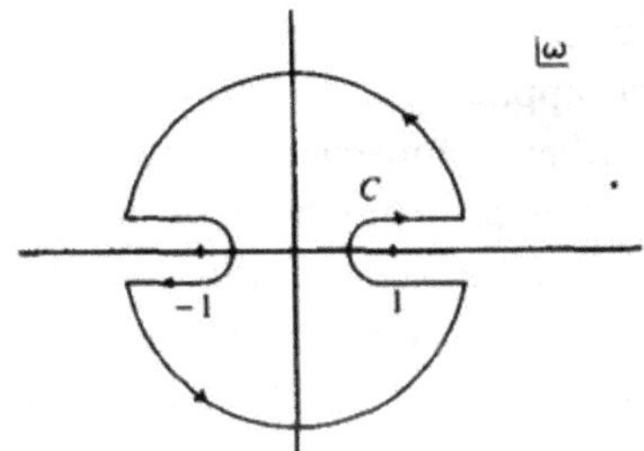

Fig. 5.2.1. The path C of the Cauchy integration of Eq.(5.2.9) in the complex-ω plane with ω defined in Eq.(5.2.8).

Equating Eqs.(5.2.9) with (5.2.10) we find [Chr 72]

$$\int_0^1 d\xi\, \xi^{n-2} F_L(\xi, Q^2) = \sum_i A^{(i)}_{L,n}\, \tilde{C}^{(i)}_{L,n}(Q^2)\ , \tag{5.2.11}$$

$$\int_0^1 d\xi\, \xi^{n-2}\, F_2(\xi, Q^2) = \sum_i A^{(i)}_{2,n}\, C^{(i)}_{2,n}(Q^2)\ , \tag{5.2.12}$$

where $Q^2 = -q^2$ and *structure functions* F_L and F_2 are defined through

$$W_{\mu\nu} = e_{\mu\nu}\omega F_L + d_{\mu\nu}\omega F_2\ . \tag{5.2.13}$$

The structure functions F_L and F_2 are related to W_1 and W_2 which were previously defined in Sec. 4.1.2, i.e.,

$$\begin{aligned} F_L(\xi, Q^2) &= -W_1(\nu, Q^2) + (1 + \nu^2/Q^2)\, W_2(\nu, Q^2)\ , \\ F_2(\xi, Q^2) &= \nu W_2(\nu, Q^2)/(2M)\ . \end{aligned} \tag{5.2.14}$$

The structure function F_L is called the *longitudinal structure function* since it corresponds to the total cross section for the longitudinal photon off protons [Han 63]. Equations (5.2.11) and (5.2.12) are called the *moment sum rules* for the structure functions. By employing the sum rules (5.2.11) and (5.2.12) we can make a prediction on the Q^2-dependence of the moments of the structure functions since the coefficient functions $\tilde{C}^{(i)}_{L,n}$ and $\tilde{C}^{(i)}_{2,n}$ are of short-distance nature and are calculable within the framework of perturbative QCD. The constants $A^{(i)}_{L,n}$ and $A^{(i)}_{2,n}$ which are related to the matrix elements of the composite operators are of long-distance nature and cannot be calculated within the framework of perturbative QCD. These constants, therefore, should be determined experimentally. This can be done by estimating the moment of the structure functions at a fixed Q^2, say Q_0^2, and solving Eqs.(5.2.11) and (5.2.12) for $A^{(i)}_{L,n}$ and $A^{(i)}_{2,n}$. If there exist many composite operators with the same quantum number and twist, we have a number of constants $A^{(i)}_{L,n}$ and $A^{(i)}_{2,n}$ to be determined and so we need more input data on the moments at some different values of Q^2.

From the point of view of practical applications the moment sum rules (5.2.11) and (5.2.12) are found to be rather inconvenient because these sum rules are not in the form for which the direct predictions on the structure functions can be made. One way of circumventing this unsatisfactory situation is to adopt inverse Mellin transformations to both sides of Eqs.(5.2.11) and (5.2.12). For this purpose we need to make an analytic continuation from integer n to complex n following the Carlson theorem [Tit 32]. In the following argument we omit the suffixes L and 2 in Eqs.(5.2.11) and

(5.2.12) for brevity. After performing the inverse Mellin transformations we have

$$F(\xi, Q^2) = \frac{1}{2\pi i}\int_{c-i\infty}^{c+i\infty} dn\, \xi^{1-n} \sum_i A_n^{(i)} \tilde{C}_n^{(i)}(Q^2)\,, \tag{5.2.15}$$

where c is an arbitrary real positive constant. To make the argument simpler we neglect, for the moment, operator mixing, i.e., we assume only one kind of the composite operator contributing to each moment. Then we may neglect the index i and so we have in this case, instead of Eq. (5.2.15),

$$F(\xi, Q^2) = \frac{1}{2\pi i}\int_{c-i\infty}^{c+i\infty} dn\, \xi^{1-n}\, A_n \tilde{C}_n(Q^2)\,. \tag{5.2.16}$$

The constant A_n is determined by giving the moments of $F(\xi, Q^2)$ at $Q^2 = Q_0^2$ as mentioned before, i.e.,

$$A_n = \frac{1}{\tilde{C}(Q_0^2)}\int_0^1 d\xi\, \xi^{n-2} F(\xi, Q_0^2)\,. \tag{5.2.17}$$

Inserting Eq.(5.2.17) into Eq.(5.2.16) and rearranging the integrals we have [Par 73, Eil 76]

$$F(\xi, Q^2) = \int_\xi^1 \frac{d\eta}{\eta} K\left(\frac{\eta}{\xi}, Q^2, Q_0^2\right) F(\eta, Q_0^2)\,, \tag{5.2.18}$$

where the kernel function $K(z, Q^2, Q_0^2)$ is defined by

$$K(z, Q^2, Q_0^2) = \frac{1}{2\pi i}\int_{c-i\infty}^{c+i\infty} dn\, z^{n-1} \frac{\tilde{C}_n(Q^2)}{\tilde{C}_n(Q_0^2)}\,. \tag{5.2.19}$$

Equation (5.2.18) tells us that, once the structure function for a given Q^2 is known for all $\xi (0 < \xi < 1)$ the structure function for the other values of Q^2 can be predicted by calculating Eq.(5.2.19) in perturbative QCD.

5.2.2. Renormalization-group-improved perturbation

We are now ready to make predictions on deep inelastic structure functions through the use of the moment sum rules (5.2.11) and (5.2.12) or their inverse (5.2.18). Our main task here is to calculate the coefficient functions $\tilde{C}_n^{(i)}(Q^2)$ perturbatively in QCD (suffixes L and 2 are omitted).

A key observation for carrying out the calculation is that the coefficient functions are independent of the states which sandwich the light-cone expansion (5.2.3) of the current product. In the previous subsection we sandwiched the operator relation (5.2.3) with the proton states to obtain the power-series expansion (5.2.7) of $T_{\mu\nu}$ in terms of ω. We could, instead, consider the quark (or gluon) Green function with the insertion of the current product,

$$t_{\mu\nu}(q,p) = i \int d^4x d^4x_1 d^4x_2 e^{iq\cdot x + ip\cdot(x_1 - x_2)} \langle 0|T[j_\mu(x) j_\nu(0) \psi(x_1) \bar{\psi}(x_2)]|0\rangle_t \ , \tag{5.2.20}$$

where p is the quark momentum and the suffix t indicates that the quark external lines are truncated in the above amplitude. An amplitude similar to Eq.(5.2.20) may also be defined for gluon fields. Substituting Eq.(5.2.2) with the expansion (5.2.3) into Eq.(5.2.20) we obtain (discarding the antisymmetric part)

$$t_{\mu\nu} = 2\sum_{i,n} [e_{\mu\nu} a^{(i)}_{L,n} \tilde{C}^{(i)}_{L,n}(Q^2) + d_{\mu\nu} a^{(i)}_{2,n} \tilde{C}^{(i)}_{2,n}(Q^2)]\omega^n \ , \tag{5.2.21}$$

where $a^{(i)}_{L,n}$ and $a^{(i)}_{2,n}$ are defined in the same way as in Eq.(5.2.6) through the quark Green functions with the insertion of the composite operators,

$$\int d^4x_1 d^4x_2 e^{ip\cdot(x_1 - x_2)} \langle 0|T[O^{(i)}_{L\mu_1\ldots\mu_n}(0)\, \psi(x_1)\, \bar{\psi}(x_2)]|\,0\rangle_t$$

$$= a^{(i)}_{L,n} p_{\mu_1} \cdots p_{\mu_n} + \text{terms containing } g_{\mu_i\mu_j} \ ,$$

$$\int d^4x_1 d^4x_2\, e^{ip\cdot(x_1 - x_2)} \langle 0|T[O^{(i)\lambda\rho}_{2\mu_1\ldots\mu_n}(0) \psi(x_1)\, \bar{\psi}(x_2)]|0\rangle_t$$

$$= a^{(i)}_{2,n+2} p^\lambda p^\rho p_{\mu_1} \cdots p_{\mu_n} + \text{terms containing } g_{\mu_i\mu_j} \ . \tag{5.2.22}$$

For later use we define the invariant amplitudes t_L and t_2 by

$$t_{\mu\nu} = e_{\mu\nu} t_L + d_{\mu\nu} t_2 \ , \tag{5.2.23}$$

where the antisymmetric part is neglected. From Eqs.(5.2.21) and (5.2.23) one obtains

$$t = 2 \sum_{i,n} a^{(i)}_n \tilde{C}^{(i)}_n (Q^2)\, \omega^n \ , \tag{5.2.24}$$

where suffixes L and 2 for t, $a^{(i)}_n$ and $\tilde{C}^{(i)}_n$ are omitted for simplicity. In Eq.(5.2.21) the coefficient functions $\tilde{C}^{(i)}_{L,n}$ and $\tilde{C}^{(i)}_{2,n}$ are the same as the previous ones appearing in Eq.(5.2.7) and in Eqs.(5.2.11) and (5.2.12). While the constants $A^{(i)}_{L,n}$ and $A^{(i)}_{2,N}$ in Eq.(5.2.7) are not calculable within the perturbative framework, the constants $a^{(i)}_{L,n}$ and $a^{(i)}_{2,n}$ in Eq.(5.2.21) are controllable in perturbation theory. Hence by calculating both t and $a^{(i)}_n$ in

Eq.(5.2.24) perturbatively one may obtain the coefficient functions $\tilde{C}_n^{(i)}(Q^2)$.

In the following argument we neglect operator mixing for simplicity so that the index i for $a_n^{(i)}$ and $\tilde{C}_n^{(i)}$ is unneccessary. Equation (5.2.24), without operator mixing, reads

$$t = 2\sum_n a_n \tilde{C}_n(Q^2)\,\omega^n \ . \tag{5.2.25}$$

We are supposed to calculate t and a_n perturbatively so as to obtain $\tilde{C}_n(Q^2)$ order by order in g (the renormalized QCD coupling constant). The resulting $\tilde{C}_n(Q^2)$ is given in powers of g and necessarily contains large logs of the form $\ln(Q^2/\mu^2)$ in each order where μ is the renormalization scale. Thus $\tilde{C}_n(Q^2)$ has to be improved using the renormalization group method. Before going into the details, let us first explain the strategy for calculating $\tilde{C}_n(Q^2)$ perturbatively. Let us assume that t and a_n are calculated order by order in g in the following form:

$$\begin{aligned} t &= t_0 + t_1 g^2 + \ldots, \\ a_n &= a_{0n} + a_{1n} g^2 + \ldots \end{aligned} \tag{5.2.26}$$

We would like to obtain $\tilde{C}_n$ in a power series in g and hence we write

$$\tilde{C}_n = \tilde{C}_{0n} + \tilde{C}_{1n}\,g^2 + \ldots \tag{5.2.27}$$

We insert Eqs.(5.2.26) and (5.2.27) into Eq.(5.2.25). Comparing the coefficients in each order of the resulting relations we easily find

$$t_0 = 2\sum_n a_{0n}\tilde{C}_{0n}\,\omega^n \ , \tag{5.2.28}$$

$$t_1 = 2\sum_n (a_{0n}\tilde{C}_{1n} + a_{1n}\tilde{C}_{0n})\,\omega^n \ , \tag{5.2.29}$$

$$\vdots \qquad\qquad \vdots \qquad\qquad \vdots$$

Expanding t_0 in powers of ω and comparing it with the right-hand side of Eq.(5.2.28), we obtain $\tilde{C}_{0n}$. Then we expand t_1 in powers of ω and use Eq.(5.2.29) to obtain $\tilde{C}_{1n}$. This procedure is continued to every order in g for obtaining $\tilde{C}_{kn}$ $(k = 0, 1, 2, \ldots)$.

The renormalization group equation for the coefficient function $\tilde{C}_n(Q^2)$ may be derived in the manner explained in Sec. 4.3.2,

$$[\mathscr{D} - \gamma_n(g)]\,\tilde{C}_n(Q^2) = 0 \ , \tag{5.2.30}$$

where $\mathscr{D}$ is the differential operator defined in Eq.(4.3.5) and γ_n the anomalous dimension of the composite operators $O_{\mu_1\ldots\mu_n}$ in Eq.(5.2.3). (Note that suffixes

L and 2 are suppressed in γ_n, $\tilde{C}_n$ and $O_{\mu_1\ldots\mu_n}$.) It should be noted here that the electromagnetic current $j_\mu(x)$ is conserved and has no anomalous dimension. In Eq.(5.2.30) the coefficient function $\tilde{C}_n(Q^2)$ and the anomalous dimension $\gamma_n(g)$ are assumed to be gauge-independent and so the term with the differentiation on the gauge parameter,

$$\delta(g, \alpha)\frac{\partial}{\partial x},$$

is absent in Eq.(5.2.30) though in general this term is necessary as in Eq.(3.2.22). The gauge independence of $\tilde{C}_n(Q^2)$ and $\gamma_n(g)$ in the MS (or $\overline{\text{MS}}$) scheme may be shown in the following way: Since the unrenormalized composite operator is gauge-independent, the renormalization constant Z_n of the composite operator is shown to be gauge-independent in the MS (or $\overline{\text{MS}}$) scheme according to the argument similar to the one given in Sec. 3.2.1. Hence $\gamma_n(g)$ as defined by

$$\gamma_n(g) = \frac{\mu}{Z_n}\frac{\partial Z_n}{\partial \mu}\bigg|_{g_0, m_0} \tag{5.2.31}$$

is also gauge-independent, where the suffixes L and 2 for $\gamma_n(g)$ and Z_n are suppressed, and g_0 and m_0 are the bare coupling constant and the bare quark mass, respectively. The coefficient function $\tilde{C}_n(Q^2)$ is calculated in the manner described above where t and a_n have to be calculated first to obtain $\tilde{C}_n(Q^2)$. Inspecting this procedure we note that the renormalization-scheme dependence with respect to the mass, coupling and field renormalizations just cancels out on both sides of Eq.(5.2.25) leaving the renormalization-scheme dependence of $\tilde{C}_n(Q^2)$ with respect to the renormalization of the composite operator. Since Z_n is gauge-independent in the MS (or $\overline{\text{MS}}$) scheme, we realize that $\tilde{C}_n(Q^2)$ is also gauge-independent in the MS (or $\overline{\text{MS}}$) scheme[3]. We work in the MS (or $\overline{\text{MS}}$) scheme so that the renormalization group equation for $\tilde{C}_n(Q^2)$ is given by Eq.(5.2.30).

The solution of Eq.(5.2.30) is easily found in the same way as in Sec. 4.3.2, i.e.,

$$\tilde{C}_n(Q^2) \equiv \tilde{C}_n\left(\frac{Q^2}{\mu^2}, g, m\right) = \tilde{C}_n(1, \bar{g}(t), \bar{m}(t)) \exp\left[-\int_0^t dt'\gamma_n(\bar{g}(t'))\right], \tag{5.2.32}$$

where $t = (1/2)\ln(Q^2/\mu^2)$. In the following, for simplicity, we neglect the quark masses, i.e., we set $m = \bar{m} = 0$. Equation (5.2.32) may be rewritten in the following way,

$$\tilde{C}_n(Q^2) = \tilde{C}_n(1, \bar{g}(t)) \exp\left[-\int_g^{\bar{g}} d\lambda \frac{\gamma_n(\lambda)}{\beta(\lambda)}\right], \tag{5.2.33}$$

[3] One may argue that $\tilde{C}_n(Q^2)$ and $\gamma_n(g)$ are gauge-independent also in the on-shell scheme.

where $\tilde{C}_n(1,\bar{g}) = \tilde{C}_n(1,\bar{g},0)$. If we expand $\tilde{C}_n(1,\bar{g})$, $\gamma_n(\bar{g})$ and $\beta(\bar{g})$ in powers of $\bar{g}$, i.e.,

$$\begin{aligned} \tilde{C}_n(1,\bar{g}) &= c_{0n} + c_{1n}\bar{g}^2 + \dots , \\ \gamma_n(\bar{g}) &= \gamma_{0n}\bar{g}^2 + \gamma_{1n}\bar{g}^4 + \dots , \\ \beta(\bar{g}) &= -\beta_0\bar{g}^3 - \beta_1\bar{g}^5 + \dots , \end{aligned} \tag{5.2.34}$$

we have from Eq. (5.2.33)

$$\tilde{C}_n(Q^2) = (c_{0n} + c_{1n}\bar{g}^2)\left(\frac{\bar{g}}{g}\right)^{\gamma_{0n}/\beta_0}\left(\frac{\beta_0+\beta_1\bar{g}^2}{\beta_0+\beta_1 g^2}\right)^{(\gamma_{0n}/\beta_0-\gamma_{1n}/\beta_1)/2} , \tag{5.2.35}$$

where higher order terms in $\bar{g}$ are neglected and β_0 and β_1 have already been given in Eqs.(3.4.23) and (3.4.24). The running coupling constant $\bar{g}(t)$ is expressed in terms of g and t by solving Eq.(3.4.51). Inserting Eq.(3.4.53) into Eq.(5.2.35) and expanding the resulting expression in powers of g we find

$$\tilde{C}_n(Q^2) = c_{0n} + \left(c_{1n} - \frac{1}{2}\gamma_{0n}\ln\frac{Q^2}{\mu^2}\right)g^2 + O(g^4) . \tag{5.2.36}$$

Equation (5.2.36) should be compared with Eq.(5.2.27). This gives us

$$\begin{aligned} \tilde{C}_{0n} &= c_{0n} , \\ \tilde{C}_{1n} &= c_{1n} - \frac{1}{2}\gamma_{0n}\ln\frac{Q^2}{\mu^2} , \\ \vdots \quad & \quad \vdots \quad\quad \vdots \end{aligned} \tag{5.2.37}$$

By calculating t_0, t_1, ... and a_{0n}, a_{1n}, ... we obtain C_{0n}, $\tilde{C}_{1n}$, ... through Eqs.(5.2.28), (5.2.29), ... and then with the help of Eq.(5.2.37) we obtain c_{0n}, c_{1n}, ... and γ_{0n}, γ_{1n}, ... This completes our calculation of $\tilde{C}_n(Q^2)$, i.e., we substitute the calculated results for c_{0n}, c_{1n}, ... and γ_{0n}, γ_{1n}, ... into Eq.(5.2.35) and obtain the renormalization-group-improved form of the coefficient function $\tilde{C}_n(Q^2)$. In the final expression of $\tilde{C}_n(Q^2)$ we have to use the form (3.4.58) for the running coupling constant $\bar{g}$. Expanding $\tilde{C}_n(Q^2)$ in powers of inverse logs we obtain

$$\begin{aligned} \tilde{C}_n(Q^2) = N_n\left(\ln\frac{Q^2}{\Lambda^2}\right)^{-\gamma_{0n}/2\beta_0}\Bigg[c_{0n} + \left(\beta_0\ln\frac{Q^2}{\Lambda^2}\right)^{-1} \\ \times\left\{c_{1n} + c_{0n}\frac{\beta_1}{2\beta_0}\left(\frac{\gamma_{1n}}{\beta_1} - \frac{\gamma_{0n}}{\beta_0}\right) - c_{0n}\frac{\beta_1\gamma_{0n}}{2\beta_0^2}\ln\ln\frac{Q^2}{\Lambda^2}\right\} + \dots\Bigg] , \end{aligned} \tag{5.2.38}$$

where Λ is the QCD *scale parameter* defined by Eq.(3.4.57) and N_n is the constant given by

$$N_n = \left(\frac{1 + \beta_1 g^2/\beta_0}{\beta_0 g^2}\right)^{\gamma_{0n}/2\beta_0} (1 + \beta_1 g^2/\beta_0)^{\gamma_{1n}/2\beta_1} . \tag{5.2.39}$$

Equation (5.2.38) is the basic formula for practical applications of perturbative QCD to deep inelastic scatterings.

In the above argument we showed an indirect method for calculating γ_{0n}, $\gamma_{1n}, \ldots$. Obviously there is another method of calculating the anomalous dimension $\gamma_n(\bar{g})$, in which we deal with the renormalization constant Z_n of the composite operators $O_{\mu_1 \ldots \mu_n}$ and compute $\gamma_n(\bar{g})$ directly through the definition (5.2.31). The relevant gauge-invariant composite operator with the lowest twist, $\tau_n = 2$ [for the definition of τ_n, see Eq.(4.3.2)], is, e.g.,

$$O_{\mu_1 \ldots \mu_n} = i^{n-1} S\bar{\psi}\gamma_{\mu_1} D_{\mu_2} \ldots D_{\mu_n} \psi + \text{terms with } g_{\mu_i \mu_j}, \tag{5.2.40}$$

where D_μ is the covariant derivative defined in Eq.(2.1.46), the factor i^{n-1} is attached for the later convenience in deriving the Feynman rule for vertices with the composite operator and S denotes the symmetrization of indices, i.e., for the tensor $f_{\mu_1 \ldots \mu_n}$

$$Sf_{\mu_1 \ldots \mu_n} \equiv \frac{1}{n!} (f_{\mu_1 \ldots \mu_n} + \text{permutations}) . \tag{5.2.41}$$

The above symmetrization is necessary since we are dealing with unpolarized scatterings. In Eq.(5.2.40) the terms with $g_{\mu_i \mu_j}$ are required to make the composite operator $O_{\mu_1 \ldots \mu_n}$ have definite spin. That the operator (5.2.40) is of the lowest twist is seen as follows: Consider the composite operator made only of quark fields $\psi(x)$ whose canonical dimension is 3/2. Since the increase of the number of the quark fields in the composite operator results in the increase of the canonical dimension of the composite operator, the number of the quark fields involved must be a minimum, i.e. two, in order to keep the twist as small as possible. Hence the expected form of the minimum-twist composite operator is

$$\bar{\psi}\Lambda_{\mu_1 \ldots \mu_n} \psi ,$$

with $\Lambda_{\mu_1 \ldots \mu_n}$ to be determined. The tensor $\Lambda_{\mu_1 \ldots \mu_n}$ may be composed of γ_μ, ∂_μ and A^a_μ. By gauge invariance, ∂_μ and A^a_μ appear only in the combination specified by D_μ as given by Eq.(2.1.46). For the combination γ_μ and D_μ in $\Lambda_{\mu_1 \ldots \mu_n}$, γ_μ can appear only once because any multiple of γ_μ always reduces to a single γ_μ due to the symmetrization S. Thus we are left with an operator of the form (5.2.40).

In the following subsections we present practical calculations of the relevant parameters in deep inelastic scatterings.

5.2.3. Flavor structure of composite operators

In order to single out dominant contributions to the right-hand side of Eqs.(5.2.11) and (5.2.12), it is necessary to find the composite operators of the lowest twist which are relevant to these moment sum rules. For this purpose we consider the property of the composite operators with respect to the flavor group.

As we are interested in deep inelastic scatterings, the quark masses may be safely neglected and so the flavor symmetry SU(N_f) turns out to be exact. The electromagnetic current $j_\mu(x)$ belongs to the singlet and adjoint representation of SU(N_f) [for $N_f = 2, j_\mu(x)$ behaves as the third component of the isovector of SU(2)]. Hence the product of two currents $j_\mu(x) j_\nu(0)$ consists of the singlet, adjoint, ... representations (for $N_f = 2$, $3 \otimes 3 = 1 \oplus 3 \oplus 5$) and the composite operators appearing in the light-cone expansion belong to these representations. As we have argued before, the lowest twist for the composite operators is two and the twist-two operators are made of two field operators. Since the quark field ψ and the gluon field A^a_μ belong to the fundamental and singlet representation of SU(N_f), respectively, the composite operators made up of two of these fields belong to the singlet or adjoint representation. Explicitly these twist-two *composite operators* are given by [Gro 74, Geo 74] (see also [Pol 74, Pet 79, Gey 79])

$$O^F_{\mu_1 \dots \mu_n} = i^{n-1} S \bar{\psi} \gamma_{\mu_1} D_{\mu_2} \dots D_{\mu_n} \psi \ , \tag{5.2.42}$$

$$O^{Fi}_{\mu_1 \dots \mu_n} = i^{n-1} S \bar{\psi} \gamma_{\mu_1} D_{\mu_2} \cdots D_{\mu_n} t^i \psi \ , \tag{5.2.43}$$

$$O^V_{\mu_1 \dots \mu_n} = i^{n-2} S F^{a_1}_{\mu_1 \lambda} D^{a_1 a_2}_{\mu_2} \dots D^{a_{n-2} a_{n-1}}_{\mu_{n-1}} F^{a_{n-1} \lambda}_{\mu_n} \tag{5.2.44}$$

where we omitted the terms containing $g_{\mu_i \mu_j}$ (the so-called trace terms), $t^i (i = 1, 2, \dots, N_f^2 - 1)$ are the generators of the flavor SU(N_f) group, and D^{ab}_μ is the matrix element of the covariant derivative D_μ in the adjoint representation of the color SU(N_c) group [see Eq.(2.3.96)]. Operators O^F and O^V in Eqs.(5.2.42) and (5.2.44), respectively are singlet under the flavor SU(N_f) and are called the singlet composite operators. Operator O^{Fi} in Eq.(5.2.43) belongs to the adjoint representation of SU(N_f) and is called the nonsinglet composite operator.

As twist-two composite operators we have here two singlet operators and one nonsinglet operator. Hence there occurs *operator mixing* among the singlet operators in the renormalization process while there is no mixing for the nonsinglet operator. We would like to separate the contributions of the singlet and nonsinglet operators to the moment sum rules. For this purpose we consider the moment sum rules for the sum and difference of the proton and neutron structure functions,

$$\int_0^1 d\xi\, \xi^{n-2} F^{\mathrm{p}\pm\mathrm{n}}(\xi, Q^2) = \sum_i A_n^{(i)\mathrm{p}\pm\mathrm{n}} \tilde{C}_n^{(i)}(Q^2) , \tag{5.2.45}$$

where the indices L and 2 are omitted, and the superfix p $\pm$ n implies the sum and difference of the relevant quantities for the proton and neutron,

$$f^{\mathrm{p}\pm\mathrm{n}} = f^{\mathrm{p}} \pm f^{\mathrm{n}} . \tag{5.2.46}$$

If the number of the flavors is two ($N_f = 2$), the flavor symmetry is SU(2), i.e., isospin symmetry. In this case, $A_n^{(i)\mathrm{p}+\mathrm{n}}$ picks out the singlet (isoscalar) composite operators and $A_n^{(i)\mathrm{p}-\mathrm{n}}$ the nonsinglet (isovector) composite operator. This can be seen as follows: The twist-two piece P of the operator-product expansion (5.2.3) consists of the isoscalar (P_0) and isovector (P_i) part, i.e.,

$$P = P_0 + P_i t_i , \tag{5.2.47}$$

with $t_i = \tau_i/2$ (τ_i are the Pauli matrices). Sandwiching Eq.(5.2.47) with the $t_3 = 1/2$ (proton) and $t_3 = -1/2$ (neutron) states we find

$$\langle\tfrac{1}{2}|P|\tfrac{1}{2}\rangle + \langle -\tfrac{1}{2}|P| -\tfrac{1}{2}\rangle = 2P_0 ,$$

$$\langle\tfrac{1}{2}|P|\tfrac{1}{2}\rangle - \langle -\tfrac{1}{2}|P| -\tfrac{1}{2}\rangle = P_3 .$$

Hence the combinations of the operator matrix elements such as $A_n^{(i)\mathrm{p}\pm\mathrm{n}}$ correspond to the isoscalar and isovector parts, respectively. Since there exist mainly u- and d-quarks inside the nucleon and the contribution of the s-, c-, b- and t-quark to the matrix elements of the composite operator is negligible,[4] the above statement may be generalized, to a good approximation, to the case of the arbitrary flavor number N_f. Thus we can isolate the contribution of the flavor singlet as well as nonsinglet composite operator to the moment sum rule by taking the combinations (5.2.45).

[4] In the region of very small ξ, the presence of a considerable amount of s- and c-quarks is expected both theoretically and experimentally.

5.2.4. The nonsinglet part

In the present subsection we concentrate on the nonsinglet operator so that we need not worry about operator mixing. The relevant moment sum rule is given by Eq.(5.2.45) with superscript p-n appended and in which no summation on i is necessary as we have only one kind of composite operator O^{Fi} of Eq.(5.2.43).

Our previous simple argument for the calculation of the coefficient function $\tilde{C}_n(Q^2)$ is directly applicable here. According to Eq.(5.2.38) we have to compute c_{0n}, c_{1n}, ... and γ_{0n}, γ_{1n}, ... in order to obtain the renormalization-group-improved perturbation series for $\tilde{C}_n(Q^2)$. Note that β_0, β_1 and β_2 were already given in Eqs.(3.4.7), (3.4.21) and (3.4.22) respectively.

Let us summarize the algorithm for calculating the c's and γ's which was explained in Sec. 5.2.2:

1. Calculate perturbatively $t_{\mu\nu}$ as defined by Eq.(5.2.20) and obtain t_L and t_2.

2. Calculate perturbatively the Green functions with insertion of composite operators O_L and O_2 as defined by Eq.(5.2.22) and obtain $a_{L,n}$ and $a_{2,n}$.

3. Compute $\tilde{C}_{0n}$, $\tilde{C}_{1n}$, ... by inserting the above t's and a's into Eqs. (5.2.28), (5.2.29), ...

4. Obtain the c's and γ's by using Eq.(5.2.37). An alternative way of determining γ's is given by the procedure described at the end of Sec. 5.2.2.

Note here that we need only the Green functions with quark external lines $t^q_{\mu\nu}$ and in particular their difference $t^{u-d}_{\mu\nu} = t^u_{\mu\nu} - t^d_{\mu\nu}$ for the u- and d- quarks since we are concerned with the nonsinglet composite operators in this subsection. As will be seen in the next subsection the discussion of the singlet composite operators necessitates the Green function with gluon external lines $t^G_{\mu\nu}$ in addition to the ones with quark external lines.

The Feynman diagrams contributing to $t^q_{\mu\nu}$ in the lower orders are shown in Fig.5.2.2. It is also necessary to calculate the Green function with insertion of the composite operator as given in Eq.(5.2.22). In Fig. 5.2.3 we present the lower-order diagrams contributing to the amplitudes $a_{2,n}$ and $a_{L,n}$. The constant c_{0n} may be determined by calculating the $t_{\mu\nu}$ and a_n in the tree approximation as depicted in Figs. 5.2.2(a) and 5.2.3(a), respectively, while c_{1n} and γ_{0n} are determined by estimating the one-loop contribution to $t_{\mu\nu}$ and a_n corresponding to the diagrams shown in Figs. 5.2.2(b) and 5.2.3(b), respectively. The two-loop, three-loop, ... diagrams are relevant to determining (c_{2n}, γ_{1n}), (c_{3n}, γ_{2n}), ...

Let us first calculate c_{0n} and γ_{0n} in order to see the leading-order form of the coefficient function $\tilde{C}_n(Q^2)$. In the following, just for technical simplicity, we

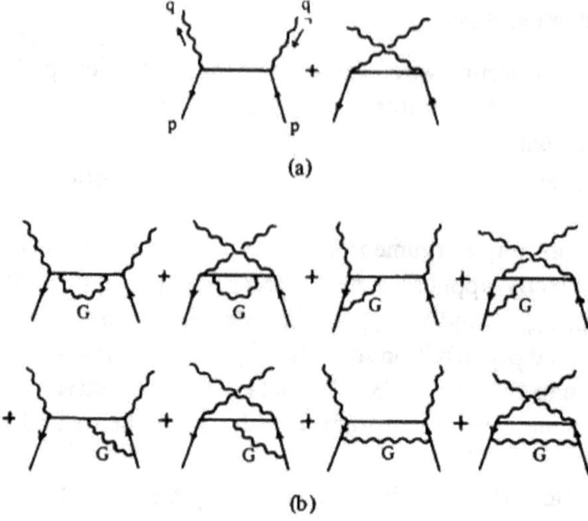

Fig. 5.2.2. The lowest (a) and one-loop (b) contributions to $t^q_{\mu\nu}$ the virtual Compton amplitude off quarks (5.2.20). The external wavy line represents the photon while the internal wavy line with G represents the gluon.

deal with the amplitudes $t_{\mu\nu}$ and a_n for which the external quarks are on their mass shell[5] and their polarizations are averaged over: Instead of $t_{\mu\nu}$ and a_n defined by Eqs.(5.2.20) and (5.2.22), respectively, we shall consider amplitudes

$$t_{\mu\nu}(q,p) = i\int d^4x\, e^{iq\cdot x}\langle p|\mathrm{T}[j_\mu(x)j_\nu(0)]|p\rangle\ , \tag{5.2.48}$$

$$a_n p_{\mu_1}\cdots p_{\mu_n} = \langle p|O^{Fi}_{\mu_1\ldots\mu_n}(0)|p\rangle\ , \tag{5.2.49}$$

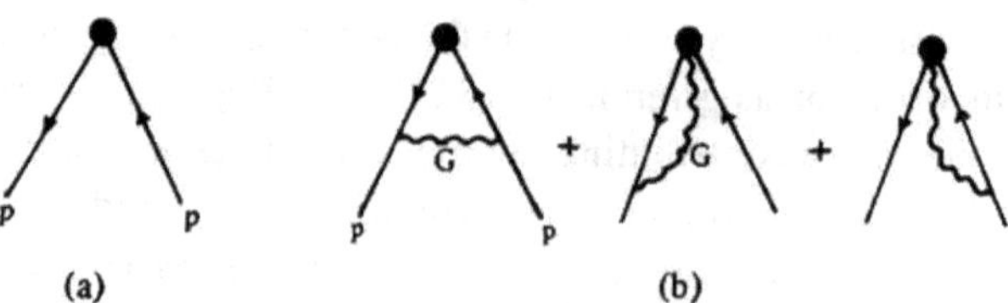

Fig. 5.2.3. The lowest (a) and one-loop (b) contributions to the quark Green functions with the insertion of the composite operators (5.2.22).

[5] Although there is no free quark state according to quark confinement, we regard the quarks as if they were on the mass shell. This approximation is allowed since we are working in perturbation theory.

where $|p\rangle$ is the quark state with momentum p and averaging over polarizations of the quarks should be understood just as we did in Eq.(4.1.27). Note that in Eq.(5.2.49) the trace terms involving $g_{\mu_i\mu_j}$ are neglected.

The contribution of the tree diagrams (Fig. 5.2.2(a)) to $t_{\mu\nu}$ denoted by $t^0_{\mu\nu}$ is given by

$$t^0_{\mu\nu} = Q_q^2 \frac{1}{2} \sum_\lambda \bar{u}_\lambda(p) \left[\gamma_\mu \frac{1}{m - \not{p} - \not{q}} \gamma_\nu + \gamma_\nu \frac{1}{m - \not{p} + \not{q}} \gamma_\mu \right] u_\lambda(p) \,, \quad (5.2.50)$$

where $u_j(p)$ is the quark spinor, m the quark mass and Q_q the charge of the quark q in units of e. In the deep inelastic region for which $|q^2| \to \infty$ with $\omega = 2p{\cdot}q/(-q^2)$ fixed, the quark masses may be neglected and Eq.(5.2.50) reduces to

$$t^0_{\mu\nu} = Q_q^2 \left(\frac{\omega}{1-\omega} - \frac{\omega}{1+\omega} \right) d_{\mu\nu} = 2Q_q^2 \sum_{n=2,4\ldots} \omega^n d_{\mu\nu} \,, \quad (5.2.51)$$

where $d_{\mu\nu}$ is defined in Eq.(5.2.8). Obviously we find from Eq.(5.2.51)

$$t_{20} = Q_q^2 \frac{2\omega^2}{1-\omega^2} \,, \qquad t_{L0} = 0. \quad (5.2.52)$$

In order to compute the quark matrix element of the composite operator O^{Fi} defined in Eq.(5.2.49), we need the Feynman rule for the operator insertion vetex. The composite operator O^{Fi} defined by Eq.(5.2.43) may be expanded in the following form

$$O^{Fi}_{\mu_1 \ldots \mu_n} = i^{n-1} S \bar{\psi} [\gamma_{\mu_1} \partial_{\mu_2} \ldots \partial_{\mu_n} + \gamma_{\mu_1} (-igT^a A^a_{\mu_2}) \partial_{\mu_3} \ldots \partial_{\mu_n} + \ldots] t^i \psi \,. \quad (5.2.53)$$

Each term in the expansion (5.2.53) contributes a primitive vertex to the operator matrix element (5.2.49) and the Feynman rule for the first two terms reads as follows [Flo 77],

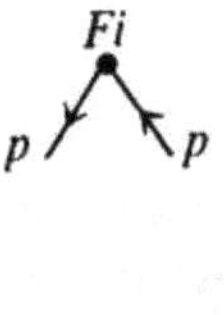

Fi, p, p

$$= S \gamma_{\mu_1} p_{\mu_2} \cdots p_{\mu_n} t^i \,, \quad (5.2.54)$$

Fi, p′ k,μ,a p

$$= S(-gT^a \gamma_{\mu_1} g_{\mu_2\mu} p'_{\mu_3} \cdots p'_{\mu_n} + gT^a \gamma_{\mu_1} p_{\mu_2} g_{\mu_3\mu} p'_{\mu_4} \cdots p'_{\mu_n} + \cdots$$
$$+ (-1)^{n-1} gT^a \gamma_{\mu_1} p_{\mu_2} \cdots p_{\mu_{n-1}} g_{\mu_n\mu}) t^i \,. \quad (5.2.55)$$

The similar Feynman rule for the other vertices

. . .

may be derived in a straightforward way. In the next subsection we shall need the Feynman rule for the other composite operators O^F and O^V and so we present it here. The rule for the operator O^F is essentially the same as the one in Eqs.(5.2.54) and (5.2.55) if the flavor generator t^i is removed. The rule for the gluon composite operator O^V reads [Flo 78]

V

$$= 2\delta_{ab}\, S(g_{\mu\lambda}\, k_{\mu_1} - g_{\mu\mu_1}\, k_\lambda)\, k_{\mu_2} \dots k_{\mu_{n-1}}\, (g_{\nu\mu_n}\, k^\lambda - g^\lambda_\nu\, k_{\mu_n}) \ ,$$

$k,\mu,a \quad k,\nu,b$

(5.2.56)

and similar but more complicated rule may be obtained for the vertices

. . .

Applying the rule (5.2.54) we easily find

$$\langle p|O^{Fi}_{\mu_1\dots\mu_n}|p\rangle = 2\langle t^i\rangle_q\, p_{\mu_1} \cdots p_{\mu_n} \ , \tag{5.2.57}$$

where the quark mass term is neglected and $\langle t^i\rangle_q$ is the matrix element of t^i for quark q. Since we have only one operator O^{Fi} available, O_L and O_2 are essentially the same. Hence the tree contribution to $a_{L,n}$ and $a_{2,n}$ denoted by $a^0_{L,n}$ and $a^0_{2,n}$, respectively is given by

$$a^0_{L,n} = a^0_{2,n} = 2\langle t^i\rangle_q \ . \tag{5.2.58}$$

Taking the $u-d$ difference of Eqs.(5.2.52) and (5.2.58) and using Eqs.(5.2.28) and (5.2.37) we have

$$\tilde{C}^0_{2,n} = c^0_{2,n} = 1 \ , \tag{5.2.59}$$

$$\tilde{C}^0_{L,n} = c^0_{L,n} = 0 \ , \tag{5.2.60}$$

where superscript 0 designates the tree contribution and the normalization of $\tilde{C}^0_{2,n}$ is readjusted to 1 as the overall factor is irrelevant.

To obtain γ_{0n} we have to calculate the one-loop divergent part of $t_{\mu\nu}$ or a_n. Here we follow the method mentioned at the end of Sec. 5.2.2. Using the Feynman rule (5.2.54-55) we compute the operator matrix element (5.2.49) in the one-loop order represented by Fig. 5.2.3(b). The loop integrations may be performed by the standard technique of dimensional regularization. We have [Gro 73a, Geo 74, Bai 74] for the one-loop contribution to the operator matrix element (5.2.49)

$$\langle p|O^{Fi}_{\mu_1\ldots\mu_n}(0)|p\rangle_{1\text{-loop}}$$

$$= -\langle t^i\rangle_q \frac{g^2}{16\pi^2} C_F \left[1 - \frac{2}{n(n+1)} + 4\sum_{j=2}^{n}\frac{1}{j}\right]\frac{1}{\varepsilon} p_{\mu_1} \ldots p_{\mu_n} + \text{finite part}, \tag{5.2.61}$$

where C_F is the constant given in Eq.(2.4.7) and $\varepsilon = (4 - D)/2$. Through Eq.(5.2.61) we obtain the renormalization constant for the nonsinglet composite operator O^{Fi} to the one-loop order in the MS scheme,

$$Z^F_n = 1 - \frac{g^2}{16\pi^2} C_F K_n \frac{1}{\varepsilon}, \tag{5.2.62}$$

with

$$K_n = 1 - \frac{2}{n(n+1)} + 4\sum_{j=2}^{n}\frac{1}{j}. \tag{5.2.63}$$

It is now straightforward to calculate the *anomalous dimension* γ^F_n of the composite operator O^{Fi} to the one-loop order in the MS scheme following the definition (5.2.31),[6]

$$\gamma^F_n(g) = \frac{g^2}{8\pi^2} C_F K_n + O(g^4). \tag{5.2.64}$$

Hence we have

$$\gamma^F_{0n} = \frac{1}{8\pi^2} C_F K_n. \tag{5.2.65}$$

With $\beta_0, c^0_{2,n}, c^0_{L,n}$ and γ^F_{0n} given by Eqs.(3.4.7), (5.2.59), (5.2.60) and (5.2.65), respectively we finally have in the leading order

$$\tilde{C}_{2,n}(Q^2) = N_n \left(\ln\frac{Q^2}{\Lambda^2}\right)^{-d_{0n}}, \tag{5.2.66}$$

with

$$\tilde{C}_{L,n}(Q^2) = 0, \tag{5.2.67}$$

$$d_{0n} = \frac{\gamma^F_{0n}}{2\beta_0}. \tag{5.2.68}$$

[6] Note that g depends on the renormalization scale μ as in Eq. (3.2.1.).

According to Eq.(5.2.67) all the moments of the nonsinglet longitudinal structure function $F_L^{p-n}(\xi, Q^2)$ vanish in the leading order. This means that the structure function $F_L^{p-n}(\xi, Q^2)$ itself vanishes. This fact corresponds to the parton model prediction (1.3.21), known as the *Callan-Gross relation* [Cal 69] which originates from the spin-1/2 nature of quarks.

It is straightforward but very tedious to push forward the approximation up to the next-to-leading order. What is important in calculating the parameters c_{1n} and γ_{1n} is to perform all the calculations consistently in the same renormalization scheme. If, for instance, c_{1n} is computed in the scheme different from the one adopted in calculating γ_{1n}, these parameters should not be combined together to compose the next-to-leading-order expression of the coefficient function $\tilde{C}_n(Q^2)$[7]. In the leading order the parameters c_{on}, γ_{on} and β_0 are, respectively, independent of the renormalization scheme chosen and hence there exists no such problem.

To calculate c_{1n} for the nonsinglet part we have to take into account the diagrams in Fig.5.2.2(b) as well as the ones in Fig.5.2.3(b). The calculation of c_{1n} in the MS (and also $\overline{\text{MS}}$) scheme was performed by Bardeen et al. [Bar 78]. (See also [Alt 78].) The two-loop nonsinglet anomalous dimension γ_{1n} was computed in the MS scheme by Floratos, Ross and Sachrajda [Flo 77] and its analytic expression was greatly simplified by Gonzalez-Arroyo, Lopez and Yndurain [Gon 79]. Combining these results on c_{1n} and γ_{1n} together with that on β_1 [Jon 74, Cas 74a] we obtain the next-to-leading-order QCD prediction on the coefficient functions $\tilde{C}_n(Q^2)$. The phenomenological implication of the result will be presented in Sec. 5.2.6. (For calculations in other schemes, see e.g. [Cal 77, deR 77].)

The constant c_{1n} has also been calculated in other renormalization schemes, e.g. the on-shell and off-shell subtraction scheme. The references to these calculations may be found in [Bar 78, Mut 79]. The constant c_{2n} for the longitudinal coefficient function $\tilde{C}_{L,n}(Q^2)$ has been computed in the MS (and $\overline{\text{MS}}$) scheme in [Duk 82, Dev 84, Cou 82, 83].

DIGRESSION

QCD CORRECTIONS TO CURRENT-ALGEBRA SUM RULES: The nonsinglet composite operator (5.2.43) for $n = 1$,

$$O_\mu^{Fi} = \bar{\psi}\gamma_\mu t^i \psi \ , \tag{5.2.69}$$

is nothing but the conserved flavor current if quark masses are neglected. Hence the operator (5.2.69) has no anomalous dimension as was shown before in Sec. 3.5,

$$\gamma_1^F = 0 \ . \tag{5.2.70}$$

[7] It will be shown in Sec. 5.3.2 that β_1 is independent of the renormalization scheme employed to calculate it.

Since Eq.(5.2.70) is an exact relation, it holds to all orders of the perturbation series. Its consequence on the coefficient function $\tilde{C}_n(Q^2)$ for $n = 1$ is found from Eq.(5.2.38) to be

$$\tilde{C}_1(Q^2) = N_1\left[c_{01} + \frac{c_{11}}{\beta_0 \ln(Q^2/\Lambda^2)} + \dots\right] . \tag{5.2.71}$$

Unfortunately in deep inelastic electron-nucleon scatterings the moment of the structure functions for odd n vanishes according to crossing symmetry and so no meaningful prediction can be derived from Eq.(5.2.71). In deep inelastic neutrino-nucleon scatterings, however, the odd-n moment of the structure functions survives under crossing symmetry and is subject to the prediction (5.2.71). Here the leading term in Eq.(5.2.71) corresponds to the current algebra sum rule while the next-to-leading term represents the QCD correction to the sum rule.

A typical example is the *Adler sum rule* [Adl 66] which corresponds to the $n = 1$ moment of $F_2^{\bar{\nu}}(\xi, Q^2) - F_2^{\nu}(\xi, Q^2)$ where $F_2^{\bar{\nu}}$ and F_2^{ν} are the F_2 structure functions for the deep inelastic antineutrino- and neutrino-nucleon scatterings, respectively. After some calculation one finds for the above moment

$$c_{11} = 0 . \tag{5.2.72}$$

Hence we recognize that the Adler sum rule does not receive any QCD correction, i.e.,

$$\int_0^1 \frac{d\xi}{\xi}[F_2^{\bar{\nu}}(\xi, Q^2) - F_2^{\nu}(\xi, Q^2)] = 2 . \tag{5.2.73}$$

On the other hand the other well-known sum rules such as the Bjorken backward sum rule [Bj 67], Gross-Llewellyn-Smith sum rule [Gro 69] and Bjorken polarized sum rule [Bj 66] are subject to the QCD correction. The correction terms for these sum rules were computed in [Bar 78, Alt 78, Kod 79] and may be tested experimentally.

5.2.5. The singlet part

To the singlet combination of the structure functions contribute two types of composite operators, O^F and O^V, defined in Eqs.(5.2.42) and (5.2.44). Thus we face the problem of operator mixing. The moment sum rule for the singlet combination is given by Eq.(5.2.45) with superscript p+n where the index i refers to F and V. The coefficient function $\tilde{C}_n^{(i)}(Q^2)$ as a solution of the renormalization group equation is presented in Eq.(4.3.19).

We write the perturbative expansion of the anomalous dimension matrix $\gamma_n(\bar{g})$ and the coefficient function $\tilde{C}_n^{(i)}(1, \bar{g})$ as follows,

$$\gamma_n(\bar{g}) = \gamma_{0n}\bar{g}^2 + \gamma_{1n}\bar{g}^4 + \dots , \tag{5.2.74}$$

$$\tilde{C}_n^{(i)}(1, \bar{g}) = c_{0n}^i + c_{1n}^i \bar{g}^2 + \dots . \tag{5.2.75}$$

Substituting Eqs.(5.2.74) and (5.2.75) into Eq.(4.3.19) and keeping only the leading order we obtain

$$\tilde{C}_n^{(i)}(Q^2) = [(\bar{g}/g)^{\gamma_{0n}/\beta_0}]^{ij} c_{0n}^j , \tag{5.2.76}$$

where $i(j) = F, V$ indicating the quark and gluon composite operator, respectively and the quark masses are neglected in Eq.(5.2.76).

In order to obtain c^i_{0n} we have to calculate $t^i_{\mu\nu}$ and $\langle i|O^{(i)}|i\rangle$ in the tree approximation where $t^F_{\mu\nu}$ is defined by Eq.(5.2.48), $t^V_{\mu\nu}$ is also given in the same way but with the gluon state, $\langle F|O^F|F\rangle$ obtained from Eq.(5.2.49) but with the quark singlet composite operator, and $\langle V|O^F|V\rangle$ similarly but with the gluon state and so on. We immediately see that c^F_{0n} is the same as the one for the nonsinglet case in the previous subsection. Hence

$$c^F_{0n} = \begin{cases} 1 & \text{for } \tilde{C}_{2,n} \,, \\ 0 & \text{for } \tilde{C}_{L,n} \,. \end{cases} \tag{5.2.77}$$

Since the gluon does not couple with the photon to lowest order, we find

$$c^V_{0n} = 0 \,. \tag{5.2.78}$$

Inserting Eqs.(5.2.77) and (5.2.78) in Eq.(5.2.76) we obtain to the leading order

$$\tilde{C}^{(i)}_{2,n}(Q^2) = [(\bar{g}/g)^{\gamma_{0n}/\beta_0}]^{iF} \,, \tag{5.2.79}$$

$$\tilde{C}^{(i)}_{L,n}(Q^2) = 0 \,. \tag{5.2.80}$$

The anomalous dimension matrix γ_{0n} has yet to be calculated in order to complete the leading-order prediction. We write

$$\gamma_{0n} = \begin{bmatrix} \gamma^{FF}_{0n} & \gamma^{FV}_{0n} \\ \gamma^{VF}_{0n} & \gamma^{VV}_{0n} \end{bmatrix} . \tag{5.2.81}$$

The relevant diagrams for the calculation of the matrix elements in Eq.(5.2.81) are listed in Fig.5.2.4. We compute the renormalization constant matrix Z_n in one-loop order and by way of the definition (4.3.17) we have

$$\gamma^{FF}_{0n} = \frac{C_F}{8\pi^2}\left[1 - \frac{2}{n(n+1)} + 4\sum_{j=2}^{n}\frac{1}{j}\right] , \tag{5.2.82}$$

$$\gamma^{FV}_{0n} = -\frac{N_f T_R}{2\pi^2}\,\frac{n^2+n+2}{n(n+1)(n+2)} , \tag{5.2.83}$$

$$\gamma^{VF}_{0n} = -\frac{C_F}{4\pi^2}\,\frac{n^2+n+2}{n(n^2-1)} , \tag{5.2.84}$$

$$\gamma^{VV}_{0n} = \frac{1}{2\pi^2}\left[C_G\left(\frac{1}{12} - \frac{1}{n(n-1)} - \frac{1}{(n+1)(n+2)} + \sum_{j=2}^{n}\frac{1}{j}\right) + \frac{N_f T_R}{3}\right] , \tag{5.2.85}$$

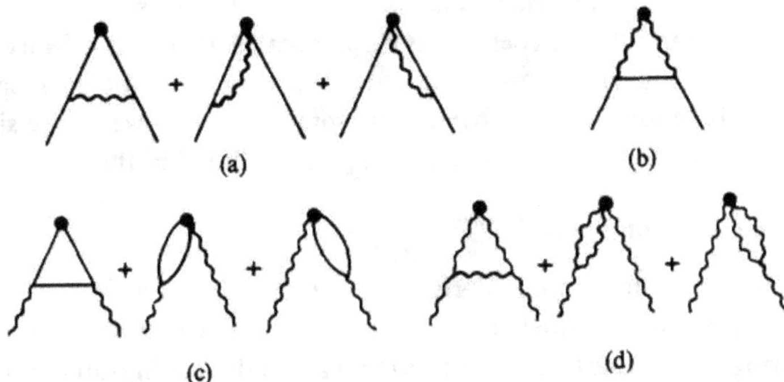

Fig. 5.2.4. The one-loop diagrams needed to calculate the quark Green functions of the (a) quark and (b) gluon composite operators, and the gluon Green functions of the (c) quark and (d) gluon composite operators. By computing the renormalization constant matrix Z_n one obtains the anomalous dimension γ_{0n}.

where C_F, C_G and T_R are defined by Eqs.(2.4.7) and (2.5.133), and N_f is the number of the quark flavors.

Substituting Eqs.(5.2.79) and (5.2.80) into Eq.(5.2.45) and suppressing superscript p + n, we have

$$\int_0^1 d\xi\, \xi^{n-2} F_2(\xi, Q^2) = \sum_i A_{2,n}^{(i)} [(\bar{g}/g)^{\gamma_{0n}/\beta_0}]^{iF} , \qquad (5.2.86)$$

$$\int_0^1 d\xi\, \xi^{n-2} F_L(\xi, Q^2) = 0 . \qquad (5.2.87)$$

For the purpose of comparing the above prediction (5.2.86) with experimental data, it is more convenient to diagnonalize the anomalous dimension matrix γ_{0n} so that Eq.(5.2.86) is rewritten as

$$\int_0^1 d\xi\, \xi^{n-2} F_2(\xi, Q^2) = A_n^+ \left(\ln\frac{Q^2}{\Lambda^2}\right)^{-\lambda_n^+} + A_n^- \left(\ln\frac{Q^2}{\Lambda^2}\right)^{-\lambda_n^-} , \qquad (5.2.88)$$

where $A_n^\pm$ are constants composed of $A_{2,n}^{(i)}$, γ_{0n}^{ij} and β_0, and $\lambda_n^\pm$ are given by

$$\lambda_n^\pm = \frac{1}{4\beta_0}\left[\gamma_{0n}^{FF} + \gamma_{0n}^{VV} \pm \sqrt{(\gamma_{0n}^{FF} + \gamma_{0n}^{VV})^2 + 4\gamma_{0n}^{FV}\gamma_{0n}^{VF}}\right] . \qquad (5.2.89)$$

It is not difficult to go beyond and obtain the next-to-leading order result but the calculation turns out to be quite tedious. The anomalous dimension matrix γ_{1n} was calculated by Floratos, Ross and Sachrajda [Flo 78] and its simpler analytic expression was found by Gonzalez-Arroyo, Lopez and Yndurain [Gon 80]. The coefficients c^i_{1n} were calculated by Bardeen et al. [Bar 78]. The resulting next-to-leading-order expression for the moment of the structure functions looks rather cumbersome. It can, however, be simplified considerably by suitable manipulations [Flo 78, Bar 79b, Bur 80].

5.2.6. Experimental tests

With the aid of the formulas derived in Sections 5.2.4 and 5.2.5 we can make critical tests of the predictions of perturbative QCD in deep inelastic scatterings. Our basic formulas are, in the case of the leading-order prediction, Eqs.(5.2.66), (5.2.67), (5.2.87) and (5.2.88) which exhibit the Q^2 dependence of the structure-function moments for deep inelastic electron-nucleon scatterings. Similar but more complicated formulas for the next-to-leading-order prediction are also at hand. The free parameter inherent in the theory is *scale parameter* Λ which corresponds to the coupling constant g. Though the operator matrix elements A_n appearing in the moment sum rules are, in principle, calculable in a nonperturbative way, there is no established method to obtain them at present. Hence we leave them as parameters to be determined from experimental data.

First of all we consider the Q^2 dependence of the running coupling constant $\bar{\alpha}_S(Q^2)$ calculated previously in Eqs.(3.4.53) and (3.4.58) where $\bar{\alpha}_S = \bar{g}^2/(4\pi)$. Once the value of the parameter Λ is fixed, a definite prediction on the behavior of $\bar{\alpha}_S$ will be made by applying these formulas. In Fig. 5.2.5 the Q^2

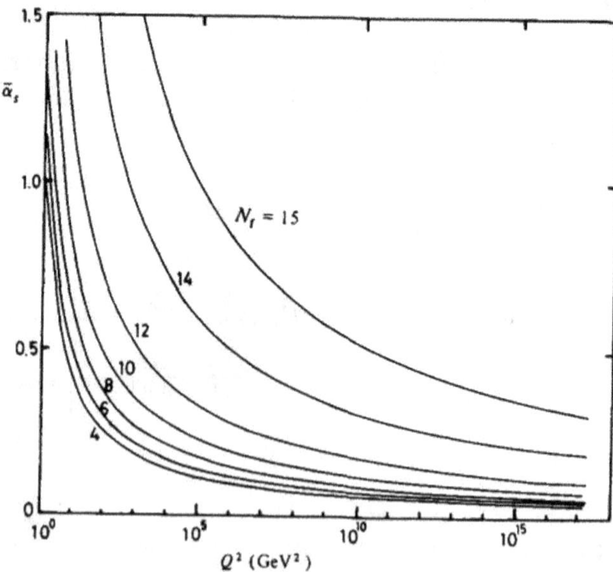

Fig. 5.2.5. The Q^2 dependence of the running coupling parameter $\bar{\alpha}_S$ in the leading order for various values of the flavor number N_f where the scale parameter Λ is chosen to be 0.5 GeV.

variation of $\bar{\alpha}_S$ in the leading order given by Eq.(3.4.53) is shown as a function of the flavor number N_f assuming $\Lambda = 0.5$ GeV. We find in Fig. 5.2.5 that $\bar{\alpha}_S$ is sufficiently small in the region around $Q^2 = 50\ \text{GeV}^2$ if $N_f \leq 6$. It should be noted that even with $N_f \simeq 12$ the running coupling constant $\bar{\alpha}_S$ stays small for Q^2 as large as $10^{15}\ \text{GeV}^2$. In Fig. 5.2.6 the magnitude of the next-to-leading-order correction to $\bar{\alpha}_S$ is shown. Two curves in Fig. 5.2.6 represent, respectively, the leading-order prediction for $\bar{\alpha}_S$ [Eq.(3.4.53)] and the prediction including the next-to-leading-order correction [Eq.(3.4.58)] for which the scale parameter Λ is set equal to 0.5 GeV. As seen in this figure, the correction is considerably large suggesting the importance of the next-to-leading-order correction.

The Q^2 dependence of the nonsinglet coefficient functions $\tilde{C}_{2,n}(Q^2)$ is shown in Fig. 5.2.7 using formula (5.2.38) in which the correction up to the next-to-leading order is taken into account in the $\overline{\text{MS}}$ scheme. Similar curves for the singlet coefficient functions may be drawn by employing the corresponding results discussed in Sec. 5.2.5. The curves as shown in Fig. 5.2.7 are to be compared with the data on the structure-function moments to test the validity of the theory.

A convenient compilation of eN and μN data is given by Duke and Roberts [Duk 80] (see also [Duk 85]) who constructed the moments of structure

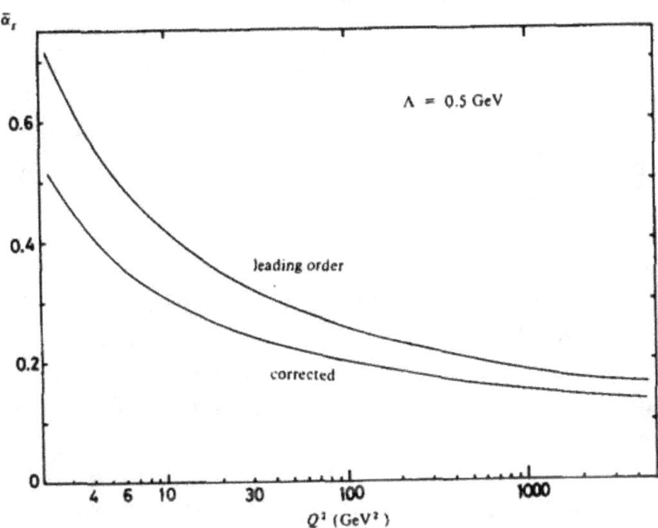

Fig 5.2.6. The Q^2 dependence of $\bar{\alpha}_S$ with $N_f = 4$ and $\Lambda = 0.5$ GeV. The upper curve corresponds to the leading-order result for $\bar{\alpha}_S$ and the lower curve corresponds to the next-to-leading-order result for $\bar{\alpha}_S$.

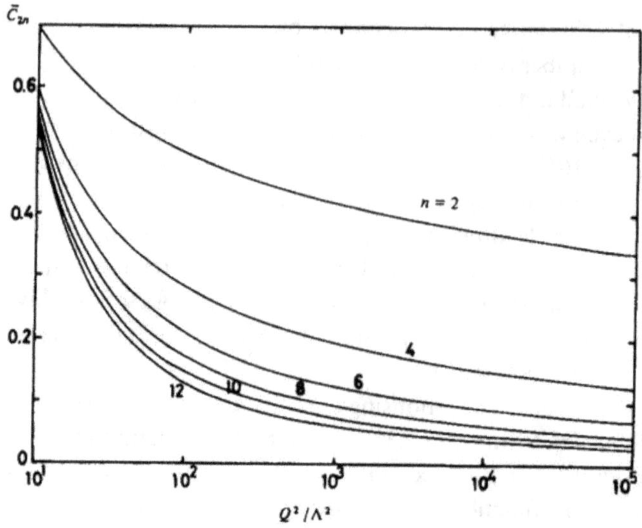

Fig. 5.2.7. The Q^2 dependence of the nonsinglet coefficient functions $\tilde{C}_{2,n}(Q^2)$ for $n = 2, 4, 6, 8, 10, 12$ with $N_f = 4$.

functions through the use of these data. The moments from [Duk 80] are shown in Fig. 5.2.8. In obtaining the above moments these authors took into account the modification of the moment formulas (5.2.11) and (5.2.12) due to the effect of the finite nucleon mass. This modified moment is called the Nachtmann moment about which we shall not go into detail [Nac 73, Geo 76].

The above data on the structure-function moments are fitted excellently by our theoretical predictions given in Fig. 5.2.7 with the value of Λ suitably determined. It is most important to note here that our prediction on the Q^2 dependence of the structure-function moments, in general, depends on the renormalization scheme adopted in the computation as will be fully explained in Sec. 5.3. Hence the value of the scale parameter Λ determined by fitting the data may vary since the form of the structure-function moment employed in fitting the data depends on the choice of the renormalization scheme. Accordingly it is necessary for us to specify the renormalization scheme when we fit the experimental data to determine the value of Λ. We have to state clearly the renormalization scheme to which the value of Λ refers. Thus, for example, $\Lambda_{\overline{MS}}$ is the value of Λ determined by fitting the data with the formula obtained in the $\overline{\text{MS}}$ scheme. The most commonly used in perturbative QCD are $\Lambda_{\overline{MS}}$ and Λ_{MOM} where Λ_{MOM} is the scale parameter determined in the momentum subtraction scheme à la Celmaster and Gonzalves [Cel 79, 79a].

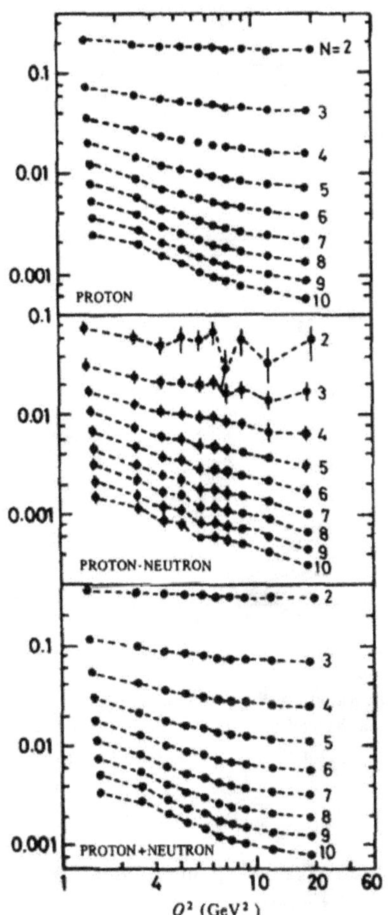

Fig. 5.2.8. The moments of the proton structure function and the difference and sum of the proton and neutron structure functions respectively. The moments are constructed by using the compilation of μN and eN data by Duke and Roberts (Duk 80).

As will be seen in Sec. 5.3.3 the leading-order prediction on the structure-function moments is scheme-independent and hence the scale parameter Λ determined by the leading-order fit is irrelevant to the renormalization-scheme problem. Beyond the leading order, however, the scheme problem should be always kept in mind.

By fitting the data of Fig. 5.2.8 with the leading-order predictions of Eqs.(5.2.66) and (5.2.88) the scale parameter Λ is determined to be

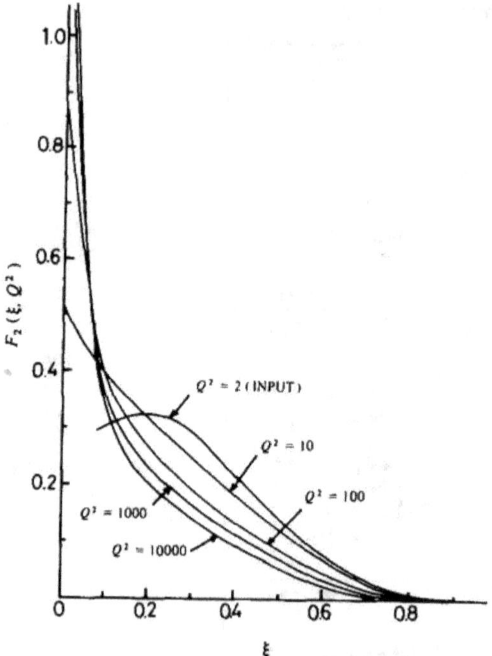

Fig. 5.2.9. The Q^2 evolution of the structure function $F_2(\xi, Q^2)$ for the e-p deep inelastic scattering. The structure function is calculated in the leading order for $N_f = 4$ and $\Lambda = 0.5$ GeV starting with the input at $Q^2 = 2$ GeV^2.

$$\Lambda \simeq 0.8 \text{ GeV} . \tag{5.2.90}$$

If we include the next-to-leading-order corrections to the moments and fit the same data as above, we obtain in the $\overline{\text{MS}}$ scheme

$$\Lambda_{\overline{\text{MS}}} \simeq 0.4 \text{ GeV} . \tag{5.2.91}$$

Both of the leading-order and next-to-leading-order fit with these values of the scale parameter reproduce the data very well. This means that to some extent the higher-order effect may be practically taken care of by the redefinition of Λ.

As already mentioned before in Sec. 5.2.1 it is more covenient to deal directly with the structure functions instead of handling their moments. By using Eq.(5.2.18) with Eq.(5.2.19) one can convert our QCD prediction on the coefficient functions $\bar{C}_n(Q^2)$ into the one for the structure functions $F(\xi, Q^2)$ once the functional form of $F(\xi, Q^2)$ in ξ is given at a fixed Q^2, say $Q^2 = Q_0^2$. In Fig. 5.2.9 we present an example of such calculations to leading order: With $F_2(\xi, Q^2)$ given at $Q^2 = 2$ GeV^2, the structure function $F_2(\xi, Q^2)$ for other values of Q^2 is completely determined where we assume $\Lambda = 0.5$ GeV. In this calculation the target is assumed to be the proton and so the combination of the singlet and nonsinglet part is considered. As seen in Fig. 5.2.9 the small ξ region is strongly enhanced with the increase of Q^2. In the parton picture the structure function $F_2(\xi, Q^2)$ is interpreted as a distribution function of partons in their momentum fraction ξ (parton momentum in units of the nucleon momentum) when measured with the space resolution of order $1/Q$ [Par 76]. Hence the above tendency in Fig. 5.2.9 suggests that the copious wee-parton production takes place as Q^2 increases, where by the *wee parton*[8] we mean a constituent of the nucleon (i.e., quark and gluon) with small momentum fraction $\xi \ll 1$. This tendency in our theoretical prediction clearly explains a remarkable feature of the experimental data which will emerge in a moment.

In Figs. 5.2.10 we give the data on $F_2(\xi, Q^2)$ for e-p and μ-p scatterings at three typical values of Q^2 which was compiled by Kato and Shimizu [Kat 80, 80a, 83]. Obviously the data shows the behavior suggested by the theoretical prediction: the increase of F_2 in the small ξ region and the decrease of F_2 in the large ξ region as Q^2 increases. More quantitatively one may predict the behavior of $F_2(\xi, Q^2)$ once one fixes Λ and $F_2(\xi, Q_0^2)$ with Q_0^2 a certain fixed value of Q^2. We determine $F_2(\xi, Q_0^2)$ as a function of ξ by fitting the data at $Q_0^2 = 5\,\mathrm{GeV}^2$ and then predict $F_2(\xi, Q^2)$ for other values of Q^2 leaving Λ as a free parameter. We use the leading-order form for $\tilde{C}_{2,n}(Q^2)$ and, of course, take

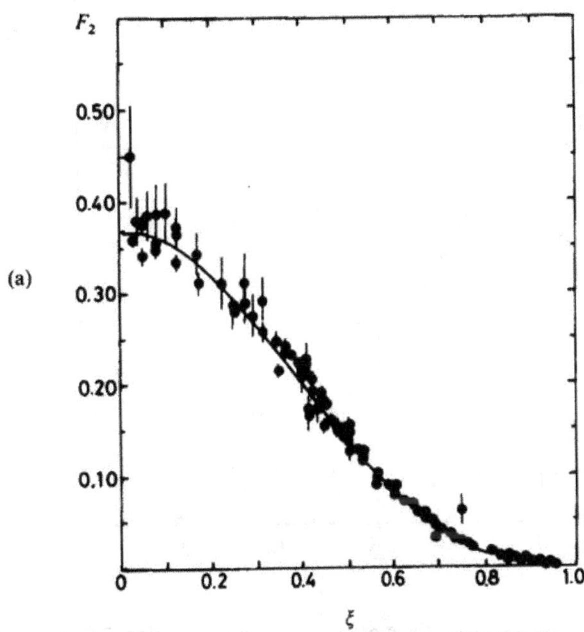

Fig. 5.2.10.

[8] According to R.P. Feynman [Fey 69, 69a].

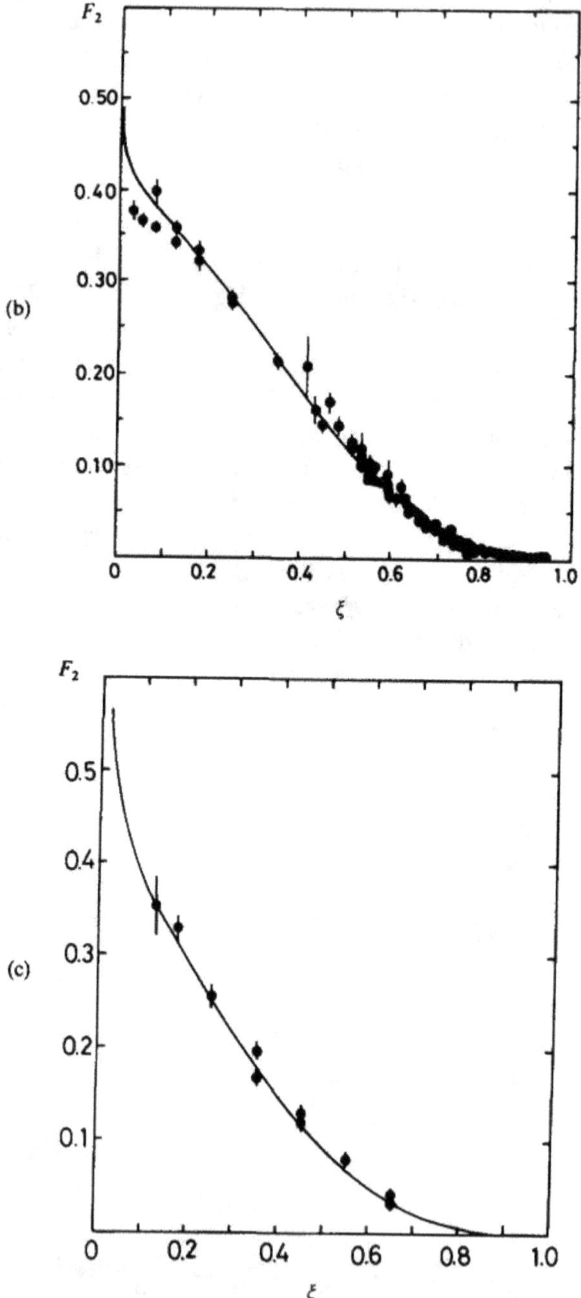

Fig. 5.2.10. Experimental data on the e-p and μ-p structure function $F_2(\xi, Q^2)$ at (a) $8 < Q^2 < 10$, (b) $14 < Q^2 < 20$ and (c) $50 < Q^2 < 100$ (GeV^2). The solid curves represent theoretical predictions as described in the text.

operator mixing into account to obtain the coefficient function for the proton. The predictions for $Q^2 = 9, 17, 75\,\text{GeV}^2$ are plotted in Figs. 5.2.10 (a), (b) and (c), respectively, with $\Lambda = 0.5\,\text{GeV}$. The agreement between the theoretical prediction and the data is satisfactory.

Comparisons of the theoretical prediction with the data including the next-to-leading-order expression for $\tilde{C}_n(Q^2)$ may be performed in a similar manner. Practically, however, the comparison turns out to be very cumbersome. For details see, for example, [Cur 80, Fur 80, 82, Flo 81, Her 81, Dev 83].

The QCD predictions for deep inelastic neutrino-nucleon scatterings are obtained exactly in the same way as in the electron (muon)-nucleon case and may be compared with experimental data. Such comparisons have been performed extensively and the satisfactory agreement between the prediction and the data has been obtained with the same value of the scale parameter Λ as that in the electron (muon)-nucleon case (see, e.g., [Bur 80]).

So far we have restricted ourselves to the discussion on the contribution of the lowest-twist composite operators to the coefficient functions $\tilde{C}_n(Q^2)$. Higher-twist operators, i.e., operators with twist 3,4, ..., give contributions to $\tilde{C}_n(Q^2)$ with a power of $1/Q^2$ higher than that for the twist-two operators. Hence the effect of the higher-twist composite operators is less important in the phenomenological analysis when Q^2 is sufficiently large. For the theoretical analysis of the effect of the higher-twist composite operators, see, e.g., [Oka 83].

5.3. RENORMALIZATION-SCHEME DEPENDENCE

5.3.1. Renormalization-scheme dependence of perturbative predictions

In quantum chromodynamics we are concerned with the Green functions of gluons and quarks,

$$\langle 0|\mathrm{T}[A_\mu A_\nu \dots \psi_\alpha \psi_\beta \dots \bar{\psi}_{\alpha'} \bar{\psi}_{\beta'} \dots]|0\rangle \ . \tag{5.3.1}$$

and the Green functions with insertion of composite operators O,

$$\langle 0|\mathrm{T}[O\, A_\mu A_\nu \dots \psi_\alpha \psi_\beta \dots \bar{\psi}_{\alpha'} \bar{\psi}_{\beta'} \dots]|0\rangle \ . \tag{5.3.2}$$

These Green functions develop divergences characteristic of perturbative calculations. The divergences are disposed of by the standard renormalization procedure, i.e., by the redefinition of the coupling constant g, the mass m, the fields A_μ and ψ, and the composite operator O in terms of the coupling renormalization constant Z_g, the mass renormalization constant Z_m, the field renormalization constants Z_3 and Z_2, and the operator renormalization constant Z_o, respectively. Since the renormalization procedure is not a unique process, the Green functions (5.3.1) and (5.3.2) are, in general, dependent on the way how the divergences are removed, i.e., the Green functions are renormalization-scheme-dependent.

The physical quantities derived from these Green functions, however, should be independent of the renormalization scheme employed so that unique physical predictions are extracted. So far, we have seen the e^+e^- annihilation total cross section and the coefficient functions in the light-cone expansion. Although the values of these physical quantities are renormalization-scheme-independent, their expression in terms of parameters g and m may vary depending on the renormalization scheme adopted. In perturbation theory the parameter g plays a special role and the above property leads us to an important consequence.

If the physical quantity $f(g)$ is expanded in powers of g,

$$f(g) = f_0 + f_1 g^2 + f_2 g^4 + \dots \, , \tag{5.3.3}$$

the coefficients $f_0, f_1, f_2 \dots$ are, in general, renormalization-scheme-dependent since the coupling constant g is renormalization-scheme-dependent. (Here and in the following we neglect the effect of the mass m for simplicity, i.e., we set $m = 0$, and we suppress in $f(g)$ physical variables such as momentum.) The reason why the coupling constant g is scheme-dependent stems from the fact that it is a Green function at specific values of its arguments. Though the coefficients f_0, f_1, f_2, ... as well as the coupling constant g are scheme-dependent, the whole perturbative series (5.3.3) when summed to all orders is scheme-independent. In the practical application of perturbation theory, however, we truncate the series (5.3.3) at a certain order to obtain an approximate expression of the physical quantity $f(g)$. Then the quantity $f(g)$ itself turns out to be scheme-dependent in the neglected orders. This is the source of the *renormalization-scheme dependence* of perturbative predictions on the physical quantity. (For a review, see, e.g. [Har 83].) It should be emphasized here that the renormalization-scheme dependence of the perturbative prediction on physical quantities is a general phenomenon not specific to quantum chromodynamics. The reason why we are concerned with this phenomenon particularly in quantum chromodynamics is that in QCD the renormalization-scheme dependence of the physical quantity in perturbation theory causes a non-negligible difference in the physical prediction since the coupling parameter $\alpha_S = g^2/(4\pi)$ is not sufficiently small and the size of the neglected orders is, in general, not very small. In the case of quantum electrodynamics the coupling parameter $\alpha = e^2/(4\pi)$ is sufficiently small for a very wide range of the renormalization scale and so the problem of renormalization-scheme dependence is practically irrelevant.

Let us look into the scheme-dependence problem in a more quantitative way. We consider the renormalization scheme different from the one adopted in Eq.(5.3.3) and denote by g' the renormalized coupling constant in this scheme. The physical quantity $f(g)$ may be written as $f'(g')$ in this new scheme. Of course we require that

$$f'(g') = f(g) \ . \tag{5.3.4}$$

Expanding $f'(g')$ in powers of g' we obtain

$$f'(g') = f'_0 + f'_1 g'^2 + f'_2 g'^4 + \dots \tag{5.3.5}$$

Since two coupling constant g and g' are to be related to each other by a finite renormalization

$$g = z_g(g')g' = (1 + z_0 g'^2 + z_1 g'^4 + \dots) g' \ , \tag{5.3.6}$$

we see that Eq.(5.3.3) may be re-expanded in powers of g',

$$f(g) = f_0 + f_1 g'^2 + (f_2 + 2z_0 f_1) g'^4 + \dots \tag{5.3.7}$$

Using Eqs.(5.3.4), (5.3.5) and (5.3.7) we conclude that

$$f'_0 = f_0, \quad f'_1 = f_1, \quad f'_2 = f_2 + 2z_0 f_1 \ , \dots \tag{5.3.8}$$

Hence only f_0 and f_1 are scheme-independent while f_2, f_3, ... are scheme-dependent. If we truncate the series at the order g^N and g'^N in Eqs.(5.3.3) and (5.3.5), respectively, there remains the difference of order g^{N+2},

$$[f(g)]_N - [f(g')]_N = O(g^{N+2}), \tag{5.3.9}$$

where $[f(g)]_N$ denotes the perturbation series truncated at the order g^N. This difference in the physical prediction gives rise to serious problems in the phenomenological applications of QCD. We have already seen some examples of the phenomenological consequences of the scheme dependence in Sections 5.1.3 and 5.2.6.

A realistic way of circumventing this problem may be to try to make the size of the neglected orders (5.3.9) as small as possible by varying the renormalization scheme. In this way one may find the most suitable renormalization scheme for guaranteeing the identity (5.3.4) up to the order under consideration. Practically, however, we have no means for estimating the size of the neglected orders (5.3.9) without calculating these terms. The next best thing we can do may be to compare the calculated highest-order term with the lower order ones and find the best renormalization scheme for which the ratio of the highest-order term to the lower ones is minimized so that the fast convergence of the perturbation series is apparently guaranteed. According to the analyses in deep inelastic scatterings, e^+e^- annihilations, $\gamma-\gamma$ scatterings and heavy quarkonium decays, one finds that the $\overline{MS}$ *scheme* [Bar 78] shows better apparent convergence than the *MS scheme* [t Ho 73] and moreover the *MOM scheme* [Cel 79, 79a, Bar 79c, Bra 81] seems to be better than the $\overline{\text{MS}}$ scheme. It is also possible to adjust renormalization scheme so that all the higher-order terms are absorbed in the definition of g and are absent in the perturbative expansion [Gru 80].

Another possible criterion for selecting a renormalization scheme is the one proposed by Stevenson [Ste 81, 81a]. In this criterion it is required that an ideal truncated perturbation series $[f(g)]_N$ has to share the property possessed by the full series $f(g)$, i.e., the property of renormalization-scheme independence. In other words the best renormalization scheme is determined by the requirement that $[f(g)]_N$ be least sensitive to a variation of renormalization schemes, i.e.,

$$\frac{\partial[f(g)]_N}{\partial(RS)} = 0 \ , \tag{5.3.10}$$

where by (RS) we mean parameters which label renormalization schemes. The best renormalization scheme determined by the Stevenson's criterion (5.3.10) may, in general, vary from process to process, e.g., the best scheme for e^+e^- annihilation processes may differ from the one for deep inelastic scatterings. Moreover the best scheme in the Stevenson's criterion does not necessarily correspond to the scheme obtained by the criterion of the fastest convergence mentioned previously.

As we discussed in Sec 3.1.1 the renormalization ambiguity is due to the arbitrariness of the renormalization scale and the renormalization scheme. While the renormalization scale is well-parametrized by the mass scale μ, the labeling of renormalization schemes is not a simple matter. A possible way of labeling renormalization schemes [Ste 81a, Abe 82, Pet 82] is to use the expansion parameters of the β function $\beta_2, \beta_3, \beta_4, \ldots$, where the β_i's are defined through

$$\beta(g) = -\beta_0 g^3 - \beta_1 g^5 - \beta_2 g^7 + \ldots \ , \tag{5.3.11}$$

and we assume that β_0 and β_1 are renormalization-scheme-independent as will be shown in the next subsection. If we choose $\beta_2, \beta_3, \ldots$ as the scheme parameters, we may write down an explicit expression for the differentiation in Eq.(5.3.10),

$$\frac{\partial}{\partial(RS)} = \frac{\partial}{\partial\beta_i} \ , \quad i = 2, 3, \ldots, N \ . \tag{5.3.12}$$

5.3.2. β function and anomalous dimensions

The renormalization group functions, i.e. the β *function* and the *anomalous dimensions*, are not directly measurable quantities and are renormalization scheme dependent. In perturbation theory, however, we can show that the first two terms of the expansion of the β function and the first term for the anomalous dimensions are scheme independent [Gro 76, Sch 79, Vla 79a]. To see this we define the β function and anomalous dimensions in two different schemes,

$$\beta(g) = \mu \frac{dg}{d\mu}\,, \qquad \beta'(g') = \mu \frac{dg'}{d\mu}\,, \tag{5.3.13}$$

$$\gamma(g) = \mu \frac{d}{d\mu} \ln Z(g,\mu)\,, \quad \gamma'(g') = \mu \frac{d}{d\mu} \ln Z'(g',\mu), \tag{5.3.14}$$

where $\gamma(g)$ is an anomalous dimension either of a local field or of a composite operator with $Z(g, \mu)$ the renormalization constant of the local field or the composite operator. Noting Eq.(5.3.6) and writing

$$\beta(g) = -\beta_0 g^3 - \beta_1 g^5 - \beta_2 g^7 + \ldots\,, \tag{5.3.15}$$

we find

$$\beta'(g') = \beta(g)/(dg/dg')\,, \tag{5.3.16}$$

$$= -\beta_0 g'^3 - \beta_1 g'^5 - [(3z_0^2 - 2z_1)\beta_0 + 2z_0\beta_1]g'^7 + \ldots \tag{5.3.17}$$

Hence we realize that the first two coefficients β_0 and β_1 are scheme independent.

The renormalization constants $Z(g, \mu)$ and $Z'(g', \mu)$ may be related through the finite renormalization $\zeta(g')$ in such a way that

$$Z(g,\mu) = \zeta(g')\, Z'(g',\mu)\,. \tag{5.3.18}$$

Inserting Eq.(5.3.18) in Eq.(5.3.14) we obtain

$$\gamma'(g') = \gamma(g) - \beta'(g') \frac{d}{dg'} \ln \zeta(g')\,. \tag{5.3.19}$$

We assume the following perturbative expansions for $\gamma(g)$ and $\zeta(g')$,

$$\gamma(g) = \gamma_0 g^2 + \gamma_1 g^4 + \ldots\,, \tag{5.3.20}$$

$$\zeta(g') = 1 + \zeta_0 g'^2 + \zeta_1 g'^4 + \ldots \tag{5.3.21}$$

Using Eqs.(5.3.6), (5.3.17), (5.3.20) and (5.3.21) in Eq.(5.3.19) we find

$$\gamma'(g') = \gamma_0 g'^2 + (\gamma_1 + 2z_0\gamma_0 + 2\zeta_0\beta_0)g'^4 + \dots \tag{5.3.22}$$

Thus we notice that only the first coefficient γ_0 is scheme-independent.

As may be seen from the relations (5.3.17) and (5.3.19) the following properties are scheme-independent [Gro 76] and so are physically meaningful.

1. The existence of a zero of $\beta(g)$ and the sign of $d\beta/dg$ at zero, i.e., the appearance of a fixed point and its stability property (infrared or ultraviolet stable).
2. The value of $\gamma(g)$ at the fixed point.

5.3.3. e^+e^- annihilations and deep inelastic scatterings

The e^+e^- annihilation total cross section is given by Eq.(4.1.56) and is related to the R-ratio defined by Eq.(5.1.1) through Eq.(4.1.59). The R-ratio has already been given in the MS and $\overline{\text{MS}}$ scheme up to order $\bar{\alpha}_S^2$ in Sec. 5.1.3. The perturbative expansion of R as given in Eq.(5.1.44) is of the form (5.3.3). As we have discussed in Sec 5.3.1, the first two coefficients f_0 and f_1 in Eq.(5.3.3) are scheme independent. Hence we realize that the first two coefficients in Eq.(5.1.44) have to be independent of the renormalization scheme. It is the coefficient A in Eq.(5.1.44) which depends on the renormalization scheme adopted. We have already seen in Eq.(5.1.49) that the $\overline{\text{MS}}$ scheme offers the better apparent convergence of the perturbation series (5.1.44) than the MS scheme. The MOM scheme in the Landau gauge provides us with the coefficient [Cel 80a]

$$A = -2.19 + 0.16N_f \quad (\text{MOM}) \ . \tag{5.3.23}$$

Comparing this value with the others in Eq.(5.1.49), we see that the criterion of the faster apparent convergence seems to favor the MOM scheme. It should be noted that the optimization method of Stevenson [Ste 81a] works well also in the discussion of e^+e^- annihilation cross sections.

In deep inelastic scatterings we consider the Q^2 dependence of the moments of the structure functions. The relevant quantities calculable in perturbative QCD are coefficient functions $\tilde{C}_n(Q^2)$. In calculating $\tilde{C}_n(Q^2)$ we are required to perform an extra renormalization in addition to the ordinary renormalization with respect to the coupling constant, mass and fields. It is the renormalization of the composite operators appearing in the operator-product expansion of current products. This feature is particularly unique to the discussion of deep inelastic scattering. We have here a new type of the renormalization-scheme dependence originating in the composite-operator renormalization. The renormalization-scheme dependence of this type leads to an arbitrariness in the definition of operator matrix elements or equivalently of the coefficient functions $\tilde{C}_n(Q^2)$. It is sometimes called the *factorization-scheme dependence* [Pol 82, Cel 82, Nak 83] since this scheme dependence occurs in connection with the factorization property associated with the operator-product expansion. In the parton language the operator matrix element is interpreted as a parton distribution function [Kod 78, Bau 78] and hence we realize that the factorization-scheme dependence gives rise to an arbitrariness in the definition of the parton distribution function.

The phenomenological importance of the *renormalization-scheme dependence* of physical predictions in perturbation theory was first noticed by Bace [Bac 78] in connection with the determination of the scale parameter Λ and was also recognized by Bardeen et al. [Bar 78] in the full QCD analysis of deep inelastic structure functions. This observation on the scheme dependence was reformulated in a more transparent way by Celmaster and Gonsalves [Cel 79a].

As is suggested by Eq.(5.2.35) the general form of the nonsinglet coefficient function $\tilde{C}_n(Q^2)$ in the renormalization-group-improved perturbation is given by

$$\tilde{C}_n(Q^2) = (\bar{\alpha}_S)^{d_n}(r_n^0 + r_n^1\bar{\alpha}_S + r_n^2\bar{\alpha}_S^2 + \ldots) \ , \tag{5.3.24}$$

where $\bar{\alpha}_S = \bar{g}^2/(4\pi)$, $d_n = \gamma_{0n}/(2\beta_0)$, and r_n^0, r_n^1, ... are given in terms of β_i, γ_{in} and c_{in} with $i = 0, 1, 2, \ldots$ By changing the renormalization scheme to the new one marked by the prime, we have the new coupling constant g' as given in Eq.(5.3.6). The coupling parameter $\bar{\alpha}_S$ in the new scheme will be denoted by $\bar{\alpha}'_S$ and the relation corresponding to Eq.(5.3.6) is written as

$$\bar{\alpha}_S = (1 + a_0\bar{\alpha}'_S + a_1\bar{\alpha}'^2_S + \ldots)\,\bar{\alpha}'_S \ . \tag{5.3.25}$$

Substituting Eq.(5.3.25) into Eq.(5.3.24) we have

$$\tilde{C}_n(Q^2) = (\bar{\alpha}'_S)^{d_n}[r_n^0 + (r_n^1 + d_n a_0 r_n^0)\bar{\alpha}'_S + \ldots] \ . \tag{5.3.26}$$

From Eq.(5.3.26) we conclude that r_n^0 is scheme independent while r_n^1 is scheme dependent.

If our argument is restricted to the first nonleading order, the change of the scheme expressed by Eq.(5.3.25) may be restated in terms of the redefinition of the scale parameter Λ,

$$\Lambda = k\Lambda' \ , \tag{5.3.27}$$

with k a numerical constant. In fact, inserting Eq.(5.3.27) into Eq.(3.4.58), we find

$$\bar{\alpha}_S = \frac{1}{4\pi\beta_0 \ln(Q^2/\Lambda'^2)}\left[1 + \frac{\ln k^2}{\ln(Q^2/\Lambda'^2)} + \ldots\right]$$
$$= \bar{\alpha}'_S\,[1 + \bar{\alpha}'_S\,(4\pi\beta_0)\ln k^2 + \ldots] \;. \tag{5.3.28}$$

We compare Eq.(5.3.28) with Eq.(5.3.25) and find

$$a_0 = 4\pi\beta_0 \ln k^2 \;. \tag{5.3.29}$$

It should be noted that beyond the first order the parameters $a_1, a_2, \ldots$ cannot be expressed only in terms of k and the scheme parameters should be taken into account. Applying the change of the scale (5.3.27) to Eq.(5.2.38), we easily find that

$$\tilde{C}_n(Q^2) = N_n\left(\ln\frac{Q^2}{\Lambda'^2}\right)^{-d_n}\left[c_{0n} + \left(\beta_0 \ln\frac{Q^2}{\Lambda'^2}\right)^{-1}(c_{1n} - c_{0n}\gamma_{0n}\ln k + \ldots) + \ldots\right] . \tag{5.3.30}$$

Hence we recognize that c_{0n} is scheme independent while c_{1n} is scheme dependent. The scheme dependence of c_{1n} is characterized by an amount $-c_{0n}\gamma_{0n}\ln k$ due to the scale change.

As an example we consider the case of the $\overline{\text{MS}}$ and MS schemes. We subtract only the pole $1/\varepsilon$ in the MS scheme while in the $\overline{\text{MS}}$ scheme we subtract the combination $1/\varepsilon - \gamma + \ln(4\pi)$. These two schemes give rise to the two different definitions of the renormalization constant for the coupling constant g. In the one-loop order, these renormalization constants read

$$Z_{\text{MS}} = 1 - Ag_{\text{MS}}^2 \frac{1}{\varepsilon} \;,$$
$$Z_{\overline{\text{MS}}} = 1 - Ag_{\overline{\text{MS}}}^2 \left(\frac{1}{\varepsilon} - \gamma + \ln(4\pi)\right) , \tag{5.3.31}$$

where the constant A is a known constant appearing in Eq.(3.4.5) and $g_{\text{MS}}(g_{\overline{\text{MS}}})$ is the renormalized coupling constant defined by Eq.(3.2.23). According to Eq.(3.2.23), Eq.(5.3.31) may be rewritten to one-loop order such that

$$Z_{\text{MS}} = 1 - Ag_0^2\left(\frac{1}{\varepsilon} + \ln\frac{\mu_0^2}{\mu_{\text{MS}}^2}\right)$$
$$Z_{\overline{\text{MS}}} = 1 - Ag_0^2\left(\frac{1}{\varepsilon} + \ln\frac{\mu_0^2}{\mu_{\overline{\text{MS}}}^2} - \gamma + \ln(4\pi)\right) . \tag{5.3.32}$$

We realize in Eq.(5.3.32) that up to one-loop order the MS and $\overline{\text{MS}}$ scheme are related to each other by changing the renormalization scale in such a way that

$$\mu_{\overline{\text{MS}}}^2 = 4\pi e^{-\gamma}\mu_{\text{MS}}^2 \ . \tag{5.3.33}$$

The above relation together with the definition of the scale parameter (3.4.57) leads us to the relation

$$\Lambda_{\overline{\text{MS}}}^2 = 4\pi e^{-\gamma}\,\Lambda_{\text{MS}}^2 \ . \tag{5.3.34}$$

Since $4\pi\exp(-\gamma) = 7.056\ldots$, we have

$$\Lambda_{\overline{MS}} = 2.66\Lambda_{\text{MS}} \ . \tag{5.3.35}$$

Comparing Eq.(5.3.35) with Eq.(5.3.27) and applying Eq.(5.3.28) to the present example we find

$$\bar{\alpha}_S^{\overline{\text{MS}}} = \bar{\alpha}_S^{\text{MS}}\,(1 + a_0\,\bar{\alpha}_S^{\text{MS}} + \ldots) \ , \tag{5.3.36}$$

$$a_0 = 4\pi\beta_0\,[\ln(4\pi) - \gamma] \ . \tag{5.3.37}$$

5.4. JETS

5.4.1. $q\bar{q}$ jets in e^+e^- annihilations

We shall show in the present section that hadronic jet phenomena are dominated by short-distance effects so that perturbative QCD may be safely applied to the discussion of jet processes. For this purpose we present the proof of the cancellation of the infrared divergences in the jet cross sections since the infrared divergence reflects the long-distance nature of QCD.

Here we confine ourselves to hadronic jets arising from the quark-antiquark pair production in e^+e^- annihilations. In order that the hadronic jets follow from the quark-antiquark pair, the quark and antiquark are required not to loose too much energy in their direction by the emission of gluons and quark pairs. In other words it is necessary to show that most of the annihilation energy is deposited along the direction of the quark and antiquark. The first step of the argument is to give a precise definition of the jet cross section and then to prove that the dangerous infrared (soft and collinear) divergences cancel out in the cross section guaranteeing that the QCD correction is controllable in its size. The jet generated by the above QCD mechanism is often called the *Sterman-Weinberg jet* [Ste 77] or the *QCD jet*.

We have calculated in Sec. 5.1 the total cross section of e^+e^- annihilations to which the QCD diagrams shown in Fig. 5.1.1 contribute. The contribution up to order g^2 was represented in Fig. 5.1.2. Here in the present section we discuss the jet contribution up to the same order as above. (For a review see, e.g. [Dok 80, Kon 80, Web 82, Hag 83, Kra 84].) The diagrams contributing to jets are the same as in Fig. 5.1.2 though the kinematical region in the final state is restricted. We define the two-jet event in such a way that most of the available energy $\sqrt{s}$, i.e., $(1 - \Delta)\sqrt{s}$ with $\Delta \ll 1$, is deposited on two cones of small half angle δ(see Fig. 5.4.1). We consider the angular distribution of the final quarks, i.e., the differential cross section $d\sigma/d\Omega$ in the solid angle Ω specified by polar and azimuthal angles, θ and ϕ, respectively. The Born contribution $(d\sigma/d\Omega)_B$ corresponding to Fig. 5.1.2 (a) is given according to Eq.(2.3.159) by

$$\left(\frac{d\sigma}{d\Omega}\right)_B = \frac{\alpha^2}{4s} \sum_i Q_i^2 (1 + \cos^2\theta) \ . \tag{5.4.1}$$

The virtual (one-loop) contribution $(d\sigma/d\Omega)_V$ corresponding to Fig. 5.1.2 (b) and (c) is directly obtained in parallel with the previous argument in deriving Eq.(5.1.17),

$$\left(\frac{d\sigma}{d\Omega}\right)_V = A_V \left(\frac{d\sigma}{d\Omega}\right)_B \ , \tag{5.4.2}$$

where A_V is given by Eq.(5.1.18).

Now comes the real-gluon-emission contribution $(d\sigma/d\Omega)_R$ corresponding to Fig. 5.1.2(d) which occupies the major part of the argument in the rest of the present subsection. The differential cross section $(d\sigma/d\Omega)_R$ is defined in a similar way as in Eq.(5.1.20) by

$$\int \left(\frac{d\sigma}{d\Omega}\right)_R d\Omega = \frac{1}{8s} \int_R \prod_{i=1}^{3} \frac{d^{D-1}k_i}{(2\pi)^{D-1} 2k_{i0}} (2\pi)^D \delta^D \left(\sum_{i=1}^{3} k_i - p_1 - p_2\right) F_R \ , \tag{5.4.3}$$

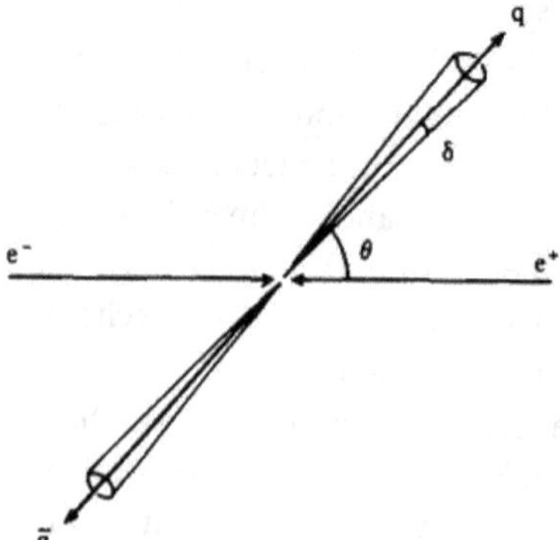

Fig. 5.4.1. The sharp cones necessary to define two-jet events.

where F_R is given by Eq.(5.1.21) and the integral region R is specified by the following conditions: The emitted gluon is either soft (i.e. $x_3 \leq \Delta$) or collinear to one of the quarks (i.e., $\theta_{13}, \theta_{23} < 2\delta$) where x_3 is defined by Eq.(5.1.31) and $\theta_{13}\,(\theta_{23})$ is the angle between the gluon and the quark (antiquark) as shown in Fig. 5.4.2. It should be noted here that the restriction of the phase space to R in Eq.(5.4.3) corresponds to taking the degenerate state of the quark and gluon which will be introduced in Chap. 6. According to the general theorem explained there, the Kinoshita-Lee-Nauenberg theorem, the infrared divergences (soft and collinear) are known to cancel out in the following sum,

$$\frac{d\sigma}{d\Omega} = \left(\frac{d\sigma}{d\Omega}\right)_B + \left(\frac{d\sigma}{d\Omega}\right)_V + \left(\frac{d\sigma}{d\Omega}\right)_R . \tag{5.4.4}$$

It is convenient to define new variables ζ_1 and ζ_2 through

$$\zeta_1 = \frac{1}{2}(1 - \cos\theta_{13}) = \frac{1 - x_2}{x_1 x_3} , \tag{5.4.5}$$

$$\zeta_2 = \frac{1}{2}(1 - \cos\theta_{23}) = \frac{1 - x_1}{x_2 x_3} , \tag{5.4.6}$$

where use has been made of the relation

$$\begin{aligned}(k_1 + k_3)^2 &= \frac{1}{4} x_1 x_3 s(1 - \cos\theta_{13}) \\ &= (q - k_2)^2 = s(1 - x_2) \text{ , etc.}\end{aligned} \tag{5.4.7}$$

Since $\theta_{13}, \theta_{23} \leq 2\delta$ in R, we have

$$\zeta_1, \zeta_2 \leq \sin^2\delta . \tag{5.4.8}$$

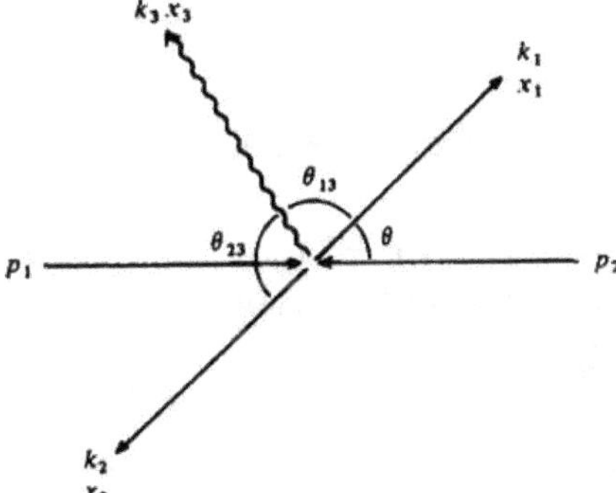

Fig. 5.4.2. The schematic view of the $e^+e^- \to q\bar{q}G$ event where p_1, p_2, k_1, k_2 and k_3 are the momentum of the electron, position, quark, antiquark and gluon respectively and x_1, x_2 and x_3 are energy fractions defined by Eq.(5.1.31). The variable $\theta_{13}(\theta_{23})$ is the angle between the directions of the emitted gluon and quark (antiquark).

Only two variables are independent among five variables $x_1, x_2, x_3, \zeta_1, \zeta_2$ and so we choose ζ_1 and x_3 as independent variables. Then the region R is specified by

$$\frac{1-\sin^2\delta}{1-x_3(2-x_3)\sin^2\delta} \le \zeta_1 \le \sin^2\delta, \qquad 0 \le x_3 \le \Delta\,, \tag{5.4.9}$$

where the lower bound of ζ_1 comes from the condition $\zeta_2 < \sin^2\delta$. In Fig. 5.4.3 the kinematical region R given by Eq. (5.4.9) is shown in the $\zeta_1 - x_3$ plane.

Neglecting terms of order δ^2 we express $(d\sigma/d\Omega)_R$ in the form of an angular distribution with respect to the quark direction,

$$\left(\frac{d\sigma}{d\Omega}\right)_R = \frac{3}{16\pi}(1+\cos^2\theta)\alpha^2\alpha_s C_F\left(\sum_i Q_i^2\right)\frac{2}{s}\left(\frac{4\pi\mu}{s}\right)^{2\varepsilon}$$

$$\times\frac{(1-\varepsilon)^2}{(3-2\varepsilon)\Gamma(2-2\varepsilon)}K_R\,, \tag{5.4.10}$$

where

$$K_R = \int_R\left[\prod_{i=1}^{3}(1-x_i)^{-\varepsilon}dx_i\right]\delta\left(2-\sum_{i=1}^{3}x_i\right)\rho\,, \tag{5.4.11}$$

$$\rho = \frac{x_1^2+x_2^2-\varepsilon x_3^2}{(1-x_1)(1-x_2)}\,. \tag{5.4.12}$$

Note that the slight deviation from the above angular distribution is expected if the precise calculation is made [Hag 81]. Probably the easiest way of calculating K_R in Eq.(5.4.11) may be the following: We first note that

$$K_R = K - K_{\bar{R}}\,, \tag{5.4.13}$$

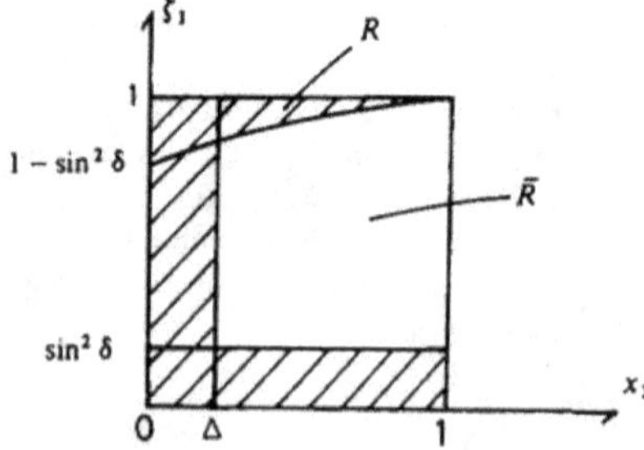

Fig. 5.4.3. The kinematical regions R and $\bar{R}$ in the $\zeta_1 - x_3$ plane.

where K is the quantity corresponding to K_R integrated over the whole phase space and is already given in Eq.(5.1.35), and $K_{\bar{R}}$ is obtained by integrating the integrand of Eq.(5.4.11) over the region $\bar{R}$ which is obtained by eliminating R from the whole phase space as shown in Fig.5.4.3. Since there is no infrared singularity in the region $\bar{R}$, we may put $\varepsilon = 0$ in $K_{\bar{R}}$ and then the calculation turns out to be straightforward. Changing the variables to ζ_1 and x_3 we obtain

$$K_{\bar{R}} = \int_{\Delta}^{1} dx_3 x_3(1 - x_3) \int_{\zeta_{\min}}^{\zeta_{\max}} d\zeta_1 \frac{\rho}{(1 - x_3\zeta_1)^2} , \tag{5.4.14}$$

where $\zeta_{\max}$ and $\zeta_{\min}$ are the maximum and minimum value of ζ_1 given in Eq.(5.4.9), and ρ defined in Eq.(5.4.12) is rewritten in terms of ζ_1 and x_3 as

$$\rho = \frac{(1 - x_3)^2 + (1 - x_3(2 - x_3)\zeta_1)^2}{x_3^2(1 - x_3)\zeta_1(1 - \zeta_1)} . \tag{5.4.15}$$

After some calculation we find

$$K_{\bar{R}} = 2\left(4\ln\delta \ln\Delta + 3\ln\delta - \frac{7}{4} + \frac{\pi^2}{3}\right) + O(\delta, \Delta) . \tag{5.4.16}$$

Hence we have

$$K_R = 4B(1-\varepsilon, 2-2\varepsilon)\, B(1-\varepsilon, 1-\varepsilon)$$
$$\times \left(\frac{1}{\varepsilon^2} - \frac{3}{\varepsilon} + \frac{17}{4} - \frac{\pi^2}{3} - 4\ln\delta \ln\Delta - 3\ln\delta\right) . \tag{5.4.17}$$

Exactly in the same way as in Eq.(5.1.37) we calculate the Born contribution to $d\sigma/d\Omega$ in D dimensions which results in

$$\left(\frac{d\sigma}{d\Omega}\right)_B = \frac{\alpha^2}{4s} \sum_i Q_i^2 (1 + \cos^2\theta) \left(\frac{4\pi}{s}\right)^{\varepsilon} \frac{3(1 - \varepsilon)\Gamma(2 - \varepsilon)}{(3 - 2\varepsilon)\Gamma(2 - 2\varepsilon)} . \tag{5.4.18}$$

Using Eqs.(5.4.10), (5.4.17) and (5.4.18) we finally obtain

$$\left(\frac{d\sigma}{d\Omega}\right)_R = \left(\frac{d\sigma}{d\Omega}\right)_B \frac{\alpha_s}{\pi} C_F \left(\frac{4\pi\mu^2}{s}\right)^{\varepsilon}$$
$$\times \frac{\cos\pi\varepsilon}{\Gamma(1 - \varepsilon)} \left(\frac{1}{\varepsilon^2} + \frac{3}{2\varepsilon} + \frac{13}{2} - \frac{\pi^2}{3} - 4\ln\delta \ln\Delta - 3\ln\delta\right) . \tag{5.4.19}$$

Combining Eqs.(5.4.1), (5.4.2) and (5.4.19) together we obviously see that the infrared divergences just cancel out, and find

$$\frac{d\sigma}{d\Omega} = \left(\frac{d\sigma}{d\Omega}\right)_B \left[1 - \frac{\alpha_s}{\pi} C_F\left(4\ln\delta\ \ln\Delta + 3\ln\delta + \frac{\pi^2}{3} - \frac{5}{2}\right)\right] . \tag{5.4.20}$$

According to Eq.(5.4.20) we realize that the order-α_S correction to the two jets from the quark pair is controllable in size within the framework of perturbation theory and the hadronic two jets approximately in the directions of the quark and antiquark are expected with the jet angular distribution being the same as that of the quark and antiquark. Thus the angular distribution of the hadronic two jets reflects the spin-1/2 nature of the constituents.

A similar argument as above may be made for hadronic jets originating from gluon sources [Shi 78, 79, Ein 78]. It has been shown that the same infrared cancellation as in the quark jets takes place in the gluon jets. Hence the hadronic jet from the gluon should also be observed experimentally. In fact clear signals of three jets from the quark, antiquark and gluon have been observed in $e^+ e^-$ annihilation processes [Bar 87].

5.4.2. Higher order effects

In the previous subsection we considered the Sterman-Weinberg $q\bar{q}$ jet only up to order α_S. The two-jet cross section to this order is obviously renormalization-scheme independent. In order to see the convergence property of the perturbation series for the jet cross section it is necessary to compute at least a term of order α_S^2 for the $q\bar{q}$ jet, i.e., the two-loop correction to the amplitude with the $q\bar{q}$ final state, the one-loop correction to the amplitude with the $q\bar{q}G$ final state and the tree contribution to the amplitude with the $q\bar{q}GG$ and $q\bar{q}q\bar{q}$ final state. Such calculation was performed in [Lam 85] and it was confirmed that the order-α_S^2 correction to the two-jet cross section is sufficiently small in the $\overline{\text{MS}}$ scheme so that the fast convergence of the perturbation series is guaranteed.

Another important example of the higher-order Sterman-Weinberg jets is the $q\bar{q}G$ three jets. In this process even the tree contribution to the jet cross section is of order α_S. Hence, to discuss feasibility of the $q\bar{q}G$ three jets, we have to consider the order-α_S^2 correction to the tree contribution, i.e., the one-loop correction to the amplitude with the $q\bar{q}G$ final state and the tree contribution to the amplitude with the $q\bar{q}GG$ and $q\bar{q}q\bar{q}$ final state. The calculation of this type has been carried out by several groups [Ell 80, 81, Fab 80, 82, Ver 81, Gut 84]. (See also [Ali 79, Kun 81].) It is found that the order-α_S^2 term is infrared-finite but large in its magnitude in the $\overline{\text{MS}}$ scheme. This means that we are required to make the resummation of the large terms to all orders.

It is worth mentioning here that an estimate of higher order effects in perturbative QCD just as the above Sterman-Weinberg jets usually requires an enormous amount of large-scale calculations. In particular the combinatorics due to the gauge couplings and the trace operation for the Dirac matrices make the calculation very cumbersome and tedious. Under this circumstance it seems wise to adopt computer-assisted manipulation. There exist some useful algebraic manipulation programs for the above purpose. The most commonly used in high energy physics are REDUCE developed by Hearn [Hea 85] and SCHOONSCHIP provided by Veltman and revised by Strubbe [Str 74]. Since algebraic parts of higher order calculations are routine work with well-defined computational rules, the algebraic manipulation programs as above may be essential in pushing the perturbative calculation to yet higher orders.

In the present section we confined ourselves to the jets in e^+e^- annihilations. It is, however, possible to generalize the Sterman-Weinberg argument to other processes such as e-N scatterings and hadron collisions. We shall not discuss these processes in this book.

5.4.3. Experimental observation

According to our QCD analyses given in Section 5.4.1 and 5.4.2 we conclude that the clear signal of hadronic two and three jets stemming from the quark-antiquark pair and the quark-antiquark-gluon system, respectively, is expected to be observed experimentally. We, however, have not shown how the hadronic jets are generated in the direction of the original quarks and gluons. The formation of hadron clusters as fragments of the quark and gluon is obviously a nonperturbative phenomenon and is beyond the scope of perturbative QCD. It is thus necessary to find a phenomenological way of describing the process in which the quarks and gluons fragment into hadrons. There are many possible models to parametrize the nonperturbative hadronization process of the quarks and gluons. Hence description of hadronic jets following from the original quarks and gluons is model dependent. Typical examples of the hadronization models are [Yam 85]:

1. The independent fragmentation model [Hoy 79, Ali 80] in which each parton (quark or gluon) is assumed to fragment independently into hadrons on the basis of the mechanism adopted by Field and Feynman [Fie 78].

2. The string-fragmentation model [Sjo 82, And 83] where in the hadronization process strings are assumed to stretch between partons in the direction of color flow according to the string picture of quark confinement. Most of the hadrons produced in a jet are subject to subsequent decays. In any

model the formation of the hadronic jet is regarded as a stochastic branching process originating from the quark or gluon which is originally produced by the perturbative QCD mechanism. The whole process of forming the hadron jets may be simulated by Monte Carlo programs using computers.

We show in Fig. 5.4.4 examples of the hadronic two and three jets generated by one of such Monte Carlo methods. They were obtained by the VENUS off-line group at TRISTAN (electron-positron collider in KEK, Ko-Enerugi-Kenkyujo) using the program EPOCS [Kat 84]. Here total incident center-of-mass energy of the electron and positron is set equal to 29 GeV for the two jets and 30 GeV for the three jets.

Experimentally such hadronic jets in e^+e^- annihilations have been observed at SLAC and DESY. The first observation of the two jets was reported in 1975 at SLAC [Han 75, 82] while the three jets were confirmed in 1979 at DESY [Bar 79d, Ber 79, Bra 79, Bar 80]. Shown in Fig. 5.4.5 are typical examples of the two and three jets observed by the JADE group at DESY ($\sqrt{s} = 29$ GeV for the two jets and $\sqrt{s} = 31$ GeV for the three jets). Here trajectories of particles in the figure are reconstructed from the computer memory of counter detection signals where the full lines represent charged particles and the dotted ones indicate photons. Obviously we recognize that the above jets are quite similar in their appearance to the Monte Carlo simulation data of the QCD hadronic jets previously shown in Fig. 5.4.4.

Further detailed and more quantitative analyses of jets may be carried out with the use of physically measurable quantities characteristic to jets such as thrust [Far 77, de R 78], sphericity [Bj 70], spherocity [Geo 77], energy correlation [Bas 78, 78a, 79] and so on. In this book we shall not go into detail regarding analyses of experimental data in terms of these quantities. We, however, mention that the measurements of the angular distribution of the jet axis in the two-jet events are in accord with the $1 + \cos^2\theta$ behavior given by Eqs. (5.4.20) and (5.4.1) suggesting the spin-1/2 nature of the fundamental constituents. (For a review, see [Bar 87].)

5.5. FACTORIZATION AND THE DRELL-YAN PROCESS

The e^+e^- annihilations and e^+e^- inelastic scatterings in the lowest order of electromagnetic interactions proceed through the one-photon and two-photon exchanges, respectively, as depicted in Fig. 5.5.1. The total cross sections for these processes are related to the imaginary part of the photon propagator and of the two-photon scattering amplitude[9] respectively. In other words these processes are described by the subprocesses where only

[9] We do not discuss the two-photon process in this book though it is one of the important QCD laboratories. For a review see, e.g., [Sas 83, Bar 85].

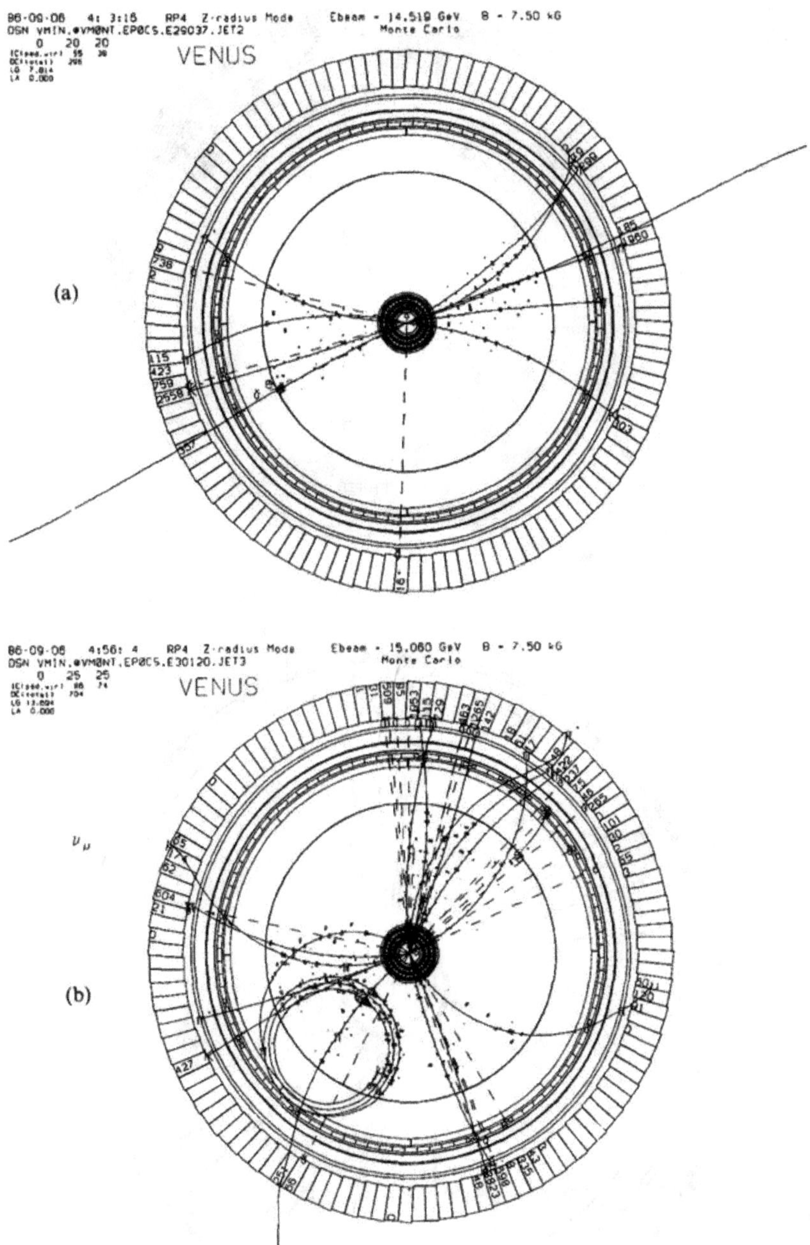

Fig. 5.4.4. The hadronic two (a) and three (b) jets generated by the Monte Carlo simulation starting with the QCD results on the $q\bar{q}$ and $q\bar{q}G$ cross sections respectively. By courtesy of the VENUS off-line group at TRISTAN. The total incident energies are chosen to be $\sqrt{s}$ = 29 GeV for (a) and $\sqrt{s}$ = 30 GeV for (b).

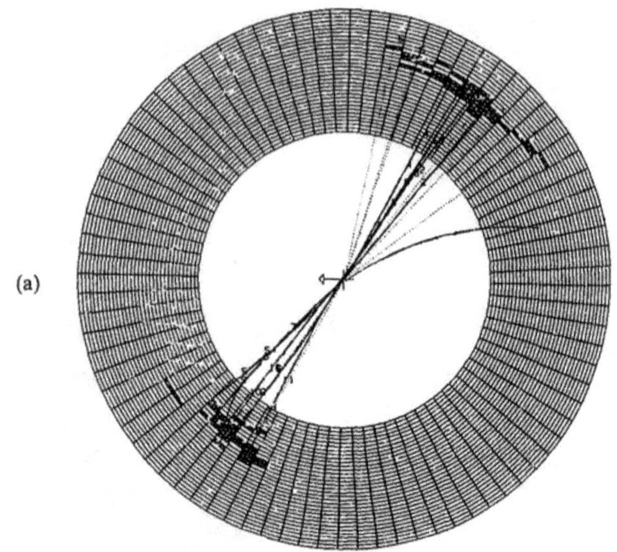

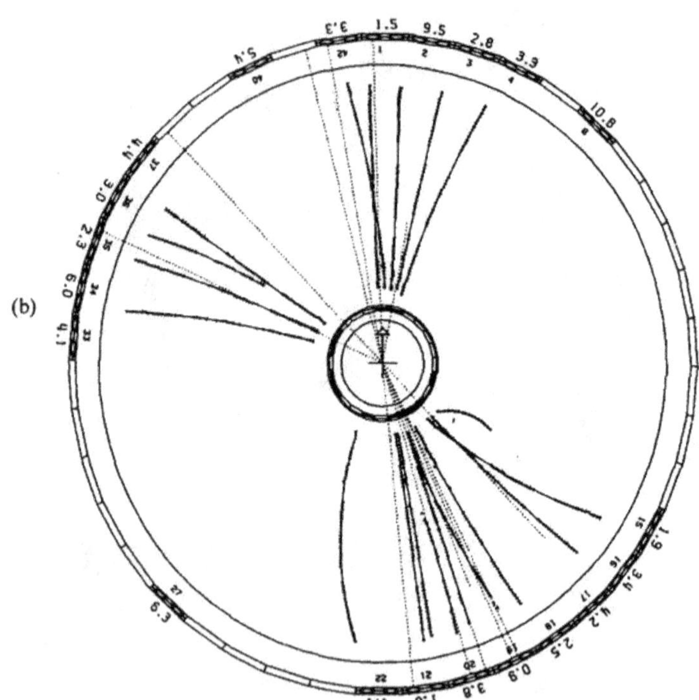

Fig. 5.4.5. Experimental data of two (a) and three (b) jets obtained by the JADE group at DESY where $\sqrt{s} = 29$ GeV for (a) and $\sqrt{s} = 31$ for (b).

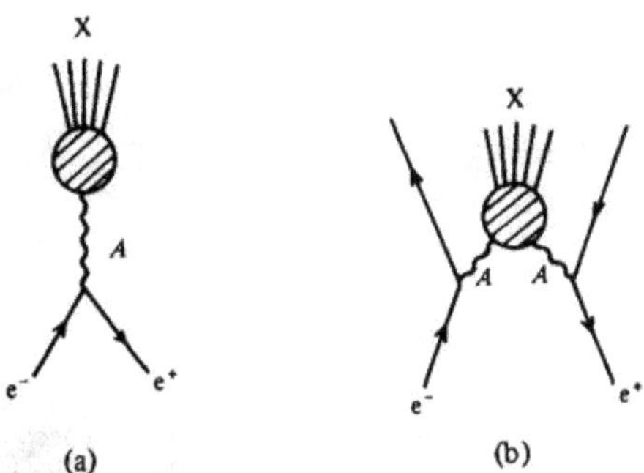

Fig. 5.5.1. The e^+e^- annihilation (a) and e^+e^- inelastic scattering (b) where wavy lines with A represent photons.

photons participate in the initial states (no hadrons in the initial states). No complication associated with the hadron structure comes about in these processes. Hence at high energies the above cross sections are purely of short-distance nature and are by themselves amenable to perturbative calculations in QCD.

The deep inelastic lepton-hadron scattering depicted in Fig. 5.5.2 is not of purely short-distance nature, but, thanks to the operator-product expansion, its cross section splits into a short-distance piece (coefficient function) and a long-distance piece (operator matrix element) as was discussed in Sec. 5.2. The short-distance piece is within the reach of perturbative QCD while the long-distance piece requires nonperturbative treatment and is out of the scope of perturbative QCD. It is usually determined in a phenomenological way. The above factorization of the cross section into two pieces may be expressed diagrammatically as in Fig.5.5.3. The subdiagram marked by $A_q(A_G)$ in Fig. 5.5.3 corresponds to the quark (gluon) operator matrix element which, in the parton language, is the quark (gluon) distribution function inside the target hadron. The subdiagram denoted by $C_q(C_G)$ corresponds to the coefficient which is the photon-quark (gluon) hard scattering cross section.

Let us turn our attention to the following reaction called the *Drell-Yan process* [Dre 70, 71],

$$N + N \to l^+ + l^- + X\ , \tag{5.5.1}$$

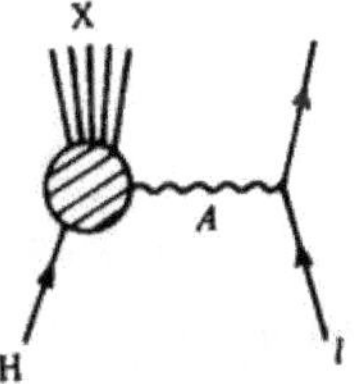

Fig. 5.5.2. The deep inelastic lepton-hadron scatterings.

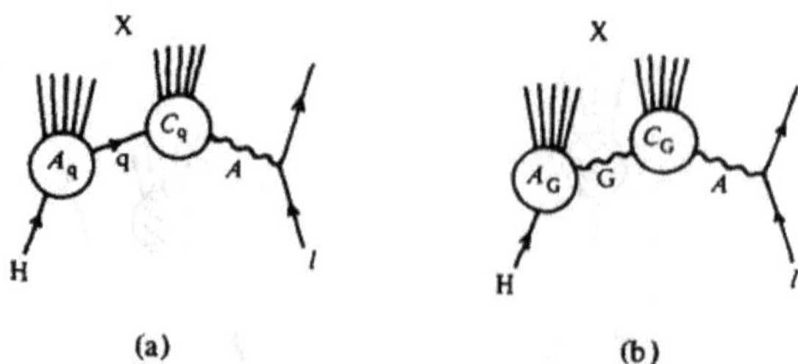

Fig. 5.5.3. The diagrammatic implication of the factorization in deep inelastic scatterings where $C_q(C_G)$ denotes the coefficient function for the quark (gluon) composite operator and $A_q(A_G)$ represents the quark (gluon) operator matrix element.

where N represents the nucleon, $l^{\pm}$ the charged lepton (usually the muon) and X the unobserved hadrons. The diagrammatic expression of the Drell-Yan process is given in Fig. 5.5.4. The cross section for this process is given in the lowest order of the electromagnetic interaction[10] by

$$\sigma = \frac{1}{2\sqrt{s(s-4M^2)}}\frac{1}{4}\sum_{\text{pol}}\int\frac{d^3k_1}{(2\pi)^3 2k_{10}}\frac{d^3k_2}{(2\pi)^3 2k_{20}}$$

$$\times \sum_{\mathrm{X}}(2\pi)^4\,\delta^4(k_1+k_2+p_{\mathrm{X}}-p_1-p_2)|\langle l^+\, l^-\mathrm{X}|T|\mathrm{NN}\rangle|^2\ , \quad (5.5.2)$$

where $s = (p_1 + p_2)^2$, M is the nucleon mass, the summation runs over all polarizations of the initial nucleons and the final leptons, and

$$\langle l^+ l^- \mathrm{X}|T|\mathrm{NN}\rangle = \bar{u}(k_1)e\gamma_\mu v(k_2)\frac{g^{\mu\nu}}{(k_1+k_2)^2}\langle \mathrm{X}|ej_\nu(0)|\mathrm{NN}\rangle\ , \qquad (5.5.3)$$

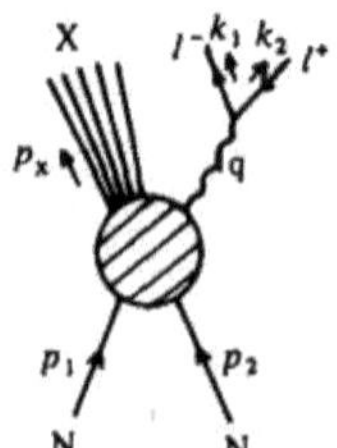

Fig. 5.5.4. The Drell-Yan process in the lowest order of the electromagnetic interaction.

[10] We neglect possible contributions of the Z boson to the cross section.

with $j_\mu(x)$ the electromagnetic current. Inserting Eq.(5.5.3) into Eq.(5.5.2) and performing the lepton polarization sum as in Eq.(4.1.25) we have

$$\sigma = \frac{2e^4}{\sqrt{s(s-4M^2)}} \int \frac{d^3k_1}{(2\pi)^3 2k_{10}} \frac{d^3k_2}{(2\pi)^3 2k_{20}} \frac{L^{\mu\nu} W_{\mu\nu}}{(k_1 + k_2)^4} , \tag{5.5.4}$$

where the lepton polarization tensor $L^{\mu\nu}$ is defined in a similar way as in Eq.(4.1.23) and is given by Eq.(4.1.25) with $k'(k)$ replaced by $k_1(k_2)$, and the hadronic tensor $W_{\mu\nu}$ is defined by

$$\begin{aligned} W_{\mu\nu} &= \sum_X (2\pi)^4 \delta^4(k_1 + k_2 + p_X - p_1 - p_2) \frac{1}{4} \sum_{\text{pol}} \langle \text{NN}|j_\mu(0)|X\rangle \langle X|j_\nu(0)|\text{NN}\rangle , \\ &= \int d^4x \, e^{-i(k_1+k_2)\cdot x} \langle p_1 p_2 | j_\mu(x) j_\nu(0) | p_1 p_2 \rangle , \end{aligned} \tag{5.5.5}$$

where the polarization sum runs only over the nucleon polarizations and we used the previously introduced notation of Eq.(4.1.27). After performing the integration in the relative momentum $k_1 - k_2$ in the lepton center-of-mass frame we obtain for the dilepton invariant mass (squared) distribution

$$\frac{d\sigma}{dq^2} = \frac{1}{\sqrt{s(s-4M^2)}} \frac{4\pi\alpha^2}{3q^2} W(\tau, q^2) , \tag{5.5.6}$$

where $\alpha = e^2/(4\pi)$, $\tau = q^2/s$ and

$$W(\tau, q^2) = \frac{1}{(2\pi)^4} \int d^4k \, \theta(k_0) \, \delta(k^2 - q^2) \, (-g^{\mu\nu} W_{\mu\nu}) , \tag{5.5.7}$$

with $k = k_1 + k_2$.

As seen in Eqs.(5.5.6) and (5.5.7) the Drell-Yan cross section is described by the hadronic tensor $W_{\mu\nu}$ defined in Eq.(5.5.5). This hadronic tensor differs from a similar one for deep inelastic scatterings (Eq.(4.1.28)) in the sense that the state sandwiching the current product $j_\mu(x) j_\nu(0)$ is not the one-nucleon state but the two-nucleon state. Hence the simple-minded application of the operator-product expansion of the form (4.2.76) does not lead to the separation of the short-distance piece from the hadronic tensor (5.5.5). It should, however, be noted that in our proof of the operator product expansion (Sec. 4.2.4) we essentially dealt with Green functions rather than operators and the factorization of the short-distance piece from the long-distance piece was actually proven for the Green functions. Thus it seems to be possible to prove the separation of the short-distance part in the hadronic tensor by simply generalizing the method employed in Sec. 4.2.4. That this is actually the case can be shown in a rigorous manner [Gup 79].

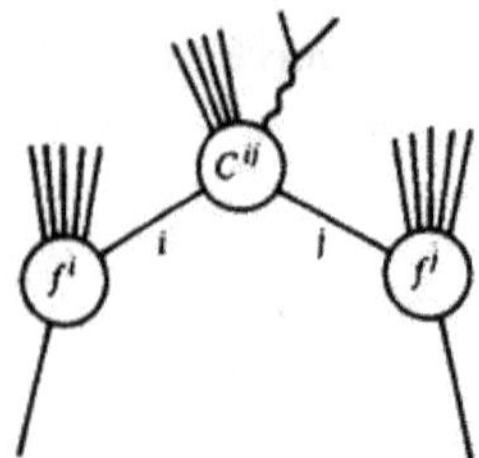

Fig. 5.5.5. The factorization of the Drell-Yan cross section corresponding to Eq.(5.5.8).

Motivated by the parton-model argument [Dre 70, 71] we are tempted to assume that $W(\tau, q^2)$ in Eq.(5.5.7) is decomposed into three factors,

$$W(\tau, q^2) = \sum_{i,j} \int_{\tau}^{1} \frac{d\xi_1}{\xi_1} f^i(\xi_1, q^2) \int_{\tau/\xi_1}^{1} \frac{d\xi_2}{\xi_2} f^j(\xi_2, q^2)\, C^{ij}(\tau/\xi_1\xi_2, \bar{g}) \ , \qquad (5.5.8)$$

where $f^i(\xi, q^2)$ is the parton distribution function inside the nucleon for parton i with momentum fraction ξ and $C^{ij}(\xi, q^2)$ the hard-scattering cross section of partons i and j. Here index i refers to the quark q, antiquark $\bar{q}$ and gluon G. Equation (5.5.8) implies that the Drell-Yan cross section is factorized into three parts as depicted in Fig. 5.5.5. That Eq.(5.5.8) is the factorization similar to the operator-product expansion can be seen more clearly [Gup 79] if we take the moment of Eq.(5.5.8) with respect to τ, i.e.,

$$W_n(q^2) = \sum_{i,j} f_n^i(q^2) f_n^j(q^2)\, C_n^{ij}(\bar{g}) \ , \qquad (5.5.9)$$

where $W_n(q^2), f_n^i(q^2)$ and $C_n^{ij}(\bar{g})$ are defined by

$$W_n(q^2) = \int_0^1 d\tau\, \tau^{n-1}\, W(\tau, q^2) \ ,$$

$$f_n^i(q^2) = \int_0^1 d\xi\, \xi^{n-1} f^i(\xi, q^2) \ ,$$

$$C_n^{ij}(\bar{g}) = \int_0^1 d\xi\, \xi^{n-1}\, C^{ij}(\xi, \bar{g}) \ . \qquad (5.5.10)$$

Equation (5.5.9) may be regarded as the counterpart of the operator-product expansion with $f_n^i(q)$ and $f_n^j(q^2)$ two operator matrix elements respectively and $C_n^{ij}(\bar{g})$ the coefficient function.

In fact the *factorization* of the form (5.5.9) has been proven in two different methods: the BPHZ-type method which is a simple generalization of the method used in Sec. 4.2.4 [Mue 78, 81, Gup 79] and the mass-singularity method which is based on the factorization of mass-singularities in the cross section (the mass singularity may spoil perturbative calculations) [Ell 78, 79, Ama 78, 78a, Lib 78]. The above proof is quite general so that it applies not only to the Drell-Yan process but also to other processes like $e^+e^- \to$ hadron + X and hadron + hadron → hadron (large p_T) + X. We shall not go into detail of the proof any further but give a brief historical survey on the proof.

The first suggestion on the factorization of the form (5.5.8) was made in [Pol 77, 77a] on the basis of the method of separating out mass singularities. There the proof of the factorization formula (5.5.8) was given in the lowest nontrivial (one-loop) order. The proof was extended to the two-loop order in [Sac 78]. On account of the large logs appearing in the above perturbative calculations it seems necessary to sum up these logs to all orders in order to make a reliable proof. The proof in this direction was promoted by many authors in the leading log order [Dok 78, 78a, Ame 78, Kaz 78, 79a, Lle 78, Fra 79, Efr 80]. It was then under immediate investigation to complete the proof in a rigorous manner to all orders. The complete proof based on the method of separating mass singularities was given in [Lib 78, Ama 78a, Ell 78, 79] and the one based on the straightforward generalization of the operator-product expansion was developed in [Mue 78, Gup 79]. The equivalence of these two methods is discussed in [Hum 81, 82].

Once the factorization of the form (5.5.8) (or (5.5.9)) is established, we are free to extract the short-distance piece $C^{ij}(\xi, \bar{g})$ (or $C_n^{ij}(\bar{g})$) which may be calculated perturbatively in QCD for large q^2. Just as in the case of deep inelastic lepton-hadron scatterings the coefficient function $C_n^{ij}(\bar{g})$ obeys the renormalization group equation and is computed in the framework of the renormalization-group-improved perturbation. In the leading order only the function $C_n^{ij}(\bar{g})$ with $(i, j) = (q, \bar{q})$ and $(\bar{q}, q)$ survives where q and $\bar{q}$ represent the quark and antiquark respectively. Hence the dominant contribution to the Drell-Yan cross section (5.5.6) comes from the $q\bar{q}$ annihilation diagram for $C_n^{ij}(\bar{g})$ as depicted in Fig. 5.5.6. In the next-to-leading order the gluons take part in the game and so the combinations $(i, j) = (q, \bar{q}), (\bar{q}, q)$ $(q, G), (G, q)$, $(\bar{q}, G), (G, \bar{q}), (G, G)$ should be taken into account where G designates the gluon. The calculation of $C_n^{ij}(\bar{g})$ (or $C^{ij}(\xi, \bar{g})$) in the next-to-leading order was performed in [Alt 78, 79, Kub 79, Con 79, Har 79, 80, 83a, Hum 79, Sch 80].

Fig. 5.5.6. The lowest-order contribution to C_n^{ij}.

One may substitute the calculated result for $C_n^{ij}(\bar{g})$ (or $C_n^{ij}(\xi, \bar{g})$) into Eq.(5.59) or (5.5.8)) and use the parton distribution function $f^i(\xi, q^2)$ determined in the deep inelastic electron (muon-)-nucleon scatterings. One then makes theoretical predictions on the Drell-Yan cross section (5.5.6). It is worth noting here that the proof of the factorization was once endangered owing to the effect of soft gluons emitted from the initial quarks [Dor 80]. Fortunately, however, it is known that the factorization is unaffected by the soft-gluon effect [Col 82] and moreover the parton distribution function in Eq.(5.5.8) remains the same as that in deep inelastic scattering [Lin 83] (see also [Bod 81, Mue 82]).

We shall not go into the details on the confrontation of the theoretical predictions with experimental data on the Drell-Yan processes. This topic is neatly reviewed, e.g., in [Ber 82, Alt 82, Bar 87].

CHAPTER

SIX

INFRARED DIVERGENCES

Although infrared divergences are of long-distance nature, they often play an essential role in the verification of the validity of the perturbative treatment of short-distance phenomena. Relying on a one-loop example we explain how the infrared (soft as well as collinear) divergence emerges and verify its cancellation by the real emission process. A general proof of the cancellation of soft divergences is given within the framework of quantum electrodynamics. The absence of infrared divergences in off-shell Green functions is shown in massless renormalizable field theories (the Kinoshita-Poggio-Quinn theorem). It is also shown that infrared divergences cancel out in the physical cross section if the degeneracy of the states is introduced (the Kinoshita-Lee-Nauenberg theorem).

6.1. ONE-LOOP EXAMPLE

6.1.1. Origin of infrared divergences

As will be explained here in a one-loop example, the infrared (soft) divergence may emerge if the theory under investigation includes a massless field like the photon in QED and the gluon in QCD. If the massless field couples with the other massless field or with itself, a further divergence is generated which is called the mass singularity (collinear divergence). In this book we make comprehensive use of the term, infrared divergence, so that the above two types of divergences are generically called the *infrared divergence*. The reason why the above two types of divergences are called soft and collinear divergences, respectively, will become evident soon.

To standardize our terminology we classify infrared divergences such that:

$$\text{infrared divergence}\begin{cases}\text{infrared divergence (soft divergence)} & \lambda \to 0,\\ \text{mass singularity (collinear divergence)} & \lambda\ (\text{and } m) \to 0.\end{cases}$$

Here λ and m designate the masses of the relevant particles of the theory which are expected to vanish. Here λ is assumed to be the fictitious mass of the gauge field while m is the mass of the matter field. Since both of the two types of infrared divergences emerge as consequences of vanishing masses, we may regard the infrared divergence as the mass singularity in a broader sense.

As a typical example we consider the one-loop QCD correction to the $e^-e^+ \to q\bar{q}$ cross section which was already discussed in Sec. 5.1.2. There we employed dimensional regularization to regulate both the ultraviolet and infrared divergences. It is, however, rather hard to see the physical significance of the infrared divergence if one uses the method of dimensional regularization. In the present section we adopt dimensional regularization only to regularize the ultraviolet divergence. We introduce a small fictitious mass λ for the gluon[1] to suppress the soft divergence and keep quarks massive with mass m so that the collinear divergence is absent.

To order g^2 of the $e^-e^+ \to q\bar{q}$ cross section contribute the diagrams shown in Fig. 6.1.1. The factor Z_2 appearing in the figure takes care of the field renormalization for the external quark lines which is necessary to define the renormalized S-matrix element [see Eq. (3.1.21)]. It should be noted that, since the factor $\sqrt{Z_2}$ corresponds to the self-energy part associated with the external quark (antiquark) line, we may re-express Fig. 6.1.1 in the form of Fig. 6.1.2 to order g^2. The first diagram in Fig. 6.1.2 corresponds to σ_B and the rest of the diagrams corresponds to $\tilde{\sigma}_V$ defined in Sec. 5.1.2.

The Born cross section σ_B is the same as before, i.e., Eq. (5.1.9), provided we neglect the terms of order m^2/s.

The virtual gluon contribution σ_V is given by Eq. (5.1.10). Here F_V takes the same expression as Eq. (5.1.11) if $\not{k}_1$ and $\not{k}_2$ are replaced by $\not{k}_1 + m$ and $\not{k}_2 - m$, respectively, while Λ_μ is different from Eq. (5.1.12) and is given by

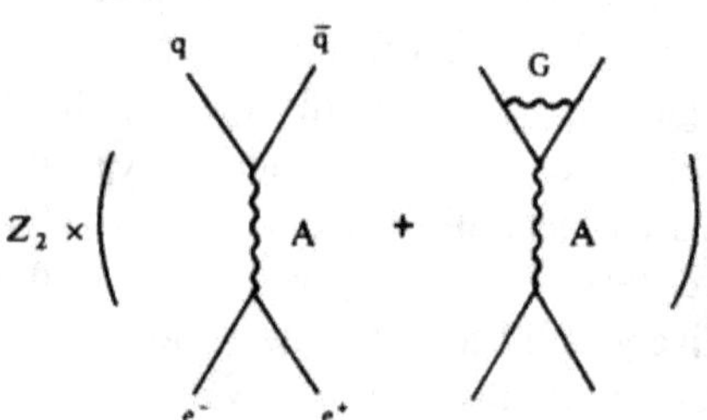

Fig. 6.1.1. Feynman diagrams contributing to the renormalized S-matrix element to order g^2 for the process $e^-e^+ \to q\bar{q}$. The factor Z_2 indicates the field renormalization constant for the external quark lines, and A and G represent the photon and gluon respectively.

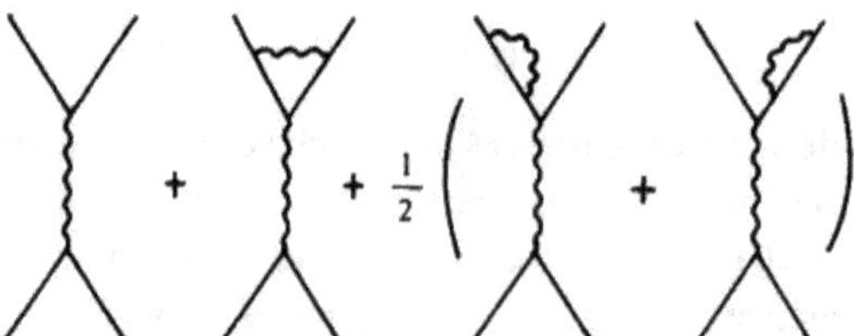

Fig. 6.1.2. Feynman diagrams for the process $e^-e^+ \to q\bar{q}$ to order g^2 in which the factor Z_2 in Fig. 6.1.1 is reexpressed as self-energy insertions to the external quark lines.

[1] By introducing the gluon mass λ we break the gauge invariance to order λ^2.

$$\Lambda_\mu = g^2 C_F \int \frac{d^D k}{(2\pi)^D i} \frac{1}{k^2 - \lambda^2} \gamma_\rho \frac{1}{m - \not{k} + \not{k}_1} \gamma_\mu \frac{1}{m - \not{k} - \not{k}_2} \gamma^\rho \, . \tag{6.1.1}$$

Recall that k_1 and k_2 are the momenta of the quark and antiquark, respectively (see Fig. 5.1.3). As in Sec. 5.1.2, the UV-divergence-free virtual-gluon contribution $\tilde{\sigma}_V$ is obtained by adding $(Z_2^2 - 1)\,\sigma_B$ to σ_V.

Since the infrared divergences in $\tilde{\sigma}_V$ essentially come from the loop integral in Λ_μ, given in Eq. (6.1.1) as $\lambda \to 0$ and $m \to 0$, we shall examine in the following the integral in Λ_μ to see the physical origins of the infrared divergences. We first let $\lambda \to 0$ and observe what happens in the integral,

$$\Lambda_\mu = g^2 C_F \int \frac{d^4 k}{(2\pi)^4 i} \frac{N_\mu}{k^2 (k^2 - 2k\cdot k_1)(k^2 + 2k\cdot k_2)} \, , \tag{6.1.2}$$

where

$$N_\mu = \gamma_\rho (m + \not{k} - \not{k}_1) \gamma_\mu (m + \not{k} + \not{k}_2) \gamma^\rho \, . \tag{6.1.3}$$

Here we ignore the ultraviolet divergence in Eq. (6.1.2) since we are only interested in the infrared divergences and are aware of its cancellation by the term $(Z_2^2 - 1)\,\sigma_B$. Noting that

$$\frac{1}{k^2} = \frac{1}{2\omega} \left(\frac{1}{k_0 - \omega + i\varepsilon} - \frac{1}{k_0 + \omega - i\varepsilon} \right) , \tag{6.1.4}$$

with $\omega = |\mathbf{k}|$ and $i\varepsilon$ the infinitesimal imaginary part ($\varepsilon > 0$), and performing the k_0-integration in Eq. (6.1.2) on the complex k_0-plane, we obtain

$$\Lambda_\mu = \frac{g^2 C_F}{8(2\pi)^3} \int_0^{2\pi} d\phi \int_0^{\pi} \sin\theta \, d\theta \int_0^{\infty} \frac{d\omega}{\omega} \frac{N_\mu}{(k_{10} - |\mathbf{k}_1| \cos\theta)(k_{20} + |\mathbf{k}_2| \cos\theta)} \, , \tag{6.1.5}$$

where $k_{i0} = \sqrt{\mathbf{k}_i^2 + m^2}$ $(i = 1, 2)$ and θ is the angle between $\mathbf{k}_1$ and $\mathbf{k}$ in the center-of-mass frame of the final quark and antiquark as depicted in Fig. 6.1.3. Note that k_0 is replaced by ω in N_μ of Eq. (6.1.5). Since $N_\mu \neq 0$ as $\omega \to 0$, Eq. (6.1.5) obviously diverges logarithmically as a consequence of the low momentum singularity $1/\omega$. The divergence emerges owing to the low-momentum gluon which is called the *soft gluon* and so it is usually called the *soft divergence*. We next let $m \to 0$. Then $k_{i0} = |\mathbf{k}_i| = \omega_i$ $(i = 1, 2)$ and hence

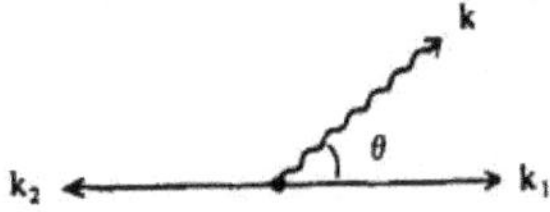

Fig. 6.1.3. The gluon momentum $\mathbf{k}$ in the center-of-mass frame of the quark-antiquark system where $\mathbf{k}_1$ and $\mathbf{k}_2$ are momenta of the quark and antiquark respectively.

$$\Lambda_\mu = \frac{g^2 C_F}{8(2\pi)^3\, \omega_1\omega_2} \int_0^\infty \frac{d\omega}{\omega} \int_0^{2\pi} d\phi \int_0^\pi \sin\theta\, d\theta \frac{N_\mu}{1-\cos^2\theta}\,. \qquad (6.1.6)$$

We now encounter a divergence in the angular integral in addition to the one in the ω integral. The new divergence comes from the integral regions $\cos\theta \sim 1$ (i.e. $\theta \sim 0$) and $\cos\theta \sim -1$ (i.e. $\theta \sim \pi$). The case $\theta = 0$ ($\theta = \pi$) corresponds to the configuration in which the gluon momentum $\mathbf{k}$ in Fig. 6.1.3 is parallel to the quark momentum $\mathbf{k}_1$ (antiquark momentum $\mathbf{k}_2$). In this sense, the above divergence is called the *collinear divergence*. Sometimes this divergence is also called a mass singularity in the narrow sense because it appears as a result of vanishing quark mass.

6.1.2. Regularization

As will be shown later in Sections 6.2 and 6.3 the infrared divergence cancels out in suitably defined physical quantities. It is, however, necessary to introduce regularization for the purpose of making each infrared-divergent integral in the physical quantity mathematically well-defined. The regularization is of course removed after the cancellation of the infrared divergence is achieved in the physical quantity.

The most natural regularization may be to introduce fictitious masses for massless particles in the theory since the infrared divergence originates from the presence of massless particles and may be regarded basically as the mass singularity in the broader sense. We have already seen a concrete example of this regularization method in Sec. 6.1.1. In this regularization scheme the infrared divergences show up as singularities in the mass parameters λ and m for their vanishing limit. We may call this method mass regularization. The introduction of the fictitious mass λ for gauge fields naturally causes the breakdown of the guage symmetry to order λ^2. Of course the gauge symmetry is recovered in physical quantities after the removal of the regularization.

The other natural way of regularizing infrared-divergent integrals is to eliminate (cut-off) potentially dangerous soft and/or collinear integral regions. In the previous example in Sec. 6.1.1 we, instead of introducing λ and m, could have eliminated the small ω region ($0 \le \omega < \omega_m$) in the ω-integral and the regions around $\theta = 0$ ($0 \le \theta < \theta_m$) and $\theta = \pi$ ($\pi - \theta_m < \theta \le \pi$) in the angular integral. Then the infrared divergence would have manifested itself as singularities in ω_m and θ_m for their vanishing limit. The physical meaning of this regularization is transparent. It is, however, rather hard to maintain covariance in this regularization scheme since we are forced to use a specific frame for adopting this scheme.

Dimensional regularization is yet another scheme for regularizing infrared divergences [Gas 73, Mar 75, And 79, Aok 82]. This scheme had already been employed in Sec. 5.1.2. It is known to give much simpler expressions for regularized results and to respect gauge and Lorentz invariance. The infrared divergence shows up as a pole in the space-time dimension D. The sign of the residue of the pole in D for the infrared divergence may be shown to be opposite to that for the corresponding ultraviolet divergence. A typical example has already been discussed in the Digression at the end of Sec. 2.5.5. There it was shown that

$$\frac{\Gamma(D/2)}{\pi^{D/2}i}\int\frac{d^Dq}{(-q^2)^\alpha}=\frac{1}{D/2-\alpha}-\frac{1}{D'/2-\alpha},$$

where D (regularizing the infrared divergence) is set equal to D' (regularizing the ultraviolet divergence) in the finite terms.

It is not difficult to find the correspondence rule for relating mass regularization and dimensional regularization. For soft divergences the rule reads

$$\ln\frac{\lambda^2}{m^2}\leftrightarrow\Gamma(\varepsilon)\left(\frac{4\pi\mu^2}{m^2}\right)^\varepsilon=\frac{1}{\varepsilon}-\gamma+\ln\frac{4\pi\mu^2}{m^2}, \qquad (6.1.7)$$

where $\varepsilon=(4-D)/2$ and μ is the mass scale of the coupling constant.

As an example let us make an explicit computation of the vertex function Λ_μ in these two regularization methods. In the mass regularization, Λ_μ is given by Eq. (6.1.1). We perform the loop-momentum integration after applying a Feynman parametrization to Eq. (6.1.1) and obtain

$$\Lambda_\mu=\frac{\alpha_s C_F}{4\pi}\int_0^1 dx\int_0^{1-x}dy\,[2(1-\varepsilon)^2\,\Gamma(\varepsilon)\,(4\pi\mu^2/K)^\varepsilon\gamma_\mu$$

$$-\gamma_\rho(m+\not{q}_1)\gamma_\mu(m-\not{q}_2)\gamma^\rho K^{-1}], \qquad (6.1.8)$$

where ε is set equal to zero in the finite term, and q_1, q_2 and K are given by

$$q_1=(1-x)k_1+yk_2, \qquad q_2=xk_1+(1-y)k_2,$$
$$K=-xyq^2+(x+y)^2m^2+(1-x-y)\lambda^2, \qquad (6.1.9)$$

with $q=k_1+k_2$. Using the on-shell condition for the final quark and antiquark, $\bar{u}(k_1)(\not{k}_1-m)=0$ and $(\not{k}_2+m)v(k_2)=0$, respectively, we obtain

$$\gamma_\rho (m + \not{k}_1) \gamma_\mu (m - \not{k}_2) \gamma^\rho = 2m^2 [a\gamma_\mu + 2b(k_1 - k_2)_\mu/m]$$
$$+ \text{ terms antisymmetric in } x \text{ and } y \,, \tag{6.1.10}$$

where

$$a = -1 + (1 - x - y)(3 - x - y) - (1 - x)(1 - y) q^2/m^2 \,,$$
$$b = (x + y)(1 - x - y)/2 \,. \tag{6.1.11}$$

Note that the terms antisymmetric in the exchange of x and y do not contribute to the integral (6.1.8). Inserting Eq. (6.1.10) into Eq. (6.1.8) and performing the x and y integration we find

$$\Lambda_\mu = \frac{\alpha_S C_F}{4\pi} \left[\gamma_\mu \left(\frac{1}{\varepsilon} - \gamma + \ln \frac{4\pi\mu^2}{m^2} + \frac{v^2 + 1}{v} \ln \frac{v + 1}{v - 1} \ln \frac{\lambda^2}{m^2} + F(v) \right) - \frac{(k_1 - k_2)_\mu}{2m} \frac{v^2 - 1}{v} \ln \frac{v + 1}{v - 1} \right] , \tag{6.1.12}$$

where

$$v = \sqrt{1 - 4m^2/q^2} \,,$$

$$F(v) = \left(3v - \frac{v^2 + 1}{2v} \ln \frac{4v^2}{v^2 - 1} \right) \ln \frac{v + 1}{v - 1} + \frac{v^2 + 1}{v} \left[\mathrm{Sp}\left(\frac{v + 1}{2v} \right) - \mathrm{Sp}\left(\frac{v - 1}{2v} \right) \right] , \tag{6.1.13}$$

with $\mathrm{Sp}(z)$ the Spence function defined by[2]

$$\mathrm{Sp}(z) = - \int_0^z dt \frac{\ln(1 - t)}{t} \,. \tag{6.1.14}$$

In dimensional regularization we use Eq. (6.1.1) with $\lambda = 0$ and regularize both the UV and IR divergence in arbitrary space-time dimension D. After performing the loop-momentum integration with the Feynman parametrization, we have

[2] See the Digression at the end of Sec. 6.1.2 for more details.

$$\Lambda_\mu = \frac{\alpha_S C_F}{4\pi}(4\pi\mu^2)^\varepsilon \int_0^1 dx \int_0^{1-x} dy [2(1-\varepsilon)^2 \Gamma(\varepsilon') K^{-\varepsilon} \gamma_\mu$$

$$- \gamma_\rho (m + \not{q}_1) \gamma_\mu (m - \not{q}_2) \gamma^\rho K^{-1-\varepsilon}] , \qquad (6.1.15)$$

where ε' simply distinguishes the UV divergence from the IR divergence and is set equal to ε in the finite terms, and

$$K = (x + y)^2 m^2 - xyq^2 . \qquad (6.1.16)$$

Again we use the Dirac equation but this time working in arbitrary dimension D to obtain

$$a = -1 + (1 - x - y)(3 - x - y) - (1 - x)(1 - y) q^2/m^2$$

$$+ \varepsilon(-(x + y)^2 + xyq^2/m^2) ,$$

$$b = (x + y)(1 - (1 - \varepsilon)(x + y))/2 . \qquad (6.1.17)$$

We are then able to perform the x and y integrations in Eq. (6.1.15). The result is

$$\Lambda_\mu = \frac{\alpha_S C_F}{4\pi}\left(\frac{4\pi\mu^2}{m^2}\right)^\varepsilon \Gamma(1+\varepsilon)\left[\gamma_\mu\left(\frac{1}{\varepsilon'} + \frac{1}{\varepsilon}\frac{v^2+1}{v}\ln\frac{v+1}{v-1} + F(v)\right)\right.$$

$$\left. - \frac{(k_1 - k_2)_\mu}{2m}\frac{v^2-1}{v}\ln\frac{v+1}{v-1}\right] . \qquad (6.1.18)$$

Comparing Eq. (6.1.18) with Eq. (6.1.12) we find the correspondence rule (6.1.7).

Finally we summarize the above three regularization schemes in the form of the table below.

regularization	soft divergence	collinear divergence
mass	gluon (photon) mass λ	quark (electron) mass m
cut-off	momentum cut ω_m	angle cut θ_m
dimensional	space-time dimension D	space-time dimension D

Digression

THE SPENCE FUNCTION: It is well-known [t Ho 79] that all the one-loop integrals can be performed resulting in elementary functions except for one special function, the *Spence function*. In particular for massive theories the Spence function inevitably occurs at the one-loop level while for massless theories the Spence function reduces to elementary functions or numerical constants. The multi-loop integrals require more special functions which are generalizations of the Spence function.

In the following we present the properties and some useful formulas of the Spence function [Lew 58, Aok 82]. The definition of the Spence function is given in Eq. (6.1.14): $\mathrm{Sp}(z)$ is an analytic function of z with a branch cut running from $z = 1$ to ∞. Equation (6.1.14) may be rewritten in the form,

$$\mathrm{Sp}\,(z) = \int_0^1 dt \frac{\ln t}{t - 1/z} . \tag{6.1.19}$$

The following relations hold among the Spence functions with different arguments,

$$\mathrm{Sp}\,(z) + \mathrm{Sp}\,(1 - z) = \frac{\pi^2}{6} - \ln z \ln (1 - z) ,$$

$$\mathrm{Sp}\left(-\frac{1}{z}\right) + \mathrm{Sp}(-z) = -\frac{\pi^2}{6} - \frac{1}{2}(\ln z)^2 . \tag{6.1.20}$$

For $|z| \le 1$, a series expansion in powers of z is possible for the Spence function,

$$\mathrm{Sp}\,(z) = \sum_{n=1}^{\infty} \frac{z^n}{n^2} . \tag{6.1.21}$$

It reduces to known numerical constants for some special values of the argument z:

$$\mathrm{Sp}\,(0) = 0, \qquad \mathrm{Sp}\,(1) = \frac{\pi^2}{6} ,$$

$$\mathrm{Sp}\,(-1) = -\frac{\pi^2}{12} , \qquad \mathrm{Sp}\left(\frac{1}{2}\right) = \frac{\pi^2}{12} - \frac{1}{2}(\ln 2)^2 . \tag{6.1.22}$$

6.1.3. Cancellation of infrared divergences

In QCD the gluon with very low momentum is called the soft gluon (correspondingly the soft photon in QED). The soft gluon is the source of the soft divergence which was discussed in Sec. 6.1.1. In the soft-gluon limit $\omega \to 0$ the four-momentum squared k^2 of the gluon vanishes and so the virtual soft gluon cannot be distinguished from the real free gluon.[3] This means that, whenever we deal with the virtual-gluon correction to the final $q\bar{q}$ state as in Sec. 6.1.1, the associated processes with an indefinite number of soft gluons should be taken into account. Thus the physical process is not just that involving a single $q\bar{q}$ final state but should be composed of an assembly of the final states $q\bar{q}$, $q\bar{q}G$, $q\bar{q}GG$, ... For this physical assembly of the final states the

[3] As we are working in perturbation theory the confinement problem is out of the scope of our argument.

soft divergences are expected to cancel out. In one-loop order the expected cancellation of the soft divergences will take place between the virtual-gluon correction to the cross section with the q$\bar{q}$ final state and the cross section with the q$\bar{q}$G final state.

Intuitively the soft gluon tends to lose its particle nature and the particle-number representation becomes inadequate for the state with soft gluons. Hence the physical process in which soft gluons participate should be defined as an assembly of the states with an indefinite number of soft gluons.

A similar but more elaborate argument can be made for the collinear divergence as will be given in Sec. 6.3.

In the present subsection we show the cancellation of the soft divergences in $e^-e^+ \to q\bar{q}$ at the one-loop level by using the mass regularization scheme. The cancellation of the collinear as well as soft divergences has been demonstrated in Sec. 5.1.2 by using dimensional regularization.

We have just observed in the preceding subsections the occurrence of the infrared divergences in the virtual gluon contribution σ_V to the $e^-e^+ \to q\bar{q}$ cross section. The divergences appear in the vertex part Λ_μ given in Eq. (6.1.12) where the collinear divergence for $m \to 0$ is not clearly seen while the soft divergence for $\lambda \to 0$ is manifest as $\ln \lambda^2$. To see the collinear divergence more clearly we let $|m^2/q^2| \ll 1$ in Eq. (6.1.12) and find

$$\Lambda_\mu = \frac{\alpha_S C_F}{4\pi}\left[\gamma_\mu\left(\frac{1}{\varepsilon} - \gamma + \ln\frac{4\pi\mu^2}{m^2} + 2\ln\frac{-q^2}{m^2}\ln\frac{\lambda^2}{m^2}\right.\right.$$
$$\left.\left. + \left(3 - \ln\frac{-q^2}{m^2}\right)\ln\frac{-q^2}{m^2} + \frac{\pi^2}{3}\right)\right], \qquad (6.1.23)$$

where we made use of the formula $\mathrm{Sp}(1) = \pi^2/6$. Equation (6.1.23) reveals the structure of both the soft and collinear divergence in an obvious manner. Inserting Eqs. (6.1.12) or (6.1.23) into Eq. (5.1.11) and using Eq. (5.1.10) we obtain σ_V. In order to get the finite cross section $\tilde{\sigma}_V$ we have to compute Z_2 in the mass regularization scheme. The quark-field renormalization constant Z_2 is calculated through the quark self-energy part $\Sigma_{ij}(q)$ whose one-loop expression is given in Eq. (A. 25). We adopt the on-shell renormalization condition for the determination of Z_2,

$$\left.\frac{\partial \Sigma_{ij}(p)}{\partial \not{p}}\right|_{\not{p}=m} = 0\ , \qquad (6.1.24)$$

in which the fictitious gluon mass λ is introduced to regularize the soft divergence. We then obtain

$$Z_2 = 1 + \frac{\alpha_S C_F}{4\pi}\left(-\frac{1}{\varepsilon} + \gamma - \ln\frac{4\pi\mu^2}{m^2} - 4 - 2\ln\frac{\lambda^2}{m^2}\right). \qquad (6.1.25)$$

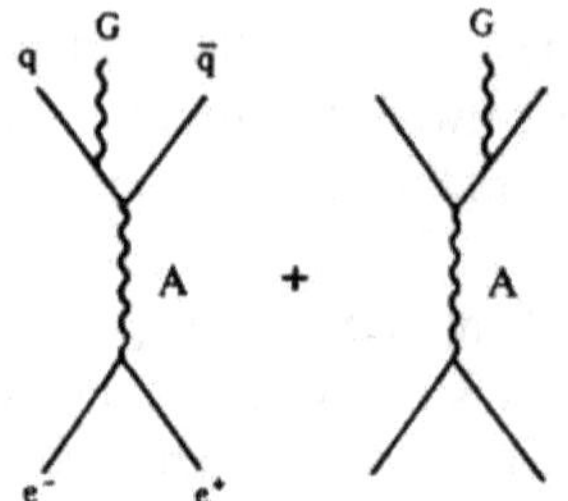

Fig. 6.1.4. Feynman diagrams contributing to the soft-gluon emission cross section σ_R to order g^2.

Using Eqs. (6.1.12) and (6.1.25) we find

$$\tilde{\sigma}_V = \frac{\alpha_S C_F}{2\pi}\sigma_B\left[-4 + \left(\frac{v^2+1}{v}\ln\frac{v+1}{v-1} - 2\right)\ln\frac{\lambda^2}{m^2} + F(v)\right]. \tag{6.1.26}$$

It should be noted that the term with $(k_1 - k_2)_\mu$ in Eq. (6.1.12) does not make any contribution to σ_V and therefore to $\tilde{\sigma}_V$.

According to the previous argument the soft divergence present in Eq. (6.1.26) is expected to be cancelled by the divergence which develops in the soft-gluon emission cross section σ_R (see Fig. 6.1.4). The calculation of σ_R proceeds in a similar way as in Sec. 5.1.2. The cross section σ_R is given by Eq. (5.1.20) with F_R defined by Eq. (5.1.21) in which $\not{k}_1$ and $\not{k}_2$ are replaced by $\not{k}_1 + m$ and $\not{k}_2 - m$ respectively. Here $S_{\mu\nu}$ in Eq. (5.1.21) is now given by

$$S_{\mu\nu} = \gamma_\mu \frac{1}{m - \not{k}_1 - \not{k}_3}\gamma_\nu + \gamma_\nu \frac{1}{m + \not{k}_2 + \not{k}_3}\gamma_\mu . \tag{6.1.27}$$

We define the soft-gluon region by $0 \le k_{30} \le \omega_m$ with $\omega_m \ll m$. The upper bound of the soft-gluon energy, ω_m, should be determined by the experimental condition that the single-quark (antiquark) state cannot be distinguished from the quark (antiquark) state accompanied by the soft gluon of energy less than ω_m. In other words, ω_m is the energy resolution of the relevant experimental apparatus to detect the gluon as an independent particle. For the soft gluon emitted from the quark (antiquark) we have $k_{30}/m \ll 1$ (and hence $k_{30} \ll k_{10}$, k_{20}) and

$$(\not{k}_1 + m)\, S_{\lambda\mu}\,(\not{k}_2 - m) = \left(-\frac{k_{1\lambda}}{k_1\cdot k_3} + \frac{k_{2\lambda}}{k_2\cdot k_3}\right)(\not{k}_1 + m)\,\gamma_\mu\,(\not{k}_2 - m) + O\left(\frac{k_{30}}{m}\right), \tag{6.1.28}$$

where $k_1^2 = k_2^2 = m^2$ and $k_3^2 = 0$ are taken into account. Using Eq. (6.1.28) for $S_{\lambda\mu}$ and a similar relation for S_ν^λ we find for the soft emitted gluons,

$$F_R = \left(\frac{k_{1\lambda}}{k_1 \cdot k_3} - \frac{k_{2\lambda}}{k_2 \cdot k_3}\right)^2 \left(\sum_i Q_i^2\right) \frac{e^4}{q^4} g^2 C_F \operatorname{Tr}[\not{p}_2 \gamma^\mu \not{p}_1 \gamma^\nu]$$
$$\times \operatorname{Tr}[(\not{k}_1 + m)\gamma_\mu (\not{k}_2 - m)\gamma_\nu] \,. \tag{6.1.29}$$

We insert Eq. (6.1.29) into Eq. (5.1.20) and recall Eqs. (2.3.150) and (2.3.151) to obtain

$$\sigma_R = g^2 C_F I \sigma_B \,, \tag{6.1.30}$$

where we have set $k_3 \simeq 0$ in the delta function and

$$I = \int_S \frac{d^3k}{(2\pi)^3\, 2k_0} \left(\frac{k_{1\mu}}{k_1 \cdot k} - \frac{k_{2\mu}}{k_2 \cdot k}\right)^2 , \tag{6.1.31}$$

with the integral region S given by

$$\lambda \le k_0 \left(= \sqrt{\mathbf{k}^2 + \lambda^2}\right) \le \omega_m \,. \tag{6.1.32}$$

Note here that we introduced the small fictitious gluon mass λ to regularize the possible infrared divergence in the integral (6.1.31).

The calculation of the integral (6.1.31) goes as follows. Equation (6.1.31) is rewritten as

$$I = \frac{1}{2(2\pi)^3} \int_\lambda^{\omega_m} |\mathbf{k}|\, dk_0 \int d\Omega \left[\frac{2m^2 - q^2}{(k_1 \cdot k)(k_2 \cdot k)} + \frac{m^2}{(k_1 \cdot k)^2} + \frac{m^2}{(k_2 \cdot k)^2}\right] , \tag{6.1.33}$$

where Ω represents the solid angle to the direction $\mathbf{k}$ and the terms of order k_0/m are neglected. The angular integrals in Eq. (6.1.33) may be performed so that

$$\int \frac{d\Omega}{(k_i \cdot k)^2} = \frac{4\pi}{m^2 \mathbf{k}^2 + \lambda^2 k_{i0}^2} \cdot \quad (i = 1, 2), \tag{6.1.34}$$

$$\int \frac{d\Omega}{(k_1 \cdot k)(k_2 \cdot k)} = \int d\Omega \int_0^1 \frac{dx}{[x k_1 \cdot k + (1 - x) k_2 \cdot k]^2} \,,$$
$$= 4\pi \int_0^1 \frac{dx}{[m^2 - x(1 - x) q^2]\, \mathbf{k}^2 + \lambda^2 k_{10}^2} \,. \tag{6.1.35}$$

Substituting Eqs. (6.1.34) and (6.1.35) into Eq. (6.1.33) and performing the remaining integrations in k_0 and x, we obtain

$$I = \frac{1}{2(2\pi)^2}\left(\frac{v^2+1}{v}\ln\frac{v+1}{v-1} - 2\right)\ln\frac{4\omega_m^2}{\lambda^2} + \text{finite terms.} \tag{6.1.36}$$

We find by combining Eqs. (6.1.30) and (6.1.36),

$$\sigma_R = \frac{\alpha_S C_F}{2\pi}\sigma_B\left[\left(\frac{v^2+1}{v}\ln\frac{v+1}{v-1} - 2\right)\ln\frac{4\omega_m^2}{\lambda^2} + \text{finite terms}\right]. \tag{6.1.37}$$

We see clearly that the soft divergences in Eqs. (6.1.26) and (6.1.37) cancel out, resulting in the finite expression,

$$\begin{aligned}\sigma &= \sigma_B + \tilde{\sigma}_V + \sigma_R\,,\\ &= \sigma_B\left[1 + \frac{\alpha_S C_F}{2\pi}\left(\left(\frac{v^2+1}{v}\ln\frac{v+1}{v-1} - 2\right)\ln\frac{4\omega_m^2}{m^2} + \text{finite terms}\right)\right].\end{aligned} \tag{6.1.38}$$

It is worth mentioning here that a diagrammatic interpretation may be given to the above cancellation mechanism for the soft divergence. To see this we first note that the cross section for the q$\bar{\text{q}}$ final state, $\sigma_B + \tilde{\sigma}_V\,[= \sigma_B + \sigma_V + (Z_2^{\frac{1}{2}} - 1)\,\sigma_B]$, which is obtained by squaring the amplitude corresponding to Fig. 6.1.2, is represented diagrammatically by the sum of the three parts, (a), (b) and (c), in Fig. 6.1.5. Here Fig. 6.1.5 (a) corresponds to σ_B, (b) to σ_V and (c) to $(Z_2^{\frac{1}{2}} - 1)\,\sigma_B$, respectively. Looking back on how the soft divergence came about in $\tilde{\sigma}_V$, we realize that the part[4] of the soft divergence in Eq. (6.1.26),

$$\frac{v^2+1}{v}\ln\frac{v+1}{v-1}\ln\frac{\lambda^2}{m^2}\,, \tag{6.1.39}$$

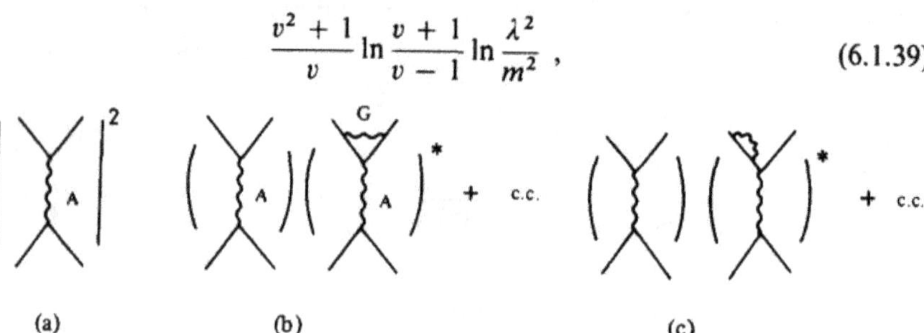

Fig. 6.1.5. Diagrammatic expression of three terms in the e$^-$e$^+$ → q$\bar{\text{q}}$ total cross section to order g^2: (a) corresponds to σ_B, (b) to σ_V and (c) to $(Z_2^{\frac{1}{2}} - 1)\,\sigma_B$ where c.c. represents the complex conjugate of the preceding term.

[4] For simplicity we omit the overall factor $\alpha_S C_F \sigma_B/(2\pi)$.

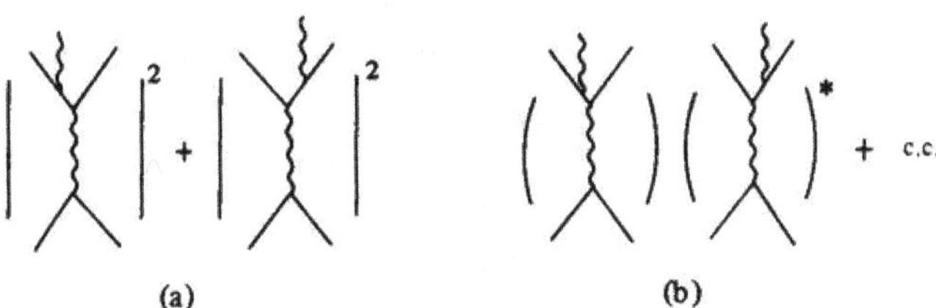

Fig. 6.1.6. Diagrammatic expression of two terms in the $e^-e^+ \to q\bar{q}G$ total cross section σ_R to order g^2: (a) represents the absolute square term and (b) the cross term where c.c. indicates the complex conjugate.

originates from σ_V and so corresponds to Fig. 6.1. 5(b) while the other part

$$-2\ln\frac{\lambda^2}{m^2}\,, \tag{6.1.40}$$

comes from $(Z_2^{\frac{1}{2}} - 1)\,\sigma_B$ and corresponds to Fig. 6.1.5 (c). Thus we may set the following correspondence as far as the soft divergence is concerned

q $\bar{q}$ q G $\bar{q}$

$$(\ \ )(\ \ )^* + \text{c.c.} \longleftrightarrow \text{Eq. (6.1.39)}, \tag{6.1.41}$$

$$(\ \ )(\ \ )^* + \text{c.c.} \longleftrightarrow \text{Eq. (6.1.40)}, \tag{6.1.42}$$

Here we neglected the electron-position vertex part and the photon line in the above diagrams.

The cross section σ_R is obtained by squaring the amplitude corresponding to Fig. 6.1.4 and hence is given by the sum of two terms corresponding to Fig. 6.1.6(a) and (b), respectively. By looking into the details of the soft divergences in σ_R given in Eq. (6.1.37) we recognize that the quantity,

$$\frac{v^2+1}{v}\ln\frac{v+1}{v-1}\ln\frac{4\omega_m^2}{\lambda^2}\,, \tag{6.1.43}$$

emerges from the first term in Eq. (6.1.33) while the other part,

$$-2\ln\frac{4\omega_m^2}{\lambda^2}\,, \tag{6.1.44}$$

originates from the second and third terms in Eq. (6.1.33). Hence Eqs. (6.1.43) and (6.1.44) correspond to Fig. 6.1.6(a) and (b), respectively. Thus we arrive at the following correspondence rule,

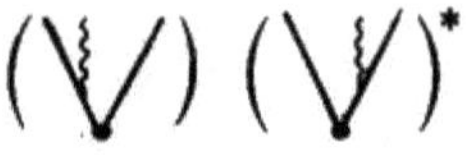

$$+\ \text{c.c.} \leftrightarrow \text{Eq. (6.1.43)}\ , \tag{6.1.45}$$

$$\leftrightarrow \text{Eq. (6.1.44)}\ . \tag{6.1.46}$$

The above correspondence rule that we derived may be re-expressed in simpler form by using the notion of a cut diagram. A cut diagram is one with a dotted horizontal line (cut line) (see Fig. 6.1.7) such that each propagator cut by the dotted (cut) line is replaced by its imaginary part. We notice that the cut

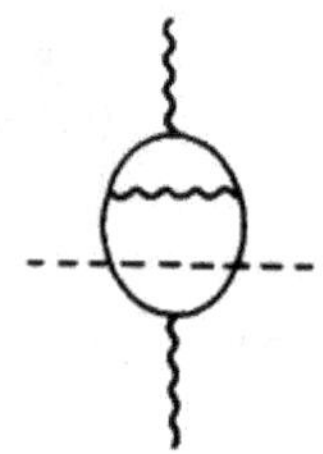

Fig. 6.1.7. An example of the cut diagrams in e^-e^+ annihilations.

diagram of Fig. 6.1.7 is essentially equivalent to the first diagram in Eq. (6.1.41). We rewrite the correspondence rule by the use of cut diagrams in the following way,

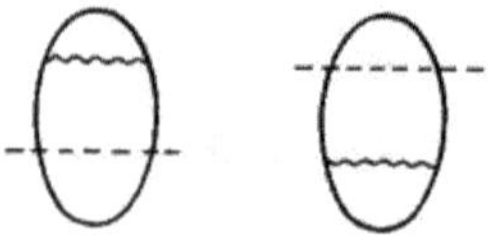

$$\leftrightarrow \text{Eq. (6.1.39)}\ , \tag{6.1.47}$$

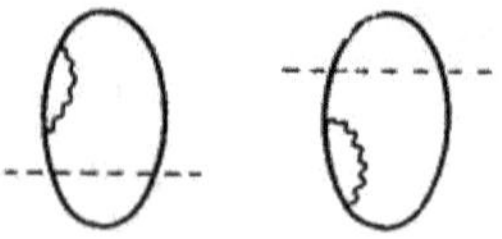

$$\leftrightarrow \text{Eq. (6.1.40)}\ , \tag{6.1.48}$$

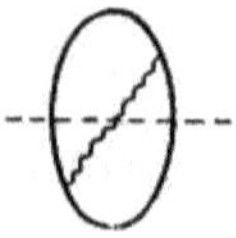

$$\leftrightarrow \text{Eq. (6.1.43)}\ , \tag{6.1.49}$$

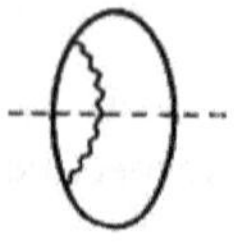

$$\leftrightarrow \text{Eq. (6.1.44)}\ . \tag{6.1.50}$$

We know that the cancellation of the soft divergences takes place between Eqs. (6.1.39) and (6.1.43) and also between Eqs. (6.1.40) and (6.1.44). These canceling pairs correspond to the different cuts of the same diagram. Thus we conclude that the soft divergences cancel out if they correspond respectively to the cut diagrams given as different cuts of the same diagram [Kin 50, 62, Nak 58, And 81]. We shall see that this result is a manifestation of the Kinoshita-Poggio-Quinn theorem which will be explained in Sec. 6.3.2.

6.2. PROOF OF THE SOFT-PHOTON CANCELLATION IN QED

6.2.1. Soft-photon emission

As was already mentioned in Sec. 6.1.3, the physically observed processes are always accompanied by the emission of an indefinite number of soft gluons (photons) and the infrared divergence should be absent in the cross section for a physically observed process (physical cross section). Thus the infrared divergence present in the cross section for the process with no soft gluon (photon) emission is expected to disappear in the physical cross section. That this is actually the case was first shown by Bloch and Nordsieck [Blo 37, Nor 37] in quantum electrodynamics (QED) and is referred to as the *Bloch-Nordsieck theorem*. Since then there have followed many works on this problem. (For a comprehensive review, see [Jau 55]. See also [Nak 58].) The complete proof of the Bloch-Nordsieck theorem is found in [Yen 61, Gra 73]. In quantum chromodynamics the same mechanism as above is expected to work. The situation in QCD is, however, quite different from that in QED since the collinear divergence, in addition to the soft divergence, occurs in QCD even if the quark masses are kept nonvanishing. The reason is that, due to the presence of the triple gluon coupling, the gluon can decay into two collinear gluons resulting in the collinear divergence (mass singularity). In spite of this situation it is believed that the infrared divergence in QCD cancels out order by order in perturbation theory if the physical cross section is considered.[5]

In Sec. 6.3 we shall present a more general and rather formal proof of the cancellation of the infrared divergence to all orders in perturbation theory for a large class of field theories with massless fields. It seems, however, to be instructive to give, before going to Sec. 6.3, a more intelligible proof based on the direct calculation of the divergence. In this section we confine ourselves to QED and present an all-order proof of the cancellation of soft divergences by following Weinberg's lucid method [Wei 65, Nas 78]. An outline of the proof is

[5] See Sec. 6.3.1 for more details and references.

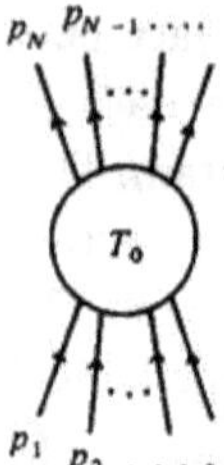

Fig. 6.2.1. The diagram with N external electron lines corresponding to the transition amplitude T_0.

as follows: We consider a transition amplitude (truncated connected Green function with external lines on their mass shell) with N external electrons and n external soft photons.[6] We assume that the transition amplitude is summed to all orders. In the total cross section obtained from this transition amplitude we extract, as a multiplicative factor, the soft divergence originating from the external soft photons and then show that this divergent factor is annihilated by the other factor coming from the internal soft photons. In the present subsection (Sec. 6.2.1) we argue the extraction of the soft divergence corresponding to the external soft photon lines in the total cross section. The soft divergence due to the internal photon lines will be discussed in the following subsection (Sec. 6.2.2.).

Consider first the transition amplitude with N external electron lines (no photon lines), T_0, whose diagrammatic expression is given in Fig. 6.2.1. Attaching a soft photon of momentum q and polarization vector $\varepsilon(q)$ to one of the outgoing electrons of momentum p and Dirac spinor $u(p)$ turns out to be the replacement of $\bar{u}(p)$ by

$$\bar{u}(p)\, e\not{\varepsilon}(q) \frac{1}{m - \not{p} - \not{q}} \simeq -\, e\frac{p\cdot\varepsilon(q)}{p\cdot q}\, \bar{u}(p)\ . \tag{6.2.1}$$

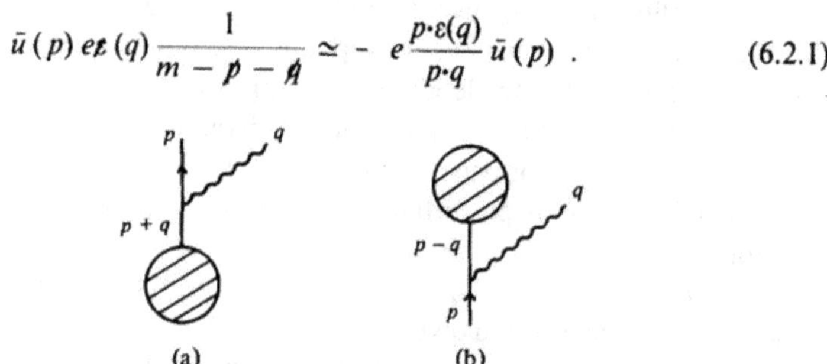

Fig. 6.2.2. Attachment of a soft photon to (a) the final external electron line and (b) the initial external electron line.

[6] For simplicity we deal with the amplitude with electron external lines (no positron lines). The generalization to include external positron lines is straightforward.

For the incoming electron we have, instead of Eq. (6.2.1),

$$\frac{1}{m - \not{p} + \not{q}} e\not{\varepsilon}(q)\, u(p) \simeq e\frac{p\cdot\varepsilon(q)}{p\cdot q} u(p)\ . \tag{6.2.2}$$

The above replacement is depicted in Fig. 6.2.2. Thus an attachment of one soft photon to the external electron line amounts to the multiplication of the extra factor

$$e\eta \frac{p\cdot\varepsilon(q)}{p\cdot q}\ , \tag{6.2.3}$$

where η is the signature factor defined by

$$\eta = \begin{cases} +1 & \text{for incoming electron,} \\ -1 & \text{for outgoing electron.} \end{cases} \tag{6.2.4}$$

For the N-electron amplitude, T_0, the extra factor for the attachment of one soft photon reads

$$\sum_{i=1}^{N} e\eta_i \frac{p_i\cdot\varepsilon(q)}{p_i\cdot q}\ . \tag{6.2.5}$$

With this extra factor we generate from T_0 the amplitude with N external electrons and one soft photon attached to the external electron lines. The transition rate (or the total cross section) corresponding to this new amplitude shares the same soft-divergence structure with the transition rate obtained from the amplitude T_1 with N external electrons and one soft photon, and hence we also denote this new amplitude by T_1. It should be noted here that the soft photon attached to internal electron lines does not create any soft divergence. The reason for this is the following: The attachment of the soft photon to the electron propagator $1/(m - \not{p})$ is equivalent to replacing the propagator by (see Fig. 6.2.3)

$$\frac{1}{m - \not{p} + \not{q}} e\not{\varepsilon}(q) \frac{1}{m - \not{p}} = e\frac{m + \not{p} - \not{q}}{m^2 - p^2 + 2p\cdot q} \not{\varepsilon}(q) \frac{m + \not{p}}{m^2 - p^2}\ . \tag{6.2.6}$$

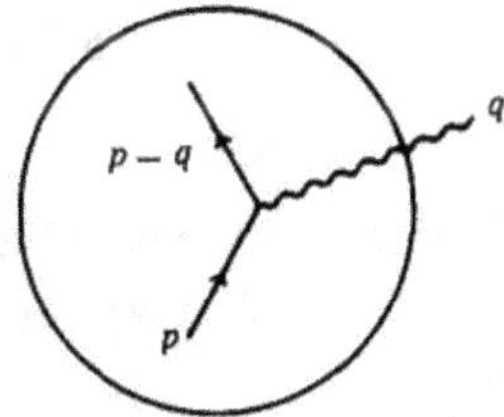

Fig. 6.2.3. Attachment of a soft photon to an internal electron line.

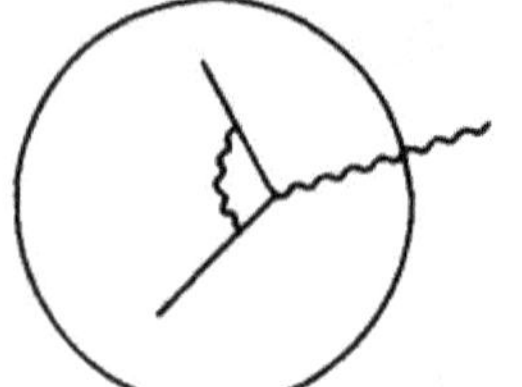

Fig. 6.2.4. An example of the diagram relevant to the overlapping soft divergence.

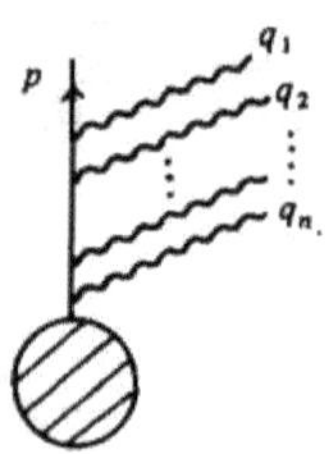

Fig. 6.2.5. The n soft photons attached to an external electron line.

Since the electron is not on its mass shell, we have $m^2 - p^2 \neq 0$, and hence the extra factor in Eq. (6.2.6) does not create any singularity for $q \to 0$. This proves that no new soft divergence is generated through the attachment of the soft photon to the internal electron line. It should also be noted that the so-called overlapping soft divergence cancels out in a gauge invariant set of subdiagrams [Yen 61]. Here by the overlapping soft divergence we mean the divergence generated by an interplay of two soft photons. Shown in Fig. 6.2.4 is an example of a diagram responsible for an overlapping soft divergence.

We would like to generalize the above rule (6.2.5) for one soft photon to the case of n soft photons. For this purpose we first consider the case of attaching n soft photons to an outgoing electron line as depicted in Fig. 6.2.5. The extra factor for this soft-photon attachment is easily obtained by the observation that

$$\begin{aligned} &\sum_{\text{perm}} \bar{u}(p)\, e\not{\varepsilon}_1 \frac{1}{m-\not{p}-\not{q}_1}\, e\not{\varepsilon}_2 \frac{1}{m-\not{p}-\not{q}_1-\not{q}_2} \cdots \frac{1}{m-\not{p}-\not{q}_1-\ldots-\not{q}_n}\,, \\ &\simeq \sum_{\text{perm}} e^n \frac{p\cdot\varepsilon_1}{-p\cdot q_1} \frac{p\cdot\varepsilon_2}{-p\cdot(q_1+q_2)} \cdots \frac{p\cdot\varepsilon_n}{-p\cdot(q_1+\ldots+q_n)}\, \bar{u}(p)\,, \\ &= (\eta e)^n \frac{p\cdot\varepsilon_1}{p\cdot q_1} \frac{p\cdot\varepsilon_2}{p\cdot q_2} \cdots \frac{p\cdot\varepsilon_n}{p\cdot q_n}\, \bar{u}(p)\,, \end{aligned} \tag{6.2.7}$$

where $\varepsilon_i = \varepsilon(q_i)$ with $i = 1, 2, \ldots, n$ and by "perm" we mean all the possible permutations of index $(1, 2, \ldots, n)$. Note that in deriving Eq. (6.2.7) we have used the formula

$$\sum_{\text{perm}} \frac{1}{p\cdot q_1} \frac{1}{p\cdot(q_1+q_2)} \cdots \frac{1}{p\cdot(q_1+\ldots+q_n)}$$

$$= \frac{1}{p\cdot q_1} \frac{1}{p\cdot q_2} \cdots \frac{1}{p\cdot q_n} . \tag{6.2.8}$$

The validity of this result follows from mathemathical induction. In fact for $n = 1$ the formulà is trivial and for $n = 2$ we easily see that it holds, i.e.,

$$\frac{1}{p\cdot q_1} \frac{1}{p\cdot(q_1+q_2)} + \frac{1}{p\cdot q_2} \frac{1}{p\cdot(q_2+q_1)} = \frac{1}{p\cdot q_1} \frac{1}{p\cdot q_2} . \tag{6.2.9}$$

Then we proceed to prove the formula for $n + 1$ assuming that it holds for n.

It is now straightforward to obtain the replacement rule for n soft photons attached to N electron lines with all the possible partitions $(n_1, n_2, \ldots, n_N)$ of soft photons $(\sum_{i=1}^{N} n_i = n)$. A typical diagram with this soft photon attachment is shown in Fig. 6.2.6. The extra factor for this attachment reads

$$\sum_{\substack{i=1 \\ \Sigma n_i = n}}^{N} \prod_{l=1}^{n_i} e\eta_i \frac{p_i\cdot\varepsilon_l}{p_i\cdot q_l} = \prod_{l=1}^{n} \sum_{i=1}^{N} e\eta_i \frac{p_i\cdot\varepsilon_l}{p_i\cdot q_l} . \tag{6.2.10}$$

As we have argued before in the case of a one-soft-photon attachment, the amplitude T_0 multiplied by the extra factor (6.2.10) gives rise to the same soft divergence in the cross section (or transition rate) as the one for the amplitude T_n with N electrons and n soft photons. Since we are only interested in the divergence structure in the soft region, we may write

$$T_n = T_0 \prod_{l=1}^{n} \sum_{i=1}^{N} e\eta_i \frac{p_i\varepsilon_l}{p_i\cdot q_l} . \tag{6.2.11}$$

The transition rate w_n for emitting n soft photons is given by

$$w_n = \frac{1}{n!} \int_R \left(\prod_{l=1}^{n} \frac{d^3 q_l}{(2\pi)^3\, 2q_{l0}} \right) \sum_{\text{pol}} |T_n|^2 , \tag{6.2.12}$$

Fig. 6.2.6. A typical diagram with attachment of n soft photons to N external electron lines.

where "pol" represents the polarization of the soft photon and R denotes the soft-photon region,

$$R = \left\{ q_1, \ldots, q_n \ ; \quad \sum_{l=1}^{n} q_{l0} \leq \omega_m \right\} . \tag{6.2.13}$$

After taking the polarization sum in Eq. (6.2.12) we obtain

$$w_n = |T_0|^2 \frac{1}{n!} \int_R \prod_{l=1}^{n} \left[\frac{d^3 q_l}{(2\pi)^3 2q_{l0}} \sum_{i,j=1}^{N} e^2 \eta_i \eta_j \frac{-p_i \cdot p_j}{(p_i \cdot q_l)(p_j \cdot q_l)} \right] . \tag{6.2.14}$$

The physical transition rate w for the N-electron process is then given by adding all w_n:

$$w = \sum_{n=0}^{\infty} w_n = |T_0|^2 \sum_{n=0}^{\infty} \frac{1}{n!} \int_R \prod_{l=1}^{n} \left(\frac{dq_{l0}}{q_{l0}} f \right) , \tag{6.2.15}$$

where

$$f = \int \frac{d\Omega}{2(2\pi)^3} \sum_{i,j=1}^{N} e^2 \eta_i \eta_j \frac{-p_i \cdot p_j}{(p_i \cdot q)(p_j \cdot q)} q_0^2 . \tag{6.2.16}$$

Here $d\Omega = \sin\theta d\theta d\phi$ and θ and ϕ are angles describing $\mathbf{q}$. Note that the quantity f defined by Eq. (6.2.16) is independent of q_0 and is a function of $p_i \cdot p_j$. This property may be explicitly seen by performing the integration in Eq. (6.2.16) under the application of the Feynman parametrization, i.e.,

$$f = \frac{e^2}{(2\pi)^2} \sum_{i,j=1}^{N} (-\eta_i \eta_j p_i \cdot p_j) \int_0^1 \frac{dx}{m^2 - x(1-x)(p_i - p_j)^2} . \tag{6.1.17}$$

The expression (6.2.15) for the physical transition rate w can be rewritten as

$$w = |T_0|^2 \sum_{n=0}^{\infty} \frac{f^n}{n!} \int \frac{dq_{10}}{q_{10}} \cdots \frac{dq_{n0}}{q_{n0}} \theta\left(\omega_m - \sum_{l=1}^{n} q_{l0} \right) . \tag{6.2.18}$$

Using the following representation for the step function,

$$\begin{aligned} \theta(\omega - \omega') &= \frac{-1}{2\pi i} \int_{-\infty}^{\infty} dx \frac{e^{-i(\omega - \omega')x}}{x + i\varepsilon} , \\ &= \frac{1}{\pi} \int_{-\infty}^{\infty} dx \frac{\sin \omega x}{x + i\varepsilon} e^{i\omega' x} , \end{aligned} \tag{6.2.19}$$

with $\varepsilon > 0$, an infinitesimal constant, we find

$$w = |T_0|^2 \frac{1}{\pi} \int_{-\infty}^{\infty} dx \frac{\sin \omega x}{x + i\varepsilon} \exp\left(f \int_{\lambda}^{\omega_m} \frac{dq_0}{q_0} e^{iq_0 x} \right), \qquad (6.2.20)$$

where λ is the fictitious photon mass employed to regularize the soft divergence. We note the identity

$$\int_{\lambda}^{\omega_m} \frac{dq_0}{q_0} e^{iq_0 x} = \ln \frac{\omega_m}{\lambda} + \int_{\lambda}^{\omega_m} \frac{dq_0}{q_0} (e^{iq_0 x} - 1) \; . \qquad (6.2.21)$$

The second term on the right-hand side of Eq. (6.2.21) is finite for $\lambda \to 0$ and so we may set $\lambda = 0$ there. We finally have

$$w = |T_0|^2 (\omega_m/\lambda)^f F(f) \; , \qquad (6.2.22)$$

where

$$\begin{aligned} F(f) &= \frac{1}{\pi} \int_{-\infty}^{\infty} dx \frac{\sin \omega x}{x + i\varepsilon} \exp\left[f \int_0^{\omega_m} \frac{dq_0}{q_0} (e^{iq_0 x} - 1) \right], \\ &= 1 - \frac{\pi^2}{12} f^2 + \ldots \end{aligned} \qquad (6.2.23)$$

Since f is of order e^2, the infrared factor $(\omega_m/\lambda)^f$ may be expanded in a perturbative series such that

$$(\omega_m/\lambda)^f = 1 + f \ln \frac{\omega_m}{\lambda} + \frac{1}{2} f^2 \ln^2 \frac{\omega_m}{\lambda} + \ldots \qquad (6.2.24)$$

Equation (6.2.24) gives us an order-by-order expression of the soft divergence coming from the soft-photon emission.

6.2.2. Virtual photons

We turn our attention to the transition amplitude T_0 which has no external photon line. Its external lines consist only of electron lines. The photon lines participate only as internal lines in T_0 which are potentially responsible for the soft divergence through loop integrals.

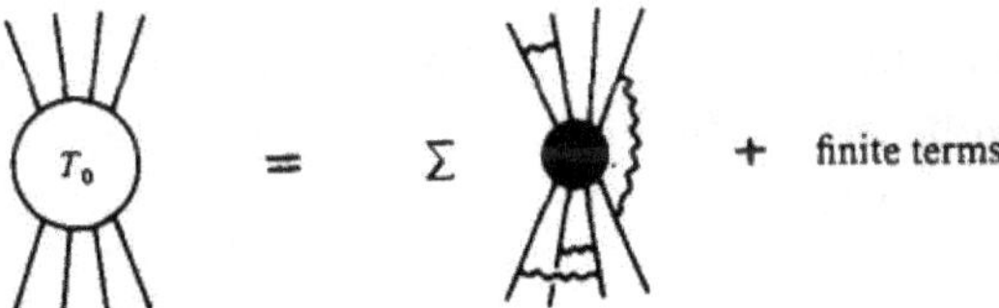

Fig. 6.2.7. Diagrammatic expression of the fact that the soft divergence emerges only through the internal photon lines connecting two external electron lines. The dark blob indicates the part of the full diagram having no infrared divergence and is called the core diagram.

$= t_0$

Fig. 6.2.8. The core diagram with N external electron lines denoted by t_0.

Fig. 6.2.9. Diagrams which do not exhibit soft divergences: (a) the soft photon connecting an external and internal electron line and (b) the soft photon joined to two internal electron lines.

Fig. 6.2.10. A diagram giving rise to an overlapping soft divergence.

It will be shown that the soft divergence shows up only when a photon line contributing to a loop integral connects two external electron lines. The diagrammatic expression of this statement is given in Fig. 6.2.7. That portion of the diagram indicated by a dark blob is called the core diagram which is infrared-finite. We denote by t_0 this core diagram which has N external electron lines as shown in Fig. 6.2.8. The statement above is in fact justified since internal photon lines other than those connecting two external electron lines do not give rise to the soft divergence: The diagrams of the type given in Fig. 6.2.9 are infrared-finite because internal electron lines are off their mass shell and serve as infrared cut-offs. Also the overlapping soft divergence coming from the diagrams of the type given in Fig. 6.2.10 is cancelled by the other soft divergence existing in its partners in the gauge-invariant subset of diagrams.

We denote by t_n the transition amplitude which is the N-electron core diagram with the insertion of n internal photon lines attached to external electron lines. We have

$$T_0 = \sum_{n=0}^{\infty} t_n + \text{finite terms.} \tag{6.2.25}$$

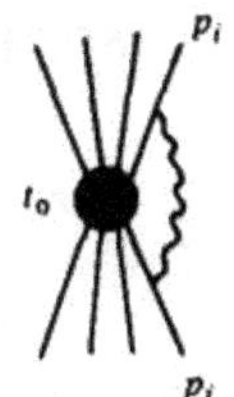

Fig. 6.2.11. The diagram obtained by inserting a virtual photon line between two external electron lines of the core diagram t_0.

We would like to obtain an explicit expression for t_n. For this purpose we first note that an insertion of a soft virtual photon between two external electron lines of the core diagram t_0 as depicted in Fig. 6.2.11 gives rise to an amplitude of the form,

$$\int_V \frac{d^4q}{(2\pi)^4 i}\bar{u}(p_i)\,e\gamma_\mu \frac{1}{m-\not{p}_i+\not{q}} M \frac{1}{m-\not{p}_j+\not{q}} e\gamma_\nu\, u(p_j)\frac{g^{\mu\nu}}{q^2}$$

$$\simeq \int_V \frac{d^4q}{(2\pi)^4 i} e^2 \frac{p_i\cdot p_j}{q^2(p_i\cdot q)(p_j\cdot q)}\bar{u}(p_i)\,Mu(p_j)\ , \tag{6.2.26}$$

where V represents the soft-photon region $\lambda \le q_0 \le \Lambda$ and $\bar{u}(p_i)\,Mu(p_j) = t_0$ the amplitude corresponding to the core diagram. The Feynman gauge is employed here. Each time we insert the soft virtual photon between two external electrons we have the extra multiplicative factor

$$\int \frac{d^4q}{(2\pi)^4 i}\frac{e^2}{q^2}\frac{-\eta_i\eta_j p_i\cdot p_j}{(p_i\cdot q)(p_j\cdot q)}\ . \tag{6.2.27}$$

It thus follows that

$$t_1 = t_0 \int \frac{d^4q}{(2\pi)^4 i}\frac{e^2}{q^2}\frac{1}{2}\sum_{i,j=1}^{N}\frac{-\eta_i\eta_j p_i\cdot p_j}{(p_i\cdot q)(p_j\cdot q)}\ . \tag{6.2.28}$$

Note that in Eq. (6.2.28) we need the factor 1/2 to correct for the overcounting in the summation on i and j. Similarly we repeat the above procedure until we obtain t_n,

$$t_n = t_0 \frac{1}{n!}\left[\int \frac{d^4q}{(2\pi)^4 i}\frac{e^2}{q^2}\frac{1}{2}\sum_{i,j=1}^{N}\frac{-\eta_i\eta_j p_i\cdot p_j}{(p_i\cdot q)(p_j\cdot q)}\right]^n , \tag{6.2.29}$$

where the factor $1/n!$ serves to correct the overcounting in estimating the number of ways that the n internal photon lines are attached to external electron lines. Inserting Eq. (6.2.29) into Eq. (6.2.25) results in

$$T_0 = t_0\, e^{g/2}\ , \tag{6.2.30}$$

where we neglected the finite terms in Eq. (6.2.25) and introduced

$$g = \int \frac{d^4q}{(2\pi)^4 i} \frac{e^2}{q^2} \sum_{i,j=1}^{N} \frac{-\eta_i \eta_j p_i \cdot p_j}{(p_i \cdot q)(p_j \cdot q)} . \tag{6.2.31}$$

Squaring Eq. (6.2.30) one finds

$$|T_0|^2 = |t_0|^2 \, e^{\mathrm{Re}\, g} . \tag{6.2.32}$$

Recalling that the infinitesimal imaginary part is present in the denominator of the propagators we find

$$\frac{1}{q^2} \to \frac{1}{q^2 + i\varepsilon} = \frac{\mathscr{P}}{q^2} - i\pi\delta(q^2) , \tag{6.2.33}$$

with $\mathscr{P}$ indicating the principal part. We calculate Re g,

$$\begin{aligned} \mathrm{Re}\, g &= \int \frac{d^4q}{(2\pi)^4} e^2 \pi \delta(q^2) \sum_{i,j=1}^{N} \frac{\eta_i \eta_j p_i \cdot p_j}{(p_i \cdot q)(p_j \cdot q)} \\ &= -\int_\lambda^\Lambda \frac{dq_0}{q_0} f = -f \ln \frac{\Lambda}{\lambda} , \end{aligned} \tag{6.2.34}$$

where f is given by Eq. (6.2.16). The final result is

$$|T_0|^2 = |t_0|^2 \, (\lambda/\Lambda)^f . \tag{6.2.35}$$

Here the factor $(\lambda/\Lambda)^f$ may be expanded in the following perturbation series,

$$(\lambda/\Lambda)^f = 1 + f \ln \frac{\lambda}{\Lambda} + \frac{1}{2} f^2 \ln^2 \frac{\lambda}{\Lambda} + \dots \tag{6.2.36}$$

6.2.3. Cancellation

It is now straightforward to show the cancellation of the soft divergences occurring in the real photon emission and the virtual photon exchange, respectively.

The physical transition rate w is given by Eq. (6.2.22) where the infrared factor $(\omega_m/\lambda)^f$ coming from the real photon emission is explicitly factored out. On the other hand another infrared factor $(\lambda/\Lambda)^f$ originating from the virtual photon effect is extracted from $|T_0|^2$ in Eq. (6.2.35). Inserting Eq. (6.2.35) into Eq. (6.2.22) we confirm the finiteness of the physical transition rate as $\lambda \to 0$, i.e.,

$$w = |t_0|^2 \, (\omega_m/\Lambda)^f \, F(f) . \tag{6.2.37}$$

This completes our all-order proof of the cancellation of the soft divergences in QED.

If we expand Eqs. (6.2.22) and (6.2.35) in a perturbation series, we have

$$w = |t_0|^2 \left[1 + f\left(\ln \frac{\lambda}{\Lambda} + \ln \frac{\omega_m}{\lambda} \right) + \frac{1}{2} f^2 \left(\ln^2 \frac{\lambda}{\Lambda} + \ln^2 \frac{\omega_m}{\lambda} \right.\right.$$
$$\left.\left. + 2 \ln \frac{\lambda}{\Lambda} \ln \frac{\omega_m}{\lambda} - \frac{\pi^2}{6} \right) + \dots \right]. \tag{6.2.38}$$

Equation (6.2.38) exhibits explicitly an order-by-order cancellation of the soft divergences for $\lambda \to 0$.

In the above argument we confined ourselves to the soft-photon region and were indifferent to the ultraviolet divergence. It should be understood that we worked in the cut-off theory as far as the ultraviolet region is concerned. Thus what is shown here is that the soft cancellation takes place in the cut-off theory. We expect the renormalization procedure not to spoil the cancellation mechanism.

For the proof of the soft cancellation we needed to sum up all the contribution of soft-photon emissions. In the low momentum region the photon tends to lose its particle nature and therefore the particle-number representation is inadequate to deal with soft photons. In the discussion of soft photons it is much more convenient to consider the superposition of states with an arbitrary number of photons. An example of such state is the so-called coherent state. The problem of the soft divergence may be formulated in a transparent way by appealing to the notion of the coherent state. (For a review and earlier references, see, e.g., [Pap 76].)

Finally it is to be remarked that the higher-order corrections to the scattering amplitude for the electron off charged particles yield a divergent imaginary part which comes from the soft region and is not cancelled. However, this soft divergence can be shown to be exponentiated and to result in a simple phase factor. Hence this divergence disappears in the physical cross section [Dal 51, Pap 76].

6.3. GENERAL ARGUMENTS FOR INFRARED CANCELLATIONS

6.3.1. Infrared divergence in QCD

In the last section the proof of the cancellation of the soft divergences in QED was given on general grounds for all orders of perturbation theory. In QCD the soft cancellation was demonstrated in the specific process $e^- e^+ \to q\bar{q}$ at the one-loop level in Sec. 6.1.3. For the same process it was also shown in Sec. 5.1.2 that the collinear as well as soft divergences cancel out in the one-loop order. The general feature involving the infrared structure in the above

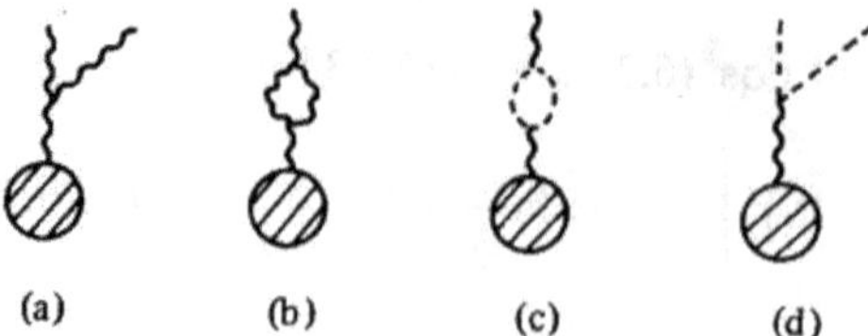

Fig. 6.3.1. An example of a set of diagrams relevant to the cancellation of the soft (or collinear) divergences arising from the triple-gluon vertex: (a) the external gluon splitting into two soft (or collinear) gluons, (b) the external gluon line with insertion of a gluon loop, (c) the external gluon line with insertion of a ghost loop and (d) the external gluon splitting into two soft (or collinear) FP-ghosts.

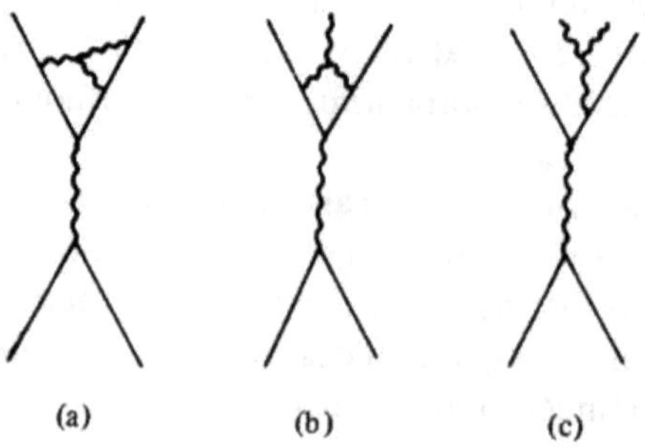

Fig. 6.3.2. Examples of diagrams in two-loop order relevant to the $e^-e^+ \to q\bar{q}$ process.

process at the one-loop level is practically the same as the one in QED since the triple gluon coupling characteristic to QCD does not come into play at the one-loop order in this process. Because of the presence of the triple gluon (and quartic gluon) coupling the infrared structure of QCD is supposed to be quite different from that of QED. For example the soft divergence due to the soft gluon emission from an external gluon line [Fig. 6.3.1.(a)] may reveal, in higher orders, a feature widely different from that in QED. The above soft divergence is expected to be cancelled by the one due to the virtual gluon. One such virtual diagram in the one-loop order is depicted in Fig. 6.3.1(b). In the covariant gauge the gluon loop appearing in Fig. 6.3.1(b) is always accompanied by the ghost loop of Fig. 6.3.1(c). This, in turn, means that we have to take into account the soft ghost emission in the covariant gauge as in Fig. 6.3.1 (d). A further complication in QCD lies in the fact that the collinear divergence may exist from the outset in addition to the soft divergence even if the quarks are kept massive. The reason for this is that a gluon can decay into two collinear gluons through the triple gluon coupling. Beyond the one-loop order in $e^-e^+ \to q\bar{q}$ the triple gluon coupling takes part in the game. Some of the relevant diagrams in the two-loop order are shown in Fig. 6.3.2.

In spite of the above-mentioned difference in the infrared nature between QCD and QED, it is expected that general theorems such as the Kinoshita-Poggio-Quinn theorem and the Kinoshita-Lee-Nauenberg theorem are operative in QCD as far as perturbation theory is concerned. To date no examples in QCD have been found to conflict with the above two theorems. We review, in the following, some important examples investigated in the first several orders in which the cancellation of the infrared divergence is explicitly demonstrated.

The one-loop calculation in the process $e^-e^+ \to q\bar{q}$ as given in Sec. 6.1.3 may be pushed forward to the two-loop order in a straightforward manner. Here the triple gluon coupling comes into play as illustrated in Fig. 6.3.2. The soft divergence is shown by an explicit calculation to cancel out in the physical cross section which takes into account the soft-gluon emissions [App 76, 77]. This cancellation may be thought of as a special case of the Kinoshita-Poggio-Quinn theorem and may correspond to the QCD version of the Bloch-Nordsieck theorem. However, it will be seen shortly that the Bloch-Nordsieck theorem breaks down in a class of QCD processes for nonleading twists. The proof of the soft cancellation in $e^-e^+ \to q\bar{q}$ is extended to all orders in [Kra 77] in accord with the general consequence of the Kinoshita-Poggio-Quinn theorem.

The quark-quark (antiquark) scattering $qq \to qq$ ($q\bar{q} \to q\bar{q}$) to the one-loop order has been discussed by many authors [Yao 76, Tyb 76, Sac 76, Sub 77, Don 78, Kon 78]. It has been shown that the soft divergence in $qq \to qq$ ($q\bar{q} \to q\bar{q}$) at one-loop level cancels (if the initial quarks are color-averaged and the final color-sum is performed) with the soft-gluon emission process $qq \to qqG$ ($q\bar{q} \to q\bar{q}G$). This result is again in conformity with the Bloch-Nordsieck theorem. It should, however, be noted that this is an artifact of the one-loop approximation. The breakdown of the Bloch-Nordsieck theorem in this process at two-loop level was pointed out in [Car 81].

The gluon-quark scattering $Gq \to Gq$ has also been analyzed in the one-loop order [Yao 76, Sug 77] and it was found that the soft cancellation takes place in spite of the presence of the triple gluon coupling.

Quark scattering off the colorless QCD potential is a means of measuring the chromodynamic form factor of quarks. The cross section for this process exhibits a soft divergence. The divergence up to two loops can be shown to cancel if the process is combined with the one involving soft-gluon emission [Kor 76, Fre 76, Lib 79]. A similar argument applies also to gluon scattering by a colorless gluon source $F^a_{\mu\nu} F^{a\mu\nu}$ [Alv 77].

The Drell-Yan processes with quarks in the initial state,

$$qq \to \gamma + X \ , \qquad q\bar{q} \to \gamma + X \ , \tag{6.3.1}$$

with X representing quarks and soft gluons have been investigated up to the two-loop order where the cross section is averaged over the color of the initial quarks. The Bloch-Nordsieck theorem would suggest that the soft divergence is absent in these inclusive cross sections. Doria, Frenkel and Taylor [Dor 80] (see also [DiL 81, Yos 81]), however, discovered that the nonleading soft divergence at two-loop level (single log divergence) cannot be cancelled by the soft-gluon emission effect. This gives a counter example to the Bloch-Nordsieck theorem. A similar phenomenon as above was observed by [Gan 81] in quark scatterings off the colored QCD potential. Soon after, it was recognized that, if the degeneracy of the initial quarks with the incoming soft gluons is taken into account, the above divergence washes out [And 81, Ito 81, Yos 81a]. This conclusion is nothing but the Kinoshita-Lee-Nauenberg theorem applied to QCD. The same conclusion as above was deduced by using the coherent-state method [Nel 81].

One of the concerns arising from the breakdown of the Bloch-Nordsieck theorem in QCD is that the factorization property of short-distance effects in inclusive processes is endangered. However one should note that the noncanceling nonleading soft divergence found by Doria, Frenkel and Taylor [Dor 80] in the process (6.3.1) does not occur in the leading (minimum)-twist term of the cross section, i.e., it occurs only in the term suppressed by $1/Q^2$ where Q^2 is the momentum of the virtual photon squared [Fre 84]. Hence the previous arguments on perturbative QCD given in Chapter 5 are unaffected since they are based on the leading-twist approximation. When we wish to deal with nonleading-twist effects, we are confronted with this problem. It may be a cure for this problem to use the soft degenerate state instead of the single quark state. In this connection it is of some interest to see whether the finite part is affected by the use of the soft degenerate state. In the case of the Drell-Yan process (6.3.1) it has been shown in the one-loop order that the finite part is unaffected [Mut 81].

In summary we see that in spite of the breakdown of the Bloch-Nordsieck theorem, the other theorems (Kinoshita-Poggio-Quinn and Kinoshita-Lee-Nauenberg theorem) seem to be still valid in QCD. This situation prompts us to review these two theorems in the succeeding subsections.

6.3.2. The Kinoshita-Poggio-Quinn theorem

In general in a theory containing at least one massless field the S-matrix element with massive and/or massless external lines suffers from some infrared divergence. By associating soft and/or collinear massless particles with the external lines we can construct degenerate states. The physical cross section or transition rate defined through the use of the degenerate states is known to be infrared-safe. This is the Kinoshita-Lee-Nauenberg theorem which will be discussed in the next subsection. When the degenerate states in the initial state are unnecessary and only the soft divergence is present, the theorem reduces to the Bloch-Nordsieck theorem. Thus the theorem includes the Bloch-Nordsieck theorem as a special case.

If the external lines are off their mass shells, we are led to the general Green functions rather than the S-matrix. Roughly speaking the external momenta which are off the mass shell serve as infrared cut-offs for the Green functions and hence they are expected to be free from infrared divergences. This observation, if stated in a more precise manner, leads us to the following theorem [Kin 62, 76, Pog 76, Ste 76].

Kinoshita-Poggio-Quinn theorem: The proper (one-particle-irreducible) Green functions with Euclidean nonexceptional external momenta are free of the infrared divergence in massless renormalizable field theories.[7] An extension to the case where some of the fields acquire mass is straightforward.

Let us first explain the terminology. By Euclidean momentum p we mean a four-momentum with positive definite metric $p^2 = p_0^2 + p_1^2 + p_2^2 + p_3^2$. It is related to the space-like Minkowskian momentum k by the identification

$$k_0 = ip_0 , \quad k_1 = p_1 , \quad k_2 = p_2 , \quad k_3 = p_3 , \tag{6.3.2}$$

so that $k^2 = -p^2$. In order to give the definition of the nonexceptional momenta it is necessary to introduce the notion of the exceptional momenta. A set of momenta $\{p_i; i = 1, 2, 3, ..., n\}$ is said to be exceptional if at least one partial sum of them vanishes, i.e.,

$$\sum_{i \in I} p_i = 0 , \tag{6.3.3}$$

where I is a subset of $\{1, 2, 3, ..., n\}$. The nonexceptional momenta are those belonging to a set of momenta in which none of the partial sums of the momenta vanishes[8].

Since all the external momenta of the Green function under investigation are Euclidean, any Feynman integral for loop momenta in the Green function may be converted into an Euclidean integral by simply applying the Wick rotation. This allows us to use the naive power-counting method in order to

[7] As an underlying theory we have in mind QCD with massless quarks or QED with massless electrons.

[8] The overall sum of the momenta, of course, vanishes in conformity with the energy-momentum conservation.

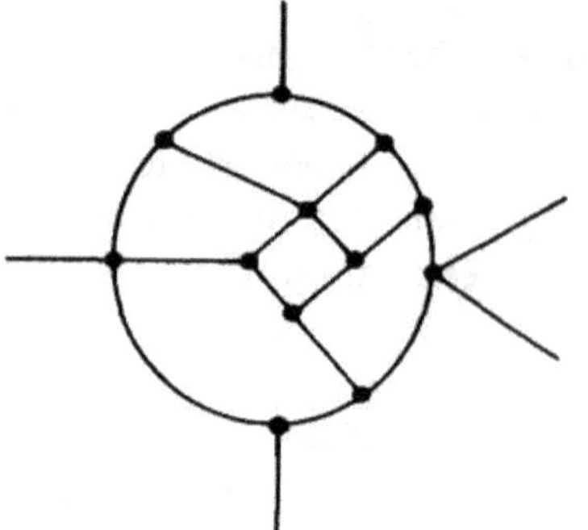

Fig. 6.3.3. An example (F_5^6) of diagrams corresponding to the Green function F_n^l which is defined in the text.

see whether the infrared divergence emerges or not. Moreover the possible collinear divergences originating from the massless external lines are avoided by choosing nonexceptional momenta.

In proving the theorem we follow an intuitive method given by Poggio and Quinn [Pog 76, 77]. Before going into the proof of the theorem we would like to show the following lemma since it will be fully employed in the main proof.

Lemma: The renormalized two-, three- and four-point proper Green functions $F_n(p_i)$ with $n = 2, 3, 4$ and $i = 1, 2, \ldots, n$ which are truncated (amputated) behave for $\eta \to 0$ as

$$F_n(\eta p_i) \to \eta^{4-n-\varepsilon}\,, \quad \text{for } n = 2, 3, 4, \tag{6.3.4}$$

where by $\eta^{-\varepsilon}$ we mean[9] the term which grows more slowly than any power of $1/\eta$ as $\eta \to 0$ (i.e., the term consisting of $\ln \eta$) and the momenta p_i are Euclidean and nonexceptional.

Proof of the lemma: We prove the lemma by induction, i.e., we assume Eq. (6.3.4) to hold up to the $(l-1)$-loop order and then show that it also holds at the l-loop level. Obviously at zero-loop (tree) level the formula (6.3.4) is satisfied with $\varepsilon = 0$. The Green function $F_n^l(p_i)$ at l-loop level may be expressed as a Feynman integral for the product of propagators and vertices F_n^k $(n = 2, 3, 4)$ with $k \leq l-1$ (the proper two-, three- and four-point Green functions up to the $(l-1)$-loop order). A typical example of the diagrams expressing F_n^l is given in Fig. 6.3.3. For the l-loop diagram corresponding to F_n^l we denote by

[9] The form of the term $\eta^{-\varepsilon}$ may differ depending on whether $n = 2, 3$ or 4. Note that the lemma is essentially equivalent to the Weinberg theorem [Wei 60].

n_I, v_2, v_3 and v_4 the numbers of internal lines and vertices F_2^k, F_3^k, F_4^k ($k \le l-1$) respectively. We have in accordance with Eq. (2.5.8) or (2.5.9),[10]

$$2n_I + n = 2v_2 + 3v_3 + 4v_4 \ . \tag{6.3.5}$$

We also have, corresponding to Eq. (2.5.10),

$$l = n_I - v_2 - v_3 - v_4 + 1 \ . \tag{6.3.6}$$

On the other hand we note, by simple power counting, that

$$F_n^l(\eta p_i) \sim \int \left(\prod_{j=1}^{l} d^4q \right) (q^2)^{-n_I} (F_2^k)^{v_2} (F_3^k)^{v_3} (F_4^k)^{v_4} \sim \eta^{a_n} \ , \tag{6.3.7}$$

where the exponent a_n is given by

$$a_n = 4l - 2n_I + (2 - \varepsilon)\, v_2 + (1 - \varepsilon)\, v_3 - \varepsilon v_4 \ . \tag{6.3.8}$$

Using Eqs. (6.3.5) and (6.3.6) we find

$$a_n = 4 - n - \varepsilon(v_2 + v_3 + v_4) \ . \tag{6.3.9}$$

This completes our proof of the lemma.[11]

We are now ready to get on to the proof of the Kinoshita-Poggio-Quinn theorem. Let us consider again the diagram of Fig. 6.3.3 as an example. According to momentum conservation all the external momenta which are nonvanishing flow through internal lines along a certain path which may contain branchings and loops. Of course this path depends on the choice of the set of the loop momenta. An example of the path is shown in Fig. 6.3.4 by

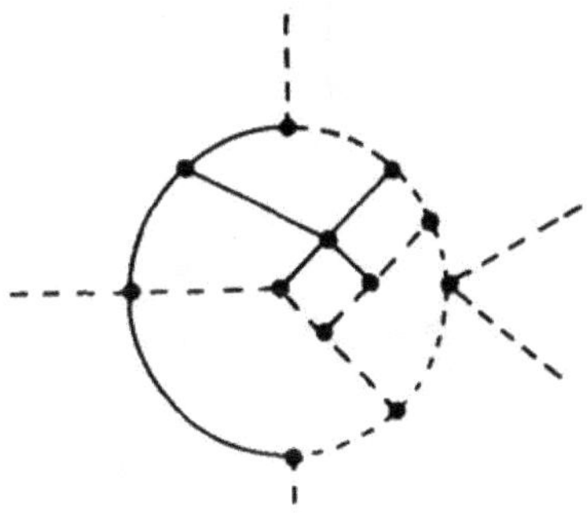

Fig. 6.3.4. The path (dotted lines) showing the flow of external momenta in Fig. 6.3.3.

[10] For simplicity we consider only bosons. The extension to include fermions is straightforward [Pog 76].
[11] As ε is infinitesimal, $\varepsilon(v_2 + v_3 + v_4)$ is also infinitesimal.

dotted lines. Since an internal line carrying the external momenta does not contribute to the infrared divergence (i.e., the external momenta play the role of infrared cut-offs), we may disregard, as far as the infrared divergence is concerned, those internal lines that belong to the above-mentioned path. Hence we reduce this path to a point. For example the diagram of Fig. 6.3.4 is converted to the one in Fig. 6.3.5. In Fig. 6.3.5 only the part of the diagram surrounded by the dotted line is relevant to the infrared divergence. In general the diagram G with n external lines can be always reduced to one like the diagram of Fig. 6.3.5 as shown in Fig. 6.3.6(b). The part of the diagram G denoted by $\bar{G}$ in Fig. 6.3.6(b) has m external lines and is responsible for the possile infrared divergence. We call $\bar{G}$ the reduced diagram. In order to see the infrared property of F_n^l, corresponding to the diagram G, it suffices to examine the infrared behavior of the Green function $\bar{F}_m$ corresponding to the reduced diagram $\bar{G}$. We apply to $\bar{F}_m$ the power-counting rule as in the proof of the lemma. Denoting the numbers of internal lines, loops and vertices F_2^k, F_3^k, F_4^k $(k \leq l-1)$ in $\bar{G}$ by $\bar{n}_I$, $\bar{l}$, $\bar{v}_2$, $\bar{v}_3$ and $\bar{v}_4$, respectively, we have

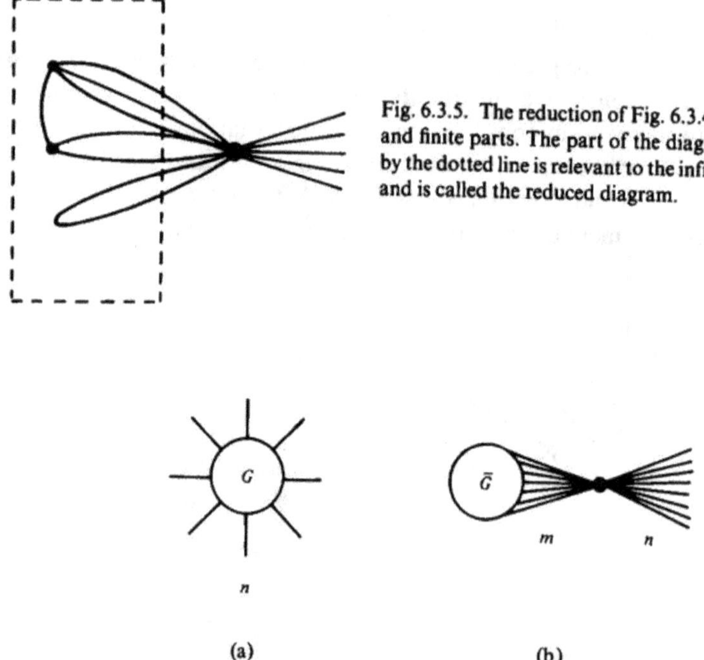

Fig. 6.3.5. The reduction of Fig. 6.3.4 to its divergent and finite parts. The part of the diagram surrounded by the dotted line is relevant to the infrared divergence and is called the reduced diagram.

Fig. 6.3.6. The reduced diagram $\bar{G}$ in (b) is obtained from the diagram G in (a) by discarding the lines carrying the external momenta.

$$2\bar{n}_I = 2\bar{v}_2 + 3\bar{v}_3 + 4\bar{v}_4 + m \,, \tag{6.3.10}$$

$$\bar{l} = \bar{n}_I - \bar{v}_2 - \bar{v}_3 - \bar{v}_4 + 1 \,. \tag{6.3.11}$$

Note that the Green function $\bar{F}_m$ is nontruncated and so m appears on the right-hand side in Eq. (6.3.10). Then the exponent $\bar{a}_m$, defined analogously to Eq. (6.3.7), for $\bar{F}_m$ reads

$$\begin{aligned}\bar{a}_m &= 4\bar{l} - 2\bar{n}_I + (2-\varepsilon)\,\bar{v}_2 + (1-\varepsilon)\,\bar{v}_3 - \varepsilon\bar{v}_4 \,, \\ &= m - \varepsilon(\bar{v}_2 + \bar{v}_3 + \bar{v}_4) \,.\end{aligned} \tag{6.3.12}$$

Since $m \geq 2$, we find that $\bar{a}_m \geq 2$. This means that $\bar{F}_m$ and consequently F^l_n has no infrared divergence. QED.

One of the important applications of the Kinoshita-Poggio-Quinn theorem may be found in the straightforward verification of the infrared cancellation in the e^+e^-annihilation total cross sections. This cancellation has already been discussed in Sections 5.1.2 and 6.1.3 at the one-loop level through explicit calculation. Here we show that the infrared cancellation in the above total cross section is a direct consequence of the Kinoshita-Poggio-Quinn theorem.

Consider the Green function for electromagnetic current $j_\mu(x)$ defined by

$$\pi_{\mu\nu}(q) = i \int d^4x\, e^{iq\cdot x} \langle 0\,|\mathrm{T}[j_\mu(x)\, j_\nu(0)]|0\rangle \,. \tag{6.3.13}$$

According to the theorem, the Green function $\pi_{\mu\nu}(q)$ is free from infrared divergences if the photon momentum q is space-like, $q^2 < 0$. Let us regard $\pi_{\mu\nu}(q)$ as a function of complex q_0 with $\mathbf{q}$ fixed. It then has a physical cut along the real q_0-axis for $q_0^2 > \mathbf{q}^2$ corresponding to the time-like q, $q^2 > 0$. Since we have no singularity except for the cut, we can make an analytic continuation of $\pi_{\mu\nu}(q)$ into the time-like region provided that q_0 is not on the cut. We conclude that, for real time-like q, $\pi_{\mu\nu}(q_0 + i\varepsilon)$ and $\pi_{\mu\nu}(q_0 - i\varepsilon)$ are infrared-finite with ε a positive infinitesimal quantity. Hence the absorptive part of $\pi_{\mu\nu}(q)$, defined by

$$\mathrm{Abs}\,\pi_{\mu\nu}(q) = \frac{\pi_{\mu\nu}(q_0 + i\varepsilon) - \pi_{\mu\nu}(q_0 - i\varepsilon)}{2i} \,, \tag{6.3.14}$$

is also infrared-finite. Using Eq. (6.3.13) we can easily show that

$$\mathrm{Abs}\,\pi_{\mu\nu}(q) = \frac{1}{2} w_{\mu\nu} \,, \tag{6.3.15}$$

where $w_{\mu\nu}$ is given by Eq. (4.1.53). Accordingly the R-ratio which is related to $w_{\mu\nu}$ through Eq. (5.1.1), i.e,

$$R = -\frac{2\pi}{q^2} g^{\mu\nu} w_{\mu\nu} , \tag{6.3.16}$$

is free from infrared divergences.

6.3.3. The Kinoshita-Lee-Nauenberg Theorem

Consider quantum electrodynamics. We have seen that an electron accompanied by an arbitrary number of soft photons cannot be distinguished from a single electron. The states with an electron and soft photons belong to the same energy-eigenstate as that of a single electron in the limit of vanishing photon energy. These states are said to be degenerate with the single-electron state and are called (soft) *degenerate states*. This degeneracy is the origin of the soft divergence in transition rates. As we have already seen, the soft divergence cancels out in the total transition rates, i.e., the transition rates summed over the final states. This is the Bloch-Nordsieck theorem.

In Sec. 6.3.1 we mentioned that the Bloch-Nordsieck theorem in general breaks down in quantum chromodynamics. It has, however, been recognized that the summation on the initial degenerate states as well as over the final states serves to eliminate the soft divergence in QCD. This is a specific example of the Kinoshita-Lee-Nauenberg theorem [Kin 62, Lee 64] which will be discussed thoroughly in this subsection.

We have also seen in QCD that a massless quark accompanied by an arbitrary number of collinear gluons cannot be distinguished from a single quark. States with a massless quark and gluons are degenerate with the single-quark state and are called *(collinear) degenerate states*. The collinear degeneracy is the origin of the collinear divergence of transition rates. It is known that the collinear divergence does not necessarily cancel out in the total transition rate. To ensure its cancellation we need a general theorem, the Kinoshita-Lee-Nauenberg theorem [Kin 62, Lee 64]. The theorem asserts that the collinear divergence cancels out if the transition rate is summed over all the initial degenerate states as well as the final ones. The theorem also holds for the soft divergence as we have just remarked above. Stated in the most general form the theorem reads as follows.

Kinoshita-Lee-Nauenberg theorem: In a theory with massless fields, transition rates are free of the infrared (soft and collinear) divergence if the summation over the initial and final degenerate states is carried out.

In order to prove the theorem we need some preparatory arguments. We shall follow the simple argument given by Lee and Nauenberg [Lee 64, 81] who make essential use of the unitarity of the S-matrix in showing the absence of the infrared divergence in each order of perturbation series. Note here that Kinoshita [Kin 62, 76] directly deals with the parametric representation of Feynman integrals and presents a straightforward proof of the cancellation of the infrared divergence.

Let us first derive a convenient expression of the S-matrix in the interaction representation. In this representation the Schrödinger equation for state $|\Psi(t)>$ reads

$$i\frac{\partial}{\partial t}|\Psi(t)> = g\hat{H}_1(t)|\Psi(t)> \ , \tag{6.3.17}$$

with

$$\hat{H}_1(t) = e^{iH_0 t} H_1 e^{-iH_0 t} \ , \tag{6.3.18}$$

where H_0 and gH_1 are the free and interaction Hamiltonian respectively and g the coupling constant. We define a unitary operator $U(t, t')$ describing the time evolution of the state $|\Psi(t)>$,

$$|\Psi(t)> = U(t, t')|\Psi(t')> \ , \tag{6.3.19}$$

with the property

$$U(t, t')\,U(t', t'') = U(t, t'') \ . \tag{6.3.20}$$

Using Eqs. (6.3.17) and (6.3.19) we obtain

$$i\frac{\partial}{\partial t}U(t, t') = g\hat{H}_1(t)\,U(t, t') \ . \tag{6.3.21}$$

Solving Eq. (6.3.21) we find

$$U(t, t') = \mathrm{T}\exp\left[-i\int_{t'}^{t} d\lambda g\hat{H}_1(\lambda)\right] \ , \tag{6.3.22}$$

where T stands for the time-ordered product. The S-matrix is defined by

$$S = U(\infty, -\infty) \ . \tag{6.3.23}$$

Using the property (6.3.20) we have

$$S = U(0, \infty)^{\dagger}\,U(0, -\infty) \ . \tag{6.3.24}$$

Sandwiching Eq. (6.3.24) between states $|a>$ and $|b>$ and squaring the resulting expression, we deduce a useful representation of the transition rate for $a \to b$,

$$|<b|S|a>|^2 = \sum_{ij} (R^+_{bij})^* R^-_{aij} , \tag{6.3.25}$$

where

$$R^\pm_{aij} = \langle i|U(0, \pm\infty)\,|a\rangle^* \langle j|U(0, \pm\infty)|a\rangle . \tag{6.3.26}$$

The representation (6.3.25) is convenient since it factorizes the transition rate into two parts, one depending only on the final state $|b>$ and the other depending only on the initial state $|a>$. This allows us to discuss the infrared divergences due to the initial degeneracy and final degeneracy separately.

Let us consider the perturbative expansion of the matrix element of $U(0, \pm\infty)$. Expanding Eq. (6.3.22) in powers of g and noting Eq. (6.3.18) we find

$$\langle i|U(0, \pm\infty)|j\rangle = \delta_{ij} - \frac{gH_{1ij}}{E_i - E_j \pm i\varepsilon} + O(g^2) , \tag{6.3.27}$$

where the Kronecker delta δ_{ij} indicates that all the quantum numbers for the state $|i>$ are the same as those for the state $|j>$. The infinitesimal constant $\varepsilon(>0)$ is introduced to ensure the convergence of the integral in Eq. (6.3.22) for $t' \to \pm\infty$. We use the notation

$$H_{1ij} = \langle i|H_1|j\rangle , \tag{6.3.28}$$

and E_i is the eigenvalue of H_0 for the state $|i>$. Substituting Eq. (6.3.27) into Eq. (6.3.26) we have

$$R^\pm_{aij} = \delta_{ia}\delta_{ja} - \frac{gH_{1ia}{}^*}{E_i - E_a \mp i\varepsilon}\delta_{ja} - \frac{gH_{1ja}}{E_j - E_a \pm i\varepsilon}\delta_{ia} + O(g^2) . \tag{6.3.29}$$

The matrix element H_{1ia} (H_{1ja}) describes the virtual transition from the state $|a>$ to $|i>$ ($|j>$). For example, in the scattering of an electron by an external source in QED, the state $|a>$ may be the electron state $|e>$ while $|i>$ ($|j>$) may be the state of an electron and a photon $|e\gamma>$. If there occurs the degeneracy between the states $|a>$ and $|i>$ ($|j>$), we have $E_i = E_a$ ($E_j = E_a$) and hence $R^\pm_{aij}$ diverges. A typical example of such degeneracy in the above QED case emerges from the soft configuration (γ is soft in $|e\gamma>$) or in the collinear configuration (γ is collinear to the massless e in $|e\gamma>$). In this case the divergence of $R^\pm_{aij}$ corresponds to the infrared divergence. Although the above degeneracy is not merely of infrared nature, we shall confine ourselves to infrared divergences in the following arguments.

The theorem asserts that in order to eliminate the divergence in Eq. (6.3.29) one should perform a summation on the initial and final degenerate states $|a>$ and $|b>$. Let us denote by $D(E)$ a set of all the states with energy E. Then according to the theorem, the quantity

$$\sum_{a\in D(E)} \sum_{b\in D(E)} |<b|S|a>|^2 , \tag{6.3.30}$$

is free from the infrared divergences. In other words on account of Eq. (6.3.25),

$$R^{\pm}_{ij}(E) = \sum_{a\in D(E)} R^{\pm}_{aij} , \tag{6.3.31}$$

has no infrared divergence.[12] In fact, by applying the degenerate-state sum of Eq. (6.3.31) to Eq. (6.3.29), we find to order g,

$$R^{\pm}_{ij}(E) = \begin{cases} 0 & \text{if } i, j \bar{\in} D(E) , \\ -\dfrac{gH_{1ij}{}^*}{E_i - E \mp i\varepsilon} & \text{if } i \bar{\in} D(E), j \in D(E) , \\ -\dfrac{gH_{1ji}}{E_j - E \pm i\varepsilon} & \text{if } i \in D(E), j \bar{\in} D(E) , \\ \delta_{ij} & \text{if } i, j \in D(E) . \end{cases} \tag{6.3.32}$$

Hence $R^{\pm}_{ij}(E)$ [and consequently Eq. (6.3.30)] are infrared-finite to order g.

The above argument up to order g can be generalized to any order of perturbation theory. In the remaining part of this subsection we shall devote ourselves to showing that $R^{\pm}_{ij}(E)$ does not suffer from infrared divergences for any order of perturbation theory. We first note that the total Hamiltonian H may be diagonalized by using the unitary operator $U(0, \pm\infty)$, i.e.,

$$U^{\dagger} H U = \hat{H}_0 , \tag{6.3.33}$$

where $\hat{H}_0$ is a diagonal operator. According to the unitarity of U we may rewrite Eq. (6.3.33) in the form $U\hat{H}_0 = HU$ and hence we have

$$[U, \hat{H}_0] = (H - \hat{H}_0)\, U = (gH_1 + \Delta)\, U , \tag{6.3.34}$$

where

$$\Delta = H_0 - \hat{H}_0 . \tag{6.3.35}$$

[12] Here $\in$ and $\bar{\in}$ indicate "included" and "not included."

Note that Δ is a diagonal operator.[13] Let us assume that Δ, U and $R^{\pm}_{ij}$ are calculated in perturbation theory, i.e.,

$$\Delta = \sum_n g^n \Delta_n \ , \tag{6.3.36}$$

$$U = \sum_n g^n U_n \ , \tag{6.3.37}$$

$$R = \sum_n g^n R_n \ , \tag{6.3.38}$$

where, for simplicity, indices i, j, and the superfix $\pm$ are omitted in Eq. (6.3.38) for $R^{\pm}_{ij}$ and $R^{\pm}_{n,ij}$. Inserting Eq. (6.3.37) into Eq. (6.3.26) and using Eq. (6.3.31) we obtain

$$R^{\pm}_{ij}(E) = \sum_{rs} g^{r+s} \sum_{a \in D(E)} \langle i|U_r(0, \pm\infty)\,|a\rangle^* \langle j|U_s(0, \pm\infty)\,|a\rangle \ . \tag{6.3.39}$$

Hence we have

$$R^{\pm}_{n,ij}(E) = \sum_r \sum_{a \in D(E)} \langle i|U_r(0, \pm\infty)\,|a\rangle^* \langle j|U_{n-r}(0, \pm\infty)|a\rangle \ . \tag{6.3.40}$$

The following lemma holds.

Lemma: If $\langle i|\Delta_n|i\rangle$ is free from infrared divergences for $n \leq N$, then $R^{\pm}_{n,ij}$ is infrared-finite for $n \leq N + 1$.

Once we prove the above lemma, it immediately follows that Eq. (6.3.30) is infrared-finite and hence the Kinoshita-Lee-Nauenberg theorem is verified.

We use induction to prove the lemma: For $n = 1$ the statement was already proven [see Eq. (6.3.32)]. Assuming $R^{\pm}_{n,ij}$ to be finite for $n \leq M$ $(M < N + 1)$ we shall show that $R^{\pm}_{M+1,ij}$ is infrared-finite. We deal with the following three cases separately.

(1) $i \in D(E)$

Here j may or may not be included in $D(E)$. By sandwiching Eq. (6.3.34) between states $|i\rangle$ and $|a\rangle$ we obtain

$$(E_a - E_i)\, U_{ia} = g \sum_k H_{1ik}\, U_{ka} + \Delta_i U_{ia} \ , \tag{6.3.41}$$

[13] In the case of QED the one-electron matrix element of Δ is nothing but the mass shift δm of the electron.

where $\Delta_i = \langle i|\Delta|i\rangle$ and

$$U_{ia} = \langle i|U(0, \pm\infty)|a\rangle \ . \tag{6.3.42}$$

Since $E_i \neq E_a$, Eq. (6.3.41) can be divided by $(E_a - E_i)$ to give a nontrivial relation among perturbative coefficients $U_{n,ia} = \langle i|U_n|a\rangle$. In fact, by substituting Eqs. (6.3.36) and (6.3.37) into Eq. (6.3.41), we find

$$U_{r,ia} = \frac{1}{E_a - E_i}\left[\sum_k H_{1ik}\, U_{r-1,ka} + \sum_s \Delta_{s,i}\, U_{r-s,ia}\right] , \tag{6.3.43}$$

where $\Delta_{s,i} = \langle i|\Delta_s|i\rangle$. Combining Eq. (6.3.43) with Eq. (6.3.40) we have

$$R^{\pm}_{n,ij}(E) = \frac{1}{E - E_i}\left[\sum_k H_{1ik}\, R^{\pm}_{n-1,kj}(E) + \sum_{s=1}^{n-1} \Delta_{s,i}\, R^{\pm}_{n-s,ij}(E)\right] . \tag{6.3.44}$$

From Eq. (6.3.44) we immediately recognize that R_n is expressed in terms of $R_m (m < n)$ and $\Delta_m\, (m < n)$. Thus we conclude that, if $R_n\, (n \leq M)$ is infrared-finite, R_{M+1} is too.

(2) $j \notin D(E)$

Here i may or may not be contained in $D(E)$. This case is essentially the same as the previous case (1) according to the hermiticity property,

$$R^{\pm}_{n,ij}(E)^* = R^{\pm}_{n,ji}(E) \ . \tag{6.3.45}$$

(3) $i, j \in D(E)$

In this case Eq. (6.3.43) turns out to be useless. We use the unitarity relation $U^\dagger U = 1$ which implies

$$\sum_r U^\dagger_r\, U_{n-r} = 0 \quad (n \neq 0) \ . \tag{6.3.46}$$

This may be rewritten as

$$\sum_r \sum_{a \in D(E)} U_{r,ia}{}^* U_{n-r,ja} + \sum_r \sum_{a \notin D(E)} U_{r,ia}{}^* U_{n-r,ja} = 0 \ . \tag{6.3.47}$$

Hence we have

$$R^{\pm}_{M+1,ij}(E) = -\sum_{r=0}^{M+1} \sum_{a \notin D(E)} U_{r,ia}{}^* U_{M+1-r,ja} . \tag{6.3.48}$$

Equation (6.3.48) implies that $R^{\pm}_{M+1,ij}(E)$ is expressible as a sum of $R^{\pm}_{M+1,ij}$ with the states $|i>$ and $|j>$ nondegenerate with the state of energy E. According to the argument in (1) these $R^{\pm}_{M+1,ij}$ are infrared-finite and hence Eq. (6.3.48) is finite. This completes the proof of the lemma.

It should be noted that in the above proof we essentially used the unitarity of the operator U. Since the matrix element of the unitary operator is bounded, no infrared divergence can appear in this matrix element. The theorem asserts the stronger statement that at each order of perturbation series the infrared divergence is absent.

In the Kinoshita-Lee-Nauenberg theorem it is necessary to perform a summation on the initial degenerate states as well as the final ones. This means that the initial states in the Kinoshita-Lee-Nauenberg theorem may be treated just like the final states for a total cross section in the Bloch-Nordsieck theorem. We saw in Sec. 6.1.3 that the Bloch-Nordsieck cancellation can be viewed diagrammatically as the compensation of the soft divergences appearing in different cuts of the same diagram. Corresponding to this diagrammatic interpretation of the Bloch-Nordsieck cancellation we may consider the diagram which is obtained by putting together the initial as well as final external lines of original diagrams relevant to the cross section under consideration. A set of diagrams for which the cancellation of the infrared divergence takes place is generated by cutting the above-mentioned diagram in all possible ways. A diagram of this type was introduced by Kinoshita [Kin 50, 62] and Nakanishi [Nak 58] and is sometimes called the Kinoshita diagram [And 81].

APPENDIX A

ONE-LOOP CONTRIBUTION TO SUPERFICIALLY DIVERGENT FEYNMAN AMPLITUDES IN QCD

We present the details of the calculations for obtaining the one-loop results exhibited in Sec. 2.5.5. The calculations will be fully given on the gluon, ghost and quark self-energy parts and the ghost-gluon and quark-gluon vertices, while for the three- and four-gluon vertices only original papers will be quoted since the calculation is too lengthy to be presented here. In the following calculations, for simplicity, we omit the suffix r on the parameters g_r, m_r and α_r, which indicates that these parameters are renormalized ones. We do not think that this omission leads to any confusion.

(1) *Gluon self-energy part* $\Pi^{ab}_{\mu\nu}(k)$

The diagrams to be considered are given in Fig. 2.5.15. The gluon-loop, gluon-tadpole-loop, ghost-loop, quark-loop and counter-term contributions to $\Pi^{ab}_{\mu\nu}$ will be marked by suffixes G, T, FP, F and C respectively. We shall now calculate those five contributions separately. Following the Feynman rules we find for $\Pi^{ab}_{G\mu\nu}(k)$

$$\Pi^{ab}_{G\mu\nu}(k) = \frac{1}{2!}\int\frac{d^Dq}{(2\pi)^Di}(-i)\,gf^{acd}\,V_{\mu\lambda\rho}(-k, q+k, -q)\frac{d^{\lambda\kappa}(q+k)}{(q+k)^2}$$
$$\times\frac{d^{\rho\sigma}(q)}{q^2}(-i)\,gf^{bdc}\,V_{\nu\sigma\kappa}(k, q, -q-k)\,. \tag{A.1}$$

After some algebra we obtain

$$\Pi^{ab}_{G\mu\nu}(k) = \frac{g^2}{2}f^{acd}f^{bcd}\int\frac{d^Dq}{(2\pi)^Di}\frac{1}{q^2(q+k)^2}[A_{\mu\nu} + (1-\alpha)\,B_{\mu\nu} + (1-\alpha)^2C_{\mu\nu}], \tag{A.2}$$

$$A_{\mu\nu} = (2q^2+2k\cdot q+5k^2)\,g_{\mu\nu} + (4D-6)q_\mu q_\nu + (2D-3)\,(q_\mu k_\nu+q_\nu k_\mu)$$
$$+ (D-6)\,k_\mu k_\nu\,, \tag{A.3}$$

$$B_{\mu\nu} = -\frac{(q^2+2k\cdot q)^2}{q^2}g_{\mu\nu} + \frac{q^2+2k\cdot q-k^2}{q^2}q_\mu q_\nu$$
$$+\frac{q^2+3k\cdot q}{q^2}(q_\mu k_\nu + q_\nu k_\mu) - k_\mu k_\nu + (q\to q+k,\, k\to -k), \tag{A.4}$$

$$C_{\mu\nu} = (k^2q_\mu - k\cdot qk_\mu)(k^2q_\nu - k\cdot qk_\nu)/[q^2(q+k)^2]. \tag{A.5}$$

Using the formulas (2.5.178–180) found in the exercise in Sec. 2.5.5, integrations in Eq. (A.2) can be performed to give

$$\Pi^{ab}_{G\mu\nu}(k) = \frac{g^2}{2(4\pi)^{2-\varepsilon}}\delta_{ab}C_G(-k^2)^{-\varepsilon}\frac{\Gamma(\varepsilon)B(2-\varepsilon,2-\varepsilon)}{1-\varepsilon}$$
$$\times[(19-12\varepsilon)k^2g_{\mu\nu} - 2(11-7\varepsilon)k_\mu k_\nu$$
$$+(k^2g_{\mu\nu}-k_\mu k_\nu)(3-2\varepsilon)(1-\alpha)(2(1-4\varepsilon)+(1-\alpha)\varepsilon)], \tag{A.6}$$

where $\varepsilon = (4-D)/2$ and we used Eq.(2.5.133) to rewrite $f^{acd}f^{bcd}$. Here clearly we see that the gluon-loop contribution $\Pi^{ab}_{G\mu\nu}$ alone does not satisfy the requirement of gauge invariance (the Ward-Takahashi identity) given by Eq. (2.5.134). To satisfy this requirement we need an additional contribution from the FP-ghost loop.

Before calculating the ghost-loop contribution we consider the gluon-tadpole-loop contribution $\Pi^{ab}_{T\mu\nu}(k)$ which reads

$$\Pi^{ab}_{T\mu\nu}(k) = \frac{1}{2!}\int\frac{d^Dq}{(2\pi)^Di}(-g^2W^{abcd}_{\mu\nu\lambda\rho})\,\delta_{cd}\frac{d^{\lambda\rho}(q)}{q^2}. \tag{A.7}$$

Equation (A.7) may be rewritten as

$$\Pi^{ab}_{T\mu\nu}(k) = g^2C_G\delta_{ab}\int\frac{d^Dq}{(2\pi)^Di}\frac{1}{q^2}\left[-(D-1)g_{\mu\nu} + (1-\alpha)\left(g_{\mu\nu} - \frac{q_\mu q_\nu}{q^2}\right)\right],$$
$$= g^2C_G\delta_{ab}g_{\mu\nu}\frac{D-1}{D}(-D+1-\alpha)\int\frac{d^Dq}{(2\pi)^Di}\frac{1}{q^2}. \tag{A.8}$$

As discussed in the Digression at the end of Sec. 2.5.5, we can set, in the sense of the dimensional regularization,

$$\int\frac{d^Dq}{q^2} = 0,$$

and hence we obtain

$$\Pi^{ab}_{T\mu\nu}(k) = 0. \tag{A.9}$$

Note here that, if the gluons were massive, Eq. (A.9) would not follow and $\Pi^{ab}_{T\mu\nu}(k)$ would receive nonvanishing contribution.

We next calculate the ghost-loop contribution $\Pi^{ab}_{FP\mu\nu}(k)$ to the gluon self-energy part. It is given by

$$\Pi^{ab}_{FP\mu\nu}(k) = -\int\frac{d^Dq}{(2\pi)^Di}(-i)\,gf^{acd}\,q_\mu\frac{-1}{(q+k)^2}(-i)\,gf^{bdc}(q+k)_\nu\frac{-1}{q^2}. \tag{A.10}$$

Using again the formulas (2.5.179) and (2.5.180), we find

$$\Pi^{ab}_{FP\mu\nu}(k) = \frac{g^2}{2(4\pi)^{2-\varepsilon}}\delta_{ab}C_G(-k^2)^{-\varepsilon}\frac{\Gamma(\varepsilon)B(2-\varepsilon,2-\varepsilon)}{1-\varepsilon}$$
$$\times\{k^2g_{\mu\nu}+2(1-\varepsilon)k_\mu k_\nu\}. \tag{A.11}$$

If we add Eq. (A.11) to Eq. (A.6), we obtain a gauge invariant expression as is expected:

$$\Pi^{ab}_{G\mu\nu}(k)+\Pi^{ab}_{FP\mu\nu}(k) = \frac{g^2}{(4\pi)^{2-\varepsilon}}\delta_{ab}C_G(-k^2)^{-\varepsilon}\frac{\Gamma(\varepsilon)B(2-\varepsilon,2-\varepsilon)}{1-\varepsilon}$$
$$\times(k^2g_{\mu\nu}-k_\mu k_\nu)[2(5-3\varepsilon)+(1-\alpha)(1-4\varepsilon)$$
$$\times(3-2\varepsilon)+(1-\alpha)^2\frac{\varepsilon}{2}(3-2\varepsilon)]. \tag{A.12}$$

We finally calculate the quark-loop contribution $\Pi^{ab}_{F\mu\nu}(k)$. It is given by

$$\Pi^{ab}_{F\mu\nu}(k) = -N_f\int\frac{d^Dp}{(2\pi)^Di}\mathrm{Tr}\left[g\gamma_\mu T^a\frac{1}{m-\not p-\not k}g\gamma_\nu T^b\frac{1}{m-\not p}\right], \tag{A.13}$$

where N_f is the number of quark flavors and the trace Tr refers to both gamma matrices γ_μ and SU(N) generators T^a. Using Eq. (2.4.74) after Feynman parametrization and applying Eq. (2.5.133), we obtain

$$\Pi^{ab}_{F\mu\nu}(k) = -4g^2T_RN_f\delta_{ab}\int_0^1dx\int\frac{d^Dp'}{(2\pi)^Di}\frac{N_{\mu\nu}}{(-p'^2+K)^2}, \tag{A.14}$$

where

$$N_{\mu\nu} = \left[m^2+x(1-x)k^2+\left(\frac{2}{D}-1\right)p'^2\right]g_{\mu\nu}-2x(1-x)k_\mu k_\nu, \tag{A.15}$$

$$K = m^2-x(1-x)k^2. \tag{A.16}$$

We then use Eq. (2.4.76) to obtain

$$\Pi^{ab}_{F\mu\nu}(k) = -\frac{8g^2}{(4\pi)^{2-\varepsilon}}\delta_{ab}T_R N_f \Gamma(\varepsilon)(k^2 g_{\mu\nu} - k_\mu k_\nu) \times \int_0^1 dx\, x(1-x)[m^2 - x(1-x)k^2]^{-\varepsilon}. \tag{A.17}$$

Hence the quark-loop contribution is gauge-invariant by itself.

The counter-term contribution is readily obtained through the Feynman rules given in Sec. 2.5.5,

$$\Pi^{ab}_{C\mu\nu}(k) = (Z_3 - 1)\delta_{ab}(k_\mu k_\nu - k^2 g_{\mu\nu}). \tag{A.18}$$

We now add up all the one-loop contributions to the gluon self-energy part to obtain

$$\Pi^{ab}_{\mu\nu}(k) = \delta_{ab}(k_\mu k_\nu - k^2 g_{\mu\nu})\,\Pi(k^2), \tag{A.19}$$

$$\begin{aligned}\Pi(k^2) = \frac{g^2}{(4\pi)^{2-\varepsilon}}(-k^2)^{-\varepsilon}\Gamma(\varepsilon)B(2-\varepsilon, 2-\varepsilon)\Bigg[-\frac{C_G}{1-\varepsilon}\Bigg\{10 - 6\varepsilon \\ + (1-\alpha)(3-2\varepsilon)\left(1 - 4\varepsilon + \frac{\varepsilon}{2}(1-\alpha)\right)\Bigg\} \\ + \frac{8T_R N_f}{B(2-\varepsilon, 2-\varepsilon)}\int_0^1 dx\, x(1-x)\left(x(1-x) - \frac{m^2}{k^2}\right)^{-\varepsilon}\Bigg] \\ + Z_3 - 1.\end{aligned} \tag{A.20}$$

Expanding Eq. (A.20) in ε we find the following expression for $\Pi(k^2)$,

$$\begin{aligned}\Pi(k^2) = -\frac{g_0^2}{(4\pi)^2}C_G\left[\left(\frac{13}{6} - \frac{\alpha}{2}\right)\left(\frac{1}{\varepsilon} - \gamma - \ln\frac{-k^2}{4\pi\mu^2}\right) + \frac{31}{9} - (1-\alpha) \right. \\ \left. + \frac{(1-\alpha)^2}{4}\right] + \frac{g_0^2}{(4\pi)^2}T_R N_f \frac{4}{3}\left[\frac{1}{\varepsilon} - \gamma - \ln\frac{-k^2}{4\pi\mu^2} \right. \\ \left. - 6\int_0^1 dx\, x(1-x)\ln\left(x(1-x) - \frac{m^2}{k^2}\right)\right] + Z_3 - 1,\end{aligned} \tag{A.21}$$

where $g = g_0\mu^\varepsilon$ with g_0 the renormalized dimensionless coupling constant (note that we have been omitting the suffix r for renormalized quantities).

(2) *FP-ghost self-energy part* $\tilde{\Pi}^{ab}(k)$

The relevant diagrams are depicted in Fig. 2.5.16. Following the Feynman rules we have

$$\tilde{\Pi}^{ab}(k) = \int \frac{d^D q}{(2\pi)^D i} (-i) g f^{cad} k_\mu \frac{-1}{(q+k)^2} (-i) g f^{cdb} (q+k)_\nu \frac{d^{\mu\nu}(q)}{q^2} + (\tilde{Z}_3 - 1)\delta_{ab} k^2 . \tag{A.22}$$

Straightforward application of Eqs. (2.5.179) and (2.5.180) leads to the result

$$\tilde{\Pi}^{ab}(k) = \delta_{ab} k^2 \left[- \frac{g^2}{(4\pi)^{2-\varepsilon}} C_G (-k^2)^{-\varepsilon} \Gamma(\varepsilon)\, B(2-\varepsilon, 1-\varepsilon) \times \left(1 + (1-\alpha)\left(\frac{1}{2} - \varepsilon\right)\right) + \tilde{Z}_3 - 1 \right] \tag{A.23}$$

$$= \delta_{ab} k^2 \left[- \frac{g_0^2}{(4\pi)^2} C_G \frac{1}{2} \left\{ \left(1 + \frac{1-\alpha}{2}\right) \times \left(\frac{1}{\varepsilon} - \ln \frac{-k^2}{4\pi\mu^2} - \gamma + 2\right) - 1 + \alpha \right\} + \tilde{Z}_3 - 1 \right] + O(\varepsilon),$$

where $g = g_0 \mu^\varepsilon$. Taking the divergent part of Eq. (A.24) we obtain Eq. (2.5.136).

(3) *Quark self-energy part* $\sum^{ij}(p)$

Corresponding to Fig. 2.5.17 we have

$$\sum^{ij}(p) = \int \frac{d^D k}{(2\pi)^D i} g\gamma_\mu T^a_{il} \frac{1}{m - \not{k} - \not{p}} g\gamma_\nu T^a_{lj} \frac{d^{\mu\nu}(k)}{k^2} + [(Z_2 - 1)\not{p} - (Z_2 Z_m - 1) m]\delta_{ij} . \tag{A.25}$$

The calculation for massless quarks ($m = 0$) has been performed in Sec. 2.4.2. Here we repeat the calculation keeping $m \neq 0$. Using Eq. (2.4.7) and applying the Feynman parametrization method we find

$$\begin{aligned}\sum^{ij}(p) = & -\delta_{ij} C_F \frac{g^2}{(4\pi)^{2-\varepsilon}} \Gamma(\varepsilon) \int_0^1 dx [2\{(2-\varepsilon)m - (1-\varepsilon)(1-x)\not{p}\} K^{-\varepsilon} \\ & + (1-\alpha)(1-x)\{(-(2-\varepsilon)m + (1-\varepsilon)(1-x)\not{p} \\ & + 2(2-\varepsilon)x\not{p})K^{-\varepsilon} + \varepsilon x^2 (m + (1-x)\not{p})p^2 K^{-1-\varepsilon}\}] \\ & + \delta_{ij}[(Z_2 - 1)\not{p} - (Z_2 Z_m - 1)m],\end{aligned} \tag{A.26}$$

where $K = xm^2 - x(1-x)p^2$. Expanding Eq. (A.26) in ε we obtain Eq. (2.5.138).

(4) *Three-gluon vertex* $\Lambda^{abc}_{\mu\nu\lambda}(k_1, k_2, k_3)$

The relevant one-loop diagrams are shown in Fig. 2.5.18. The calculation is straightforward but tedious. Here we have no space to present the whole calculation and only refer to the original paper [Cel 79a] (see also [Pas 80]).

(5) *Ghost-gluon vertex* $\tilde{\Lambda}^{abc}_{\mu}(k, p, p')$

We calculate two one-loop diagrams in Fig. 2.5.19. Let us call contributions of these two diagrams $\tilde{\Lambda}_1$ and $\tilde{\Lambda}_2$ so that

$$\tilde{\Lambda}^{abc}_{\mu} = \tilde{\Lambda}^{abc}_{1\mu} + \tilde{\Lambda}^{abc}_{2\mu} + (\tilde{Z}_1 - 1)(-i)gf^{abc}p_\mu, \tag{A.27}$$

where

$$\begin{aligned}\tilde{\Lambda}^{abc}_{1\mu}(k,p,p') = &\int\frac{d^Dq}{(2\pi)^D i}(-i)gf^{a'bc'}p_\nu\frac{-1}{(q+p)^2}(-i)gf^{ac'b'}\\ &\times(q+p)_\mu\frac{-1}{(q+p')^2}(-i)gf^{a'b'c}(q+p')_\lambda\frac{d^{\nu\lambda}(q)}{q^2},\end{aligned} \tag{A.28}$$

$$\begin{aligned}\tilde{\Lambda}^{abc}_{2\mu}(k,p,p') = &\int\frac{d^Dq}{(2\pi)^D i}(-i)gf^{c'ba'}p_\nu\frac{-1}{(q+p')^2}(-i)gf^{b'a'c}\\ &\times(q+p')_\lambda\frac{d^{\lambda\rho}(q)}{q^2}(-i)gf^{ac'b'}V_{\mu\sigma\rho}(k,q-k,-q)\\ &\times\frac{d^{\nu\sigma}(q-k)}{(q-k)^2}.\end{aligned} \tag{A.29}$$

Using the formula (B.11) in Appendix B we rewrite Eq. (A.28) to find

$$\begin{aligned}\tilde{\Lambda}^{abc}_{1\mu}(k,p,p') = &-i\frac{g^3}{2}C_G f^{abc}\int\frac{d^Dq}{(2\pi)^D i}\frac{(q+p)_\mu}{(q+p)^2q^2(q+p')^2}\\ &\times\left[p\cdot(q+p') - (1-\alpha)\frac{q\cdot p q\cdot(q+p')}{q^2}\right].\end{aligned} \tag{A.30}$$

We would like to calculate the divergent part in Eq. (A.30) and hence we keep only integrands contributing to the divergences,

$$\tilde{\Lambda}^{abc}_{1\mu}(k,p,p') = -i\alpha\frac{g^3}{2}C_G f^{abc}p^\nu\int\frac{d^Dq}{(2\pi)^D i}\frac{q_\mu q_\nu}{(q+p)^2q^2(q+p')^2} + \text{finite}. \tag{A.31}$$

The integration in Eq. (A.31) may be easily performed to give

$$\tilde{\Lambda}^{abc}_{1\mu}(k,p,p') = -igf^{abc}p_\mu \frac{g^2}{(4\pi)^2} C_G \frac{\alpha}{8\varepsilon} + \text{finite}. \tag{A.32}$$

In a similar way we have

$$\tilde{\Lambda}^{abc}_{2\mu}(k,p,p') = -igf^{abc}p_\mu \frac{g^2}{(4\pi)^2} C_G \frac{3\alpha}{8\varepsilon} + \text{finite}. \tag{A.33}$$

Hence we finally obtain

$$\tilde{\Lambda}^{abc}_{\mu}(k,p,p') = -igf^{abc}p_\mu \left[\frac{g^2}{(4\pi)^2} C_G \frac{\alpha}{2\varepsilon} + \tilde{Z}_1 - 1\right] + \text{finite}. \tag{A.34}$$

(6) *Quark-gluon vertex* $\Lambda^{aij}_{F\mu}(k,p,p')$

The one-loop contribution to the quark-gluon vertex is shown in Fig. 2.5.20. Denoting the contributions of the first two diagrams in Fig. 2.5.20 by Λ_{F1} and Λ_{F2} we have

$$\Lambda^{aij}_{F\mu}(k,p,p') = \Lambda^{aij}_{F1\mu}(k,p,p') + \Lambda^{aij}_{F2\mu}(k,p,p') + (Z_{1F}-1)g\gamma_\mu T^a_{ij}, \tag{A.35}$$

$$\Lambda^{aij}_{F1\mu}(k,p,p') = g^3(T^bT^aT^b)_{ij}\int \frac{d^Dq}{(2\pi)^D i}\gamma_\lambda \frac{1}{m-\not{p}-\not{q}}\gamma_\mu \frac{1}{m-\not{p}'-\not{q}}\gamma_\rho \frac{d^{\lambda\rho}(q)}{q^2}, \tag{A.36}$$

$$\Lambda^{aij}_{F2\mu}(k,p,p') = -ig^3 f^{abc}(T^bT^c)_{ij}\int \frac{d^Dq}{(2\pi)^D{}_i}\gamma_\rho \frac{1}{m-\not{p}'-\not{q}}\gamma_\sigma \frac{d^{\rho\nu}(q-k)}{(q-k)^2}$$
$$\times V_{\mu\nu\lambda}(k,q-k,-q)\frac{d^{\sigma\lambda}(q)}{q^2}. \tag{A.37}$$

We first note that two color factors in Eqs. (A.36) and (A.37) are given by

$$\begin{aligned} -if^{abc}T^bT^c &= -if^{abc}(if^{bcd}T^d + T^cT^b) \\ &= \frac{1}{2}f^{abc}f^{bcd}T^d = \frac{1}{2}C_GT^a, \end{aligned} \tag{A.38}$$

$$\begin{aligned} T^bT^aT^b &= T^b(if^{abc}T^c + T^bT^a) \\ &= (C_F - C_G/2)T^a. \end{aligned} \tag{A.39}$$

Equations (A.36) and (A.37) may be rewritten in the following forms,

$$\Lambda^{aij}_{F1} = \frac{g^3}{(4\pi)^{D/2}}(T^b T^a T^b)_{ij}[I_{1\mu} + (1-\alpha)J_{1\mu}], \tag{A.40}$$

$$\Lambda^{aij}_{F2\mu} = \frac{g^3}{(4\pi)^{D/2}}(-i)f^{abc}(T^b T^c)_{ij}[I_{2\mu} + (1-\alpha)J_{2\mu} + (1-\alpha)^2 K_{2\mu}], \tag{A.41}$$

where I_1, J_1, I_2, J_2 and K_2 are given by

$$I_{1\mu} = \int \frac{d^Dq}{\pi^{D/2}i}\frac{1}{q^2}\gamma_\lambda \frac{1}{m-\not{q}-\not{p}}\gamma_\mu \frac{1}{m-\not{q}-\not{p}'}\gamma^\lambda \tag{A.42}$$

$$J_{1\mu} = -\int \frac{d^Dq}{\pi^{D/2}i}\frac{1}{q^4}\not{q}\frac{1}{m-\not{q}-\not{p}}\gamma_\mu \frac{1}{m-\not{q}-\not{p}'}\not{q}, \tag{A.43}$$

$$I_{2\mu} = \int \frac{d^Dq}{\pi^{D/2}i}\frac{1}{q^2(q-k)^2}\gamma^\nu \frac{1}{m-\not{q}-\not{p}'}\gamma^\lambda V_{\mu\nu\lambda}(k, q-k, -q), \tag{A.44}$$

$$J_{2\mu} = -\int \frac{d^Dq}{\pi^{D/2}i}\frac{1}{q^2(q-k)^2}\left[(\not{q}-\not{k})\frac{1}{m-\not{q}-\not{p}'}\frac{(q-k)^\nu}{(q-k)^2}\gamma^\lambda \right.$$
$$\left. + \gamma^\nu \frac{q^\lambda}{q^2}\frac{1}{m-\not{q}-\not{p}'}\not{q}\right] V_{\mu\nu\lambda}(k, q-k, -q), \tag{A.45}$$

$$K_{2\mu} = \int \frac{d^Dq}{\pi^{D/2}i}\frac{1}{q^4(q-k)^4}(\not{q}-\not{k})\frac{1}{m-\not{q}-\not{p}'}\not{q}(q-k)^\nu q^\lambda V_{\mu\nu\lambda}(k, q-k, -q). \tag{A.46}$$

We now evaluate the $1/\varepsilon$ terms in the above expressions. The divergent part of $I_{1\mu}$ may be extracted in such a way that

$$I_{1\mu} = \int \frac{d^Dq}{\pi^{D/2}i}\frac{\gamma_\nu \not{q}\gamma_\mu \not{q}\gamma^\nu}{q^2(m^2-(q+p)^2)(m^2-(q+p')^2)} + \text{finite}, \tag{A.47}$$

and then we use the Feynman parametrization to obtain

$$I_{1\mu} = \frac{1}{2}\gamma_\nu\gamma_\lambda\gamma_\mu\gamma^\lambda\gamma^\nu \int_0^1 dx \int_0^{1-x} dy\, \Gamma(\varepsilon)\, L^{-\varepsilon} + \text{finite}, \tag{A.48}$$

where $\varepsilon = (4-D)/2$ and

$$L = x(m^2 - p^2) + y(m^2 - p'^2) + (xp + yp')^2. \tag{A.49}$$

Noting that $\gamma_\nu\gamma_\lambda\gamma_\mu\gamma^\lambda\gamma^\nu = (2-D)^2\gamma_\mu$ we have

$$I_{1\mu} = \frac{1}{\varepsilon}\gamma_\mu + \text{finite}. \tag{A.50}$$

In the same way we calculate the $1/\varepsilon$ terms of all the other quantities (A.43) – (A.46), i.e.

$$J_{1\mu} = -\frac{1}{\varepsilon}\gamma_\mu + \text{finite}, \tag{A.51}$$

$$I_{2\mu} = \frac{3}{\varepsilon}\gamma_\mu + \text{finite}, \tag{A.52}$$

$$J_{2\mu} = -\frac{3}{2}\frac{1}{\varepsilon}\gamma_\mu + \text{finite}, \tag{A.53}$$

$$K_{2\mu} = \text{finite}. \tag{A.54}$$

Inserting Eqs. (A.50) – (A.54) into Eqs. (A.40) and (A.41) we find

$$\Lambda^{aij}_{\mathrm{F}1\mu} = g\gamma_\mu T^a_{ij}\frac{g^2}{(4\pi)^2}(C_{\mathrm{F}} - C_{\mathrm{G}}/2)\frac{\alpha}{\varepsilon} + \text{finite}, \tag{A.55}$$

$$\Lambda^{aij}_{\mathrm{F}2\mu} = g\gamma_\mu T^a_{ij}\frac{g^2}{(4\pi)^2}C_{\mathrm{G}}\frac{3(1+\alpha)}{4\varepsilon} + \text{finite}. \tag{A.56}$$

Hence we finally have

$$\Lambda^{aij}_{\mathrm{F}\mu}(k,p,p') = g\gamma_\mu T^a_{ij}\left[\frac{g^2}{(4\pi)^2}\left(\frac{3+\alpha}{4}C_{\mathrm{G}} + \alpha C_{\mathrm{F}}\right)\frac{1}{\varepsilon} + Z_{1\mathrm{F}} - 1\right] + \text{finite}. \tag{A.57}$$

(7) *Four-gluon vertex* $\Lambda^{a_1\ldots a_4}_{\mu_1\ldots\mu_4}(k_1, \ldots, k_4)$

We shall not go into the calculation of the contribution of the one-loop diagrams in Fig. 2.5.21, because of space limitations. A considerable account of the detail may be found in [Pas 80].

APPENDIX B

USEFUL FORMULAS INVOLVING THE SU(N) GENERATORS

The SU(N) generators T^a ($a = 1, 2, \ldots, N^2-1$) are hermitian, traceless matrices which generate the closed SU(N) algebra,

$$[T^a, T^b] = if^{abc}\, T^c, \tag{B.1}$$

with f^{abc} the structure constants.

The fundamental representation is N-dimensional where the T^a satisfy an additional relation,

$$\{T^a, T^b\} = \frac{1}{N}\,\delta_{ab} + d^{abc}\, T^c, \tag{B.2}$$

which is consistent with the normalization

$$\mathrm{Tr}[T^a\, T^b] = \frac{1}{2}\delta_{ab}\,. \tag{B.3}$$

Here d^{abc} is totally symmetric in a, b and c and is given by

$$d^{abc} = 2\mathrm{Tr}[\{T^a, T^b\}\, T^c]\,. \tag{B.4}$$

According to Eqs. (B.1) and (B.2) we have

$$T^a\, T^b = \frac{1}{2N}\,\delta_{ab} + \frac{1}{2}\,d^{abc}\, T^c + \frac{1}{2}\,if^{abc}\, T^c\,. \tag{B.5}$$

The traces of products of generators T^a in the fundamental representation are calculated by using Eq. (B.5),

$$\mathrm{Tr}[T^a\, T^b\, T^c] = \frac{1}{4}\,(d^{abc} + if^{abc})\,, \tag{B.6}$$

$$\mathrm{Tr}[T^a\, T^b\, T^c\, T^d] = \frac{1}{4N}\,\delta_{ab}\,\delta_{cd} + \frac{1}{8}\,(d^{abe} + if^{abe})\,(d^{cde} + if^{cde})\,. \tag{B.7}$$

In the adjoint representation the generator T^a is a $(N^2-1) \times (N^2-1)$ matrix and its matrix element is given by

$$(T^a)_{bc} = -if^{abc}. \tag{B.8}$$

The traces of products of generators T^a yield

$$\mathrm{Tr}[T^aT^bT^c] = (N/2)\, if^{abc}, \tag{B.9}$$

$$\mathrm{Tr}[T^aT^bT^cT^d] = \delta_{ab}\delta_{cd} + \delta_{ad}\delta_{bc} + \frac{N}{4}(d^{abe}\,d^{cde} - d^{ace}\,d^{bde} + d^{ade}\,d^{bce}). \tag{B.10}$$

Equation (B.9) reduces to

$$f^{ade}\,f^{bef}\,f^{cfd} = \frac{N}{2}f^{abc}. \tag{B.11}$$

The Jacobi identities

$$[T^a, [T^b, T^c]] + \text{cyclic permutations} = 0, \tag{B.12}$$

$$[T^a, \{T^b, T^c\}] + \text{cyclic permutations} = 0, \tag{B.13}$$

lead to the relations

$$f^{abe}\,f^{cde} + f^{cbe}\,f^{dae} + f^{dbe}\,f^{ace} = 0, \tag{B.14}$$

$$f^{abe}\,d^{cde} + f^{cbe}\,d^{dae} + f^{dbe}\,d^{ace} = 0. \tag{B.15}$$

APPENDIX C

RENORMALIZATION CONSTANTS IN QCD

In quantum chromodynamics we have the following eight renormalization constants: the quark mass renormalization constant Z_m, the gluon, ghost and quark field renormalization constant Z_3, $\tilde{Z}_3$ and Z_2, the three-gluon, ghost-gluon, quark-gluon and four-gluon vertex renormalization constant Z_1, $\tilde{Z}_1$, Z_{1F} and Z_4. In presenting the explicit expressions of these constants we stay in the MS (or $\overline{\mathrm{MS}}$) scheme. Since the MS (or $\overline{\mathrm{MS}}$) scheme[1] respects the Slavnov-Taylor identity (2.5.128)[2], three of them are not independent of the other five. We choose here Z_m, Z_3, $\tilde{Z}_3$, Z_2 and $\tilde{Z}_1$ as independent constants. Then the remaining three constants Z_1, Z_{1F} and Z_4 are given in terms of the above five constants,

$$Z_1 = Z_3 \frac{\tilde{Z}_1}{\tilde{Z}_3}, \tag{C.1}$$

$$Z_{1F} = Z_2 \frac{\tilde{Z}_1}{\tilde{Z}_3} \tag{C.2}$$

$$Z_4 = Z_3 \left(\frac{\tilde{Z}_1}{\tilde{Z}_3}\right)^2 . \tag{C.3}$$

The constant Z_m was calculated by [Tar 81] and the constants Z_3, $\tilde{Z}_3$, Z_2 and $\tilde{Z}_1$ were by [Jon 74, Vla 77, Ego 78] in the MS scheme up to two loops. The resulting expressions for these constants are given as follows,

$$Z_m = 1 - \left(\frac{g_R}{4\pi}\right)^2 \frac{3C_F}{\varepsilon} + \left(\frac{g_R}{4\pi}\right)^4 \frac{C_F}{\varepsilon}\left[\frac{1}{\varepsilon}\left(\frac{9}{2}C_F + \frac{11}{2}C_G - 2T_R N_f\right) - \frac{3}{4}C_F - \frac{97}{12}C_G + \frac{5}{3}T_R N_f\right] + O(g_R^6), \tag{C.4}$$

[1] See Sec. 2.4.3.
[2] See Sec. 2.5.6.

$$Z_3 = 1+\left(\frac{g_R}{4\pi}\right)^2 \frac{1}{\varepsilon}\left[\frac{1}{2} C_G\left(\frac{13}{3} - \alpha_R\right) - \frac{4}{3}T_R N_f\right]$$
$$+ \left(\frac{g_R}{4\pi}\right)^4 \frac{1}{\varepsilon}\left[\frac{1}{2\varepsilon} C_G\left\{\frac{1}{2}C_G\left(-\frac{13}{2} - \frac{17}{6}\alpha_R + \alpha_R^2\right) + 2T_R N_f\left(1 + \frac{2\alpha_R}{3}\right)\right\}\right.$$
$$\left. + \frac{1}{8} C_G\left(\frac{59}{2} - \frac{11}{2}\alpha_R - \alpha_R^2\right) - \left(2C_F + \frac{5}{2} C_G\right) T_R N_f\right] + O(g_R^6), \tag{C.5}$$

$$\tilde{Z}_3 = 1 + \left(\frac{g_R}{4\pi}\right)^2 \frac{C_G}{\varepsilon}\frac{3-\alpha_R}{4} + \left(\frac{g_R}{4\pi}\right)^4 \frac{1}{\varepsilon}\left[\frac{1}{2\varepsilon}\left(C_G^2\frac{3\alpha_R^2-35}{16} + \alpha_R^2 C_F\right)\right.$$
$$\left. + \frac{1}{4}C_G\left\{\frac{1}{8}C_G\left(\frac{95}{3} + \alpha_R\right) - \frac{5}{3} T_R N_f\right\}\right] + O(g_R^6), \tag{C.6}$$

$$Z_2 = 1 - \left(\frac{g_R}{4\pi}\right)^2 \frac{C_F\alpha_R}{\varepsilon} + \left(\frac{g_R}{4\pi}\right)^4 \frac{C_F}{\varepsilon}\left[\frac{\alpha_R}{2\varepsilon}\left(C_G\frac{3+\alpha_R}{2} + \alpha_R^2 C_F\right)\right.$$
$$\left. - C_G\frac{25+8\alpha_R+\alpha_R^2}{8} + T_R N_f + \frac{3}{4} C_F\right] + O(g_R^6), \tag{C.7}$$

$$\tilde{Z}_1 = 1 - \left(\frac{g_R}{4\pi}\right)^2 \frac{C_G\alpha_R}{2\varepsilon} + \left(\frac{g_R}{4\pi}\right)^4 \frac{C_G^2\alpha_R}{4\varepsilon}\left[\frac{1}{\varepsilon}\left(\frac{3}{2} + \alpha_R\right) - \frac{5+\alpha_R}{4}\right] + O(g_R^6), \tag{C.8}$$

where g_R is defined in Eq. (3.2.23), α_R is the renormalized gauge parameter and $\varepsilon = (4-D)/2$. The constants C_F, T_R and C_G are defined in Eqs. (2.4.7) and (2.5.133), and N_f is the number of flavors.

APPENDIX D
FEYNMAN RULES

The Feynman rule for quantum chromodynamics is summarized in the following. For comparison the rule frequently used in other text books is presented together with our Feynman rule. As a typical example of those rules we pick out the rule employed in the book by Itzykson and Zuber [Itz 80].

		Present book	Itzykson-Zuber
Propagators			
Gluon	$a\mu$ ~~~ k ~~~ $b\nu$	$\delta_{ab}\dfrac{d_{\mu\nu}(k)}{k^2}$	$\delta_{ab}\dfrac{-id_{\mu\nu}(k)}{k^2}$
Ghost	a - - k - - b	$\delta_{ab}\dfrac{-1}{k^2}$	$\delta_{ab}\dfrac{i}{k^2}$
Quark	i — p — j	$\delta_{ij}\dfrac{1}{m-\not p}$	$\delta_{ij}\dfrac{-i}{m-\not p}$
Vertices			
3-gluons	$a_1\mu_1$, k_1, k_2, k_3, $a_2\mu_2$, $a_3\mu_3$	$-igf^{a_1a_2a_3}V_{\mu_1\mu_2\mu_3}(k_1,k_2,k_3)$	$gf^{a_1a_2a_3}V_{\mu_1\mu_2\mu_3}(k_1,k_2,k_3)$
4-gluons	$a_1\mu_1$, $a_2\mu_2$, $a_3\mu_3$, $a_4\mu_4$	$-g^2W^{a_1a_2a_3a_4}_{\mu_1\mu_2\mu_3\mu_4}$	$-ig^2W^{a_1a_2a_3a_4}_{\mu_1\mu_2\mu_3\mu_4}$
Gluon-ghost	$a\mu$, k, b, c	$-igf^{abc}k_\mu$	$gf^{abc}k_\mu$
Gluon-quark	$a\mu$, i, j	$g\gamma_\mu T^a_{ij}$	$ig\gamma_\mu T^a_{ij}$

Here $d_{\mu\nu}(k)$, f^{abc}, $V_{\lambda\mu\nu}(k,p,q)$, $W^{abcd}_{\lambda\mu\nu\rho}$ and T^a_{ij} are defined in Eqs. (2.3.133), (2.1.25), (2.3.139), (2.3.140) and (2.1.29) respectively in the text.

BIBLIOGRAPHY

Abe 73 E. S. Abers and B. W. Lee, *Phys. Reports* **9C** (1973) 1. (Sec. 2.2)

Abe 82 O. Abe, M. Haruyama and A. Kanazawa, *Prog. Theor. Phys.* **67** (1982) 1541. (Sec 5.3)

Adl 66 S. L. Adler, *Phys. Rev.* **143** (1966) 1144. (Sec. 5.2.)

Adl 69 S. L. Adler, *Phys. Rev.* **177** (1969) 2426. (Sec. 1.2)

Ahm 75 M. A. Ahmed and G. G. Ross, *Phys. Lett.* **59B** (1975) 369. (Sec. 1.1)

Ali 79 A. Ali, J. G. Körner, Z. Kunszt, J. Willrodt, G. Kramer, G. Schierholz, and E. Pietarinen, *Phys. Lett.* **82B** (1979) 285; *Nucl. Phys.* **B167** (1980) 454. (Sec. 5.4)

Ali 80 A. Ali, E. Pietarinen, G. Kramer, and J. Willrodt, *Phys. Lett.* **93B** (1980) 155. (Sec. 5.4)

Ali 82 A. Ali, *Phys. Lett.* **110B** (1982) 67. (Sec. 5.1)

Alt 74 G. Altarelli and L. Maiani, *Phys. Lett.* **52B** (1974) 351. (Sec. 1.1)

Alt 77 G. Altarelli and G. Parisi, *Nucl. Phys.* **B126** (1977) 298. (Secs. 1.1, 5.2)

Alt 78 G. Altarelli, R. K. Ellis and G. Martinelli, *Nucl. Phys.* **B143** (1978) 521; **B146** (1978) 544 (Erratum). (Secs. 1.1, 5.5)

Alt 79 G. Altarelli, R. K. Ellis and G. Martinelli, *Nucl. Phys.* **B157** (1979) 461. (Sec. 5.5)

Alt 82 G. Altarelli, *Phys. Reports* **81** (1982) 1. (Sec. 5.5)

Alv 77 A. G. Alvarez, *Nucl. Phys.* **B120** (1977) 355. (Sec. 6.3)

Ama 78 D. Amati, R. Petronzio and G. Veneziano, *Nucl. Phys.* **B140** (1978) 54. (Secs. 1.1, 4.2, 5.5)

Ama 78a D. Amati, R. Petronzio and G. Veneziano, *Nucl. Phys.* **B146** (1978) 29. (Secs. 1.1, 4.2, 5.5)

And 79 A. Andrasi and J. C. Taylor, *Nucl. Phys.* **B154** (1979) 111. (Sec. 6.1)

And 81 A. Andrasi, M. Day, R. Doria, J. Frenkel, and J. C. Taylor, *Nucl. Phys.* **B182** (1981) 104. (Secs. 6.1, 6.3)

And 83 B. Andersson, G. Gustafson, G. Ingelman, and T. Sjöstrand, *Phys. Reports* **97** (1983) 31. (Sec. 5.4)

Ani 73 S. A. Anikin, O. I. Zav'yalov and M. K. Polivanov, *Teor. Mat. Fiz.* **17** (1973) 189 *(Theor. Math. Phys.* **17** (1974) 1082). (Sec. 2.5)

Aok 82 K-I. Aoki, Z. Hioki, R. Kawabe, M. Konuma, and T. Muta, *Suppl. Prog. Theor. Phys.* No. 73 (1982). (Secs. 2.3, 6.1)

App 69	T. Appelquist, *Ann. Phys.* **54** (1969) 27. (Sec. 2.5)
App 73	T. Appelquist and H. Georgi, *Phys. Rev.* **D8** (1973) 4000. (Secs. 1.1, 5.1)
App 75	T. Appelquist and H. D. Politzer, *Phys. Rev. Lett.* **34** (1975) 43. (Sec. 1.1)
App 75a	T. Appelquist and J. Carazzone, *Phys. Rev.* **D11** (1975) 2856. (Sec. 4.2)
App 75b	T. Appelquist and H. D. Politzer, *Phys. Rev.* **D12** (1975) 1404. (Sec. 1.1)
App 76	T. Appelquist, J. Carazzone, H. Kluberg-Stern, and M. Roth, *Phys. Rev. Lett.* **36** (1976) 768; 1161 (Erratum). (Sec. 6.3)
App 77	T. Appelquist and J. Carazzone, *Nucl. Phys.* **B120** (1977) 77. (Sec. 6.3)
Ash 72	J. F. Ashmore, *Lett. Nuovo Cim.* **4** (1972) 289. (Sec. 2.4)
Bac 78	M. Bacé, *Phys. Lett.* **78B** (1978) 132. (Sec. 5.3)
Bai 74	D. Bailin, A. Love and D. V. Nanopoulos, *Lett. Nuovo. Cim.* **9** (1974) 501. (Sec. 5.2)
Bak 77	M. Baker and C. K. Lee, *Phys. Rev.* **D15** (1977) 2201. (Sec. 2.5)
Ban 81	M. Bander, *Phys. Reports* **75** (1981) 205. (Sec. 1.1)
Bar 78	W. A. Bardeen, A. J. Buras, D. W. Duke, and T. Muta, *Phys. Rev.* **D18** (1978) 3998. (Secs. 1.1, 2.4, 5.1, 5.2, 5.3)
Bar 79	W. A. Bardeen and A. J. Buras, *Phys. Rev.* **D20** (1979) 166; **D21** (1980) 2041 (Erratum). (Sec. 1.1)
Bar 79a	R. Barbieri, G. Curci, E. d'Emilio, and E. Remiddi, *Nucl. Phys.* **B154** (1979) 535. (Sec. 1.1)
Bar 79b	W. A. Bardeen and A. J. Buras, *Phys. Lett.* **86B** (1979) 61; **90B** (1980) 485 (Erratum). (Sec. 5.3)
Bar 79c	R. Barbieri, L. Caneschi, G. Curci, and E. d'Emilio, *Phys. Lett.* **81B** (1979) 207. (Secs. 2.4, 5.3)
Bar 79d	D. P. Barber et al., *Phys. Rev. Lett.* **43** (1979) 830. (Sec. 5.4)
Bar 80	W. Bartel et al., *Phys. Lett.* **91B** (1980) 142. (Sec. 5.4)
Bar 85	W. A. Bardeen, *Proc. 6th Int. Workshop on Photon-Photon Collisions.* World Scientific, 1985. (Sec. 5.5)
Bar 87	V. D. Barger and R.J.N. Phillips, *Collider Physics* (Addison-Wesley, Tokyo, 1987). (Secs. 5.4, 5.5)
Bas 78	C. L. Basham, L. S. Brown, S. D. Ellis and S. T. Love, *Phys. Rev. Lett.* **41** (1978) 1585. (Sec. 5.4)
Bas 78a	C. L. Basham, L. S. Brown, S. D. Ellis and S. T. Love, *Phys. Rev.* **D17** (1978) 2298. (Sec. 5.4)
Bas 79	C. L. Basham, L. S. Brown, S. D. Ellis and S. T. Love, *Phys. Rev.* **D19** (1979) 2018. (Sec. 5.4)
Bau 78	L. Baulieu and C. Kounnas, *Nucl. Phys.* **B141** (1978) 423. (Secs. 1.1, 5.3)
Bec 76	C. Becchi, A. Rouet and R. Stora, *Ann. Phys.* **98** (1976) 287. (Sec. 2.3)
Bel 69	J. S. Bell and R. Jackiw, *Nuovo Cim.* **60A** (1969) 47. (Sec. 1.2)

Bel 74 A. A. Belavin and A. A. Migdal, *Zh. Eksp. Teor. Fiz.* **19** (1974) 317 (*Sov. Phys. JETP* **19** (1974) 181). The factor 183/16 should be replaced by 3413 in Eq. (8). (Sec. 3.4)

Ber 66 F. A. Berezin, *The Method of Second Quantization* (translated from Russian by N. Mugibayashi and A. Jeffrey), Academic Press, 1966. (Sec. 2.2)

Ber 68 J. Bernstein, *Elementary Particles and Their Currents*, Freeman, 1968. (Sec. 1.3)

Ber 74 M. C. Bergere and J. B. Zuber, *Comm. Math. Phys.* **35** (1974) 113. (Sec. 2.5)

Ber 79 C. Berger et al., *Phys. Lett.* **86B** (1979) 418. (Sec. 5.4)

Ber 82 E. L. Berger, *Particles and Fields-1982. Proc. American Phys. Soc. Meeting*, College Park, Maryland, 1982. (Sec. 5.5)

Bj 64 B. J. Bjorken and S. L. Glashow, *Phys. Lett.* **11** (1964) 255. (Sec. 1.2)

Bj 65 J. D. Bjorken and S. D. Drell, *Relativistic Quantum Fields*. McGraw-Hill, 1965. (Sec. 3.1)

Bj 66 J. D. Bjorken, *Phys. Rev.* **148** (1966) 1467. (Sec. 5.2)

Bj 67 J. D. Bjorken, *Phys. Rev.* **163** (1967) 1767. (Sec. 5.2)

Bj 69 J. D. Bjorken, *Phys. Rev.* **179** (1969) 1547. (Sec. 1.1)

Bj 69a J. D. Bjorken and E. A. Paschos, *Phys. Rev.* **185** (1969) 1975. (Sec. 1.1)

Bj 70 J. D. Bjorken and S. Brodsky, *Phys. Rev.* **D1** (1970) 1416. (Sec. 5.4)

Blo 37 F. Bloch and A. Nordsieck, *Phys. Rev.* **52** (1937) 54. (Sec. 6.2)

Bod 81 G.T. Bodwin, S.J. Brodsky and G. P. Lepage, *Phys. Rev. Lett.* **47** (1981) 1799. (Sec. 5.5)

Boe 70 H. Boerner, *Representations of Groups* (translated by P. G. Murphy), North-Holland, 1970. (Sec. 2.4)

Bog 57 N. N. Bogoliubov and O. S. Parasiuk, *Acta Math.* **97** (1957) 227. (Sec. 2.5)

Bog 80 N. N. Bogoliubov and D. V. Shirkov, *Introduction to the Theory of Quantized Fields* (3rd ed.), John Wiley, 1980. (Secs. 2.5, 3.1, 4.1)

Bol 64 C. G. Bollini, J. J. Giambiagi and A. Gonzales Dominguez, *Nuovo Cim.* **31** (1964) 550. (Sec. 2.4)

Bol 72 C. G. Bollini and J. J. Giambiagi, *Phys. Lett.* **40B** (1972) 566; *Nuovo Cim.* **12B** (1972) 20. (Sec. 2.4)

Boy 67 T. H. Boyer, *Ann. Phys.* **44** (1967) 1. (Sec. 3.5)

Bra 67 R. A. Brandt, *Ann. Phys.* **44** (1967) 221. (Sec. 4.2)

Bra 69 R. A. Brandt, *Phys. Rev. Lett.* **23** (1969) 1260. (Sec. 4.2)

Bra 76 R. A. Brandt, *Nucl. Phys.* **B116** (1976) 413. (Sec. 2.5)

Bra 79 R. Brandelik et al., **86B** (1979) 243. (Sec. 5.4)

Bra 81 E. Braaten and J.P. Leveille, *Phys. Rev.* **D24** (1981) 1369. (Sec. 5.3)

Bre 74 E. Brezin, J. C. Le Guillou and J. Zinn-Justin, *Phys. Rev.* **B9** (1974) 1121. (Sec. 3.4)

Bre 77 P. Breitenlohner and D. Maison, *Comm. Math. Phys.* **52** (1977) 39; 55. (Sec. 2.5)

Bur 80 A. J. Buras, *Rev. Mod. Phys.* **52** (1980) 199. (Sec. 5.2)

Cal 69 C. G. Callan and D. J. Gross, *Phys. Rev. Lett.* **22** (1969) 156. (Secs. 1.3, 5.2)

Cal 70 C. G. Callan, Jr., *Phys. Rev.* **D2** (1970) 1541. (Sec. 3.2)

Cal 73 C. G. Callan, Jr., *Particle Physics, Proc. Les Houches Summer School of Theoretical Physics,* Gordon and Breach, 1973, p.216. (Sec. 3.2)

Cal 77 M. Calvo, *Phys. Rev.* **D15** (1977) 730. (Sec. 5.2)

Cam 97 C. Campagnari and M. Franklin, *Rev. Mod. Phys.* **69** (1997) 137. (Sec. 1.2)

Car 81 C. E. Carneiro, M. Day, J. Frenkel, J. C. Taylor, and M. T. Thomaz, *Nucl. Phys.* **B183** (1981) 445. (Sec. 6.3)

Cas 74 W. E. Caswell and F. Wilczek, *Phys. Lett.* **49B** (1974) 291. (Sec. 3.2)

Cas 74a W. E. Caswell, *Phys. Rev. Lett.* **33** (1974) 244. (Sec. 3.4)

Cas 82 W. E. Caswell and A. D. Kennedy, *Phys. Rev.* **D25** (1982) 392. (Sec. 2.5)

CDF 95 CDF Collaboration, *Phys. Rev. Lett.* **74** (1995) 2626. (Sec. 1.2)

Cel 79 W. Celmaster and R. J. Gonsalves, *Phys. Rev. Lett.* **42** (1979) 1435. (Secs. 1.1, 5.3)

Cel 79a W. Celmaster and R. J. Gonsalves, *Phys. Rev.* **D20** (1979) 1420. (Secs. 1.1, 2.5, 5.3, Appendix A)

Cel 80 W. Celmaster and R. Gonsalves, *Phys. Rev. Lett.* **44** (1980) 560. (Secs. 1.1, 5.1)

Cel 80a W. Celmaster and R. Gonsalves, *Phys. Rev.* **D21** (1980) 3112. (Secs. 1.1, 5.3)

Cel 82 W. Celmaster and D. Sivers, *Ann. Phys.* **143** (1982) 1. (Sec. 5.3)

Cha 75 C. Chang et al., *Phys. Rev. Lett.* **35** (1975) 901. (Sec. 1.1)

Cha 79 M. Chanowitz, M. Furman and I. Hinchliffe, *Nucl. Phys.* **B159** (1979) 225. (Sec. 2.4)

Cha 82 N. P. Chang, A. Das, D. X. Li, D. C. Xian, and X. J. Zhou, *Phys. Rev.* **D25** (1982) 1630. (Sec. 4.2)

Cha 84 M. Chaichian and N. F. Nelipa, *Introduction to Gauge Field Theories,* Springer-Verlag, 1984. (Sec. 2.2)

Che 61 G. F. Chew, *S-Matrix Theory of Strong Interactions,* Benjamin, 1961. (Sec. 1.2)

Che 66 G. F. Chew, *The Analytic S Matrix,* Benjamin, 1966. (Sec. 1.2)

Che 79 K. G. Chetyrkin, A. L. Kataev and F. V. Tkachov, *Phys. Lett.* **85B** (1979) 277. (Secs. 1.1, 2.5, 5.1)

Che 80 K. G. Chetyrkin, A. L. Kataev and F. V. Tkachov, *Nucl. Phys.* **B174** (1980) 345. (Secs. 1.1, 2.5, 5.1)

Che 82 K. G. Chetyrkin, S. G. Gorishny and F. V. Tkachov, *Phys. Lett.* **119B** (1982) 407. (Sec. 4.2)

Che 83 K. G. Chetyrkin, *Phys. Lett.* **126B** (1983) 371. (Sec. 4.2)

Che 83a K. G. Chetyrkin, S. G. Gorishny, S. A. Larkin, and F. V. Tkachov, *Phys. Lett.* **132B** (1983) 351. (Sec. 3.4)

Che 84 T-P. Cheng and L.-F. Li, *Gauge Theory of Elementary Particle Physics*, Clarendon Press, 1984. (Sec. 4.2)

Che 84a K. G. Chetyrkin and V. A. Smirnov, *Phys. Lett.* **144B** (1984) 419. (Sec. 4.2)

Chi 81 T. W. Chiu, *Nucl. Phys.* **B181** (1981) 450. (Sec. 2.5)

Chr 72 N. Christ, B. Hasslacher and A. H. Mueller, *Phys. Rev.* **D6** (1972) 3543. (Secs. 1.1, 5.2)

Cic 72 G. M. Cicuta and E. Montaldi, *Lett. Nuovo Cim.* **4** (1972) 329. (Sec. 2.4)

Clo 79 F. E. Close, *An Introduction to Quarks and Partons*, Academic Press, 1979. (Sec. 1.3)

Col 71 S. Coleman and R. Jackiw, *Ann. Phys.* **67** (1971) 552. (Sec. 3.5)

Col 72 S. Coleman, *Development in High Energy Physics, Proc. Int. School of Phys. Enrico Fermi* (ed. R. Gatto), Academic Press, 1972, p.280. (Sec. 3.2)

Col 73 S. Coleman and D. J. Gross, *Phys. Rev. Lett.* **31** (1973) 851. (Sec. 1.1)

Col 73a S. Coleman and E. Weinberg, *Phys. Rev.* **D7** (1973) 1888. (Sec. 3.4)

Col 73b S. Coleman, *Properties of the Fundamental Interactions, Proc. 1971 Int. School of Subnuclear Phys. Ettore Majorana* (ed. A. Zichichi), Editorice Compositori, 1973, p.358. (Sec. 3.2)

Col 74 J. C. Collins and A. J. Macfarlane, *Phys. Rev.* **D10** (1974) 1201. (Sec. 3.2)

Col 75 J. C. Collins, *Nucl. Phys.* **B92** (1975) 477. (Sec. 2.5)

Col 82 J. C. Collins, D. Soper and G. Sterman, *Phys. Lett.* **109B** (1982) 388; *Nucl. Phys.* **B223** (1983) 381; *Phys. Lett.* **134B** (1984) 263. (Sec. 5.5)

Col 84 J. C. Collins, *Renormalization*, Cambridge Univ. Press, 1984. (Sec. 2.5)

Con 79 A. P. Contogouris and J. Kripfganz, *Phys. Rev.* **D19** (1979) 2207. (Secs. 1.1, 5.5)

Cou 82 S. N. Coulson and R. E. Ecclestone, *Phys. Lett.* **115B** (1982) 415. (Sec. 5.2)

Cou 83 S. N. Coulson and R. E. Ecclestone, *Nucl. Phys.* **B211** (1983) 317. (Sec. 5.2)

Cre 83 M. Creutz, *Quarks, Gluons and Lattices*, Cambridge Univ. Press, 1983. (Sec. 2.4)

Cur 80 G. Curci, W. Furmanski and R. Petronzio, *Nucl. Phys.* **B175** (1980) 27. (Secs. 1.1, 5.2)

D0 95 D0 Collaboration, *Phys. Rev. Lett.* 74 (1995) 2633. (Sec. 1.2)

Dar 51 R. H. Dalitz, *Proc. Roy. Soc.* **A206** (1951) 509. (Sec. 6.2)

Dav 82 F. David, *Nucl. Phys.* **B209** (1982) 433. (Sec. 4.2)

Dav 84 F. David, *Nucl. Phys.* **B234** (1984) 237. (Sec. 4.2)

Dea 78 W. S. Deans and J. A. Dixon, *Phys. Rev.* **D18** (1978) 1113. (Sec. 4.2)

deR 74 E. de Rafael and J. L. Rosner, *Ann. Phys.* **82** (1974) 369. (Sec. 3.3)

deR 77 A. de Rujula, H. Georgi and H. D. Politzer, *Ann. Phys.* **103** (1977) 315. (Sec. 5.2)

deR 78 A. de Rujula, J. Ellis, E. G. Floratos, and M. K. Gaillard, *Nucl. Phys.* **B138** (1978) 387. (Sec. 5.4)

Dev 83 A. Devoto, D. W. Duke, J. F. Owens, and R. G. Roberts, *Phys. Rev.* **D27** (1983) 508. (Sec. 5.2)

Dev 84 A. Devoto, D. W. Duke, J. D. Kimel, and G. A. Sowell, *Phys. Rev.* **D30** (1984) 541. (Sec. 5.2)

DeW 67 B. DeWitt, *Phys. Rev.* **162** (1967) 1195; 1239. (Sec. 2.1)

DiL 81 C. Di'Lieto, S. Gendron, I. G. Halliday, and C. T. Sachrajda, *Nucl. Phys.* **B183** (1981) 223. (Sec. 6.3)

Din 79 M. Dine and J. Sapirstein, *Phys. Rev. Lett.* **43** (1979) 668. (Sec. 1.1, 5.1)

Dir 45 P. A. M. Dirac, *Rev. Mod. Phys.* **17** (1945) 195. (Sec. 2.2)

Dix 74 J. A. Dixon and J. C. Taylor, *Nucl. Phys.* **B78** (1974) 552. (Sec. 4.2)

Dok 78 Yu. L. Dokshitzer, D. I. Dyakonov and S. I. Troyan, *Materials of the 13th Winter School of the Leningrad Nucl. Phys. Inst.*, Vol. 1, p.3-89 (1978) (English trans., SLAC TRANS-183, 1978). (Sec. 5.5)

Dok 78a Yu. L. Dokshtzer, D. I. Dyakonov and S. I. Troyan, *Phys. Lett.* **78B** (1978) 290. (Sec. 5.5)

Dok 80 Yu. L. Dokshitzer, D. I. Dyakonov and S.I. Troyan, *Phys. Reports* **58** (1980) 269. (Sec. 5.4)

Don 78 P. H. Dondi, *Phys. Rev.* **D17** (1978) 1101. (Sec. 6.3)

Dor 80 R. Doria, J. Frenkel and J. C. Taylor, *Nucl. Phys.* **B168** (1980) 93. (Secs. 5.5, 6.3)

Dre 70 S. D. Drell and T. M. Yan, *Phys. Rev. Lett.* **25** (1970) 316. (Sec. 5.5)

Dre 71 S. D. Drell and T. M. Yan, *Ann. Phys.* **66** (1971) 578. (Sec. 5.5)

Duk 80 D. W. Duke and R. G. Roberts, *Nucl. Phys.* **B166** (1980) 243. (Sec. 5.2)

Duk 82 D. W. Duke, J. D. Kimel and G. A. Sowell, *Phys. Rev.* **D25** (1982) 71. (Sec. 5.2)

Duk 85 D. W. Duke and R. G. Roberts, *Phys. Reports* **120** (1985) 275. (Sec. 5.1)

Dys 49 F. J. Dyson, *Phys. Rev.* **75** (1949) 486. (Sec. 2.5)

Dys 49a F. J. Dyson, *Phys. Rev.* **75** (1949) 1736. (Sec. 2.5)

Efr 80 A. V. Efremov and A. V. Radyushkin, *Teor. Mat. Fiz.* **44** (1980) 157; 327 (*Theor. Math. Phys.* **44** (1981) 664; 774). (Sec. 5.5)

Ego 78 E. Sh. Egorian and O. V. Tarasov, *Teor. Mat. Fiz.* **41** (1978) 26 (*Theor. Math. Phys.* **41** (1979) 863). (Sec. 3.4, Appendix C)

Eil 76 G. Eilam and M. Glück, *Phys. Lett.* **61B** (1976) 85. (Sec. 5.2)

Ein 78 M. B. Einhorn and B. G. Weeks, *Nucl. Phys.* **B146** (1978) 445. (Secs. 1.1, 5.4)

Ell 77 J. Ellis, *Weak and Electromagnetic Interactions at High Energy* (eds. R. Balian and C. H. Llewellyn-Smith), North-Holland, 1977, p.1. (Sec. 4.2)

Ell 78 R. K. Ellis, H. Georgi, M. Machacek, H. D. Politzer, and G. G. Ross, *Phys. Lett.* **78B** (1978) 281. (Secs. 1.1, 4.2, 5.5)

Ell 79 R. K. Ellis, H. Georgi, M. Machacek, H. D. Politzer, and G. G. Ross, *Nucl. Phys.* **B152** (1979) 285. (Secs. 1.1, 4.2, 5.5)

Ell 80 R. K. Ellis, D. A. Ross and A. E. Terrano, *Phys. Rev. Lett.* **45** (1980) 1226. (Secs. 5.1, 5.4)

Ell 81 R. K. Ellis, D. A. Ross and A. E. Terrano, *Nucl. Phys.* **B178** (1981) 421. (Secs. 1.1, 5.1, 5.4)

Erd 54 A. Erdelyi (ed.), *Higher Transcendental Functions*, Vol. 1, McGraw-Hill, 1954. (Secs. 2.4, 5.1)

Erd 54a A. Erdelyi (ed.), *Higher Transcendental Functions*, Vol. 2, McGraw-Hill, 1954. (Sec. 4.1)

Eri 63 K. E. Eriksson, *Nuovo Cim.* **30** (1963) 1423. (Sec. 3.1)

Fab 80 K. Fabricius, I. Shmitt, G. Schierholz, and G. Kramer, *Phys. Lett.* **97B** (1980) 431. (Secs. 1.1, 5.1, 5.4)

Fab 82 K. Fabricius, G. Kramer, G. Schierholz, and I. Schmitt, *Z. Phys.* **C11** (1982) 315. (Secs. 1.1, 5.1, 5.4)

Fad 67 L. D. Faddeev and V. N. Popov, *Phys. Lett.* **25B** (1967) 29. (Secs. 1.1, 2.1, 2.3)

Fad 80 L. D. Faddeev and A. A. Slavnov, *Gauge Fields* (translated by D. B. Pontecorvo), Benjamin/Cummings, 1980. (Secs. 2.2, 2.5)

Far 77 E. Farhi, *Phys. Rev. Lett.* **39** (1977) 1587. (Sec. 5.4)

Fey 48 R. P. Feynman, *Rev. Mod. Phys.* **20** (1948) 367. (Sec. 2.2)

Fey 63 R. P. Feynman, *Acta Phys. Polon.* **24** (1963) 697. (Sec. 2.1)

Fey 69 R. P. Feynman, *High Energy Collisions*, Gordon and Breach, 1970, p.237. (Secs. 1.1, 4.2, 5.2)

Fey 69a R. P. Feynman, *Phys. Rev. Lett.* **23** (1969) 1415. (Secs. 1.1, 4.1, 4.2, 5.2)

Fey 77 R. P. Feynman, *Weak and Electromagnetic Interactions at High Energy* (eds. R. Balian and C. H. Llewellyn-Smith), North-Holland, 1977, p.121. (Sec. 2.2)

Fie 78 R. Field and R. Feynman, *Nucl. Phys.* **B136** (1978) 1. (Sec. 5.4)

Flo 77 E. G. Floratos, D. A. Ross and C. T. Sachrajda, *Nucl. Phys.* **B129** (1977) 66; **B139** (1978) 545 (Errata). (Secs. 1.1, 5.2)

Flo 78 E. G. Floratos, D. A. Ross and C. T. Sachrajda, *Nucl. Phys.* **B152** (1979) 493. (Secs. 1.1, 5.2)

Flo 81 E. G. Floratos, C. Kounnas and R. Lacaze, *Phys. Lett.* **98B** (1981) 89; 285; *Nucl. Phys.* **B192** (1981) 417. (Secs. 1.1, 5.2)

Fra 70 E. S. Fradkin and I. V. Tyutin, *Phys. Rev.* **D2** (1970) 2841. (Sec. 2.1)

Fra 79 W. R. Frazer and J. F. Gunion, *Phys. Rev.* **D19** (1979) 2447. (Sec. 5.5)

Fre 76 J. Frenkel, R. Meuldermans, I. Mohammad, and J. C. Taylor, *Phys. Lett.* **64B** (1976) 211; *Nucl. Phys.* **B121** (1977) 58. (Sec. 6.3)

Fre 84 J. Frenkel, J. G. M. Gatheral and J. C. Taylor, *Nucl. Phys.* **B233** (1984) 307. (Sec. 6.3)

Fri 70 Y. Frishman, Phys. *Rev. Lett.* **25** (1970) 966. (Secs. 4.1, 4.2)

Fri 71 H. Fritzsch and M. Gell-Mann, *Proc. Coral Gables Conf. on Fundamental Interactions at High Energy* (eds. M. D. Cin, G. J. Iverson and A. Perlmutter), Gordon and Breach, Vol. 2, p.1. (Sec. 4.2)

Fri 72 H. Fritzsch and M. Gell-Mann, *Proc. XVI Int. Conf. on High Energy Physics* (eds. J. D. Jackson and A. Roberts), Fermilab, 1972, Vol. 2, p.135. (Sec. 1.1)

Fri 73 H. Fritzsch, M. Gell-Mann and H. Leutwyler, *Phys. Lett.* **47B** (1973) 365. (Sec. 1.1)

Fri 74 Y. Frishman, *Phys. Reports* **13C** (1974) 1. (Sec. 4.2)

Fuj 81 Y. Fujii, N. Ohta and H. Taniguchi, *Nucl. Phys.* **B177** (1981) 297. (Sec. 2.4)

Fuk 49 H. Fukuda and Y. Miyamoto, *Prog. Theor. Phys.* **4** (1949) 347. (Sec. 1.2)

Fur 80 W. Furmanski and R. Petronzio, *Phys. Lett.* **97B** (1980) 437. (Secs. 1.1, 5.2)

Fur 82 W. Furmanski and R. Petronzio, *Z. Phys.* **C11** (1982) 293. (Sec. 5.2)

Gai 74 M. K. Gaillard and B. W. Lee, *Phys. Rev. Lett.* **33** (1974) 108. (Sec. 1.1)

Gan 81 V. Ganapathi and G. Sterman, *Phys. Rev.* **D23** (1981) 2408. (Sec. 6.3)

Gas 73 R. Gastmans and R. Meuldermans, *Nucl. Phys.* **B63** (1973) 277. (Sec. 6.1)

Gel 54 M. Gell-Mann and F. E. Low, *Phys. Rev.* **95** (1954) 1300. (Secs. 3.1, 3.2)

Gel 62 M. Gell-Mann, *Phys. Rev.* **125** (1962) 1067. (Sec. 1.2)

Gel 64 M. Gell-Mann, *Phys. Lett.* **8** (1964) 214. (Secs. 1.1, 1.2)

Gel 64a I. M. Gel'fand and G. E. Shilov, *Generalized Functions*, Vol. 1 (translated by E. Saletan), Academic Press, 1964. (Sec. 4.2)

Gel 72 M. Gell-Mann, *Acta Phys. Austriaca Suppl.* **9** (1972) 733. (Sec. 1.1)

Geo 74 H. Georgi and H. D. Politzer, *Phys. Rev.* **D9** (1974) 416. (Secs. 1.1, 3.4, 4.2, 5.2)

Geo 76 H. Georgi and H. D. Politzer, *Phys. Rev.* **D14** (1976) 1829. (Secs. 2.5, 3.2)

Geo 77 H. Georgi and M. Machacek, *Phys. Rev. Lett.* **39** (1977) 1237. (Sec. 5.4)

Gey 79 R. Geyer, R. Robaschik and W. Wieczorek, *Fortsch. d. Phys.* **27** (1979) 75. (Secs. 4.3, 5.2)

Gil 74 R. Gilmore, *Lie Groups. Lie Algebras and Some of Their Applications.* John Wiley, 1974. (Sec. 2.2)

Gla 61 S. L. Glashow, *Nucl. Phys.* **22** (1961) 579. (Sec. 2.4)

Gol 80 H. Goldstein, *Classical Mechanics* (2nd ed.), Addison-Wesley, 1980. (Sec. 2.2)

Gon 79 A. Gonzalez-Arroyo, C. Lopez and F. J. Yndurain, *Nucl. Phys.* **B153** (1979) 161. (Sec. 5.2)

Gon 80 A. Gonzalez-Arroyo and C. Lopez, *Nucl. Phys.* **B166** (1980) 429. (Sec. 5.2)

Gou 67 M. Gourdin, *Unitary Symmetries.* North-Holland, 1967. (Sec. 1.2)

Gra 65 I. S. Gradshteyn and I. M. Ryzhik, *Table of Integrals. Series and Products* (translated by A. Jeffrey), Academic Press, 1965. (Sec. 2.4)

Gra 73 G. Grammer and D. R. Yennie, *Phys. Rev.* **D8** (1973) 4332. (Sec. 6.2)

Gre 64 O. W. Greenberg, *Phys. Rev. Lett.* **13** (1964) 598. (Sec. 1.1)

Gri 78 V. N. Gribov, *Nucl. Phys.* **B139** (1978) 1. (Sec. 2.2)

Gro 69 D. J. Gross and C. H. Llewellyn-Smith, *Nucl. Phys.* **B14** (1969) 337. (Sec. 5.2)

Gro 71 D. J. Gross and S. B. Treiman, *Phys. Rev.* **D4** (1971) 1059. (Sec. 4.3)

Gro 73 D. J. Gross and F. Wilczek, *Phys. Rev. Lett.* **30** (1973) 1343. (Secs. 1.1, 3.4)

Gro 73a D. J. Gross and F. Wilczek, *Phys. Rev.* **D8** (1973) 3633. (Secs. 1.1, 3.4, 4.2, 5.2)

Gro 74 D. J. Gross and F. Wilczek, *Phys. Rev.* **D9** (1974) 980. (Secs. 1.1, 4.2, 4.3, 5.2)

Gro 74a D. J. Gross and A. Neveu, *Phys. Rev.* **D10** (1974) 3235. (Sec. 1.1)

Gro 76 D. J. Gross, *Methods in Field Theory,* (eds. R. Balian and J. Zinn-Justin), North-Holland, 1976, p.141. (Secs. 3.2, 5.3)

Gru 80 G. Grunberg, *Phys. Lett.* **95B** (1980) 70. (Sec. 5.3)

Gup 79 S. Gupta and A. H. Mueller, *Phys. Rev.* **D20** (1979) 118. (Sec. 5.5)

Gup 79a S. Gupta, *Phys. Rev.* **D20** (1979) 3160. (Sec. 4.2)

Gup 80 S. Gupta, *Phys. Rev.* **D21** (1980) 984. (Secs. 4.2, 5.5)

Gut 84 F. Gutbrod, G. Kramer and G. Schierholz, *Z. Phys.* **C21** (1984) 235. (Sec. 5.4)

Haa 58 R. Haag, *Phys. Rev.* **112** (1958) 669. (Sec. 4.1)

Hag 81 K. Hagiwara and T. Yoshino, *Nucl. Phys.* **B179** (1981) 347. (Sec. 5.4)

Hag 82 K. Hagiwara and T. Yoshino, *Phys. Rev.* **D26** (1982) 2038. (Sec. 3.4)

Hag 83 K. Hagiwara, *Suppl. Prog. Theor. Phys.* No.77 (1983), p.100. (Sec. 5.4)

Hah 68 Y. Hahn and W. Zimmermann, *Comm. Math. Phys.* **10** (1968) 330. (Sec. 2.5)

Han 63 L. N. Hand, *Phys. Rev.* **129** (1963) 1834. (Sec. 5.2)

Han 65 M. Han and Y. Nambu, *Phys. Rev.* **139B** (1965) 1006. (Sec. 1.1)

Han 75 G. Hanson et al., *Phys. Rev. Lett.* **35** (1975) 1609. (Sec. 5.4)

Han 82 G. Hanson et al., *Phys. Rev.* **D26** (1982) 991. (Sec. 5.4)

Har 64 Y. Hara, *Phys. Rev.* **134B** (1964) 701. (Sec. 1.2)

Har 75 H. Harari, *Phys. Lett.* **57B** (1975) 265. (Sec. 1.2)

Har 75a H. Harari, *Proceedings of Summer Inst. on Particle Phys.*, SLAC Report No.191 (1975), p.159. (Sec. 1.2)

Har 79 K. Harada, T. Kaneko and N. Sakai, *Nucl. Phys.* **B155** (1979) 169. (Secs. 1.1, 5.5)

Har 80 K. Harada and T. Muta, *Phys. Rev.* **D22** (1980) 663. (Sec. 5.5)

Har 83 M. Haruyama, *Suppl. Prog. Theor. Phys.* No.77 (1983), p.77. (Sec. 5.3)

Har 83a K. Harada, *Suppl. Prog. Theor. Phys.* No.77 (1983), p.138. (Sec. 5.5)

Hea 85 A. C. Hearn, *REDUCE User's Manual* (Version 3.2), Rand Corporation, 1985. (Sec. 5.4)

Hei 29 W. Heisenberg and W. Pauli, *Z. Phys.* **56** (1929) 1; **59** (1930) 168. (Sec. 2.2)

Hep 66 K. Hepp, *Comm. Math. Phys.* **2** (1966) 301. (Sec. 2.5)

Her 81 R. T. Herrod and S. Wada, *Phys. Lett.* **96B** (1981) 195; *Z. Phys.* **C9** (1981) 351. (Sec. 5.2)

Hey 72 A. J. G. Hey and J. E. Mandula, *Phys. Rev.* **D5** (1972) 2610. (Sec. 5.2)

Hol 74 M. J. Holwerda, W. L. van Neerven and R. P. van Royen, *Nucl. Phys.* **B75** (1974) 302. (Sec. 3.2)

Hoy 79 P. Hoyer, P. Osland, H.G. Sander, T. F. Walsh, and P. M. Zerwas, *Nucl. Phys.* **B161** (1979) 349. (Sec. 5.4)

Hum 79 B. Humpert and W. L. van Neerven, *Phys. Lett.* **84B** (1979) 327, **85B** (1979) 471 (Erratum). (Secs. 1.1, 5.5)

Hum 81 B. Humpert and W. L. van Neerven, *Phys. Lett.* **102B** (1981) 426. (Sec. 5.5)

Hum 82 B. Humpert and W. L. van Neerven, *Phys. Rev.* **D25** (1982) 2593. (Secs. 4.2, 5.5)

Iof 69 B. L. Ioffe, *Phys. Lett.* **30B** (1969) 123. (Sec. 4.1)

Ito 81 I. Ito, *Prog. Theor. Phys.* **65** (1981) 1466; **67** (1982) 1216. (Sec. 6.3)

Itz 80 C. Itzykson and J. Zuber, *Quantum Field Theory*, McGraw-Hill, 1980. (Secs. 2.5, 4.1, Appendix D)

Jac 70 R. Jackiw, R. van Royen and G. B. West, *Phys. Rev. D2* (1970) 2473. (Sec. 4.2)

Jau 55 J. M. Jauch and F. Rohrlich, *The Theory of Photons and Electrons*, Addison-Wesley, 1955. (Secs. 2.4, 6.2)

Jog 76 S. D. Joglekar and B. W. Lee, *Ann, Phys.* **97** (1976) 160. (Sec. 4.2)

Jon 74 D. R. T. Jones, *Nucl. Phys.* **B75** (1974) 531. (Sec. 3.4, Appendix C)

Kai 74 W. Kainz, W. Kummer and M. Schweda, *Nucl. Phys.* **B79** (1974) 484. (Sec. 4.2)

Kat 80 K. Kato, Y. Shimizu and H. Yamamoto, *Prog. Theor. Phys.* **63** (1980) 1295. (Sec. 5.2)

Kat 80a K. Kato and Y. Shimizu, *Prog. Theor. Phys.* **64** (1980) 703; **68** (1982) 862. (Sec. 5.2)

Kat 83 K. Kato, *Suppl. Prog. Theor. Phys.* No.77 (1983), p.237. (Sec. 5.2)

Kat 84 K. Kato and T. Munehisa, *e^+e^- Event Generator EPOCS* (in Japanese), KEK Report 84-18, 1984. (Sec. 5.4)

Kaz 78 Y. Kazama and Y. P. Yao, *Phys. Rev. Lett.* **41** (1978) 611. (Sec. 5.5)

Kaz 79 Y. Kazama and Y. P. Yao, *Phys. Rev. Lett.* **43** (1979) 1562. (Sec. 4.2)

Kaz 79a Y. Kazama and Y. P. Yao, *Phys. Rev.* **D19** (1979) 3111. (Sec. 5.5)

Kaz 80 Y. Kazama and Y. P. Yao, *Phys. Rev.* **D21** (1980) 1116; 1138. (Sec. 4.2)

Kin 50 T. Kinoshita, *Prog. Theor. Phys.* **5** (1950) 1045. (Sec. 6.1)

Kin 62 T. Kinoshita, *J. Math. Phys.* **3** (1962) 650: See also *Lectures in Theoretical Physics,* Vol. 4 (eds. W. E. Brittin, B. W. Downs and J. Downs), Interscience, 1962, p.121. (Secs. 6.1, 6.3)

Kin 76 T. Kinoshita and A. Ukawa, *Phys. Rev.* **D13** (1976) 1573. (Sec. 6.3)

Klu 75 H. Kluberg-Stern and J. B. Zuber, *Phys. Rev.* **D12** (1975) 467. (Secs. 2.5, 4.2)

Klu 75a H. Kluberg-Stern and J. B. Zuber, *Phys. Rev.* **D12** (1975) 482. (Secs. 2.5, 4.2)

Klu 75b H. Kluberg-Stern and J. B. Zuber, *Phys. Rev.* **D12** (1975) 3159. (Sec. 4.2)

Kob 73 M. Kobayashi and T. Maskawa, *Prog. Theor. Phys.* **49** (1973) 652. (Sec. 1.2)

Kod 78 J. Kodaira and T. Uematsu, *Nucl. Phys.* **B141** (1978) 497. (Secs. 1.1, 5.3)

Kod 79 J. Kodaira, S. Matsuda, T. Muta, T. Uematsu, and K. Sasaki, *Phys. Rev.* **D20** (1979) 627. (Sec. 5.2)

Kod 80 J. Kodaira, *Nucl. Phys.* **B165** (1980) 129. (Sec. 4.2)

Kon 78 W. Konetschny, *Phys. Lett.* **74B** (1978) 365. (Sec. 6.3)

Kon 80 K. Konishi, *Surv. High Energy Phys.* **2** (1980) 1. (Sec. 5.4)

Kor 76 C. P. Korthals Altes and E. de Rafael, *Phys. Lett.* **62B** (1976) 320; *Nucl. Phys.* **B125** (1977) 275. (Sec. 6.3)

Kra 77 F. G. Krausz, *Phys. Lett.* **66B** (1977) 251; *Nucl. Phys.* **B126** (1977) 340. (Sec. 6.3)

Kra 84 G. Kramer, *Springer Tracts in Mod. Phys.*, Vol. 102, Springer Verlag, 1984. (Sec. 5.4)

Kub 79 J. Kubar-Andre and F. E. Paige, *Phys. Rev.* **D19** (1979) 221. (Secs. 1.1, 5.5)

Kug 78 T. Kugo and I. Ojima, *Phys. Lett.* **73B** (1978) 459; *Prog. Theor. Phys.* **60** (1978) 1869. (Secs. 2.2, 2.3)

Kug 79 T. Kugo and I. Ojima, *Suppl. Prog. Theor. Phys.* No. 66 (1979) 1. (Secs. 2.2, 2.3)

Kun 81 Z. Kunszt, *Phys. Lett.* **99B** (1981) 429; **107B** (1981) 123. (Sec. 5.4)

Lam 85 B. Lampe and G. Kramer, DESY preprint 85-30 (1985). (Sec. 5.4)

Lan 54 L. D. Landau, A. A. Abrikosov and I. M. Khalatnikov, *Dokl. Akad. Nauk USSR* **95** (1954) 497; 773; 1177. (Sec. 3.3)

Lan 56 L. D. Landau, A. A. Abrikosov and I. M. Khalatnikov, *Nuovo Cim. Suppl. Ser. X* **3** (1956) 80. (Sec. 3.3.)

Lau 67 B. Lautrup, *Mat. Fys. Medd. Dan. Vid. Selsk.* **35** (1967) 29. (Sec. 2.3)

Lee 64 T. D. Lee and M. Nauenberg, *Phys. Rev.* **133B** (1964) 1549. (Sec. 6.3)

Lee 72 B. W. Lee and J. Zinn-Justin, *Phys. Rev.* **D5** (1972) 3121; 3137. (Sec. 2.5)

Lee 76 B. Lee, *Methods in Field Theory* (eds. R. Balian and J. Zinn-Justin), North-Holland, 1976, p.79. (Sec. 2.5)

Lee 76a C. Lee, *Phys. Rev.* **D14** (1976) 1078. (Sec. 4.2)

Lee 81 T. D. Lee, *Particle Physics and Introduction to Field Theory*, Harwood Academic, 1981. (Sec. 6.3)

Leh 54 H. Lehmann, *Nuovo Cim.* **11** (1954) 342. (Sec. 4.1)

Leh 55 H. Lehmann, K. Symanzik and W. Zimmermann, *Nuovo Cim.* **1** (1955) 205. (Sec. 2.3)

Lei 75 G. Leibbrandt, *Rev. Mod. Phys.* **47** (1975) 849. (Sec. 2.4)

Leu 70 H. Leutwyler and J. Stern, *Nucl. Phys.* **B20** (1970) 77. (Sec. 4.2)

Lew 58 L. Lewin, *Dilogarithms and Associated Functions*. MacDonald, 1958. (Sec. 6.1)

Lib 79 S. B. Libby and G. Sterman, *Phys. Rev.* **D19** (1979) 2468. (Sec. 6.3)

Lig 58 M. J. Lighthill, *Introduction to Fourier Analysis and Generalized Functions*. Cambridge Univ. Press, 1958. (Sec. 4.1)

Lin 83 W. W. Lindsay, D. A. Ross and C. T. C. Sachrajda, *Nucl. Phys.* **B222** (1983) 189. (Secs. 4.2, 5.5)

Lle 78 C. H. Llewellyn-Smith, *Acta Physica Austriaca Suppl.* **19** (1978) 331. (Sec. 5.5)

Low 70 J. Lowenstein, *Comm. Math. Phys.* **16** (1970) 265. (Sec. 4.2)

Mac 74 A. J. Macfarlane and G. Woo, *Nucl. Phys.* **B77** (1974) 91. (Secs. 1.1, 2.5, 3.4)

Mac 81 P. M. Mackenzie and G. P. Lepage, *Phys. Rev. Lett.* **47** (1981) 1244. (Sec. 1.1)

Mac 84 M. E. Machacek and M. T. Vaughn, *Nucl. Phys.* **B236** (1984) 221. (Sec. 3.4)

Mak 64 Z. Maki, *Prog. Theor. Phys.* **31** (1964) 331; 333. (Sec. 1.2)

Man 68 S. Mandelstam, *Phys. Rev.* **175** (1968) 1580; 1604. (Sec. 2.1)
Man 83 E. B. Manoukian, *Renormalization*, Academic Press, 1983. (Sec. 2.5)
Mar 75 W. J. Marciano and A. Sirlin, *Nucl. Phys.* **B88** (1975) 86. (Sec. 6.1)
Mar 78 W. Marciano and H. Pagels, *Phys. Reports* **36C** (1978) 137. (Sec. 1.1)
Mat 55 P. T. Mathews and A. Salam, *Nuovo Cim.* **2** (1955) 120. (Sec. 2.2)
Mat 80 T. Matsuki and N. Yamamoto, *Phys. Rev.* **D22** (1980) 2433. (Sec. 3.4)
Mor 60 S. Moriguchi, K. Udagawa and S. Hitotsumatsu, *Mathematical Formulae*, Vol. 3 (in Japanese), Iwanami, 1960. (Secs. 2.4, 4.1, 5.1)
Mue 78 A. H. Mueller, *Phys. Rev.* **D18** (1978) 3705. (Secs. 1.1, 4.2, 5.5)
Mue 81 A. H. Mueller, *Phys. Reports* **73** (1981) 237. (Secs. 4.2, 5.5)
Mue 82 A. H. Mueller, *Phys. Lett.* **108B** (1982) 335. (Sec. 5.5)
Mut 77 T. Muta and D. Popović, *Prog. Theor. Phys.* **57** (1977) 690. (Sec. 1.1)
Mut 79 T. Muta, *Phys. Rev.* **D20** (1979) 1232. (Sec. 5.2)
Mut 81 T. Muta and C. A. Nelson, *Phys. Rev.* **D25** (1981) 2222. (Sec. 6.3)

Nac 73 O. Nachtmann, *Nucl. Phys.* **B63** (1973) 237. (Sec. 5.2)
Nac 81 O. Nachtmann and W. Wetzel, *Nucl. Phys.* **B187** (1981) 333. (Sec. 3.4)
Nak 57 N. Nakanishi, *Prog. Theor. Phys.* **17** (1957) 401. (Sec. 2.5)
Nak 58 N. Nakanishi, *Prog. Theor. Phys.* **19** (1958) 159. (Sec. 6.1, 6.3)
Nak 63 N. Nakanishi, *J. Math. Phys.* **4** (1963) 1385. (Sec. 2.5)
Nak 66 N. Nakanishi, *Prog. Theor. Phys.* **35** (1966) 1111. (Sec. 2.3)
Nak 75 N. Nakanishi, *Quantum Theory of Fields* (in Japanese), Baifu-kan, 1975. (Sec. 2.5)
Nak 83 H. Nakkagawa and A. Niégawa, *Prog. Theor. Phys.* **70** (1983) 511. (Sec. 5.3)
Nan 79 D. V. Nanopoulos and D. A. Ross, *Nucl. Phys.* **B157** (1979) 273. (Sec. 3.4)
Nas 78 C. Nash, *Relativistic Quantum Fields*, Academic Press, 1978. (Sec. 6.2)
Nel 81 C. A. Nelson, *Nucl. Phys.* **B186** (1981) 187. (Sec. 6.3)
Nis 58 K. Nishijima, *Phys. Rev.* **111** (1958) 995. (Sec. 4.1)
Nis 77 K. Nishijima and Y. Tomozawa, *Prog. Theor. Phys.* **57** (1977) 654. (Sec. 3.2)
Nor 37 A. Nordsieck, *Phys. Rev.* **52** (1937) 59. (Sec. 6.2)
Nov 80 V. A. Novikov, M. A. Shifman, A. I. Vainshtein, and V. I. Zakharov, *Nucl. Phys.* **B174** (1980) 378. (Sec. 4.2)
Nov 84 V. A. Novikov. M. A. Shifman, A. I. Vainshtein, and V. I. Zakharov, *Phys. Reports* **116** (1984) 103. (Sec. 4.2)
Nov 85 V. A. Novikov, M. A. Shifman, A. I. Vainshtein, and V. I. Zakharov, *Nucl. Phys.* **B249** (1985) 445. (Sec. 4.2)

Ohn 73 Y. Ohnuki and S. Kamefuchi, *Prog. Theor. Phys.* **50** (1973) 258. (Sec. 1.1)

Ohn 78 Y. Ohnuki and T. Kashiwa, *Prog. Theor. Phys.* **60** (1978) 548. (Sec. 2.2)

Oka 81 H. Okada, T. Munehisa, K. Kudoh, and K. Kitani, *Phys. Lett.* **B102** (1981) 49. (Sec. 1.1)

Oka 83 M. Okawa, Suppl. *Prog. Theor. Phys.* No. 77 (1983) p.264. (Sec. 5.2)

Ovs 56 L. V. Ovsyannikov, *Dokl. Acad. Nauk SSSR* **109** (1956) 1112. (Sec. 3.1, 3.2)

Owe 78 J. F. Owens, *Phys. Lett.* **76B** (1978) 85. (Sec. 1.1)

Pan 55 W. K. H. Panofsky and M. Phillips, *Classical Electricity and Magnetism*, Addison-Wesley, 1955. (Sec. 2.1)

Pan 68 W. K. H. Panofsky, *Proc. 14th Int. Conf. on High Energy Physics*, CERN Scientific Information Service, 1968, p.23. (Sec. 1.1)

Pap 76 N. Papanicolaou, *Phys. Reports* **24C** (1979) 229. (Sec. 6.2)

Par 73 G. Parisi, *Phys. Lett.* **43B** (1973) 207. (Sec. 5.2)

Par 76 G. Parisi, *Proc. 11th Rencontre de Moriond on Weak Interactions and Neutrino Physics* (ed. J. Tran Thanh Van), 1976. (Secs. 1.1, 5.2)

Par 81 G. Parisi and Y. Wu, *Sci. Sinica* **24** (1981) 483. (Sec. 2.2)

Par 96 *Particle Data Group, Phys. Rev.* **D54** (1996) 1. The data table is updated every two years. (Sec. 1.2)

Pas 80 P. Pascual and R. Tarrach, *Nucl. Phys.* **B174** (1980) 123; **B181** (1981) 546 (Errata). (Sec. 2.5, Appendix A)

Pau 49 W. Pauli and F. Villars, *Rev. Mod. Phys.* **21.** (1949) 434. (Sec. 2.4)

Pet 79 A. Peterman, *Phys. Reports* **53** (1979) 157. (Sec. 5.2)

Pet 82 A. Peterman, *Phys. Lett.* **114B** (1982) 333. (Secs. 3.1, 5.3)

Pog 76 E. C. Poggio and H. R. Quinn, *Phys. Rev.* **D14** (1976) 578. (Secs. 3.3, 6.3)

Pog 77 E. C. Poggio, *Quark Confinement and Field Theory* (eds. D. R. Stump and D. H. Weingarten), John Wiley, 1977. (Sec. 6.3)

Pol 73 H. D. Politzer, *Phys. Rev. Lett.* **30** (1973) 1346. (Sec. 1.1)

Pol 74 H. D. Politzer, *Phys. Reports* **14** (1974) 129. (Sec. 5.2)

Pol 77 H. D. Politzer, *Phys. Lett.* **70B** (1977) 430. (Secs. 4.2, 5.5)

Pol 77a H. D. Politzer, *Nucl. Phys.* **B129** (1977) 301. (Secs. 4.2, 5.5)

Pol 82 H. D. Politzer, *Nucl. Phys.* **B194** (1982) 493. (Sec. 5.3)

Rey 81 E. Reya, *Phys. Reports* **69** (1981) 195. (Sec. 5.2)

Rit 97 T. van Ritbergen, J. A. M. Vermaseren and S. A. Larin, *Phys. Lett.* **B400** (1997) 379. (Sec. 3.4)

Sac 76 C. T. Sachrajda, *Phys. Rev.* **D14** (1976) 1072. (Sec. 6.3)

Sac 78 C. T. Sachrajda, *Phys. Lett.* **73B** (1978) 185. (Sec. 5.5)

Sak 51 S. Sakata, H. Umezawa and S. Kamefuchi, *Phys. Rev.* **84** (1951) 154. (Sec. 2.5)

Sak 52 S. Sakata, H. Umezawa and S. Kamefuchi, *Prog. Theor. Phys.* **7** (1952) 377. (Sec. 2.5)

Sal 51 A. Salam, *Phys. Rev.* **82** (1951) 217. (Sec. 2.5)

Sal 68 A. Salam, *Elementary Particle Theory* (eds. N. Svartholm), Almqvist and Wilsell, 1968, p.367. (Sec. 2.4)

Sar 74 S. Sarkar, *Nucl. Phys.* **B82** (1974) 447. (Sec. 4.2)

Sar 75 S. Sarkar and H. Strubbe, *Nucl. Phys.* **B90** (1975) 45. (Sec. 4.2)

Sas 83 K. Sasaki, *Suppl. Prog. Theor. Phys.* No. 77 (1983) 197. (Sec. 5.5)

Sch 51 J. Schwinger, *Phys. Rev.* **82** (1951) 914. (Sec. 2.2)

Sch 51a J. Schwinger, *Proc. Nat. Acad. Sci.* **37** (1951) 452; 455. (Secs. 2.2, 2.5)

Sch 61 S. S. Schweber, *An Introduction to Relativistic Quantum Field Theory*, Harper and Row, 1961. (Sec. 4.1)

Sch 79 A. N. Schellekens, *Lett. Nuovo Cim.* **24** (1979) 513. (Sec. 5.3)

Sch 80 A. N. Schellekens and W. L. van Neerven, *Phys. Rev.* **D21** (1980) 2619. (Sec. 5.5)

Shi 78 K. Shizuya and S.-H. H. Tye, *Phys. Rev. Lett.* **41** (1978) 787. (Secs. 1.1, 5.4)

Shi 79 M. A. Shifman, A. I. Vainshtein and V. I. Zakharov, *Nucl. Phys.* **B147** (1979) 385. (Sec. 4.2)

Shi 79a K. Shizuya and S.-H. H. Tye, *Phys. Rev.* **D20** (1979) 1101. (Sec. 5.4)

Shi 85 K. Shimizu, *Prog. Theor. Phys.* **74** (1985) 610. (Sec. 2.4)

Siv 76 D. Sivers, S. J. Brodsky and R. Blankenbecler, *Phys. Reports* **23C** (1976) 1. (Sec. 1.3)

Sjo 82 T. Sjöstrand, *Comp. Phys. Com.* **27** (1982) 243; **28** (1983) 229. (Sec. 5.4)

Sla 72 A. A. Slavnov, *Teor. Mat. Fiz.* **10** (1972) 153 (*Theor. Math. Phys.* **10** (1973) 99). (Sec. 2.5)

Spe 68 E. R. Speer, *J. Math. Phys.* **9** (1968) 1408. (Sec. 2.4)

Spe 69 E. R. Speer, *Generalized Feynman Amplitudes*, Princeton Univ. Press, 1969. (Sec. 2.4)

Spe 74 E. R. Speer, *J. Math. Phys.* **15** (1974) 1. (Secs. 2.4, 2.5)

Spe 76 E. Speer, *Renormalization Theory* (eds. G. Velo and A. S. Wightman), D. Reidel, 1976. See Theorem 2.14. (Sec. 2.5)

Ste 49 J. Steinberger, *Phys. Rev.* **76** (1949) 1180. (Sec. 1.2)

Ste 76 G. Sterman, *Phys. Rev.* **D14** (1976) 2123. (Sec. 6.3)

Ste 77 G. Sterman and S. Weinberg, *Phys. Rev. Lett.* **39** (1977) 1436. (Secs. 1.1, 5.4)

Ste 81 P. M. Stevenson, *Phys. Lett.* **100B** (1981) 61. (Sec. 5.3)

Ste 81a P. M. Stevenson, *Phys. Rev.* **D23** (1981) 2916. (Secs. 3.1, 5.3)

Str 74 H. Strubbe, *Comp. Phys. Com.* **8** (1974) 1. (Sec. 5.4)

Stu 53 E. C. G. Stueckelberg and A. Petermann, *Helv. Phys. Acta* **26** (1953) 499. (Secs. 3.1, 3.2)

Sug 77 A. Sugamoto, *Phys. Rev.* **D16** (1977) 1065. (Sec. 6.3)

Sym 70 K. Symanzik, *Comm. Math. Phys.* **18** (1970) 227. (Sec. 3.2)
Sym 73 K. Symanzik, *Lett. Nuovo Cim.* **6** (1973) 77. (Sec. 1.1)

Tak 57 Y. Takahashi, *Nuovo Cim.* **6** (1957) 371. (Sec. 2.5)
Tar 80 O. V. Tarasov, A. A. Vladimirov and A. Yu. Zharkov, *Phys. Lett.* **93B** (1980) 429. (Sec. 3.4)
Tar 81 R. Tarrach, *Nucl. Phys.* **B183** (1981) 384. (Sec. 3.4, Appendix C)
Tay 71 J. C. Taylor, *Nucl. Phys.* **B33** (1971) 436. (Sec. 2.5)
Tay 76 J. C. Taylor, *Gauge Theories of Weak Interactions.* Cambridge Univ. Press, 1976. (Sec. 2.3)
Thi 50 W. Thirring, *Phil. Mag.* **41** (1950) 1193. (Sec. 3.1)
tHo 71 G. 't Hooft, *Nucl. Phys.* **B33** (1971) 173. (Sec. 1.1)
tHo 72 G. 't Hooft and M. Veltman, *Nucl. Phys.* **B44** (1972) 189. (Sec. 2.4)
tHo 72a G. 't Hooft, unpublished. See, however, *Proc. Colloquium on Renormalization of Yang-Mills Fields and Applications to Particle Physics*, Marseilles, 1972 (ed. C. P. Korthals-Altes). See also *Nucl. Phys.* **B254** (1985) 11. (Sec. 1.1)
tHo 72b G. 't Hooft and M. Veltman, *Nucl. Phys.* **B50** (1972) 318. (Sec. 2.5)
tHo 73 G. 't Hooft, *Nucl. Phys.* **B61** (1973) 455. (Secs. 2.5, 5.1, 5.3)
tHo 74 G. 't Hooft and M. Veltman, *Particle Interactions at Very High Energies* (eds. D. Speiser, F. Halzen and J. Weyers), Plenum Press, 1974, p.177. (Sec. 2.5)
tHo 79 G. 't Hooft and M. Veltman, *Nucl. Phys.* **B153** (1979) 365. (Sec. 6.1)
Tit 32 E. C. Titchmarsh, *The Theory of Functions.* Oxford Univ. Press, 1932, p.186. (Sec. 5.2)
Tit 37 E. C. Titchmarsh, *Introduction to the Theory of Fourier Integrals.* Oxford Univ. Press, 1937. (Sec. 4.1)
Tka 83 F. V. Tkachov, *Phys. Lett.* **124B** (1983) 212. (Sec. 4.2)
Tyb 76 L. Tyburski, *Phys. Rev. Lett.* **37** (1976) 319; *Nucl. Phys.* **B116** (1976) 291. (Sec. 6.3)

Uem 78 T. Uematsu, *Phys. Lett.* **79B** (1978) 97. (Sec. 1.1)
Ume 56 H. Umezawa, *Quantum Field Theory.* North-Holland, 1956. (Sec. 2.5)

Ver 81 J. A. Vermaseren, K. J. F. Gaemers and S. J. Oldham, *Nucl. Phys.* **B187** (1981) 301. (Secs. 1.1, 5.1, 5.4)
Vla 75 A. A. Vladimirov, *Teor. Mat. Fiz.* **25** (1975) 335 (*Theor. Math. Phys.* **25** (1975) 1170). (Sec. 3.4)
Vla 77 A. A. Vladimirov and O. V. Tarasov, *Yad. Fiz.* **25** (1977) 1104 (*Sov. J. Nucl. Phys.* **25** (1977) 585). (Sec. 3.4, Appendix C)
Vla 78 A. A. Vladimirov, *Teor. Mat. Fiz.* **36** (1978) 271 (*Theor. Math. Phys.* **36** (1978) 732). (Sec. 3.4)

Vla 79 A. A. Vladimirov, D. I. Kazakov and O. V. Tarasov, *Zh. Eksp. Teor. Fiz.* **77** (1979) 1035 (*Sov. Phys. JETP* **50** (1979) 521). (Sec. 3.4)

Vla 79a A. A. Vladimirov and D. V. Shirkov, *Usp. Fiz. Nauk* **129** (1979) 407 (*Sov. Phys. Usp.* **22** (1979) 860). (Sec. 5.3)

War 50 J. C. Ward, *Phys. Rev.* **78** (1950) 1824. (Sec. 2.5)

War 51 J. C. Ward, *Phys. Rev.* **84** (1951) 897. (Sec. 2.5)

Wat 75 Y. Watanabe et al., *Phys. Rev. Lett.* **35** (1975) 898. (Sec. 1.1)

Web 82 B.R. Webber, *Phys. Scripta* **25** (1982) 198. (Sec. 5.4)

Wei 60 S. Weinberg, *Phys. Rev.* **118** (1960) 838. (Secs. 2.5, 6.3)

Wei 65 S. Weinberg, *Phys. Rev.* **140B** (1965) 516. (Sec. 6.2)

Wei 67 S. Weinberg, *Phys. Rev. Lett.* **19** (1967) 1264. (Sec. 2.4)

Wei 73 S. Weinberg, *Phys. Rev. Lett.* **31** (1973) 494. (Sec. 1.1)

Wei 73a S. Weinberg, *Phys. Rev.* **D8** (1973) 3497. (Secs. 2.4, 3.2)

Wei 80 S. Weinberg, *Phys. Lett.* **91B** (1980) 51. (Sec. 4.2)

Wey 18 H. Weyl, *Sitzungsberichte der Preussischen* Akad. d. Wissenschaften, 1918. (Sec. 2.1)

Wie 23 N. Wiener, *J. Math. Phys.* **2** (1923) 131. (Sec. 2.2)

Wil 64 K. Wilson, *On Products of Quantum Field Operators at Short Distances* (Cornell report, 1964). (Sec. 4.1)

Wil 69 K. G. Wilson, *Phys. Rev.* **179** (1969) 1499. (Secs. 1.1, 4.1, 4.2, 4.3)

Wil 70 K. G. Wilson, *Phys. Rev.* **D2** (1970) 1473. (Secs. 3.5, 4.2)

Wil 71 K. G. Wilson, *Phys. Rev.* **D3** (1971) 1818. (Sec. 3.1)

Wil 72 K. G. Wilson and W. Zimmermann, *Comm. Math. Phys.* **24** (1972) 87. (Sec. 4.2)

Wit 77 E. Witten, *Nucl. Phys.* **B120** (1977) 189. (Sec. 1.1)

Wit 84 E. Witten, *Phys. Rev.* **D30** (1984) 272. (Sec. 1.2)

Yam 85 H. Yamamoto, *Proc. 1985 Int. Symposium on Lepton and Photon Interactions at High Energies*, Organizing Committee of the Symposium at Yukawa Hall, Kyoto Univ., 1986. (Sec. 5.4)

Yan 54 C. N. Yang and R. L. Mills, *Phys. Rev.* **96** (1954) 191. (Sec. 1.1)

Yao 76 Y. P. Yao, *Phys. Rev. Lett.* **36** (1976) 653. (Sec. 6.3)

Yen 61 D. R. Yennie, S. C. Frautschi and H. Suura, *Ann. Phys.* **13** (1961) 379. (Sec. 6.2)

Ynd 83 F. J. Yndurain, *Quantum Chromodymanics*, Springer-Verlag, 1983. (Sec. 3.5)

Yos 81 N. Yoshida, *Prog. Theor. Phys.* **66** (1981) 269. (Sec. 6.3)

Yos 81a N. Yoshida, *Prog. Theor, Phys.* **66** (1981) 1803. (Sec. 6.3)

Zav 65 O. I. Zav'yalov and B. M. Stepanov, *Yad. Fiz.* **1** (1965) 922 (*Sov. J. Nucl. Phys.* **1** (1965) 658). (Sec. 2.5)

Zee 73 A. Zee, *Phys. Rev.* **D7** (1973) 3630. (Sec. 1.1)

Zee 73a A. Zee, *Phys. Rev.* **D8** (1973) 4038. (Secs. 1.1, 5.1)
Zim 58 W. Zimmermann, *Nuovo Cim.* **10** (1958) 597. (Sec. 4.1)
Zim 68 W. Zimmermann, *Comm. Math. Phys.* **11** (1968) 1. (Sec. 2.5)
Zim 69 W. Zimmermann, *Comm. Math. Phys.* **15** (1969) 208. (Sec. 2.5)
Zim 71 W. Zimmermann, *Lectures on Elementary Particles and Quantum Field Theory, Proc. 1970 Brandeis Summer Institute in Theor. Phys.* (eds. S. Deser, M. Grisaru and H. Pendleton), MIT Press, 1971, p.396. (Secs. 1.1, 4.1, 4.2)
Zim 73 W. Zimmermann, *Ann. Phys.* **77** (1973) 570. (Secs. 1.1, 4.1, 4.2)
Zwe 64 G. Zweig, Preprints CERN-TH 401 and 412 (1964). (Secs. 1.1, 1.2)

INDEX

www.ingramcontent.com/pod-product-compliance
Lightning Source LLC
Chambersburg PA
CBHW061614220326
41598CB00026BA/3760

* 9 7 8 9 8 1 2 7 9 3 5 4 6 *